PROGRESS IN LOW TEMPERATURE PHYSICS

XV

PROGRESS IN LOW TEMPERATURE PHYSICS

EDITED BY

W.P. HALPERIN

Chairperson, Department of Physics and Astronomy
Northwestern University, Evanston, IL, USA

VOLUME XV

ELSEVIER

Amsterdam – Boston – Heidelberg – London – New York – Oxford
Paris – San Diego – San Francisco – Singapore – Sydney – Tokyo

ELSEVIER B.V.
Radarweg 29
P.O. Box 211
1000 AE Amsterdam
The Netherlands

ELSEVIER B.V.
525 B Street, Suite 1900
San Diego
CA 92101-4495
USA

ELSEVIER Ltd
The Boulevard, Langford
Lane, Kidlington
Oxford OX5 1GB
UK

ELSEVIER Ltd
84 Theobalds Road
London
WC1X 8RR
UK

First edition 2005

ISBN-13: 978-0-444-51944-3
ISBN-10: 0-444-51944-0
ISSN: 0079-6417

∞ The paper used in this publication meets the requirements of ANSI/NISO Z39.48-1992 (Permanence of Paper).
Printed in Great Britain.

PREFACE

The present volume, the 15th in the book series *Progress in Low Temperature Physics*, continues a sequence of volumes with distinguished reviews and articles that mark significant developments in their fields. Here we have four major chapters written with such depth that they truly define their subject in four important areas of low temperature physics.

Superfluid ^{3}He has been the subject of intense theoretical and experimental study since it's first observation in 1971 by Osheroff, Richardson, and Lee. Shortly thereafter the theoretical framework was established by Leggett; all four have now received Nobel prizes for their accomplishments. The stability of the ^{3}He phases and their nucleation are among the many fascinating superfluid properties that have been carefully investigated. Fundamental questions about the nucleation mechanism were raised by Leggett more than two decades ago including the possible effects of cosmic rays. Since that time, at the Helsinki Low Temperature Laboratory, seminal experiments have been performed on defect production after rapid cooling that follow irradiation and capture of a high energy particle, such as a cosmic muon or a neutron. The resulting vortex creation is a measure of an inhomogeneous order parameter that is expected under conditions of extreme rapid cooling following intense local heating. These conditions are similar to those that may have existed after the Big Bang and which subsequently established large-scale structure in the universe. The scenario suggested by Kibble can be tested in the ^{3}He laboratory. In the first chapter of this volume, Eltsov, Krusius, and Volovik, have an extensive review of their experiments and interpretation of them in terms of the Kibble-Zurek mechanism for formation of defects in superfluid ^{3}He. These remarkable superfluid phases continue to serve as a model system for many areas of physics. Another example of ^{3}He in its role as a paradigm material is its parallel with unconventional superconductivity. It is not surprising then that this fact is abundantly referred to in the second chapter by Flouquet. His review includes a lengthy discussion of unconventional superconductivity in heavy fermion materials.

Since the last volume 10 years ago there have been discoveries of many exotic and unexpected superconducting states in new materials, including some of them in heavy fermion compounds and some, surprisingly, in ferromagnetic materials. From this work there is substantial evidence for unconventional order parameters controlled by proximity to a quantum critical point and where spin-fluctuations appear to be ubiquitous. An extensive review of this topic is presented here by Flouquet, whose group has made many advances in the understanding of these systems. His discussion includes some of the very latest results such as those from the most recently discovered compounds; for example, the cerium 115 materials, recently discovered by Sarrao, Fisk, Thompson and colleagues at Los Alamos, and the skutterudites found by Maple's group at San Diego. This chapter in the present volume will serve as an excellent guide for future work on heavy fermion physics and has extensive referencing and many new insights.

Among the quantum fluids ^{3}He has always been exceptionally important in its own right. One of its key aspects is its magnetism and the interplay between spin fluctuations and transport behavior. One way to probe this, and concomitantly to explore theoretical ideas that have been proposed to account for it, is to freeze out the spin degrees of freedom by polarizing the fluid. In the third chapter Buu, Puech, and Wolf describe elegant experiments of this kind performed at the Centre de Recherches sur les Très Basses Températures in Grenoble. They summarize their most recent results, on the highly polarized Fermi liquid obtained by rapid melting, a method originally proposed by Castaing and Nozières. From the investigations by this group and their collaborators we see how it is possible to rapidly refrigerate the polarized fluid after it has been melted from the highly polarized solid. This allows study of its behavior up to polarizations of 70% in the degenerate limit. From these experiments it is possible to discriminate between predictions of several theoretical models for the state of liquid ^{3}He, notably the *nearly localized* and *nearly ferromagnetic* models. Neither, as it turns out, can completely account for the observations.

The most important low temperature experimental variable is temperature and one of the most important techniques for its measurement is the melting curve thermometer. This subject is reviewed in the fourth chapter by Adams. The method, and the contributions made by Adams to their development, have been pivotal to low-temperature physics. The highly sensitive pressure gauge, invented by Straty and Adams, is an essential component of the technique. Adams gives a historical perspective, which is intermingled with detailed instructions on how to construct and use such thermometers. It is particularly timely that the newest proposal for a low temperature scale has recently been presented to the low temperature community and is considered to be resident in the pressure - temperature relation for melting ^{3}He. The proposed new scale is discussed in this chapter with appropriate references to earlier work.

We can look forward to many more new discoveries and research accomplishments in low-temperature physics. But it is interesting to see that in the present volume, as will very likely remain the case for the future, there is a natural entanglement of fundamental concepts covering a breadth of topics that are having impact on a wide range of complementary subjects in physics.

Bill Halperin
Evanston, March 2005

CONTENTS

VOLUME XV

CONTENTS OF PREVIOUS VOLUMES

Volumes I-VI, edited by C.J Gorter

Volume I (1955)

Volume II (1957)

Volume III (1961)

Volume VI (1970)

Volumes VII-XIII, edited by D.E. Brewer

Volume VII (1978)

Volume VIII (1982)

Volume IX (1985)

Volume XIV (1995)

CHAPTER 1

VORTEX FORMATION AND DYNAMICS IN SUPERFLUID ^{3}HE AND ANALOGIES IN QUANTUM FIELD THEORY

BY

V. B. ELTSOV

Low Temperature Laboratory, Helsinki University of Technology, Box 2200, FIN-02015 HUT, Finland and P.L. Kapitza Institute for Physical Problems, 119334 Moscow, Russia

M. KRUSIUS

Low Temperature Laboratory, Helsinki University of Technology, Box 2200, FIN-02015 HUT, Finland

and

G. E. VOLOVIK

Low Temperature Laboratory, Helsinki University of Technology, Box 2200, FIN-02015 HUT, Finland and L.D. Landau Institute for Theoretical Physics, 119334 Moscow, Russia

The formation and dynamics of topological defects of different structure has been of central interest in the study of the ^{3}He superfluids. Compared to superfluid ^{4}He-II, the variability of the important parameters with temperature and pressure is wider and many features are more ideal. This has made experimentally new approaches possible. An example is the formation of quantized vortices and other topological defects in a rapid time-dependent phase transition from the normal state to ^{3}He-B. This

Progress in Low Temperature Physics, Volume XV

ISSN: 0079-6417 DOI: 10.1016/S0079-6417(05)15001-X

vortex formation process is known as the Kibble-Zurek mechanism and is one of the central topics of this review. It demonstrates the use of ^{3}*He-B as a model system for quantum fields, with detailed control from the laboratory.*

Contents

1. Superfluid ^{3}He and quantum field theory

In recent times, condensed matter and elementary particle physics have been experiencing remarkable convergence in their developments, as many-body aspects have become increasingly more important in particle physics: the Early Universe is the ultimate application field for the theories on interacting particle systems which can be worked out for different energy regimes, or epochs of the expansion after the Big Bang. Compact astrophysical objects provide other "laboratories" with a narrower range of conditions in which to test theories. Actual collider experiments are set up to study interacting particle systems, such as the quark–gluon plasma in heavy ion collisions or the pion condensate.

Collective phenomena in interacting many-body systems is what condensed matter physics is about, but elementary particle systems bring extreme quantum behavior plus relativistic motion. So far systems, where all of these features would be of importance, are not available for direct observation. During the 1990s, it turned out that valuable analogs can nevertheless be constructed from comparisons with nonrelativistic many-body quantum systems of condensed matter, such as superconductors and helium superfluids. More recently, optically cooled alkali atom clouds have been added to this list. Here we limit our discussion to the fermionic ^{3}He superfluids.

The liquid ^{3}He phases provide attractive advantages as model systems for the study of various general concepts in quantum field theory: the liquid is composed of neutral particles in an inherently isotropic environment – no complications arise from electrical charges or a symmetry constrained by a crystalline lattice. This allows one to concentrate on the consequences from a most complex symmetry-breaking, which gives rise to a multidimensional order parameter space, but is well described by a detailed microscopic theory. Experimentally superfluid ^{3}He is devoid of extrinsic imperfections. In fact, with respect to impurities and dirt it is one of the purest of all condensed matter systems, excepting optically cooled atom clouds. The superfluid coherence length is large such that surface roughness can be reduced sufficiently to transform the container walls to almost ideally behaving boundaries. There are topologically stable defects of different dimensionality and type, which often can be detected with NMR methods with single-defect sensitivity. The study of the different intrinsic mechanisms, by which these defects are formed, provides one of the important parallels to other systems. Phase transitions of both first and second order exist, which can be utilized in the investigation of defect formation. Another example is the use of the zero-temperature ^{3}He systems to model the complicated physical vacuum of quantum field theory – the modern ether: the bosonic and fermionic excitations in ^{3}He (in the dilute limit) are in many respects similar to the excitations of the physical vacuum – the gauge bosons and elementary particles. This approach has been quite successful in constructing a physical picture of the interactions of elementary particles with strings and domain walls, as described in a recent monograph by one of the present authors (Volovik 2003).

This review describes some examples from superfluid ^{3}He research where strong connections exist to quantum field theory. In fact, some of these studies were only taken up because they can be used as laboratory models, to answer the question whether a proposed principle is physically correct or not. Our discussion has been split in two parts: the first part, section 2, deals with a detailed description of defect formation and evolution in a rapid non-equilibrium phase transition of second order, caused by the localized heating from an absorption event in ionizing radiation. This phenomenon originally became interesting because it can be thought to model phase transitions in the Early Universe. Defect formation in these transitions has been suggested as the origin for the inhomogeneous mass distribution in the present Universe. The second part, section 3 discusses concepts from vortex dynamics which have been used in section 2 to analyze vortex formation in rapid phase transitions. This discussion proceeds in more general terms and continues to highlight connections to quantum field theory.

2. Defect formation in quench-cooled superfluid transition

2.1. *Nonequilibrium phase transitions*

A rapid phase transition is generally associated with a large degree of inhomogeneity. After all, this is the process by which materials like steel or amorphous solids are prepared. We attribute this disorder to heterogeneous extrinsic influence which is usually present in any system which we study: impurities, grain boundaries, and other defects depending on the particular system. To avoid disorder and domain formation, we generally examine phase transitions in the adiabatic limit, as close to equilibrium as possible.

But suppose we would have an ideally homogeneous infinite system with no boundaries. It is rapidly cooling from a symmetric high-temperature state to a low-temperature phase of lower symmetry, which we call the broken-symmetry phase, for example by uniform expansion. What would happen in such a transition? Are defects also formed in this case? Such a measurement of a homogeneous transition as a function of the transition speed is difficult. As we shall see, ultimately it also raises the question whether the source of the precipitated inhomogeneity has been reliably identified.

2.2. *Cosmic large-scale structure*

According to the standard theory of cosmology the Universe started off in the "Big Bang" in a homogeneous state. It then rapidly cooled through a sequence of phase transitions, in which the four fundamental forces of nature separated out. Today the Universe exists in a state with inhomogeneous large-scale structure. The clumped distribution of visible mass has become most evident from galaxy surveys – maps which show that galaxies form clusters and these in turn superclusters, such as the "Great Wall", which are the largest structures discovered to date (Geller and Huchra 1989). The clustering takes the form of long chains or filaments, which are separated by large voids, regions empty of visible mass. Recent extended galaxy surveys indicate that the length scale of large-scale structure is of order 100 Mpc[1] (Peacock et al. 2001).

Another image of large-scale structure has been preserved in the cosmic microwave background radiation, from the time when the Universe had cooled to a few eV, when nuclei and electrons combined to form atoms and the Universe became transparent to photons. Since then the background radiation has cooled in the expanding Universe. It now matches to within three parts in 10^5 the spectrum of a black body at 2.728 K, as measured, e.g. with the spectrometers on board of the satellite Cosmic Background Explorer (COBE) in 1990. Later balloon-borne measurements examined the spatial distribution of the residual anisotropy with high angular resolution and amplitude sensitivity. The recent satellite Wilkinson Microwave Anisotropy Probe (WMAP) has extended this work, mapping all-sky surveys with an angular resolution better than 1° on different frequency bands, to separate out the interfering signal from our galaxy (Bennett et al. 2003, Spergel et al. 2003). Combined with other information, this anisotropy of the cosmic microwave background radiation, 30 μK in amplitude, which represents the density fluctuations in the structure of the Universe when it was only 400 000 years old, is expected to explain the mechanisms which originally seeded and led to the formation of the large-scale structure, as we observe it today.

2.3. *Kibble-Zurek mechanism*

One of the early explanations for the origin of large-scale structure was offered by Tom Kibble (1976). He suggested that the inhomogeneity was created in rapid phase transitions of the early expanding Universe. Even in a

[1] 1 Mpc = 10^6 parsec = 3.262×10^6 light years.

perfectly homogeneous transition of second-order defects can be expected to form, if the transition proceeds faster than the order parameter of the broken-symmetry phase is able to relax. In such a non-equilibrium transition the new low-temperature phase starts to form, due to fluctuations of the order parameter, simultaneously and independently in many parts of the system. Subsequently during further cooling, these regions grow together to form the new broken-symmetry phase. At the boundaries, where different causally disconnected regions meet, the order parameter does not necessarily match and a domain structure is formed.

If the broken symmetry is the $U(1)$ symmetry, then domains with different value of the order parameter phase are formed. Such a random domain structure reduces to a network of linear defects, which are vortex lines in superfluids and superconductors or perhaps cosmic strings in the Early Universe. If the symmetry break is more complicated, as is the case in ^{3}He superfluids, then defects of different dimensionality and structure may be formed.

Subsequent numerical simulations of rapidly cooled second-order phase transitions confirmed that defects were indeed forming (Vachaspati and Vilenkin 1984). In 1985, Wojciech Zurek proposed a conceptually powerful phenomenological approach how to understand a phase transition far out of equilibrium, when the outcome from the transition becomes time-dependent (Zurek 1985, 1996). He characterizes the transition speed with a quench time

$$\tau_Q = \left(\frac{1}{T_c} \left| \frac{dT}{dt} \right|_{T=T_c} \right)^{-1}, \tag{1}$$

which allows him to approximate temperature and time with a linear dependence during the thermal quench.

The quench time τ_Q is compared to the order parameter relaxation time τ, which in the time-dependent Ginzburg–Landau (TDGL) model a second-order phase transition is of the form

$$\tau(T) = \tau_0(1 - T/T_c)^{-1}. \tag{2}$$

In superfluid ^{3}He, τ_0 is of the order of $\tau_0 \sim \xi_0 / v_F$, where ξ_0 is the zero-temperature limiting value of the temperature (T) and pressure (P) dependent superfluid coherence length $\xi(T, P)$. Close to T_c, it is of the form $\xi(T, P) = \xi_0(P)(1 - T/T_c)^{-1/2}$. The second quantity, v_F, is the velocity of the quasiparticle excitations in normal Fermi liquid.

As sketched in fig. 1, this means that below T_c the order parameter coherence spreads out with the velocity $c(T) \sim \xi/\tau = \xi_0\,(1 - T/T_c)^{1/2}/\tau_0$. The freeze-out of defects occurs at $t = t_{\mathrm{Zurek}}$, when the causally disconnected

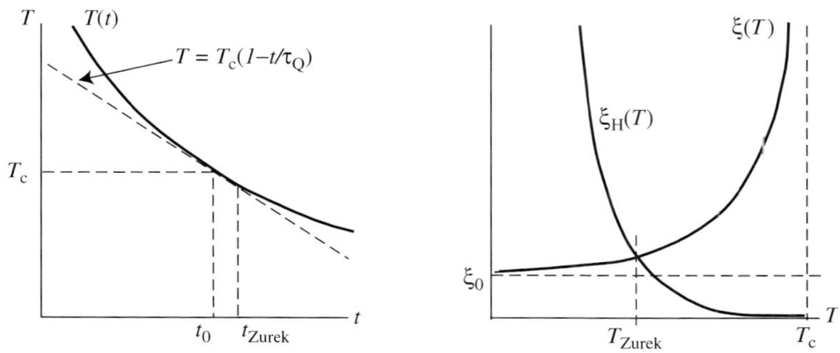

Fig. 1. Principle of KZ mechanism. Rapid thermal quench through T_c : (*Left*) Temperature $T(t)$ and its linear approximation $T = T_c(1 - t/\tau_Q)$ during the quench, as a function of time t. (*Right*) Superfluid coherence length $\xi(t)$ and order-parameter relaxation time $\tau(T)$ diverge at T_c. At the freeze-out point t_{Zurek}, when phase equilibrium is achieved, the edge of the correlated region, the causal horizon, has moved out to a distance $\xi_H(t_{Zurek})$, which has to equal the coherence length $\xi(t_{Zurek})$.

regions have grown together and superfluid coherence becomes established in the whole volume. At the freeze-out temperature $T(t_{Zurek}) < T_c$, the causal horizon has traveled the distance $\xi_H(t_{Zurek}) = \int_0^{t_{Zurek}} c(T)\,dt = \xi_0\tau_Q (1 - T_{Zurek}/T_c)^{3/2}/\tau_0$, which has to be equal to the coherence length $\xi(t_{Zurek})$. This condition establishes the freeze-out temperature $T_{Zurek}/T_c = 1 - \sqrt{\tau_0/\tau_Q}$ at the freeze-out time $t_{Zurek} = \sqrt{\tau_0\tau_Q}$, when the domain size has reached the value

$$\xi_v = \xi_H(t_{Zurek}) = \xi_0 \ (\tau_Q/\tau_0)^{1/4}. \tag{3}$$

In superfluid ^{3}He we may have $\xi_0 \sim 20$ nm, $\tau_0 \sim 1$ ns, and in the best cases a cool-down time of $\tau_Q \sim 1$ μs can be reached. From these values we expect the domain structure to display a characteristic length scale of order $\xi_v \sim 0.1$ μm. Assuming a $U(1)$-symmetry-breaking transition, where vortex lines are formed at the domain boundaries, the average inter-vortex distance and radius of curvature in the randomly organized vortex network is of the order of the domain size ξ_v. In general, we can expect a rapid quench to lead to an initial defect density (defined as vortex length per unit volume)

$$l_v = \frac{1}{a_l \xi_v^2}, \tag{4}$$

where the numerical factor $a_l \sim 1 - 100$ depends on the details of the model system. Numerical simulation calculations (Ruutu et al. 1998a) give $a_l \approx 2$ for ^{3}He–B. In this system with global $U(1)$ symmetry and negligible thermal fluctuations the Kibble-Zurek (KZ) model of defect formation can be expected to work well. However, the TDGL model is not a good description of ^{3}He–B and more rigorous time-dependent microscopic calculations are required.

2.4. *Experimental verification of KZ mechanism*

The predictions of the KZ model have been tested in numerous numerical experiments under varying conditions (cf. Bettencourt et al. 2000). Laboratory experiments are fewer and their results are less conclusive.

The first experiments were performed on liquid crystals (Chuang et al. 1991, Bowick et al. 1994), where the transition from the isotropic to the nematic phase was examined. This is a weakly first-order transition of rod-like molecules from a disordered to an ordered state, where defects of different dimensionality can be formed. The results were generally found to be consistent with the KZ interpretation. However, a liquid-crystal at room temperature is a classical system and one might question whether it is a good medium for modelling quantum field theory.

The first experiments on coherent quantum systems were started with liquid ^{4}He (Hendry et al. 1994), by releasing the pressure on a convoluted phosphor–bronze bellows so that the sample expanded from the normal liquid through the λ transition into the superfluid phase. For a mechanical pressure system the quench time is of order $\tau_Q \sim 10$ ms. The depressurization is followed by a dead time of similar length, which is required for damping down the vibrations after the mechanical shock. The vortex density after the transition is determined from the attenuation in second sound propagation through bulk liquid. Expansion of ^{4}He across the $T_\lambda(P)$ line (which has negative slope in the (P, T) phase diagram) requires careful elimination of all residual flow which might result from the expansion itself, foremost the flow out of the filling capillary of the sample cell, but also that around the convolutions of the bellows. Such flow will inevitably exceed the critical velocities, which approach zero at T_λ, and also leads to vortex nucleation (section 2.6.1). After the initial attempts, where vortices were detected but a fill line was still present, later much improved measurements failed to yield any evidence of vortex line production, in apparent contradiction with the KZ prediction (Dodd et al. 1998).

The early ^{4}He expansion experiments were analyzed by Gill and Kibble (1996), who concluded that the initially detected vortex densities were unreasonably large and must have originated from extrinsic effects, most likely created by the flow out the fill line. The absence of vorticity in the later results was shown not to be evidence against the KZ mechanism (Karra and Rivers, 1998, Rivers, 2000): although the pressure drop in expansion cooling is large, 20–30 bar, the change in relative temperature $\varepsilon = 1 - T/T_\lambda$ is small, $|\Delta\varepsilon| < 0.1$. The final temperature $\varepsilon_f < 0.06$ is below T_λ but still above the Ginzburg temperature T_{Gi}, i.e. in the regime where thermal fluctuations might be sufficiently effective to wipe out domain formation. Here T_{Gi} is defined as the temperature at which $k_B T_{Gi}$ equals the energy of a vortex loop of size ξ. However, numerical simulations by Bettencourt et al. (2000) suggest that above T_{Gi} no substantial suppression occurs in the final density of defects the current view about the ^{4}He expansion experiments holds that the vortices decay away before the observation switches on (Hendry et al. 2000).

Simultaneously with the ^{4}He expansion experiments, measurements of quite different nature were carried out in ^{3}He–B (Ruutu et al. 1996a, Bäuerle et al. 1996). Here a detailed comparison with the KZ predictions was achieved. These efforts were followed by measurements of thermal quenches through the superconducting transition in different superconducting geometries.

The most conclusive results have been obtained with high-T_c YBCO films (Maniv et al. 2003). The film is heated above its transition temperature of 90 K with a laser pulse. The hot film is then cooled from below back to the ambient temperature of 77 K. This happens laterally uniformly through the contact with an optically transparent substrate of much larger thickness and heat capacity. The maximum cooling rate was measured to be 10^8 K/s, by monitoring the electrical resistance of the film. The total net flux, which remains frozen in the pinning sites of the YBCO film after the transition, was measured with a sensing coil below the substrate and connected to a SQUID magnetometer. This flux was found to display very weak dependence on the cooling rate, as predicted by the KZ scaling theory.

As discussed by Kibble and Rajantie (2003), the important information to obtain from such measurements would be the spatial distribution of the flux quanta in the film. Spatial correlations between the fluxoids would support the KZ mechanism, if the vortices were found to display negative correlations, i.e. neighboring vortices would prefer to have opposite signs. A competing mechanism involves thermal fluctuations of the magnetic field which generates vortices during the quench cooling through T_c where the critical field vanishes: $H_{c2} \to 0$. This second mechanism is predicted to display positive correlations, i.e. vortices of same sign would tend to cluster. An analogous situation arises in ^{3}He–B which will be described below. Here at temperatures below 3 mK thermal fluctuations are not important since any energy barriers will generally automatically be orders of magnitude higher. Nevertheless, even under these conditions a phenomenon can be identified which competes with the KZ mechanism (which is a volume effect). This involves the phase boundary between the normal liquid and ^{3}He–B (a surface effect) if a phase boundary exists in the presence of flow.

The reason is again that at T_c the critical velocity for vortex formation vanishes: $v_c \to 0$.

In superconductors a measurement with spatial resolution and single-fluxoid sensitivity has been recently accomplished (Kirtley et al. 2003). Here a scanning SQUID microprobe is used to map the flux quanta trapped in an amorphous Mo_3Si film which had been lithographically patterned in ring-like elements. Unfortunately, in this measurement, the cooling rate was limited to ~20 K/s at the $T_c = 7.8$ K of the film and the detected quantized flux was concluded to originate from a thermally activated process.

These examples illustrate the difficulties in attempting to find experimental proof for the KZ process. The KZ mechanism is intuitively easy to accept, but its experimental verification is complicated and involves careful considerations. In a phase transition defect formation is more usual than its absence. In most experiments both extrinsic and intrinsic sources of vortex formation are present and compete in actual vortex generation. For instance, even slow cooling of liquid ^{4}He across the λ transition is known to produce primordial vortices which remain pinned as remanent vorticity on the walls of the container (Awschalom and Schwarz 1984). There are no claims in the literature yet that an experiment would have produced a perfectly vortex-free bulk sample of superfluid ^{4}He.

A second point to note is that the KZ model describes a second-order transition where the energy barrier, which separates the symmetric high-temperature phase from the broken-symmetry states at low temperatures, vanishes at T_c. Thus the transition becomes an instability. A first-order transition is different: here the barrier remains finite and the low-temperature phase has to be nucleated, usually by overcoming the barrier via thermal activation. Nucleation may occur in different parts of the system nearly simultaneously, depending on the properties of the system, the fluctuations, the quench time τ_Q, and the nucleation mechanism. Such a situation, which often is called bubble nucleation, also leads to domain formation and to a final state which is qualitatively similar to that expected after the second-order transition and the KZ mechanism.

In all of the experimental examples listed above, other mechanisms are also possible which might replace or operate in parallel to the KZ process. The differences are subtle, as we shall see in section 2.8 for ^{3}He–B or as has been emphasized for superconductors by Rajantie (2001). In practice, measurements differ from the original KZ model, i.e. from a homogeneous transition, which occurs simultaneously in the whole system and where there are no gradients in density or temperature. In practical measurements on bulk material, a rapid quench through the transition is often forced by strong gradients. On some level, gradients will always appear in any laboratory system and their influence has to be investigated.

Superfluid ^{3}He measurements have so far established the most quantitative comparison to the KZ scaling theory (Ruutu et al. 1996a, Bäuerle et al. 1996). A happy coincidence of many valuable features has made this possible: a reduced likelihood of defect formation from extrinsic sources, the virtual absence of remanent vorticity, the possibility to perform thermal quenches in microseconds, the presence of a pure second-order phase transition, and the availability of measuring methods to detect different types of defects, often with a resolution of one single defect. A critical discussion of these questions is presented below.

The search for firm proof of the KZ mechanism has grown to an important research field, as described in a recent collection of lectures by Kibble (2003). The reason is that it provides (at least superficially) easy interpretation of the fastest non-equilibrium phase transitions. Amusingly this question does not enjoy similar interest any more in cosmology where the problem was born. Measurements of the anisotropy in the cosmic microwave background radiation with the WMAP satellite (with an angular resolution $\sim 0.25°$ and an amplitude sensitivity $\sim 1\ \mu K$) have concluded that the angular distribution of the anisotropy expanded in its harmonic components gives a multipole expansion with a sharp dominating peak at $\ell \sim 200$, which corresponds to an angular scale $\sim 1°$ (Bennett et al. 2003, Spergel et al. 2003). This distribution can be fit with models which are based on inflationary expansion of the Early Universe, i.e. with models with an early period of exponentially accelerated expansion. In contrast, topological defects, such as cosmic strings, are expected to generate a different broader peak. These two scenarios have been the ruling contenders as explanation for the origin of large-scale structure and thus one of them has now been dismissed.

At present the KZ mechanism appears to survive only in condensed matter physics. Interestingly, one could have assumed that this verdict would also have reduced the interest in the hypothetical cosmic string, the counterpart of a vortex in superfluids and superconductors. However, as reviewed by Kibble (2004), new reasons have surfaced to revive the interest in cosmic strings and other types of strings and their creation via the KZ mechanism.

2.5. *Principle of superfluid ^{3}He experiments*

To study reliably defect formation in a quench-cooled transition, two basic requirements are the following: first, the transition can be repeated reproducibly, and second, a measurement is required to detect the defects after their formation, either before they annihilate or by stabilizing their presence with a bias field. Both of these requirements can be satisfied in ^{3}He experiments.

2.5.1. *Outline of experimental method*

A schematic illustration of one set of experiments (Ruutu et al. 1996a) is depicted in fig. 2. The top part displays the events which are controlled by the observer in the laboratory. A cylindrical container filled with superfluid ^{3}He–B is rotated at constant velocity Ω. The velocity of rotation is maintained below the critical value at which quantized vortex lines are spontaneously formed. In other words, the initial state is one of metastable vortex-free counterflow of the superfluid and normal components under constant conditions.

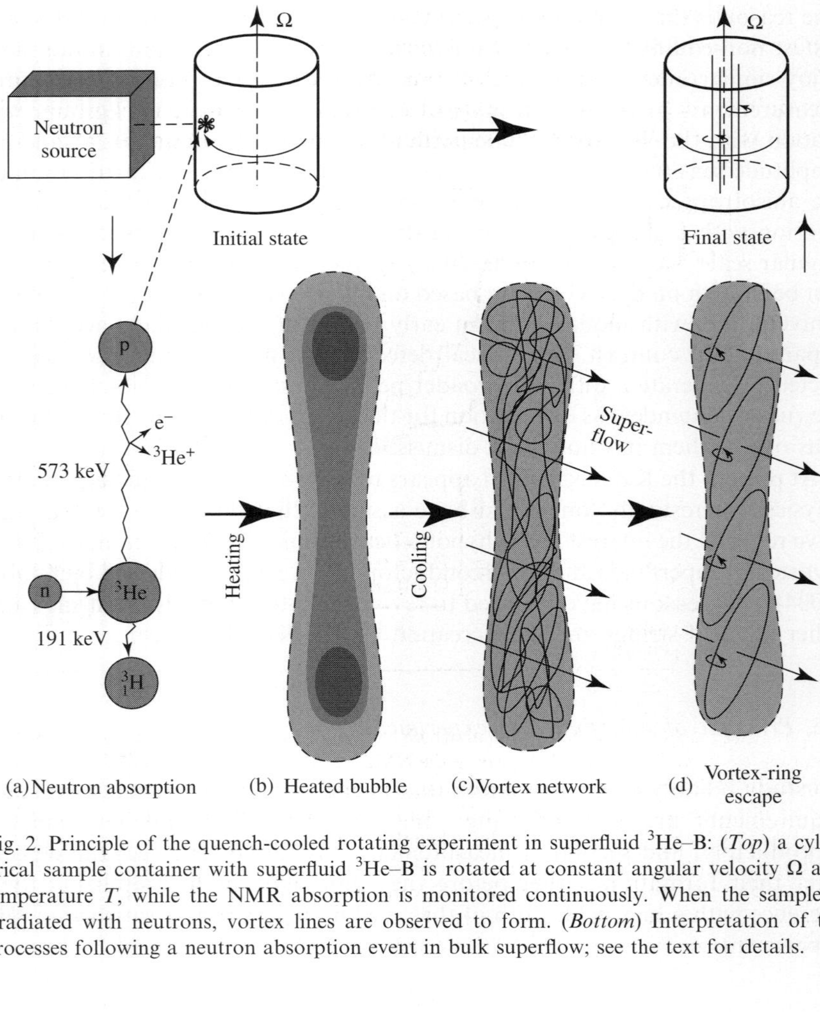

Fig. 2. Principle of the quench-cooled rotating experiment in superfluid ^{3}He–B: (*Top*) a cylindrical sample container with superfluid ^{3}He–B is rotated at constant angular velocity Ω and temperature T, while the NMR absorption is monitored continuously. When the sample is irradiated with neutrons, vortex lines are observed to form. (*Bottom*) Interpretation of the processes following a neutron absorption event in bulk superflow; see the text for details.

Next a weak source of thermal neutrons is placed in the vicinity of the ^{3}He sample. If the rotation velocity is sufficiently high, then vortices start to appear at a rate which is proportional to the neutron flux. The neutron source is positioned at a convenient distance d from the sample so that vortex lines are observed to form in well-resolved individual events. The experimental signal for the appearance of a new vortex line is an abrupt jump in NMR absorption. An example, measured at low rotation, is shown by the lower recorder trace in fig. 3.

The lower part of fig. 2 shows in more detail what is thought to happen within the superfluid. Liquid ^{3}He can be locally heated with the absorption reaction of a thermal neutron: $n + {}^3_2\text{He} \rightarrow p + {}^3_1\text{H} + E_0$, where $E_0 = 764$ keV. The reaction products, a proton and a triton, are stopped by the liquid and produce two collinear ionization tracks (Meyer and Sloan 1997). The ionized particles, electrons and ^{3}He ions, diffuse in the liquid to recombine such that 80% or more of E_0 is spent to heat a small volume with a radius $R_b \sim 50$ μm from the superfluid into the normal phase. The rest of

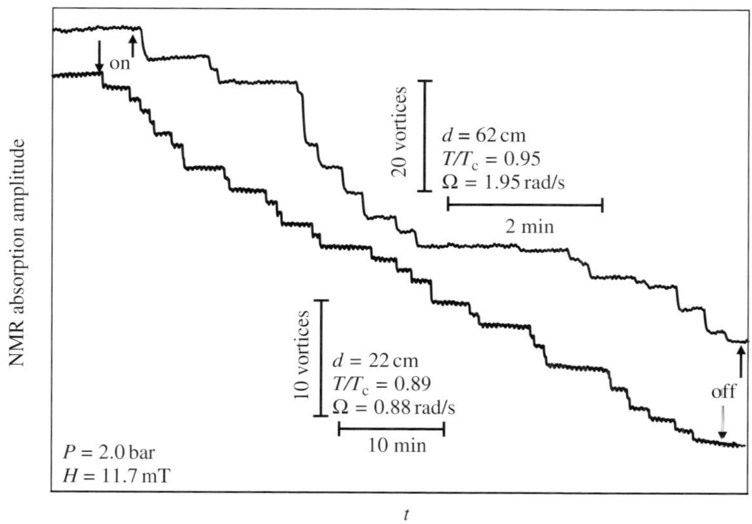

Fig. 3. Peak height of the NMR signal which monitors the velocity of the bias flow from rotation (the so-called counterflow peak in the NMR absorption spectrum of ^{3}He–B), shown as a function of time t during neutron irradiation. The *lower trace* is for a low and the *upper trace* for a high value of rotation velocity Ω. The initial state in both cases is metastable vortex-free counterflow. The vertical arrows indicate when the neutron source was turned on/off. Each step corresponds to one neutron absorption event and its height, when compared to the adjacent vertical calibration bar, gives the number of vortex lines formed per event. The distance of the neutron source from the sample is denoted by d. Copyright 1996 Institute of Physics ASCR; with kind permission from Springer Science and Business Media.

the reaction energy escapes in the form of ultraviolet emission (Stockton et al. 1971, Adams et al. 1995) and, possibly, in the form of long-living molecular $^3He_2^*$ excitations (Keto et al. 1974, Kafanov et al. 2000), which partially relax at the walls of the container.

For measurements in rotating ^{3}He, neutron absorption is an ideal heating process. For slow neutrons from a room temperature source with a Maxwellian distribution (at 20.4°C or 0.0253 eV) the velocity at the peak of the distribution is 2200 m/s. The capture cross section of the ^{3}He nucleus for these neutrons is huge: 5327 ± 10 barn $= 5.3 \times 10^{-21}$ cm^2 (Beckurts and Wirtz 1964). This means a short mean free path of about 100 μm. Thus most neutrons are absorbed within a close distance from the wall, although still within the bulk liquid, but such that the velocity of the applied counterflow is accurately specified. In other parts of the refrigerator neutron absorption is negligible and thus the temperature of the ^{3}He sample can be maintained stable during the irradiation. Finally, the cooling is fast, as the small volume heated by the neutron absorption event is embedded in the cold bulk system.

Subsequently, the normal liquid in the heated neutron bubble cools back through T_c in microseconds. The measurements in fig. 3 demonstrate that vortex lines are indeed created in this process. Vortex loops, which form within a cooling neutron bubble in the bulk superfluid, would normally contract and disappear in the absence of rotation. This decay is brought about by the inter-vortex interactions between the vortex filaments in the presence of a friction force experienced by a vortex when it moves with respect to normal component, the thermal excitations. We refer to this force as a mutual friction force. In rotation the externally applied counterflow provides a bias which causes sufficiently large loops to expand to rectilinear vortex lines and thus maintains them for later detection. In the rotating experiments in Helsinki these rectilinear vortex lines are counted with NMR methods (Ruutu et al. 1996a, 1998a, Finne et al. 2004a).

In a second series of ^{3}He experiments, performed in Grenoble (Bäuerle et al. 1996, 1998a) and Lancaster (Bradley et al. 1998), the vortices formed in a neutron absorption event are detected calorimetrically with very-low-temperature techniques. In the zero-temperature limit mutual friction becomes exponentially small and the lifetime of the vorticity might be very long, even if there is no flow. In this situation the existence of the vorticity can be resolved as a deficit in the energy balance of the neutron absorption reaction.

2.5.2. *Interpretation of ^{3}He experiments*

Several alternative suggestions can be offered on how to explain these observations. The KZ model is one of them. Whatever the mechanism, the superfluid transition in a localized bubble represents a new process for creating vortex rings. This phenomenon is associated with one of the fastest second-order phase transitions, which have been probed. It takes place in the bulk superfluid and not in immediate contact with a solid wall, where most other vortex formation processes occur in the presence of applied counterflow.

The KZ interpretation of the later stages of the experiment is contained in the two illustrations marked as (c) and (d) in fig. 2. First a random vortex network (c) is created while cooling through T_c. Next the network starts to evolve under the influence of inter-vortex interactions and the applied flow. At high temperatures the process is highly dissipative, owing to the large mutual friction force. The average inter-vortex distance increases, small loops are smoothed out, and the network becomes rarefied or "coarse grained". Also reconnections, which take place when two lines cross, contribute to the rarefaction. The applied flow favors the growth of loops with the proper winding direction and orientation. It causes sufficiently large loops to expand, while others contract or reorient themselves with respect to the flow. The final outcome is that correctly oriented large loops (d), which exceed a critical threshold size, start expanding spontaneously as vortex rings, until they meet the chamber walls. There the superfluous sections of the ring will annihilate and only a rectilinear vortex line will finally be left over. It is pulled to the center of the container, where it remains in stationary state parallel to the rotation axis, stretched between the top and bottom surfaces.

This picture applies at high temperatures in ^{3}He–B where Kelvin-wave excitations on vortices are exponentially damped and spontaneous loop formation on existing vortices does therefore not occur. In this situation the number of vortex rings, which are extracted from the neutron bubble (fig. 2d), remains conserved during the later expansion. Thus the number of rectilinear vortex lines, the end result from the neutron absorption event, characterizes the extraction process as a function of the applied bias velocity. This situation is very different from that in ^{4}He-II where mutual friction damping never reaches this high values (except extremely close to T_λ).

Very different timescales are at work in this process of vortex formation. Except for the retarded molecular excitations, the heating from the initial ionization and subsequent recombination is limited by the diffusion of charges in the liquid and takes place much faster than the thermal recovery. The initial vortex network forms during cooling through T_c, for which the relevant timescale is microseconds. The later evolution of the network and the loop escape happen again on a much slower timescale, namely from milliseconds up to seconds, since here the vortex motion is governed by the mutual-friction-dependent superfluid hydrodynamics.

2.6. *Measurement of vortex lines in* ^{3}*He–B*

There are two major phases of superfluid ^{3}He, the A and B phases (fig. 14). The neutron measurements have been performed in the quasi-isotropic ^{3}He–B. In the present context we may think of vortices in ^{3}He–B as being similar to those in superfluid ^{4}He-II, where only the $U(1)$ symmetry is broken. The order parameter outside the vortex core is of the form $\Psi = |\Psi(T)|e^{i\Phi(r)}$: the vortices are topologically stable and have a singular core, inside of which $|\Psi|$ deviates from its bulk value, while outside the phase Φ changes by $2\pi\nu$ on a closed path which encircles the core. A persistent superfluid current is trapped as a single-quantum circulation ($\nu = 1$) around the core, with the circulation quantum $\kappa = h/(2m_3) \simeq 0.0661$ mm^2/s and a superflow velocity $\mathbf{v}_{\mathrm{s,vortex}} = \kappa/(2\pi)\nabla\Phi = \kappa/(2\pi r)\hat{\phi}$.

2.6.1. *Critical velocity of vortex formation*

The energy of a single rectilinear vortex line, aligned along the symmetry axis of a rotating cylindrical container of radius R and height Z, consists primarily of the hydrodynamic kinetic energy stored in the superfluid circulation trapped around the core,

$$E_{\mathrm{v}} = \frac{1}{2}\int \rho_{\mathrm{s}} v^2_{\mathrm{s,vortex}}\, \mathrm{d}V = \frac{\rho_{\mathrm{s}}\kappa^2 Z}{4\pi}\ln\frac{R}{\xi}. \tag{5}$$

Here the logarithmic ultraviolet divergence has been cut off with the core radius, which has been approximated with the coherence length ξ. If the container rotates at an angular velocity Ω, the state with the first vortex line becomes energetically preferred when the free energy $E_{\mathrm{v}} - \Omega L_{\mathrm{z}}$ becomes negative. The hydrodynamic angular momentum from the superflow circulating around the vortex core is given by

$$L_{\mathrm{z}} = \int \rho_{\mathrm{s}} r v_{\mathrm{s,vortex}}\, \mathrm{d}V = \rho_{\mathrm{s}}\kappa R^2 Z. \tag{6}$$

One finds that $E_{\mathrm{v}} - \Omega L_{\mathrm{z}} < 0$, when the velocity ΩR of the cylinder wall exceeds the Feynman critical velocity, $v_{\mathrm{c1}} = \kappa/(2\pi R)\ \ln(R/\xi)$. With a container radius R of a few mm, the Feynman velocity is only $\sim 10^{-2}$ mm/s. However, if we can exclude remanent vorticity and other extrinsic mechanisms of vortex formation (which is generally the case in ^{3}He–B at temperatures $T > 0.6T_{\mathrm{c}}$, but not in ^{4}He-II), then vortex-free superflow will persist as a meta-stable state to much higher velocities because of the existence of a finite nucleation energy barrier.

The height of the nucleation barrier decreases with increasing velocity. In ^{4}He-II intrinsic vortex formation is observed in superflow through submicron-size orifices. Here, the barrier is ultimately at sufficiently high velocities overcome by thermal activation or at low temperatures by quantum tunneling (Packard 1998, Varoquaux et al. 1998). The other possibility is that the thermodynamic stability limit of the superfluid is reached (Andreev and Melnikovsky 2003). In ^{3}He–B the barrier is unpenetrably high in practically any situation and the flow velocity has to be increased to the point where the barrier approaches 0. Vortex formation then occurs in the form of an instability which takes place locally at some sharp asperity on the cylindrical wall of the rotating sample where the stability limit of superflow is first reached (Parts et al. 1995, Ruutu et al. 1997b). The average velocity at the cylindrical container wall we call the spontaneous critical velocity $v_c(T, P)$ (fig. 13), which now depends on surface roughness and is therefore container specific. When the instability occurs and a vortex is formed, the actual local velocity at the critical dominating asperity equals the intrinsic bulk instability velocity $v_{cb}(T, P) > v_c$.

When the flow reaches the instability value v_{cb} at the dominating sharpest protrusion, a segment of a vortex ring of size ξ is formed there. The energy of this smallest possible vortex ring is of order $E_v \sim \rho_s \kappa^2 \xi$. This energy constitutes the nucleation barrier. On dimensional grounds we may estimate that $E_v/(k_B T) \sim (\xi/a)(T_F/T)$, where $T_F = \hbar^2/(2m_3 a^2) \sim 1$ K is the degeneracy temperature of the quantum fluid and a the inter-atomic spacing. In ^{4}He-II the coherence length is of order $\xi \sim a$ and the temperature $T \sim T_F$, which means that $E_v/(k_B T) \sim 1$ and thermally activated nucleation becomes possible.

In contrast, in ^{3}He–B, the coherence length is more than 10 nm and we find $E_v/(k_B T) \sim 10^5$ at temperatures of a few mK or less. The consequence from such an enormous barrier is that both thermal activation and quantum tunneling are out of question as nucleation mechanisms. The only remaining intrinsic mechanism is a hydrodynamic instability, which develops at rather high velocity $v_{cb} \sim \kappa/\xi$. This fact has made possible the measurement of neutron-induced vortex formation as a function of the applied flow velocity $v < v_c$.

2.6.2. *Rotating states of the superfluid*

When a container with superfluid is set into rotation and the formation of vortex lines is inhibited by a high energy barrier, then the superfluid component remains at rest in the laboratory frame. Its velocity is 0 in the whole container: $v_s = 0$ (if we neglect deviation in the shape of the container from an ideal cylinder). The normal component, in contrast, corotates with the container and $v_n = \Omega r$. This state of vortex-free counterflow is called the Landau state. It corresponds to the Meissner state in superconductors, with complete flux expulsion.

In ^{4}He-II vortex-free flow with sufficiently high velocity is out of reach in most situations, because even at very low velocities remanent vorticity leads to efficient vortex formation. In ^{3}He–B the superfluid coherence length is several orders of magnitude larger and remanent vorticity can be avoided if the container walls are sufficiently smooth. Coupled with a high nucleation barrier, metastable rotating states then become possible. These can include an arbitrary number N of rectilinear vortex lines, where $N \leq N_{\rm max}$. In most experimental situations the maximum possible vortex number, $N_{\rm max}$, equals that in the equilibrium vortex state, $N_{\rm eq} \approx 2\pi\Omega R^2/\kappa$. The equilibrium vortex state is obtained by cooling the container slowly at constant Ω through $T_{\rm c}$.

If the number of lines is smaller than that in the equilibrium state, then the existing vortex lines are confined within a central vortex cluster, as shown in fig. 4. The confinement comes about through the Magnus force from the normal–superfluid counterflow, $v = v_{\rm n} - v_{\rm s}$, which circulates around

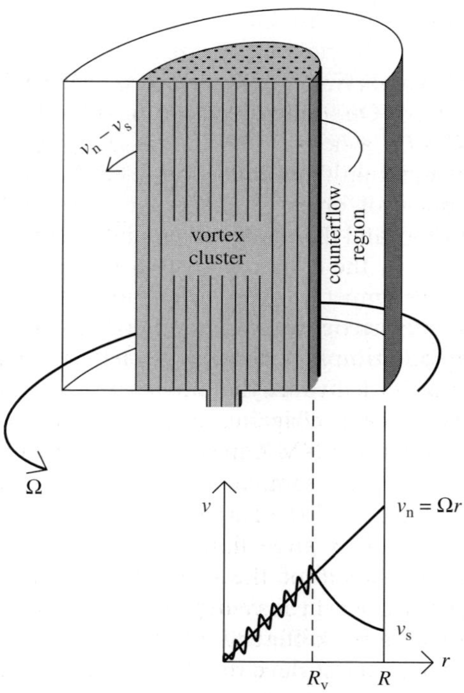

Fig. 4. (*Top*) Metastable state with a central vortex cluster in the rotating cylinder. This axially symmetric arrangement consists of rectilinear vortex lines surrounded by vortex-free counterflow. The vortex cluster corotates with the container. (*Bottom*) Radial distribution of the velocities of the superfluid component, $v_{\rm s}$, and the normal component, $v_{\rm n}$. Because of its large viscosity, the normal fraction corotates with the container in all experimentally relevant situations.

the central cluster in the vortex-free region. Within the cluster mutual repulsion keeps the lines apart such that they form an array with triangular nearest-neighbour coordination and an areal density, $n_v = 2\Omega/\kappa$, which is constant over the transverse cross section of the cluster. At this line density the cluster corotates with the container at constant Ω, like a solid body. The total number of lines is given by $N = \pi n_v R_v^2$, where R_v is the radius of the cluster.

With N vortex lines in the cluster, the superflow velocity outside has the value $v_s = \kappa N/(2\pi r)$, which is equivalent to that around a giant vortex with N circulation quanta κ (fig. 4). The counterflow velocity v vanishes inside the cluster (measured on length scales which exceed the inter-vortex distance), while outside it increases from 0 at $r = R_v$ to its maximum value at the cylinder wall, $r = R$:

$$v(R, N) = \Omega R - \frac{\kappa N}{2\pi R}. \tag{7}$$

In the absence of the neutron source, new vortex lines are only formed if $v(R)$ is increased to its container-dependent spontaneous critical value $v_c(T, P)$, by increasing the rotation velocity Ω. In neutron irradiation vortices are formed at lower velocities, but again $v(R)$ in eq. (7) is the applied counterflow velocity, the externally controlled bias, because of the short mean free path of neutron capture in liquid ^{3}He.

2.6.3. *Experimental setup*

Figure 5 shows a measuring setup for rotating NMR measurements in ^{3}He–B. The cylindrical sample is connected via an orifice to a long liquid ^{3}He column which is needed to fill the sample volume, to provide the thermal contact with the nuclear refrigeration stage, and to isolate the sample in a metal-free environment for NMR measurement. The major resistances in the thermal path between the sample and the nuclear cooling stage are the orifice, the liquid column, and the Kapitza surface resistance of the porous sintered heat exchanger in a large liquid volume below the column. The most important characteristic of the sample container is its spontaneous critical velocity $v_c(T, P)$, which may depend in complicated ways on temperature and pressure, since different sources contribute to vortex formation. These arise from (i) the surface roughness of the cylindrical walls in the volume above the orifice, (ii) sites in which remanent vortices can be trapped, and (iii) the leakage of vortices through the orifice (since the volume in contact with the rough heat exchanger surfaces will be filled with the equilibrium number of vortices already at low Ω). In practice, for a clean quartz cylinder, the most important source is some isolated surface protrusion, a localized defect or a piece of dirt, which increases the flow velocity at

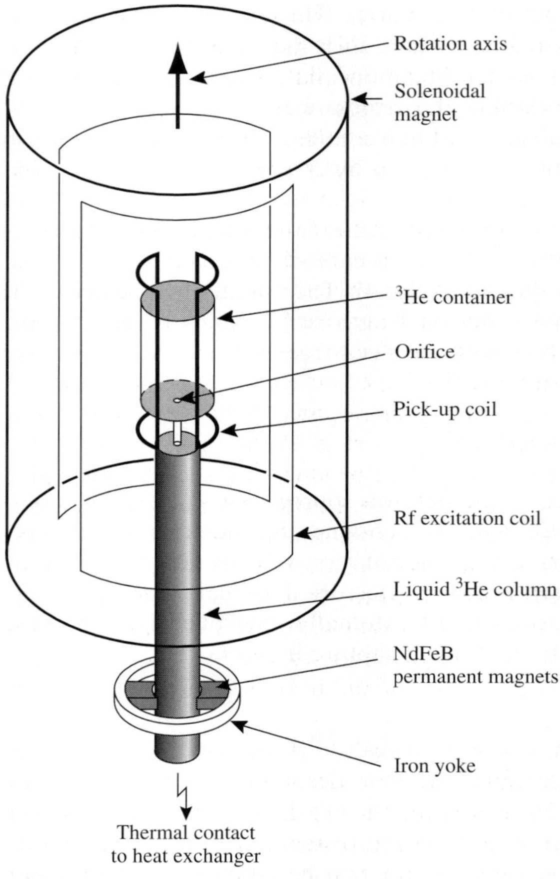

Fig. 5. Liquid ^{3}He sample container with coils for NMR measurement. The cylindrical sample on the top is connected via a narrow aperture with a long liquid ^{3}He column which serves as a thermal path to the sintered heat exchanger on the nuclear refrigeration stage. The nuclear stage is located below this structure while the mixing chamber of the precooling dilution refrigerator is above. The excitation coil and the outer magnet have not been drawn to scale (from Ruutu et al. 1997b) Copyright 1997 Plenum Publishing Coroporation; with kind permission from Springer Science and Business Media.

a sharp asperity by up to an order of magnitude (Parts et al. 1995, Ruutu et al. 1997b). To achieve a high critical velocity, the geometry and surface quality of the sample container are thus of utmost importance.

The sample container structure in fig. 5 is prepared from fused quartz. The sample cylinder itself has a radius $R = 2.5$ mm, length $Z = 7$ mm, and wall thickness 0.5 mm. An orifice of 0.5 mm diameter links it with the

60-mm-long connecting column. The long length is needed to place the sample in the middle of the NMR polarization magnet. The container is fused together from tubular and flat plate parts in an oxygen–acetylene flame. After assembly the structure is annealed in an oven, etched, and cleaned with solvents. All indications point to an average surface roughness well below 1 μm, such that it is dirt particles, and perhaps isolated localized defects, on the glass surface which ultimately control the critical velocity. For the quartz glass cylinder in fig. 5 the spontaneous critical velocity v_c was found to be a factor of 3 or more larger than the lowest velocity $v_{cn}(T, P)$ at which vortices start to appear in neutron irradiation (cf. fig. 13).

The non-invasive continuous-wave (CW) NMR measurement is conducted from the outside with a system of three orthogonal superconducting coils. The innermost is a saddle-shaped detector coil, wound from 25 μm solid Nb wire on a thin epoxy shell and mounted directly on the sample cylinder. It is thermally anchored with a Cu strip to the nuclear cooling stage. The outermost coil is a magnet which produces the axially oriented homogeneous polarization field. This end-compensated solenoid on a bronze body is fixed mechanically and thermally to the mixing chamber. The coil in the middle is used for radio frequency (rf) excitation. It consists of two turns in each half and is fixed inside the polarization magnet. All parts shown here are located inside a superconducting Nb shield, to avoid interference from the demagnetization field which is required for cooling and temperature stabilization. The Nb jacket is part of the heat shield which is fixed to the mixing chamber.

The neutron source is a weak ^{241}Am/Be specimen which is sandwiched between two paraffin moderator tiles so that a cubic box results with sides of 25 cm length. For irradiation the box is carried to the cryostat and placed at room temperature at a desired distance from the ^{3}He sample. By varying this distance the neutron flux can be adjusted to the required low value, as determined from the vortex formation rate $\dot{N}$ in fig. 8. Between measurements the source is moved a fair distance away from the cryostat (~25 m).

2.6.4. *NMR measurement*

The unusual NMR properties of the ^{3}He superfluids are a direct consequence from the Cooper pairing in states with spin $S = 1$ and orbital momentum $L = 1$. In ^{3}He–B the NMR absorption spectrum, recorded at low rf excitation with traditional transverse CW methods, is related to the spatial distribution of the order parameter orientation in the rotating cylinder. Spin–orbit coupling, although weak, is responsible for the large frequency shifts which give rise to absorption peaks shifted from the Larmor frequency (Vollhardt and Wölfle 1990). The height of the so-called *counterflow* (CF) peak, which depends on the CF velocity around the vortex cluster, can be used to determine the number of rectilinear vortex lines in the cluster.

In the top panel of fig. 6, NMR absorption spectra are shown, which have been recorded at the same Ω, but with a different number N of vortex lines

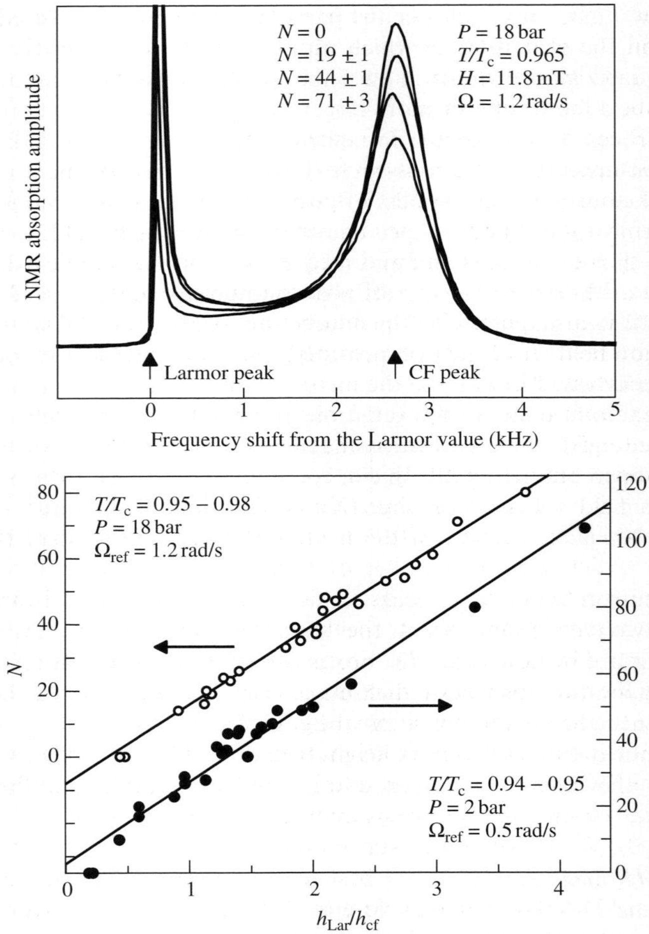

Fig. 6. NMR measurement of vortex lines in superfluid ^{3}He–B, when rotated in a long cylinder with the magnetic field oriented axially: (*Top*) NMR spectra recorded at a reference velocity $\Omega_{\text{ref}} = 1.2$ rad/s with different number of vortex lines N. The Larmor frequency is on the left and the counterflow peak on the right vertical arrow. (*Bottom*) Two calibration measurements of the number of vortex lines N as a function of the ratio of the Larmor and counterflow peak heights, $h_{\text{Lar}}/h_{\text{cf}}$, measured at two different rotation velocities Ω_{ref} (from Xu et al. 1996) Copyright 1996 Institute of Physics ASCR; with kind permission from Springer Science and Business Media.

in the vortex cluster. In the vortex-free state the large CF peak on the right increases rapidly with Ω when the orienting effect from the CF grows. The absorption intensity for this growth in peak height is shifted from the asymmetric peak on the left close to the Larmor frequency. This second peak reflects the order parameter orientation in the center of the container in the region of the vortex cluster. At fixed Ω, if more vortex lines are added, the central vortex cluster expands, the CF velocity outside the cluster is reduced, and the absorption intensity from the CF peak is shifted back to the Larmor peak (Kopu et al. 2000).

The frequency shift of the CF peak increases monotonically with decreasing temperature and is used to measure the temperature. The total integrated absorption in the NMR spectrum is proportional to the B phase susceptibility $\chi_B(T, P)$ and decreases rapidly with temperature. However, by maintaining the CF peak at fixed frequency shift, the temperature is kept stable, and the relative amount of absorption in the Larmor and CF peaks can then be used to determine the number of vortex lines. By monitoring the peak height of either of the two absorption maxima, one can detect a change in N, as illustrated in fig. 3.

In the bottom panel of fig. 6 the ratio of the two peak heights has been calibrated to give the vortex number at two different angular velocities Ω_{ref}. In both cases only a relatively small number of vortex lines $N \ll N_{\mathrm{eq}}$ is present and the CF peak is large (Xu et al. 1996). These calibration plots were measured by starting with an initially vortex-free sample rotating at Ω_{ref}, into which a given number of vortex lines N was introduced with neutron absorption reactions, as in fig. 3. After that the irradiation was stopped and the entire NMR spectrum was recorded (by sweeping the magnetic polarization field H). From the spectrum the ratio of the peak heights of the two absorption maxima is worked out and plotted against N. This gives the linear relationships in fig. 6.

The reduction in the CF peak height from the addition of one single-vortex line can be discerned with good resolution under favorable conditions (fig. 3). The optimization of this measurement has been analyzed by Kopu et al. (2000). With a small-size superconducting magnet and its modest field homogeneity ($\Delta H/H \sim 10^{-4}$), the best conditions are usually achieved at low magnetic field ($H \sim 10-20$ mT) close to T_c ($T \geqslant 0.8 T_c$) and at relatively high CF velocity ($\Omega > 0.6$ rad/s), where the CF peak is well developed (Korhonen et al. 1990). The single-vortex signal, i.e. the change in peak height per vortex line, decreases with increasing Ω and CF peak height, as shown in fig. 7. In neutron irradiation measurements, the rate of vortex line creation $\dot{N}$ is determined directly from records like the two traces in fig. 3. The plot in fig. 7 provides a yard stick for the single-vortex signal, to estimate the number of new lines if multiple lines are created in one neutron absorption event.

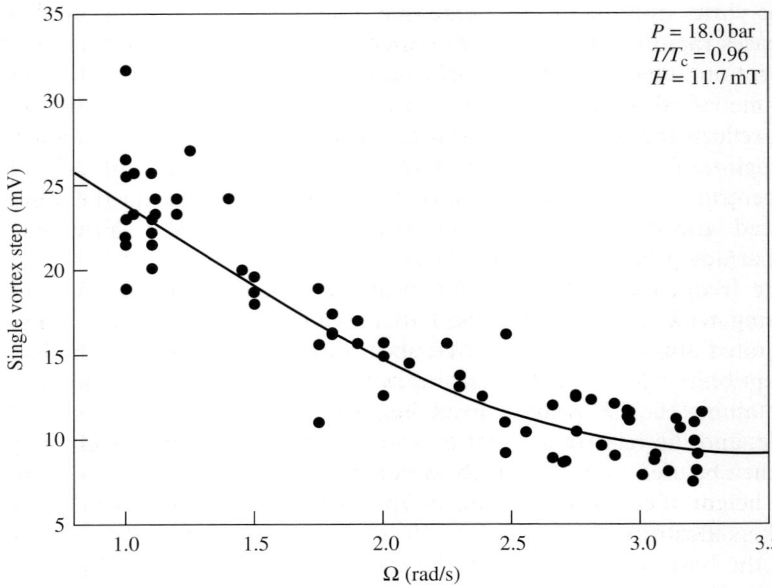

Fig. 7. Absolute magnitude of the single-vortex signal (as in fig. 3), measured in the vortex-free counterflow state at different Ω. The reduction in the height of the counterflow peak is given here in millivolts at the output of the cooled preamplifier which has a gain of ≈ 10 and is located inside the vacuum jacket of the cryostat. The scatter in the data arises from residual differences in the measuring temperature, perhaps from small differences in the order-parameter texture from one cool down to the next, and from an electronic interference signal which is generated by the rotation of the cryostat. More details on the NMR spectrometer are given by Parts et al. (1995) and Ruutu et al. (1997b).

2.7. *Vortex formation in neutron irradiation*

A measurement of vortex-line formation in neutron irradiated normal-superfluid counterflow is performed at constant ambient conditions. The externally controlled variables include the rotation velocity Ω, temperature T, pressure P, magnetic field H, and neutron flux ϕ_n. The initial state is one of vortex-free counterflow ($N = 0$). When stable conditions have been reached, a weak neutron source is placed at a distance d from the ^{3}He–B sample and the output from the NMR spectrometer is monitored, as shown in fig. 3. From this NMR absorption record as a function of time, the vortex lines can be counted which are formed during a given irradiation period.

The process evidently exhibits stochastic variation. To measure the vortex-formation rate $\mathrm{d}N/\mathrm{d}t = \dot{N}$, records with a sufficiently large number of detected neutron absorption events are analyzed to obtain a representative result. After that one of the experimental parameters is changed and a new run is performed. In this way the dependence of the vortex formation process on the external variables can be studied. It was found that $\dot{N}$ varies as a function of all the external parameters, i.e. neutron flux, rotation velocity, temperature, pressure, and magnetic field. In the following we shall first describe these empirically established dependences.

The rate of vortex-line formation is proportional to the neutron flux. The latter is varied by changing the distance of the source from the cryostat. In this way individual absorption events can be studied, which are well separated in time. A calibration plot of the measured rate $\dot{N}$, as a function of the distance between sample and source, is shown in fig. 8. By means of this plot the results can be scaled to correspond to the same incident neutron flux, for example with the source at the minimum distance $d = 22$ cm, which is given by the outer radius of the liquid He dewar.

A most informative feature is the dependence of $\dot{N}$ on rotation velocity (Ruutu et al. 1996a, c, 1998a). Rotation produces the applied bias velocity at the neutron absorption site. Since the absorption happens close to the wall of the container, the velocity of this bias flow is approximately equal to

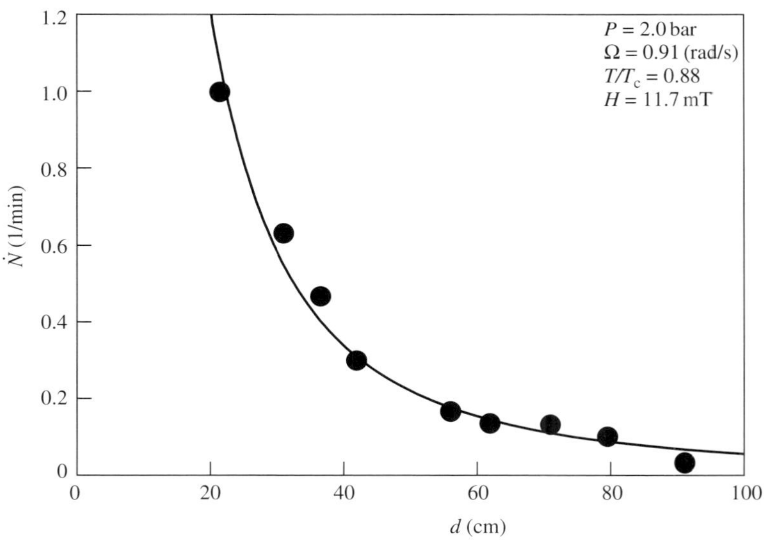

Fig. 8. Rate of vortex-line formation in neutron-irradiated counterflow, $\dot{N}$, as a function of the distance d between neutron source and ^{3}He–B sample. The initial state for all data points is vortex-free counterflow at a rotation velocity of 0.91 rad/s. The neutron irradiation time is 30 min. The fitted curve is of the form $\ln[1 + (R_s/d)^2]$, where R_s is the radius of the front surface of the paraffin moderator box (perpendicular to **d**), in which the Am–Be source is embedded.

$v(R, N)$ (eq. (7)), the counterflow velocity at the cylinder wall. We shall denote this velocity simply with v. It depends both on the angular velocity Ω and the number of vortex lines $N(t)$, which are already present in a central cluster. The bias flow provides the force which allows vortex rings to escape from the heated neutron bubble and to expand to rectilinear vortex lines, which are then preserved in the central cluster.

These features can be seen in fig. 9. It demonstrates that it is the counterflow velocity v which is the important variable governing the rate $\dot{N}$ and not, for instance, that of the normal component, $v_n(R) = \Omega R$. Here the total vortex number $N(t)$ is recorded during neutron irradiation at constant external conditions over a time span of 6 h. The irradiation is started from the vortex-free state: $N(t=0) = 0$. The rate $\dot{N}(t)$, at which vortex lines accumulate in the center of the container, is not constant: initially both the counterflow velocity and the rate of vortex formation are the highest. When more vortex lines collect in the central cluster, both the velocity v and the rate $\dot{N}$ fall off. Finally, the vortex number N approaches a saturation value, beyond which no more lines are formed. This means that there exists

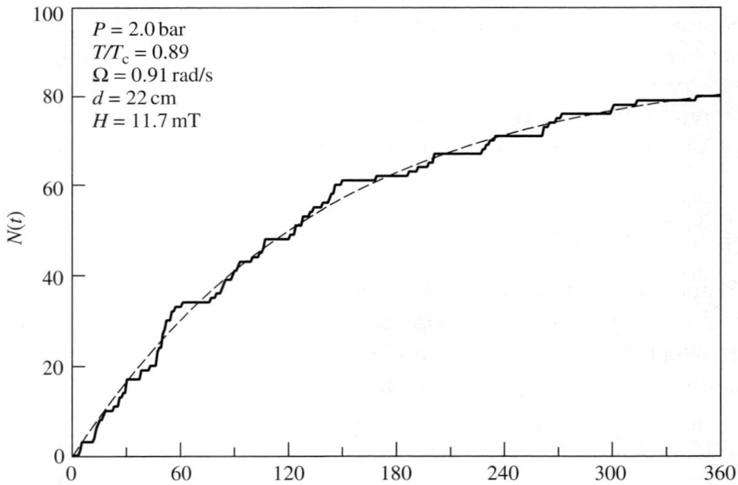

Fig. 9. Cumulative number of vortex lines $N(t)$ as a function of time t, after turning on the neutron irradiation at constant flux on an initially vortex-free ^{3}He–B sample rotating at constant Ω. The dashed curve represents a fit to eq. (10) with $\gamma = 1.1\ \text{min}^{-1}$ and $v_{cn} = 1.9$ mm/s. The result demonstrates that the rate of vortex-line formation is controlled by the counterflow velocity $v(R, N)$ and not by that of the normal component $v_n(R) = \Omega R$, which is constant during this entire measurement (from Ruutu et al. 1996c) Copyright 1996 Institute of Physics ASCR; with kind permission from Springer Science and Business Media.

a lower limit, a threshold value for the counterflow velocity v, below which the neutron absorption events produce no vortex lines at all.

The dependence of $\dot{N}$ on the counterflow velocity v is obtained from experiments like that in fig. 9. However, more efficient is a measurement of only the initial slope of the $N(t)$ record. Typically 15–30 min accumulation periods are then enough. Only a small number of vortices form during such a run compared to the number of vortices in the equilibrium state. Thus the total decrease in the counterflow velocity is small and can be accounted for by assigning the average of the velocities before and after irradiation as the appropriate value of the applied bias velocity. An example of the measured rates is shown in fig. 10.

These measurements reveal a vortex-formation rate $\dot{N}$ as a function of v which has an onset at a threshold velocity v_{cn}, followed by a rapid non-linear increase. As shown in fig. 3, close above the threshold a successful neutron absorption event produces one vortex line, but at high flow rates many lines may result. Therefore the non-linear dependence $\dot{N}(\Omega)$ arises in the following manner: initially the fraction of successful absorption events increases, i.e. of those which produce one new vortex line. Eventually this effect is limited by the neutron flux, when essentially all absorbed neutrons produce at least one new vortex. The second part of the increase is brought about by the fact that more and more lines are produced, on average, in each absorption event.

As demonstrated in fig. 11 the dependence of the rate $\dot{N}$ on the counterflow velocity v can be approximated by the empirical expression

$$\dot{N}(v) = \gamma\left[\left(\frac{v}{v_{cn}}\right)^3 - 1\right] \tag{8}$$

with γ and v_{cn} as parameters. This equation also describes the results in fig. 9: if initially v in the vortex-free state is only slightly larger than the critical velocity v_{cn}, then eq. (8) can be linearized to $\dot{N} \simeq 3\gamma(v/v_{cn} - 1)$. Thus $N(t)$ is obtained by integration from

$$\dot{N}(t) = 3\gamma\left(\frac{v(t=0) - \kappa N(t)/(2\pi R)}{v_{cn}} - 1\right) \tag{9}$$

with the solution

$$N(t) = \frac{2\pi R}{\kappa}\,[v(t=0) - v_{cn}]\left[1 - \exp\left(-\frac{3\gamma\kappa t}{2\pi R v_{cn}}\right)\right]. \tag{10}$$

This equation has been fit to the measurements in fig. 9, to give the two parameters γ and v_{cn}. The resulting values agree with those of the horizontal and vertical zero intercepts in fig. 11.

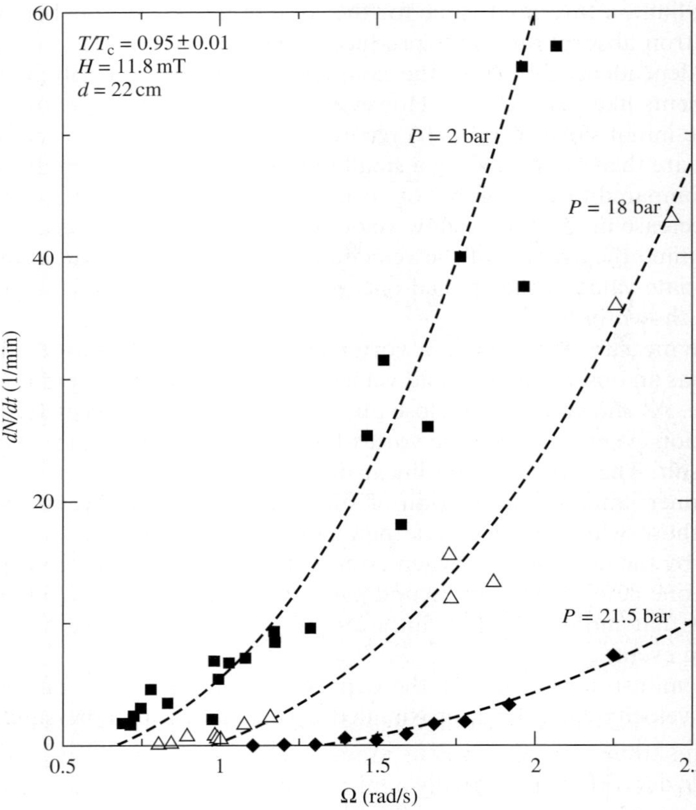

Fig. 10. Initial rate $\dot{N}$ of vortex line formation at different rotation velocities Ω. $\dot{N}$ has been measured as the average of all vortex lines formed during an accumulation period of 15–30 min. The fitted curves are of the form $\gamma[(\Omega/\Omega_{cn})^3 - 1]$ (eq. (8)). Their low velocity endpoints determine the threshold velocity $v_{cn} = \Omega_{cn} R$.

At high CF velocities a large number of vortex lines is produced in each neutron absorption event. The accumulation record for one neutron irradiation session (similar to the upper trace in fig. 3) then consists of many large steps where, in the worst case, two absorption events might even be overlapping. Nevertheless, the final total number of lines, which have been collected into the central cluster, can be determined in a few different ways and these give consistent answers. These tests have been described in fig. 12. One of them includes the use of the rate equation (8) where the multiplier γ has the value from fig. 11.

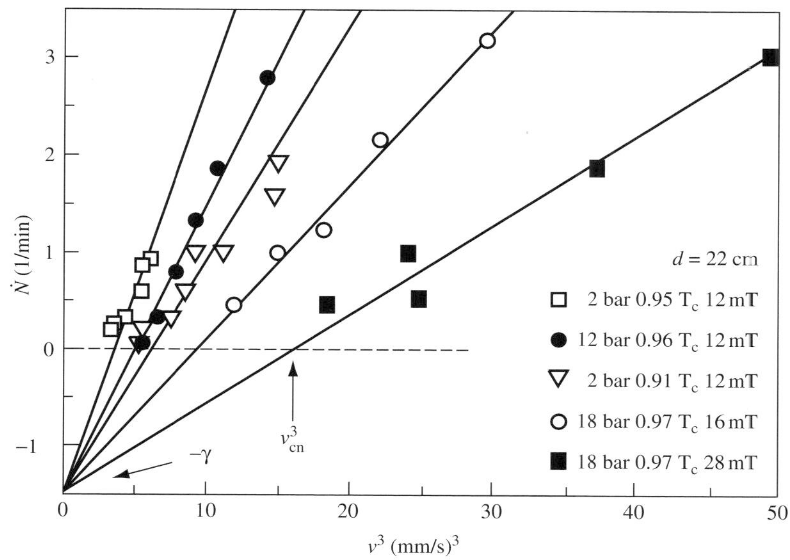

Fig. 11. Vortex-formation rate $\dot{N}$ plotted against the cube of the bias velocity v. This plot is used to determine the threshold velocity v_{cn}. A line has been fit to each set of data, measured under different external conditions, to identify the horizontal intercept v^3_{cn} and the common vertical intercept $-\gamma = -1.4\ \mathrm{min}^{-1}$ (from Ruutu et al. 1996a).

In fact, it can be guessed from fig. 11 that eq. (8) has wider applicability than is apparent from the present examples: the same equation holds universally under different externally applied conditions with all dependence on external variables contained in $v_{cn}(T, P, H)$, which is the critical or threshold velocity for vortex formation in neutron irradiation. The rate factor γ is proportional to the neutron flux but does not depend on temperature, pressure, or magnetic field. Hence $\dot{N}(v/v_{cn})$ appears to be a universal function for all measurements that have been performed in the temperature regime $T > 0.8T_c$.

To summarize, the experimental result displays two distinguishing features: (1) the cubic dependence on the bias v and (2) the universality that all dependence of the vortex formation properties on the experimental variables T, P, and H is contained in the threshold velocity v_{cn}.

2.8. *Volume or surface mechanism?*

In the following sections we examine the various experimental properties of the neutron-induced vortex formation. The analysis demonstrates that most

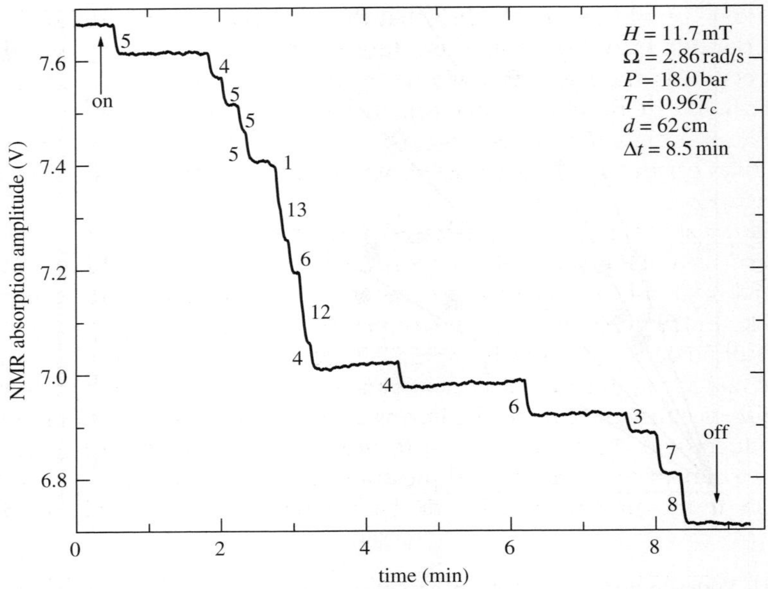

Fig. 12. Neutron irradiation and vortex-line yield at high CF velocity. The total number of lines, which are created in one irradiation session, can be determined in several ways: (1) In this example the direct count of the steps in signal amplitude yields 88 lines (cf. fig. 3). The number next to each step denotes its equivalent in rectilinear vortex lines. (2) From fig. 7 one finds that the drop in signal amplitude per vortex line is 11 mV under the present measuring conditions. The total amplitude drop of 960 mV corresponds thus to 87 vortex lines. (3) A measurement of the annihilation threshold, by successive deceleration to lower and lower Ω (Ruutu et al. 1997b), gives the total number of lines in the vortex cluster. Here this measurement yields $\Omega_v = 0.245$ rad/s. The annihilation threshold, which at $\Omega_v \gtrsim 0.2$ rad/s coincides with the equilibrium vortex state, corresponds to $N = \pi R^2 2\Omega_v \kappa^{-1}(1 - 0.18\sqrt{1\ \text{rad/s}/\Omega_v})$, as measured by Ruutu et al. (1998b). With $R = 2.45$ mm, this gives 88 vortex lines. (4) The rate equation (8), with the measured values of γ and v_{cn} from fig. 11, gives 88 lines for an irradiation time of 8.5 min (with $\Omega_{cn} = 0.75$ rad/s). Thus all of the reduction in the height of the counterflow peak (cf. fig. 6) can be reliably accounted for in many different ways.

features are consistent with the KZ mechanism – none have so far been found contradictory. But does this fact constitute solid proof of the mechanism?

Neutron-induced vortex formation in a rotating superfluid does not exactly fit the ideal KZ model – a rapid second-order transition in an infinite homogeneous system. In the heated neutron bubble there is a strong thermal gradient and a strict boundary condition applies at its exterior, imposed by the bulk superfluid state outside. The comparison of experiment and model is further complicated by the fact that the cool down occurs so fast that an extrapolation from the equilibrium state theories becomes uncertain, whether it concerns the hydrodynamics or even the superfluid state itself. (The applicability of the KZ mechanism to inhomogeneous transitions is considered in more detail in section 2.15.) Perhaps, other processes of hydrodynamic origin can be suggested which account for the experimental observations?

The most viable alternative arises from the boundary condition in the presence of the applied bias flow: it is the instability of superflow along the normal–superfluid interface in the outer peripheral regions of the neutron bubble. In the Ginzburg–Landau temperature regime the container-dependent spontaneous critical velocity $v_c(T, P)$ decreases with increasing temperature and vanishes at T_c (dashed curves in fig. 13). It is related (via surface roughness enhancement) to the intrinsic instability velocity $v_{cb}(T, P)$ of the bulk superfluid (Parts et al. 1995, Ruutu et al. 1997b), which has qualitatively a similar temperature and pressure dependence, but is larger in magnitude. In the outermost region of the neutron bubble the fluid remains in

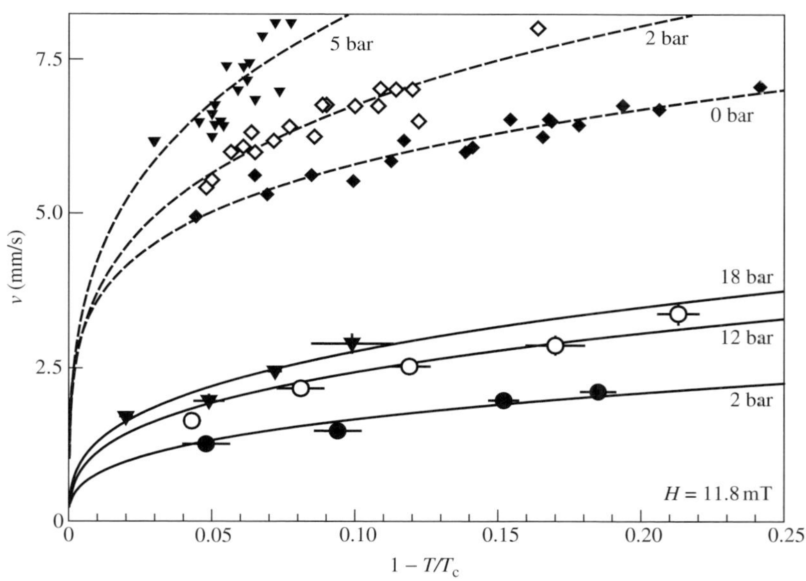

Fig. 13. Critical values of the applied bias velocity v vs. temperature at different pressures: (*Solid lines*) Threshold velocity $v_{cn} \propto (1 - T/T_c)^{1/3}$ for vortex formation in neutron irradiation (Ruutu et al. 1996a). (*Dashed lines*) Critical velocity $v_c \propto (1 - T/T_c)^{1/4}$ of spontaneous vortex formation in the same quartz glass container in the absence of the neutron source (Parts et al. 1995, Ruutu et al. 1997b).

the B phase, but is heated above the surrounding bulk temperature T_0. Consequently, if no other process intervenes, the superflow instability has to occur within a peripheral shell surrounding the hot neutron bubble where $v_{cb}(T, P)$ drops below the applied bias v. This process was discussed by Ruutu et al. (1998b) and compared to rotating measurements, but was found incompatible with the measured results. As explanation it was suggested that the KZ mechanism is inherently the fastest process for creating vortices.

The interplay between the interior volume of the neutron bubble and its boundary was examined by Aranson et al. (1999) in a rapidly cooling model system with a scalar order parameter, using the thermal diffusion equation to account for cooling and the TDGL model for order-parameter relaxation. Analytic calculations within this model showed that the normal–superfluid interface becomes unstable in the presence of superflow along the interface. Numerical calculations (Aranson et al. 1999, 2001) further confirmed that the instability develops such that vortex rings are formed. The rings encircle the bubble perpendicular to the applied flow and screen the superflow, so that the superfluid velocity (in the rotating frame) is 0 inside the bubble. The number of rings depends on the shape and size of the bubble and on the bias velocity. In this process it does not matter what the state of the liquid is inside the bubble: the rings are formed within an external cooler layer, which remains throughout the process in the B phase. After their formation, the rings start to expand and might eventually be pulled away by the Magnus force. Thus these rings could be the source for the vortex lines which are observed in the rotating measurements.

The calculations also confirmed that in addition to the superflow instability at the bubble boundary, the KZ mechanism in the bubble interior is also present – thus in this calculation both processes produce vortices. In the simulation the loops, which eventually manage to escape into the bulk bias flow, originate from the boundary while the random vortex network in the interior collapses and decays away, since the counterflow is screened by vortices at the boundary. Thus Aranson et al. confirmed the presence of the KZ mechanism, but concluded that the vortices observed in the rotating measurements are produced by the boundary instability. The TDGL treatment with one common relaxation time for both the phase and the magnitude of the order parameter is rigorously obeyed only within a narrow temperature interval from T_c down to $1 - T/T_c \sim 10^{-3}$. The bulk liquid temperature in the rotating measurements is below 0.98 T_c (see fig. 13). Another difference between experiment and calculation lies in the size of the neutron bubble: the numerical calculation is carried out on a grid with unit length smaller than the coherence length $\xi(T_0)$. To make this tractable, the reaction energy E_0 is reduced by one or two orders of magnitude. This scales down the neutron bubble radius R_b via the thermal diffusion equation, which is used to model the neutron bubble (see section 2.9.1). A smaller bubble (see eq. (14)) means a faster velocity of the phase front ($\propto 1/R_b$). At present time it is not known how reliably the calculation with its differences extrapolates into the range of the measurements.

How do we, in fact, distinguish from measurements whether it is the KZ mechanism or the boundary instability which is responsible for the observed vortex lines? Clearly the critical velocity v_{cn} does not discriminate between them. In both cases v_{cn} is the velocity of the bias flow at which the largest vortex ring which fits in the bubble is stable, as will be shown in section 2.9. However, the different nature of the two processes has important consequences for the vortex formation rate $\dot{N}(v)$. The row of vortex rings, which is produced by the boundary instability to screen the interior of the neutron bubble from the bias flow, resembles a vortex sheet. The sheet is formed in the heated peripheral shell of the neutron bubble, where the outside superfluid moves with velocity v along the bubble interface and the superfluid inside is stationary. The density of vorticity in the sheet is v/κ and the number of loops produced by one neutron absorption event is $N \propto vR_b/\kappa$, where R_b is the size of the bubble along the flow direction. Thus $N(v)$ grows linearly with the applied bias flow velocity, $N(v) \propto v$, which agrees with the result of Aranson et al. (1999). In contrast, the KZ mechanism is a volume effect, which results in the cubic dependence expressed in eq. (8) (see also section 2.12.2). The measured result in the figures of the previous section is not linear, but is consistent with the cubic dependence.

Are there other arguments emerging from the measurements which would support the KZ origin of the observed vortices? Several more can be listed in addition to the rate dependence $\dot{N}(v)$, which favor a volume effect: (i) the boundary instability is expected to be a deterministic process which in every neutron event produces the same surface density of vortex rings. Variations in the number of rings arise only owing to variations in the shape of the neutron bubble and its orientation with respect to the bias flow. (Unfortunately, the distribution of the number of rings was not studied in detail in the simulations by Aranson et al. 2001.) In contrast, the KZ mechanism produces a random vortex network from which the number of loops extracted by the bias flow varies and follows a stochastic distribution (fig. 29). (ii) The boundary instability should not be particularly sensitive to the processes inside the heated neutron bubble and only regular mass-flow vortices are expected to form in such an instability. The KZ mechanism, on the other hand, can be expected to produce all possible defects. Their presence might be either directly observed in the final state or via their influence on the evolution of the vortex network inside the neutron bubble. Such observations are discussed in section 2.10.

2.9. *Threshold velocity for vortex loop escape*

The dependence on the counterflow bias v can be studied from the threshold $v_{\rm cn}$ up to the critical limit $v_{\rm c}$ at which a vortex is spontaneously nucleated at the cylindrical wall in the absence of the neutron flux (Parts et al. 1995, Ruutu et al. 1997a). The threshold velocity $v_{\rm cn}(T, P, H)$ is one of the features which can be examined to learn more about neutron-induced vortex formation. Measurements of both critical velocities, $v_{\rm cn}$ and $v_{\rm c}$, are shown in fig. 13 as a function of temperature.

2.9.1. *Properties of threshold velocity*

By definition, the threshold $v_{\rm cn}$ represents the smallest bias velocity at which a vortex ring can escape from the neutron bubble after the absorption event. It can be connected with the bubble size in the following manner. A vortex ring of radius $r_{\rm o}$ is in equilibrium in the applied bias flow at v if it satisfies the equation

$$r_{\rm o}(v) = \frac{\kappa}{4\pi v} \ln\left(\frac{r_{\rm o}(v)}{\xi(T,P)}\right). \tag{11}$$

As explained in more detail in section 2.12.2, a ring with a radius larger than $r_{\rm o}$ will expand in the flow while a smaller one will contract. Thus the threshold or minimum velocity at which a vortex ring can start to expand towards a rectilinear vortex line corresponds to the maximum possible vortex-ring size. This must be comparable to the diameter of the heated bubble. For a simple estimate we set the vortex ring radius equal to that of a spherical neutron bubble: $r_{\rm o}(v_{\rm cn}) \sim R_{\rm b}$. (In fact, the numerical simulations to be described in section 2.14.3 suggest that $r_{\rm o}(v_{\rm cn}) \approx 2R_{\rm b}$ because of the complex convoluted shape of the largest rings in the random vortex network.)

A simple thermal diffusion model can be used to yield an order of magnitude estimate for the radius $R_{\rm b}$ of the bubble which originally was heated above $T_{\rm c}$. In the temperature range close to $T_{\rm c}$ the cooling occurs via diffusion of quasiparticle excitations out into the surrounding superfluid with a diffusion constant $D \approx v_{\rm F} l$, where $v_{\rm F}$ is their Fermi velocity and l their mean free path. The difference from the surrounding bulk temperature T_0 as a function of the radial distance r from the centre of the bubble can be calculated from the diffusion equation

$$\frac{\partial T(r,t)}{\partial t} = D\left(\frac{\partial^2 T}{\partial r^2} + \frac{2}{r}\frac{\partial T}{\partial r}\right). \tag{12}$$

With the assumption that at $t = 0$ the reaction energy E_0 is deposited at $r = 0$, the solution is given by

$$T(r,t) - T_0 \approx \frac{E_0}{C_v} \frac{1}{(4\pi Dt)^{3/2}} \exp\left(\frac{-r^2}{4Dt}\right), \tag{13}$$

where C_v is the specific heat. For now, we may assume that all the energy of the neutron absorption reaction is deposited as heat. The bubble of normal fluid, $T(r) > T_c$, first expands and reaches a maximum radius

$$R_b = \sqrt{\frac{3}{2\pi e}} \left(\frac{E_0}{C_v T_c}\right)^{1/3} (1 - T_0/T_c)^{-1/3}. \tag{14}$$

It then starts cooling and rapidly shrinks with the characteristic time $\tau_Q \sim R_b^2/D \sim 1\ \mu s$. Since v_{cn} is inversely proportional to $r_o \sim R_b$, it has the temperature dependence $v_{cn} \propto (1 - T_0/T_c)^{1/3}$. This is in agreement with the solid curves in fig. 13 which have been fitted to measurements on v_{cn}. The prefactor of these curves is in agreement with that from eqs. (14) and (11) within a factor of ~ 2, and its increase with increasing pressure is well described by the decrease in bubble size according to eq. (14), where C_v and T_c increase with pressure (Greywall 1986).

2.9.2. *Influence of ^{3}He-A on the threshold velocity*

Measurements on the threshold velocity v_{cn} offer interesting possibilities to study whether the interior of the heated bubble participates in vortex formation or not. In the left panel of fig. 14 the phase diagram of ^{3}He is shown at temperatures below 2.5 mK. In the measurements which we discussed so far, the liquid pressure has been below 21.2 bar, so that the heated bubble cools from the normal phase directly into the B phase (along trajectory b). On the top right in fig. 14 the measured v_{cn} is plotted as a function of pressure P at constant reduced temperature T/T_c. Here the pressure dependence displays an abrupt increase at about 21.2 bar, the pressure P_{PCP} of the polycritical point: it thus makes an unexpectedly large difference whether the quench trajectory follows a path denoted by a or by b! The two cooling trajectories differ in that above P_{PCP} a new phase, ^{3}He-A, is a stable intermediate phase between the normal and B phases. Consequently, although the bulk liquid is well in the B phase in all of the measurements of fig. 14, vortex formation is less than expected on the basis of extrapolations from lower pressures, when the quench trajectory crosses the stable A-phase regime.

The fitted curves in fig. 14 represent $v_{cn}(P, T)$ at pressures below P_{PCP}:

$$v_{cn} = \mathscr{A}\kappa/(4\pi R_b)\ \ln(R_b/\xi), \tag{15}$$

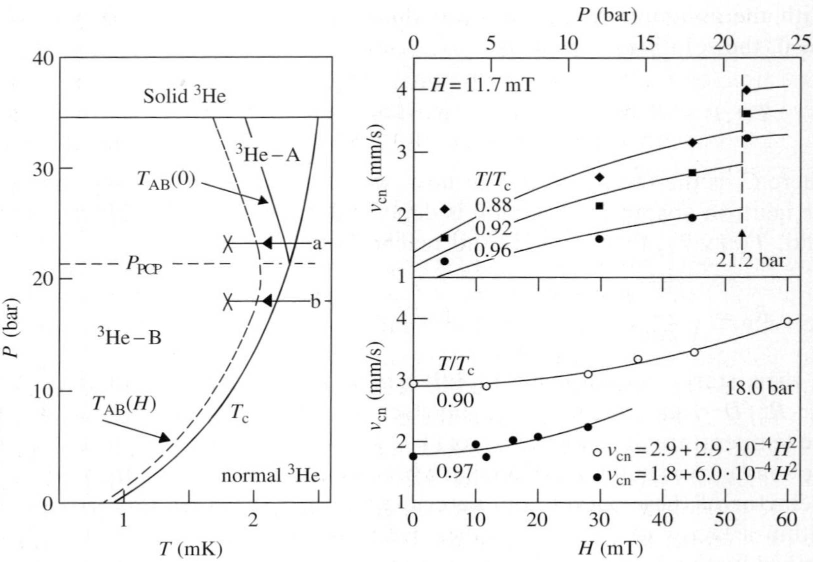

Fig. 14. Threshold velocity v_{cn} for the onset of vortex formation during neutron irradiation: (*Left*) Phase diagram of ^{3}He superfluids in the pressure vs. temperature plane, with the A $\rightarrow$ B transition at $T_{AB}(0)$ in zero field (solid line) and at $T_{AB}(H)$ in non-zero field (dashed line). Two quench trajectories, distinguishing different types of measurements on the right, are marked with (a) and (b). (*Right top*) The pressure dependence of v_{cn} displays a steep change at the pressure P_{PCP} of the polycritical point. (*Right bottom*) The magnetic field dependence of v_{cn} is parabolic, similar to that of the equilibrium state A $\rightarrow$ B transition $T_{AB}(H)$ (Reprinted figure with permission from Ruutu et al. 1998a Copyright (1998) by American Physical Society).

where R_b is obtained from eq. (14). If we assume that all of the reaction energy is transformed to heat ($E_0 = 764$ keV), then the common scaling factor $\mathscr{A}$ of the three curves has the value $\mathscr{A} = 2.1$. This is the only fitting parameter in figs. 13 and 14 where the same fit is compared to measurements as a function of both temperature and pressure. The agreement is reasonable and suggests that the spherical thermal diffusion model is not too far off. However, as seen in fig. 14, v_{cn} is clearly above the fit at $P > P_{PCP}$. Also this offset is largest at the highest temperature, which in fig. 14 is $0.96T_c$. Both features suggest that vortex formation is reduced when the relative range of A-phase stability increases over the quench trajectory below T_c.

The lower right panel of fig. 14 shows the dependence of v_{cn} on the applied magnetic field H. Below P_{PCP} the magnetic field acts to stabilize ^{3}He-A in a narrow interval from T_c down to the first-order A $\rightarrow$ B transition at $T_{AB}(P, H)$. This result confirms our previous conclusion: v_{cn} again increases when the range of A-phase stability increases. The parabolic magnetic field dependence of $v_{cn}(H)$ in the lower right panel is reminiscent of that of the equilibrium $T_{AB}(H)$ transition temperature.[2] It should be noted that the measurements in the upper right panel of fig. 14 have not been carried out in zero magnetic field but in 11.7 mT. However, as shown in the lower panel, the field dependence in the range 0–12 mT is not visible within the experimental precision.

Thus by increasing the pressure above P_{PCP} or by increasing the magnetic field we reduce the A-phase energy minimum relative to that of the B phase and both operations act to increase v_{cn}. If we go back to figs. 10 and 11, we note that both operations seem to leave the rate equation (8) and its rate constant γ unchanged. To be more careful, we conclude that in first order the changes from increased A-phase stability appear to affect only v_{cn}. This conclusion is compatible with the KZ model, as will be discussed in the following sections. What about the competing model, the superflow instability at the boundary of the heated neutron bubble, can it also account for the observations in fig. 14? At first glance this does not appear to be the case.

The superflow instability occurs in B phase in the outer periphery of the heated bubble in a shell where the temperature is close to T_c. In principle this phenomenon should not be influenced by the appearance of A phase in the interior of the bubble during its cool down to the bath temperature T_0. However, it can be affected by the presence of an A-phase shell around the hot bubble, bounded by an inner surface $T = T_c$ and an outer $T = T_{AB}$, i.e. a region where the temperature exceeds the AB transition temperature $T_{AB}(P, H)$. The B-phase bulk critical velocity is approximately equal to the pair-breaking velocity, and in the Ginzburg–Landau temperature regime it is of the form (Vollhardt et al. 1980, Kleinert 1980)

$$v_{cb}(T, P) \approx v_{c0}(P)\sqrt{1 - T/T_c},$$

where $v_{c0}(P) = 1.61\,(1 + F_1^s/3)k_B T_c/p_F$. This velocity is a smooth function of pressure and cannot account for the steep rise of v_{cn} at P_{PCP} in fig. 14 (upper right panel). The B-phase boundary instability occurs within an outer shell where the inner surface is defined by $T = T_c$ and the outer by the condition $v_{cb}(T, P) = v$. For the data in fig. 14 the latter condition puts the outer surface of this shell at $1 - T/T_c \sim 10^{-3} - 10^{-5}$, which falls inside the shell within which A phase is stable. Thus here it is the A-phase shell within which the boundary instability has to occur. Such a situation is complicated (see Parts et al. 1993, Blaauwgeers et al. 2002) and possibly one where the escape of vorticity from the A-phase shell into B phase is reduced. In first order these considerations suggest that v_{cn} increases with increasing pressure

[2]According to Tang et al. (1991), below P_{PCP}, the equilibrium $A \leftrightarrow B$ transition temperature is in first order of the form $T_{AB}(P, H) = T_c(P)(1 - bH^2)$, where $b(P) \sim (0.5 - 10) \times 10^{-6}\,(\mathrm{mT})^{-2}$.

and magnetic field, when the range of the stable A-phase regime increases over the length of the quench trajectory as a function of temperature. However, it is unclear whether in a three-phase problem – where an outer region of warm superfluid surrounds a hot interior with normal liquid, while both are embedded in a cold bath of B phase – coherence is established such that we can apply shell structure considerations. Therefore, in the absence of more quantitative estimates, we must conclude that the measurements in fig. 14 do not yet allow a preference between the KZ model or the superflow instability at the neutron bubble boundary. Clearly more extensive measurements are needed to study rapid thermal quenches in the three-phase situation, as will be emphasized in section 2.13.2.

2.10. *Other defect structures formed in neutron irradiation*

2.10.1. *Radiation-induced supercooled* A $\rightarrow$ B *transition*

The first-order transition of supercooled ^{3}He-A to ^{3}He–B is another experimentally confirmed phenomenon which is catalyzed by the interior of the neutron bubble. Usual arguments about homogeneous first-order phase transitions show that the A $\rightarrow$ B transition is forbidden (Leggett 1984). The normal phase symmetry $SO(3) \times SO(3) \times U(1)$ of liquid ^{3}He can be broken to the $U(1) \times U(1)$ symmetry of the A phase or to the $SO(3)$ symmetry of the B phase, with only a small energy difference between these two states, but separated by an extremely high energy barrier ($\sim 10^6 k_B T_c$). Nevertheless, so far it has not been reported that the transition would not have taken place ultimately, if only one cools to sufficiently low temperature.

The A $\rightarrow$ B transition has not been found to be affected by rotation (Hakonen et al. 1985) and in the absence of ionizing radiation it is generally believed to be caused by extrinsic sources, perhaps at solid surfaces. A dramatic increase in the transition probability has been shown to take place in the bulk liquid in the presence of γ radiation or thermal neutrons (Schiffer et al. 1992, Schiffer and Osheroff 1995). A generally accepted understanding of this feature is still missing, because two quite different models, the "baked Alaska" (Leggett 1992, Schiffer et al. 1995, Leggett 2002) and the KZ mechanism extended to domain wall formation (Volovik 1996, Bunkov and Timofeevskaya 1998a, b), are discussed in this context. Since only the latter is appropriate in radiation-induced vortex formation, we briefly describe it in the context of the supercooled A $\rightarrow$ B transition. Domain wall formation in a quench-cooled transition was originally suggested in the cosmological context by Kobzarev et al. (1974).

When ^{3}He-A is supercooled to low temperatures, below the thermodynamic $A \leftrightarrow B$ transition, the sample stays in the A phase for long times, perhaps even indefinitely, depending on temperature and other largely unknown requirements. The deeper the supercooling, the faster the transition into the B phase follows. With increasing magnetic field the transition probability decreases, since a magnetic field acts to stabilize the A phase. However, when neutron irradiation is turned on, the transition probability increases sharply. In such measurements the initial state is supercooled ^{3}He-A at varying temperatures and magnetic fields. Consequently here the B-phase state is enforced by the boundary condition around the hot neutron bubble: the boundary condition is not the source for the A $\rightarrow$ B transition! Rather the new B-phase state has to emerge from the interior of the bubble.

According to the KZ model, in different parts of the rapidly cooling neutron bubble the order parameter may fall initially either in the A- or B-phase energy minima. Thus a domain structure of size ξ_v (eq. (3)) is laid down, where A- and B-phase blobs form a patch work. When these blobs later grow together, then AB interfaces are formed. The lower the temperature, the lower is the B-phase energy minimum relative to that of the A phase, the larger is the proportion of B-phase blobs in the patch work, and the more likely it is that many of them manage to merge together to form one large bubble, where the AB interface exceeds the critical diameter of about 1 μm, needed for spontaneous B-phase expansion to start. This one seed is then sufficient to initiate the A $\rightarrow$ B transition in the whole sample.

The time-dependent Ginzburg–Landau calculations of Bunkov and Timofeevskaya (1998a, b) are consistent with this model. In the calculations there is no spatial dependence, an initial fluctuation is imposed at T_c into a random direction of the phase space, and the evolution towards the final state is followed. As seen in fig. 15, the energy difference between the A and B phases is small, but pressure-dependent. During rapid cool down through T_c the order parameter may settle into either of these two energy minima with a pressure-dependent probability, which is also shown in fig. 15.

Thus the extended KZ mechanism provides a simple explanation of the A $\rightarrow$ B transition probability as a function temperature and magnetic field in supercooled ^{3}He-A. This model is independent of the boundary condition, which dictates ^{3}He-A. Furthermore, in ambient conditions, where only B phase is stable, A-phase blobs ultimately shrink away and only B phase remains. Before that, however, a network of AB interfaces is created which interact with the formation of other topological defects. This feature can be used to explain the sharp increase in v_{cn} at P_{PCP} in fig. 14 (top right panel).

2.10.2. *Vortex formation, AB interfaces, and KZ mechanism*

The KZ model predicts that in different parts of the cooling neutron bubble the order parameter may fall initially either in A- or B-phase energy minima, and a patch work of AB interfaces is laid down. The relative number of

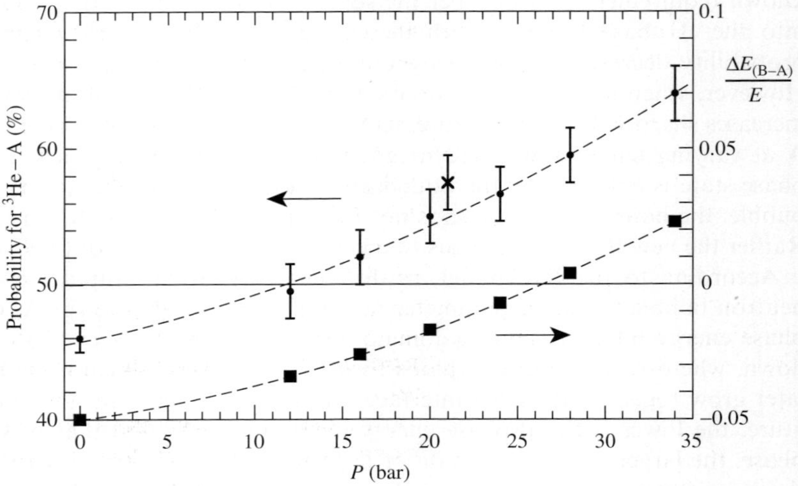

Fig. 15. Normalized energy difference, $\Delta E_{(B-A)}/E(P)$, between A and B phases (*right vertical axis*) and probability of A-phase formation in a rapid cool-down (*left vertical axis*), plotted as a function of pressure at temperatures close to T_c in the Ginzburg–Landau regime. The results have been calculated with the weak coupling β parameters ($\beta_i = -1, 2, 2, 2, -2$), except for the data point marked with a cross ($\times$) on the upper curve, for which strong-coupling parameter values have been used according to the spin fluctuation model (from Bunkov and Timofeevskaya 1998, b) Copyright 1998 Plenum Publishing Corporation; with kind permission from Springer Science and Business Media.

A- and B-phase blobs of size ξ_v is not only determined by the difference in the A- and B-phase energies, but also depends on the trajectory from the normal phase to the new energy minimum in the phase space spanned by the order-parameter components. The latter aspect gives more preference to the higher energy A state: although the A-phase is energetically not favorable in a wide pressure range, the probability of its formation has a value close to 50% even at lower pressures, as shown in fig. 15. The AB interfaces interfere with the simultaneous formation of a random vortex network, which is laid down separately in both the A and B phases. It is known from experiments on both a moving AB interface (Krusius et al. 1994) and a stationary AB interface (Blaauwgeers et al. 2002) that the penetration of vortex lines through the AB phase boundary is suppressed. We thus expect that the presence of A-phase blobs reduces the combined volume which remains available for B-phase blobs and their vortex network inside a neutron bubble. Consequently, when B-phase blobs merge to larger units, the overall volume, into which the B-phase vortex network is confined, is smaller, and B-phase vortex formation becomes impeded. This process increases the value of v_{cn}.

To conclude, the KZ mechanism extended to domain wall formation provides a unified picture of the pressure and magnetic field dependences of v_{cn} (fig. 14) and also the A $\rightarrow$ B transition in supercooled ^{3}He-A. In both cases AB interfaces appear as a new type of defect in the rapidly cooling neutron bubble. Normally, to create a sizeable bubble of B phase within bulk A liquid requires that an AB interface of large size is formed. This is a slow and energy-consuming process. In quench cooling through T_c the freeze-out of disconnected blobs with A- and B-phase-like fluctuations happens first and these are initially separated by supercooled normal liquid. The AB interfaces appear later as metastable defects when the blobs grow together. Such an interpretation suggests that the KZ mechanism is the fastest process by which defects are created, before other effects manage to switch on.

2.10.3. *Spin-mass vortex*

Unlike vortex lines, which are produced in neutron irradiation and are then partly stabilized by the applied flow, no bias is provided for maintaining AB interfaces in fig. 14. Thus in these rotating experiments, the evidence for AB interfaces remains indirect. However, there exists another topologically stable defect, which is formed in a neutron absorption event and is preserved by the bias flow for later examination. This is a combined object called the spin-mass vortex (Kondo et al. 1992, Korhonen et al. 1993).

The signature from the spin-mass vortex is explained in fig. 16, where the counterflow peak height is plotted as a function of time during neutron irradiation (as in figs. 3 and 12). This accumulation record shows one oversize step in the absorption amplitude, which coincides with a similar discontinuous jump in the out-of-phase dispersion signal. As explained in the context of fig. 12, the total number of vortex lines extracted from an irradiation session can be determined in several different ways. In the case of fig. 16 a comparison of the different line counts shows that, within the uncertainty limits, the large jump cannot represent more than a few vortex lines. As will be seen below, this is the signature from a spin-mass vortex, which in fig. 16 should be counted as one line.

The identification of the spin-mass vortex is based on its peculiar NMR spectrum (Kopu and Krusius 2001). In an axially oriented magnetic field the NMR spectrum has a textural cutoff frequency, beyond which no absorption is allowed (excepting line-broadening effects). The distinguishing feature of the spin-mass vortex is the absorption which has been pushed to the maximum possible frequency shift, beyond that of the CF peak (which corresponds to 80% of the maximum value, see fig. 17).

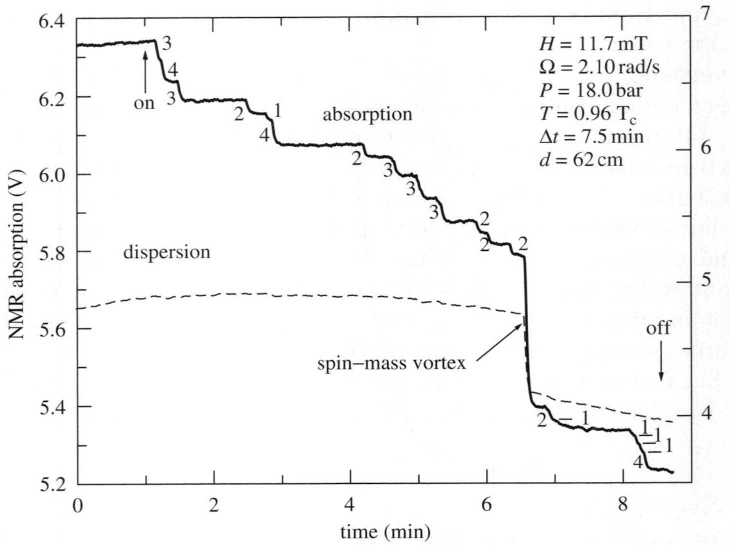

Fig. 16. Signal from spin-mass vortex in neutron irradiation. This irradiation session includes one large jump in signal amplitude. Like in fig. 12, the various methods for extracting the total number of lines give: (1) A direct count of the steps in signal amplitude yields 45 lines plus the large jump. The combined amplitude drop for these 45 lines amounts to 716 mV which translates to 16 mV/vortex, in agreement with fig. 7. With this step size the large jump of 390 mV would correspond to 25 additional lines. (2) A measurement of the annihilation threshold gives $\Omega_v = 0.155$ rad/s, which corresponds to 48 lines. (3) The rate equation (8) gives $\dot{N} =$ 6.2 lines/min or a total of 48 lines for an irradiation of 7.5 min duration with the measured threshold value of $\Omega_{cn} = 0.75$ rad/s. For these three estimates to be consistent, the large jump cannot be 25 lines. Instead it is interpreted to represent one spin-mass vortex line. In this case the size of the jump is not related to the number of lines, but to the counterflow velocity $\sqrt{v}$ (eq. (7)). It controls the length $R(1 - \sqrt{1 - v/\Omega R})$ of the soliton sheet (fig. 18). The dispersion signal also displays a simultaneous large jump, while it behaves smoothly when usual mass-current vortex lines are formed. Outside the discontinuity the dispersion signal is slowly drifting, because it is very sensitive to a residual temperature creep (Reprinted figure with permission from Eltsov et al. 2000, Copyright (2000) by the American physical Society).

Because of this shifted absorption, the CF peak is slightly shifted both in frequency and in height in the presence of a spin-mass vortex, compared to its location without the spin-mass vortex. This difference is displayed in fig. 17, where the spectrum plotted with the solid curve includes one spin-mass vortex while the dashed spectrum is without it. The small frequency shift between these two peaks is the reason why also the NMR dispersion signal is discontinuous when the spin-mass vortex is formed during the neutron irradiation in fig. 16. Since both the NMR absorption and dispersion

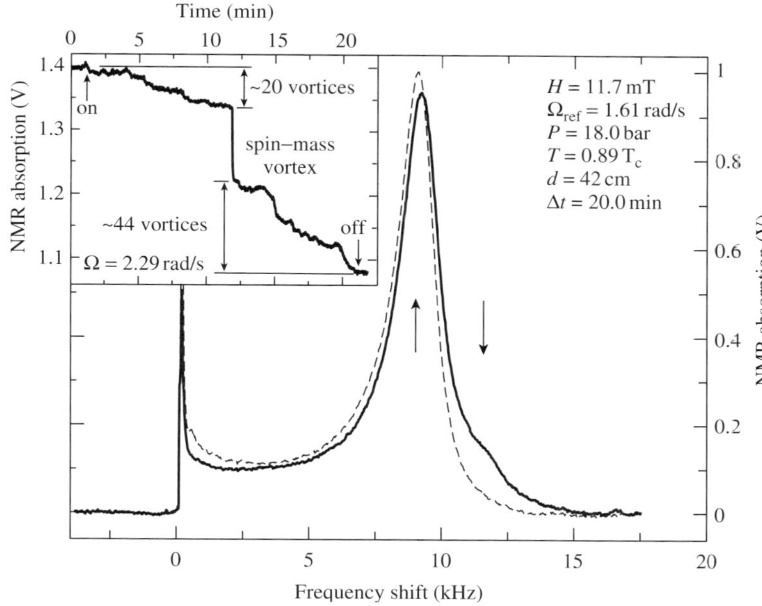

Fig. 17. NMR signal characteristics of the spin-mass vortex. (*Inset*) This 20 min irradiation session includes one large jump, in addition to vortex-line formation. The latter amounts to 64 lines, as determined from the measured annihilation threshold $\Omega_v = 0.19$ rad/s. The rate equation (8) gives $\dot{N} = 3.35$ lines/min or a total of 67 lines (with the measured $\Omega_{cn} = 1.10$ rad/s). The combined drop in signal amplitude from these lines is 206 mV, which corresponds to 3.2 mV/line. Here at a lower temperature the step size per vortex is much reduced from that in fig. 12. The single large jump would correspond to 36 additional lines, but like in fig. 16 it is interpreted to represent one spin-mass vortex. (*Main panel*) NMR spectra of the accumulated vortex sample, recorded at a lower reference velocity Ω_{ref}: the solid curve with the smaller counterflow peak is the original state after irradiation and deceleration from 2.29 to 1.61 rad/s. The dashed curve with the higher counterflow peak was traced later after cycling Ω from 1.61 to 0.20 rad/s and back, i.e. after decelerating close to, but still above the annihilation threshold of the regular vortex cluster. In the former spectrum (solid curve) the soliton sheet shifts a sizeable fraction of the absorption to the maximum possible frequency shift (at right vertical arrow). After cycling Ω the outermost spin-mass vortex (see fig. 18 top panel) is annihilated (dashed curve) and the CF peak is shifted to the value (at left vertical arrow) without the soliton sheet. This shift in the frequency of the counterflow peak creates the jump in the dispersion signal in fig. 16. Since the remaining regular cluster was not decelerated at any stage below its annihilation threshold of 0.19 rad/s, no usual vortex lines were annihilated from the dashed-curve spectrum.

signals are continuously monitored at the lock-in-amplifier output, a jump in the dispersion channel provides the immediate alert for a spin-mass vortex. In contrast, no discontinuity in the dispersion signal is observed in first order when a small number of vortex lines is added to the cluster at high bias velocity (see spectra in fig. 6). In this case some absorption intensity is shifted toward the Larmor edge to small frequency shifts, and the change in the dispersion signal at the site of the CF maximum is negligible.

The large-scale configuration, in which the spin-mass vortex appears, is shown in the top part of fig. 18. It consists of a linear object and a planar domain-wall-like soliton sheet. The line defect provides one termination for the sheet while its second end is anchored on the cylindrical wall of the container. The line defect consists of a singular core, which supports a circulating superfluid mass current, but has additionally trapped a disclination line in the B-phase order parameter. The latter feature can also be described in terms of a spin vortex which supports a persistent spin current. Thus the composite object, the spin-mass vortex, has the combined properties of a usual mass-current vortex and a spin-current vortex. A gain in the core energy of the combined structure provides the energy barrier against their dissociation. The soliton sheet is the continuation into the bulk superfluid from the breaking of the spin–orbit coupling: it separates two regions with antiparallel orientations of the B-phase anisotropy axis $\hat{\mathbf{n}}$ (lower part of fig. 18). The spin-mass vortex feels the Magnus force from the applied bias flow, but this is partly compensated by the surface tension of the soliton sheet. The equilibrium position of the spin-mass vortex is therefore at the edge of the vortex cluster (top part of fig. 18) from where it can be selectively removed by annihilation, as was done in the experiment of fig. 17.

The origin of the two defects in the spin-mass vortex can be seen from the form which the B-phase order parameter takes (Vollhardt and Wölfle 1990)

$$A_{\alpha j} = \Delta_{\mathrm{B}}(T, P) \mathrm{e}^{\mathrm{i}\Phi} R_{\alpha j}(\hat{\mathbf{n}}, \theta). \tag{16}$$

It includes the isotropic energy gap $\Delta_{\mathrm{B}}(T, P)$, the phase factor $\mathrm{e}^{\mathrm{i}\Phi}$, and the rotation matrix $R_{\alpha j}(\hat{\mathbf{n}}, \theta)$. The latter defines the rotation by which the $SO(3)_{\mathrm{L}}$ orbital and $SO(3)_{\mathrm{S}}$ spin spaces are rotated with respect to each other around the axis $\hat{\mathbf{n}}$ by the angle θ. To minimize the weak spin–orbit interaction, it is required that $\theta = \arccos(-\frac{1}{4}) \approx 104°$ homogeneously everywhere within the bulk superfluid. The mass-current vortex is associated with a 2π circulation in the phase factor while the spin current involves a disclination line in $R_{\alpha j}(\hat{\mathbf{n}}, \theta)$ field. The minimum energy configuration then becomes the structure depicted in the lower part of fig. 18. Here the volume in the bulk, where the spin–orbit interaction is not minimized and θ traverses through π, is concentrated within a soliton tail.

In ^{3}He–B weak anisotropy energies arise in an external magnetic field. They produce an extended orientational distribution, or texture, of the anisotropy axis $\hat{\mathbf{n}}$. The characteristic length scale of this texture in the conditions of the rotating NMR measurements is $\xi_H \propto 1/H{\sim}1$ mm. Since the NMR frequency is controlled locally by the orientation of $\hat{\mathbf{n}}$ with respect to

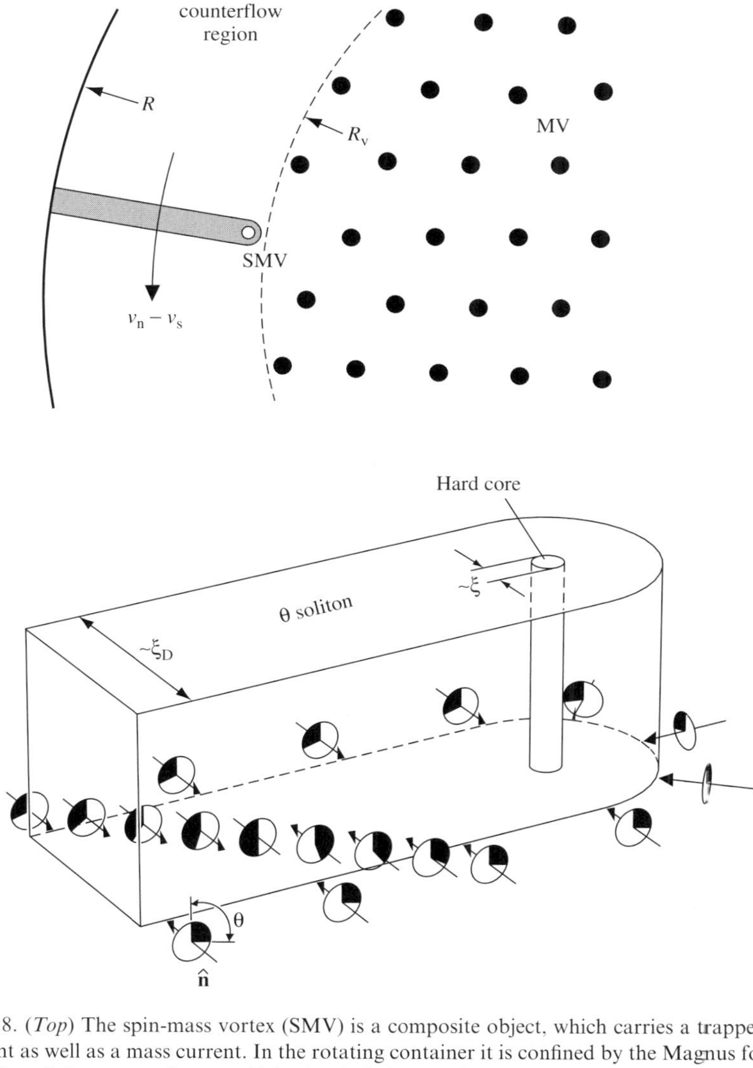

Fig. 18. (*Top*) The spin-mass vortex (SMV) is a composite object, which carries a trapped spin current as well as a mass current. In the rotating container it is confined by the Magnus force to the edge of the vortex cluster, which also includes usual mass-current vortex (MV) lines. The extension of the spin vortex into the bulk liquid is a soliton tail (gray strip). The SMV itself provides one termination line of the soliton while its second end is anchored on the cylindrical wall. (*Bottom*) Inside the soliton the angle θ changes through π while outside it is homogeneously at 104°. At the soliton surface the rotation axis $\hat{\mathbf{n}}$ is oriented perpendicular to the surface while further away it goes smoothly over into the flare-out texture of an ideal cylinder in an axially oriented magnetic field. The distortion of the $\hat{\mathbf{n}}$ texture from the flare-out configuration in the region around the soliton (Kopu and Krusius 2001) gives rise to the NMR signature of the SMV in fig. 17.

H, the NMR spectrum is an image of the $\hat{\mathbf{n}}$ texture in the cylindrical container. In an axially oriented magnetic field and an infinitely long cylinder, the texture is known to be of the "flare-out" form (Vollhardt and Wölfle 1990). In this texture there are no regions where $\hat{\mathbf{n}}$ would lie in the transverse plane perpendicular to the field **H**. In contrast, at the soliton sheet $\hat{\mathbf{n}}$ is oriented strictly perpendicular to the sheet and thus to **H**. This part of the texture is responsible for shifting NMR absorption into the region of maximum possible frequency shift and becomes thus the signature of the spin-mass vortex in the NMR spectrum (fig. 17).

The presence of the soliton sheet has many important experimental implications: it determines the location of the spin-mass vortex at the edge of the cluster and it leaves a clear signature in the NMR spectrum at high bias velocities when the sheet becomes stretched. It also affects the threshold velocity at which a spin-mass vortex might be expected to escape from the neutron bubble. This was only observed to happen well above v_{cn}, the threshold velocity for usual mass-vortex rings (fig. 13). It is explained by the fact that for spontaneous expansion a spin-mass vortex has to overcome, in addition to the line tension of the mass vortex, also the line tension of the spin vortex and the surface tension from the soliton sheet.

The spin-mass vortex was originally discovered as a defect which is formed when an A $\rightarrow$ B transition front moves slowly close to adiabatic conditions through a rotating container (Kondo et al. 1992, Korhonen et al. 1993). In this experiment the initial A-phase state is one with the equilibrium number of doubly quantized singularity-free vortex lines (skyrmions) while the final state is found to contain less than the equilibrium number of singly quantized B-phase vortex lines plus some number of spin-mass vortices. This means that A-phase vortex lines interact with the moving AB interface, they are not easily converted into B-phase vorticity, a critical value of bias flow velocity has to be exceeded before the conversion becomes possible, and even then some fraction of the conversion leads to vorticity with the additional defect in $R_{\alpha j}(\hat{\mathbf{n}}, \theta)$ (Parts et al. 1993, Krusius et al. 1994). In neutron measurements the spin-mass vortex was not observed at 2.0 bar, but at 18.0 bar. In the KZ scenario this is the pressure regime where the probability for the formation of A-phase blobs in the neutron bubble increases (fig. 15). Interestingly one might therefore surmise that even in the neutron bubble the spin-mass vortex could result from vorticity which originally was contained within A-phase blobs and which is partly transferred to the B phase when the A-phase blobs shrink away.

The two NMR spectra in fig. 17 illustrate the easy identification of the spin-mass vortex after neutron irradiation. Although the spin-mass vortex is a rare event in neutron irradiation, compared to the yield of vortex lines, its presence demonstrates that in addition to usual mass-current vortices also other order-parameter defects are created. This lends support to the discussion in section 2.9.2 that also the AB interface could be among such defects. The presence of the spin-mass vortex also limits the possible mechanisms of defect formation. An explanation in terms of the superflow instability at the neutron bubble boundary (section 2.8) is not a probable alternative: the spin-mass vortex is not normally created at the bulk critical velocity limit $v_{cb}(T, P)$ as a response to superflow.

2.11. *Vortex formation in gamma radiation*

To study the dependence of the vortex formation rate $\dot{N}$ on the dimensions of the heated neutron bubble in the ^{3}He–B bath, it would be useful to investigate alternative heating methods. In addition to thermal neutrons, β and γ radiation have frequently been used as a source of heat. In the rotating experiments a ^{60}Co γ source was experimented with, which emits γ rays at 1.17 and 1.33 MeV energy.

A γ ray, which scatters in the container wall or in the liquid ^{3}He bath, knocks off a secondary electron which gives rise to a heat release in the liquid ^{3}He bath. The interaction probability is roughly proportional to the density of the material and thus a large fraction of the secondary electrons originate from the quartz glass wall. Although the initial energies of the secondary electrons cover a broad spectrum, the variation in the effective heat release is narrower. Energetic electrons knock off additional electrons which produce their own ionization tracks. However, once the energy has dropped in the regime of several keV, the remaining ionization track is short: roughly the rate of energy loss per unit distance traveled increases inversely proportional to the electron's decreasing energy. Therefore the heating becomes concentrated within a distance $< 10\ \mu m$ of the final stopping point.

As a source for localized heating in liquid ^{3}He experiments, gamma rays suffer from several shortcomings, compared to thermal neutrons: (1) it is not possible to concentrate on single heat-release events, since many secondary electrons can be produced by a single γ ray. Thus there is more variation in these events. For instance, the end points of different ionization tracks may fall close to each other and may merge to produce a large hot bubble of variable size. (2) The events are not localized in a well-determined location within the container. (3) It is not only the liquid ^{3}He bath which preferentially absorbs the heat, but even more so the structural materials of the refrigerator. This interferes with the temperature control of the measurements: the heating produced by the γ radiation in the metal parts of the cryogenic equipment may rise to excessive levels, and radiation shielding and collimation need to be built into the experiment.

It was found that qualitatively vortex formation in γ and neutron radiation is similar. But the measurements also showed that quantitative differences are large: the threshold velocity $v_{\rm cn}$ is substantially increased and thus the yield $\dot{N}(\Omega)$ at a given rotation velocity Ω in the vortex-free state is smaller. So far no careful quantitative measurements have been carried out with γ radiation.

2.12. *Bias dependence of loop extraction*

The dependence of the vortex formation rate $\dot{N}$ on the externally applied bias v is the most tangible quantitative result from the rotating measurements. It will be described and analyzed below (Ruutu et al. 1998a). The empirical rate equation (8) is shown to be consistent with the KZ model, including the measured value of the prefactor γ.

2.12.1. *Experimental velocity dependence*

The frequency and number of vortex lines can be counted from the accumulation record (figs. 3 and 12), when the NMR absorption is monitored as a function of time during neutron irradiation. Figure 19 shows the vortex formation rate $\dot{N}$ (a), the rate of those neutron absorption events $\dot{N}_{\rm e}$ which produce at least one line and thus become observable (b), and the average number of lines produced by each absorption event (c). All three quantities are determined independently and directly from the absorption records. In fig. 19 the results are plotted as a function of the normalized bias $v/v_{\rm cn}$. To construct the plot, the horizontal axis was divided into equal bins from which all individual measurements were averaged to yield the evenly distributed data points displayed in the graphs.

The rates in fig. 19a and b are proportional to the intensity of the neutron flux, which is contained in the prefactors of the expressions given in the different panels of the figure. For counting the rates from the NMR absorption records it is vital that absorption events do not overlap, when the applied counterflow velocity is increased. Therefore three different source positions were used which were scaled to the same distance using the measured graph in fig. 8.

The rates in fig. 19 increase rapidly with the applied bias velocity v: at $v/v_{\rm cn} \approx 4.5$, close to the maximum velocity limit imposed by the spontaneous instability limit (fig. 13), there are almost no unsuccessful (or unobserved) absorption events left: $\dot{N}_{\rm e}(\infty) \approx \dot{N}_{\rm e}(4.5v_{\rm cn})$. The middle panel shows that this saturation corresponds to a flux of 20 neutrons/min. The same flux of neutrons was estimated to be absorbed in the sample based on independent measurements with commercial ^{3}He counters. Also we note that in

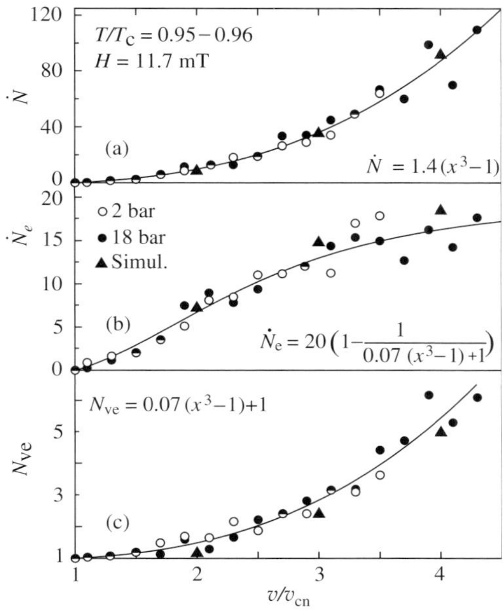

Fig. 19. Rates of vortex line formation plotted vs. normalized bias velocity $x = v/v_{cn}$: (a) total number of lines formed per minute ($\dot{N}$), (b) number of observed neutron absorption events per minute ($\dot{N}_e$), and (c) average number of lines per observed event (N_{ve}). All three quantities have been determined *independently* from discontinuities in NMR absorption, i.e. each data point is based on one or several accumulation plots like those in fig. 3. The data in the two upper plots are proportional to the neutron flux, while the data in the bottom plot are neutron-flux-independent. Solid curves are fits to the data, given by the expressions in each panel. Triangles are results from numerical simulations described in section 2.14 (Reprinted figure with permission from Ruutu et al. 1998a. Copyright (1998) by the American Physical Society).

agreement with the scaling properties expressed by the rate equation (8), measurements at the two pressures of 2 and 18 bar fall on the same universal curves.

Figure 19 demonstrates that the neutron-induced vortex formation process involves a strong stochastic element: close to the critical threshold at $v = 1.1v_{cn}$ only one neutron capture event from 40 manages to produce a sufficiently large vortex loop for spontaneous expansion. On increasing the bias flow by a factor of 4, almost all neutron capture events give rise to at least one escaping vortex loop. Secondly, the non-linear increase of $\dot{N}$ as a function of v/v_{cn} in the top panel has to be attributed to two factors: (i) the increase in the event rate $\dot{N}_e$ in the middle panel and (ii) the rapid rise in the number of lines produced per event in the bottom panel. These conclusions fit with the KZ predictions, as we shall see in the next section.

The most detailed information from the rate measurements is the dispersion into events in which a given number of lines is formed. Figure 20 displays the rates $\dot{N}_{ei}$ of events which produce $i = 1-5$ lines. These data exhibit larger scatter, owing to its statistical variation, but again after averaging one gets for each value of i a curve, which peaks at a maximum, and then trails off. With increasing value of i the curves shift to successively higher velocities. The starting points of each curve, the threshold velocities v_{cni}, are plotted in the inset. Their values increase monotonically with each consecutive value of i. This means that at and immediately above the first threshold, $v_{cn} = v_{cn1}$, only single-vortex events occur.

A surprising finding from the measurements is the total absence of a background contribution to the measured rates. A number of tests were performed to look for spontaneous events in the absence of the neutron source, to check whether a contribution from the background radiation level should be subtracted from the measured rates. For instance, a vortex-free sample was rotated for 90 min at different velocities (0.9, 1.3, and 2.1 rad/s at 2.0 bar and 0.94 T_c), but not a single event was noted.

2.12.2. *Analytic model of vortex loop escape*

An analytic calculation can be constructed for independent vortex loops in the applied flow, which explains the cubic velocity dependence of the escape rate in fig. 19. Vortex rings, which are part of the random vortex network after the quench, have a critical radius (eq. (11)) below which they contract and above which they expand spontaneously in homogeneous superflow. By comparing this radius to the typical curvature in the random network, we arrive at the desired result (Ruutu et al. 1996a).

During the quench through the superfluid transition a random vortex network is formed with a characteristic length scale of the order of ξ_v. The later evolution of the network leads to a gradual increase in this length. The average inter-vortex distance or, equivalently, the typical radius of curvature of the loops increases with time. We shall call this length $\tilde{\xi}(t)$. This "coarse-graining" process preserves the random character of the network; in other words the network remains self similar or scale-invariant. Only later a change occurs in this respect, when the loops become sufficiently large to interact with the externally applied bias flow: this causes large loops to expand, if they are oriented transverse to the flow with the correct winding direction, while small loops contract, and loops with the wrong sign of circulation tend to both contract and to change orientation. The timescale of these processes is determined by the magnitude of mutual friction force. In the Ginzburg–Landau temperature range the evolution of the network occurs in milliseconds and the expansion of the extracted loops into rectilinear lines in a fraction of a second.

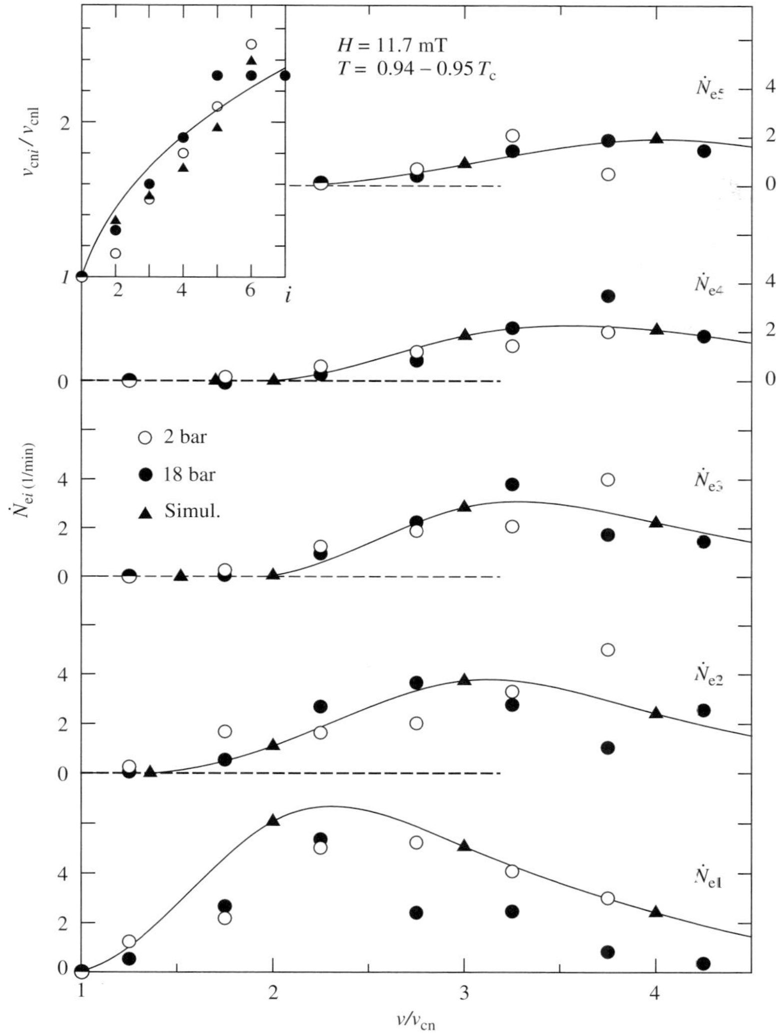

Fig. 20. Rates $\dot{N}_{ei}$ of vortex formation events, grouped according to the number of lines i formed per absorption event, plotted vs. v/v_{cn}. Triangles are results from numerical simulations described in section 2.14. Solid curves are spline fits to simulation data. (*Inset*) Threshold velocity v_{cni}/v_{cn1} for the onset of an event with i lines, plotted vs. the number of lines i. The solid curve represents the fit $v_{cni}/v_{cn1} = [2.0\,(i-1)+1]^{1/3}$ (Reprinted figure with permission from Ruutu et al. 1998a. Copyright (1998) by the American Physical Society).

In superfluid ^{3}He the large viscosity of the normal component clamps it to corotation with the container. In the rotating frame of reference we may write $v_n = 0$ and $v_s = v$. The energy of a vortex loop, which is stationary with respect to the walls, is given by (Donnelly 1991)

$$\mathscr{E} = E_{\text{kin}} + \mathbf{pv}, \tag{17}$$

where $\mathbf{v}$ is the velocity of the applied superflow. In simple configurations, the hydrodynamic kinetic energy or self-energy of the loop arises from the trapped superfluid circulation with the velocity $\mathbf{v}_{\text{s,vortex}} = (\kappa/2\pi)\nabla\Phi$,

$$E_{\text{kin}} = \frac{1}{2}\int \rho_s v_{\text{s,vortex}}^2 \, \mathrm{d}V = \varepsilon L, \tag{18}$$

and is proportional to the length L of the loop and its line tension,

$$\varepsilon = \frac{\rho_s \kappa^2}{4\pi} \ln\frac{\tilde{\xi}(t)}{\xi}. \tag{19}$$

Here we neglect the small contribution from the core energy, use intervortex spacing $\tilde{\xi}(t)$ as the upper cutoff, and the superfluid coherence length $\xi(T, P)$ as the lower cutoff for the integration in eq. (18). This equation is valid in the logarithmic approximation, when $\tilde{\xi}(t) \gg \xi(T, P)$. While the first term in eq. (17) is proportional to the length L of the loop, the second term involves its linear momentum,

$$\mathbf{p} = \int \rho_s \mathbf{v}_{\text{s,vortex}} \, \mathrm{d}V = \frac{1}{2\pi}\rho_s\kappa \int \nabla\Phi \, \mathrm{d}V = \rho_s\kappa\mathbf{S}, \tag{20}$$

where the last step follows from Gauss's theorem and involves the area S of the loop in the direction of the normal $\mathbf{S}/S$ to the plane of the loop. Thus we write for the energy of a loop

$$\mathscr{E}(L, S, t) = \rho_s\kappa\left[L\,\frac{\kappa}{4\pi}\ln\frac{\tilde{\xi}(t)}{\xi} - vS\right], \tag{21}$$

where S is the algebraic area, perpendicular to the flow and of proper winding direction. This equation expresses the balance between a contracting loop due to its own line tension, which dominates at small applied velocities, and expansion by the Magnus force from the external superflow, which dominates at high applied velocities. The divide is the equilibrium condition, which was expressed in eq. (11) and corresponds to the situation when the height of the energy barrier, which resists loop expansion, vanishes. In this extremal configuration $\mathbf{p}$ is antiparallel to $\mathbf{v}$, the loop moves with the velocity $-\mathbf{v}$ in a frame of the superfluid component, but is stationary in the rotating frame. In a more general sense, if we consider loops in the random network which still deviate from circular shape, the extremal case degenerates to a saddle point. This is because the extremum requires also minimization with respect to deviations from circular shape, i.e. the total energy is invariant under small variations of the radius of the ring, or $\delta\mathscr{E} = \delta E_{\text{kin}} + \mathbf{v}\cdot\delta\mathbf{p} = 0$.

The expansion of the vortex loop should be calculated by including the forces acting on a vortex from the normal and superfluid components. In our analytic description of vortex loop escape, we shall neglect such complexity. Instead we shall make use of three scaling relations which apply to Brownian networks and are described in more detail in section 2.14. These expressions relate the mean values in the statistical distributions of the loop diameter $\mathscr{D}$, area S, and density n to the length L of the loop:

$$\mathscr{D} = AL^{\delta}\tilde{\xi}^{1-\delta}, \quad (A \approx 0.93, \quad \delta \approx 0.47), \tag{22}$$

$$|S| = B\mathscr{D}^{2-\zeta}\tilde{\xi}^{\zeta}, \quad (B \approx 0.14, \quad \zeta \approx 0), \tag{23}$$

$$n = CL^{-\beta}\tilde{\xi}^{\beta-3}, \quad (C \approx 0.29, \quad \beta \approx 2.3). \tag{24}$$

For a Brownian random walk in infinite space the values of δ, β, and ζ are 1/2, 5/2, and 0, which will be used below.

The important simplification here is that these relations are assumed valid during the entire evolution of the network, until sufficiently large rings are extracted by the bias flow into the bulk. Using eqs. (22) and (23), we may write eq. (21) for the energy of a loop in the form

$$\mathscr{E}(\mathscr{D}, t) = \rho_{\text{s}}\kappa\mathscr{D}^2\left[\frac{\kappa}{4\pi\tilde{\xi}(t)A^2}\ln\frac{\tilde{\xi}(t)}{\xi} - vB\right]. \tag{25}$$

When the scale $\tilde{\xi}(t)$ exceeds a critical size $\tilde{\xi}_{\text{c}}(v)$, which depends on the particular value of the applied superflow velocity v,

$$\tilde{\xi}_{\text{c}}(v) = \frac{1}{A^2B}\frac{\kappa}{4\pi v}\ln\frac{\tilde{\xi}_{\text{c}}}{\xi}, \tag{26}$$

the energy in eq. (25) becomes negative and the loops in the network start expanding spontaneously. The total number of loops N_{b}, which will be extracted from one neutron bubble, can then be obtained by integrating their density from the smallest size $\tilde{\xi}_{\text{c}}$ to the upper cutoff, which is provided by the diameter of the entire network, or that of the heated bubble, $2R_{\text{b}}$:

$$N_{\text{b}} = V_{\text{b}}\int_{\tilde{\xi}_{\text{c}}}^{2R_{\text{b}}} \text{d}\mathscr{D}n(\mathscr{D}). \tag{27}$$

Here the density distribution $n(L) = C\tilde{\xi}^{-3/2}L^{-5/2}$, combined with that for the average diameter $\mathscr{D}(L) = A(L\tilde{\xi})^{1/2}$, gives $n(\mathscr{D})\,\text{d}\mathscr{D} = 2A^3C\mathscr{D}^{-4}\,\text{d}\mathscr{D}$. On inserting this into the integral (27) we obtain

$$N_{\mathrm{b}} = \frac{1}{9}\,\pi A^3 C\left[\left(\frac{2R_{\mathrm{b}}}{\tilde{\xi}_{\mathrm{c}}}\right)^3 - 1\right]. \tag{28}$$

From this equation we see that the requirement $N_{\mathrm{b}}(v_{\mathrm{cn}}) = 0$ returns us the definition of the threshold velocity $v_{\mathrm{cn}} : \tilde{\xi}_{\mathrm{c}}(v = v_{\mathrm{cn}}) = 2R_{\mathrm{b}}$. This in turn gives us, from eq. (26) for the radius of the heated bubble,

$$R_{\mathrm{b}} = \frac{1}{A^2 B}\,\frac{\kappa}{8\pi v_{\mathrm{cn}}}\,\ln\,\frac{2R_{\mathrm{b}}}{\xi(T,P)}, \tag{29}$$

which we used in section 2.9.1 to derive the temperature dependence of the threshold velocity: $v_{\mathrm{cn}} \propto (1 - T/T_{\mathrm{c}})^{1/3}$.

Equations (26) and (29) thus show that $\tilde{\xi}_{\mathrm{c}} \propto 1/v$ and $R_{\mathrm{b}} \propto 1/v_{\mathrm{cn}}$, so that we may write for the vortex-formation rate $\dot{N} = \phi_{\mathrm{n}} N_{\mathrm{b}}$ from eq. (28),

$$\dot{N} = \frac{1}{9}\,\pi A^3 C \phi_{\mathrm{n}}\left[\left(\frac{v}{v_{\mathrm{cn}}}\right)^3 - 1\right], \tag{30}$$

where ϕ_{n} is the neutron flux. This is the form of the measured cubic velocity dependence in the empiric equation (8). By inserting $A \approx 0.93$, $C \approx 0.29$ from eqs. (22) and (24), respectively, and $\phi_{\mathrm{n}} \approx 20$ neutrons/min, as determined from the saturation of the event rate $\dot{N}_{\mathrm{e}}$ in fig. 19b, we obtain for the rate factor in eq. (30) $\gamma \approx 1.6\ \mathrm{min}^{-1}$, which agrees with the experimental value in figs. 11 and 19a. We note that the cubic dependence on the applied bias flow in eq. (30) comes only from the assumption that the whole volume of the heated bubble contributes equally to the production of vortices, while the values of the prefactor γ and of the constant term in eq. (30) depend on the scaling relations (22)–(24).

The definition of the threshold velocity $v_{\mathrm{cn}i}$, which applies for an event in which i rings are formed simultaneously, is roughly consistent with the requirement $N_{\mathrm{b}}(v = v_{\mathrm{cn}i}) \approx i$. This gives $v_{\mathrm{cn}i}/v_{\mathrm{cn}} \sim i^{1/3}$, which agrees with the measured result in fig. 20.

To summarize, we note that the KZ model, combined with the simplest possible interpretation for the loop escape from a random vortex network, reproduces both the measured cubic dependence on the normalized velocity, $(v/v_{\mathrm{cn}})^3$, and the magnitude of the extraction rate.

2.13. *Neutron-induced vortex formation at low temperatures*

The low-temperature properties of superfluid ^{3}He could not have been extrapolated from measurements restricted to temperatures above $0.80T_{\mathrm{c}}$, even within the framework of the general theory. Similarly, to understand the neutron-induced vortex formation process, it is important to extend the measurements to lower temperatures. One such study has been performed with NMR down to $0.4T_c$ in a rotating sample (Finne et al. 2004a). This measurement is a continuation of the NMR work, which we have discussed so far above $0.80T_c$. Another series of measurements has been performed calorimetrically below $0.2T_c$ in a quiescent bath (Bäuerle et al. 1996, 1998a). It was originally expected that at these very low temperatures quantized vortex lines have a long life time in zero applied flow. Thus there would be sufficient time after the neutron absorption event to detect the vortices.

2.13.1. *Experimental techniques*

NMR measurement in ^{3}He–B loses rapidly in amplitude resolution on cooling to lower temperatures: the susceptibility drops with decreasing temperature and the width of the NMR spectrum increases. Thus the change of NMR absorption per one rectilinear vortex line quickly diminishes and single-vortex resolution is lost. Also below $0.6T_c$ a new phenomenon enters, namely superfluid turbulence. This was discovered in recent measurements when a few seed vortex loops were injected into rotating vortex-free flow (Finne et al. 2003). With decreasing temperature vortex damping via mutual friction dissipation drops, and the seed loops are more and more likely to become unstable in the flow. At high temperatures the loops evolve individually to rectilinear vortex lines so that the number of vortices is conserved. At low temperatures the injection is followed by a sudden turbulent burst in which a tangle of vortices is formed. In the turbulent burst the number of individual vortices proliferates and their total length exceeds that in the equilibrium array of rectilinear lines. From this tangle, after a brief transient period, the final stable state, evolves, the equilibrium vortex state. A surprisingly sharp transition as a function of temperature was found to separate the laminar high-temperature and the turbulent low-temperature responses. These changes with decreasing temperature mean that neutron absorption measurements in rotating flow are very different at low temperatures.

The rotating NMR measurements at low temperatures are performed in the setup of fig. 21. For the study of turbulence it proved fortunate that this arrangement includes two NMR detection coils at both ends of the long sample cylinder. A second bonus is the relatively high critical velocity of the container so that vortex-free superflow could be maintained to above 3.5 rad/s below $0.8T_c$ at 29.0 bar.

On cooling to lower temperatures vortex-free superflow generally becomes more and more difficult to generate and maintain. The lower section of the ^{3}He volume below the orifice is directly connected with the sintered

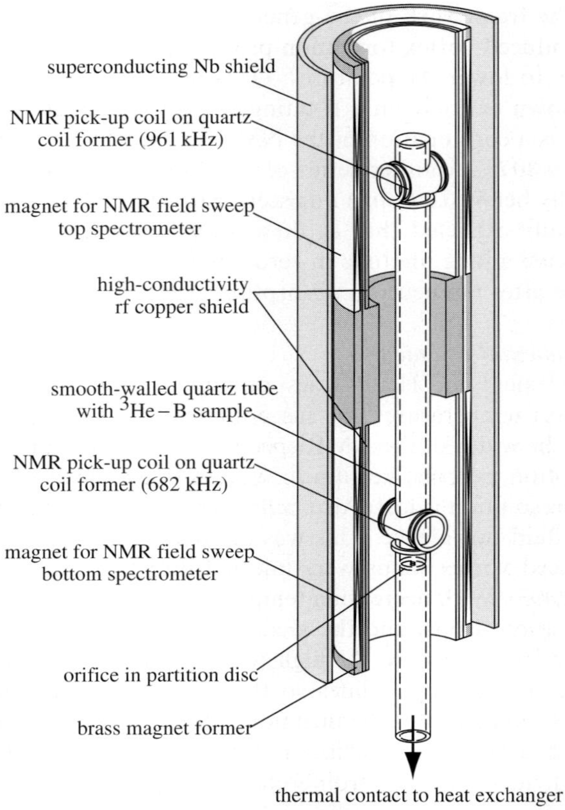

Fig. 21. ^{3}He sample with NMR measuring setup. The sample is contained in a fused quartz tube which has a diameter of 6 mm and length 110 mm. This space is separated from the rest of the liquid ^{3}He volume with a partition disc. In the center of the disc an orifice of 0.75 mm diameter provides the thermal contact to the liquid column which connects to the sintered heat exchanger on the nuclear cooling stage. Two superconducting solenoidal coil systems with end-compensation sections produce two independent homogeneous field regions with axially oriented magnetic fields. An exterior niobium cylinder provides shielding from external fields and additional homogenization of the NMR fields. The NMR magnets and the Nb shield are thermally connected to the mixing chamber of the precooling dilution refrigerator and have no solid connection to the sample container in the center. The two split-half detection coils are fixed directly on the sample container (from Finne et al. 2004a) Copyright 2004 Plenum Publishing Corporation; with kind permission from Springer Science and Business Media.

heat exchanger. This region is flooded with remanent vortices from the porous sinter already at low rotation. Generally one finds that these vortices start to leak through the orifice into the sample volume on cooling below $0.6T_c$. Surprisingly the sample tube, with which the neutron measurements were performed, seemed immune to this problem in spite of the fact that the orifice had a relatively large diameter of 0.75 mm.

Below $0.5T_c$ remnant vortices become a problem. Here successive accelerations to rotation have to be separated by extensive waiting periods at stand still, to allow the slow annihilation of the last one or two remnant vortices. If rotation is started too soon, a remnant vortex is still present and becomes easily unstable in rotation. Vortex-free rotation at velocities of order ~1 rad/s is then not possible to achieve. Just below $0.60T_c$ the waiting time is around 5 min in the container of fig. 21, but at $0.40T_c$ it is found to be 30 min or more. No procedure of rotation at very low velocity in different directions or with different deceleration rates seemed to help in driving the last vortex to annihilation and in cutting down on the waiting period.

The experimental setup in fig. 21 is equipped with two independent CW NMR spectrometers. Each spectrometer includes a split-half excitation/detection coil, wound from thin superconducting wire. The two coils are installed at both ends of the sample cylinder. Each coil is part of a high-Q tank circuit, with a resonance frequency approaching 1 MHz and $Q \sim 10^4$. Figure 22 shows a number of absorption spectra measured with the bottom spectrometer, similar to the high-temperature examples in the top panel of fig. 6. The spectra illustrate the relevant NMR line shapes and their large width at low temperatures, with varying numbers of rectilinear vortex lines in the sample. Again the peak height of the CF maximum can be calibrated to yield the number of lines N in the linear limit, when $N \ll N_{eq}$. Often enough, to determine N, one has to increase Ω to some higher reference value where the sensitivity of the CF peak is restored.

Figure 23 shows an example of a calibration where the CF peak height is measured as a function of N (or equivalently Ω_v). The measurement has been performed for small vortex clusters ($N \lesssim 100$), by measuring each calibration point in a four step process: (1) first the spectrum is recorded in the vortex-free state in the reference conditions and the CF peak height $A_0(\Omega_{ref})$ is obtained. (2) Next a large number of vortex lines is created by irradiating with neutrons (with the rotation at Ω_{ref} or higher). (3) The sample is then decelerated to the low rotation velocity $\Omega_{low} \ll \Omega_{ref}$ so that part of the vortex lines is observed to annihilate. Since the long cylinder is oriented along the rotation axis only within a precision of ~0.5°, this means that any annihilation barrier must be negligible or that at Ω_{low} the sample is in the equilibrium vortex state $\Omega_{low} = \Omega_v(N)$, with a known number of vortex lines N. (4) Finally the rotation is increased back to the reference value Ω_{ref} and the spectrum is recorded in order to measure the CF peak height $A(\Omega_{ref}, \Omega_v(N))$. In fig. 23 the reduction $\Delta A = A_0(\Omega_{ref}) - A(\Omega_{ref}, \Omega_v)$ in the CF peak height of the two spectra is plotted as a function of Ω_v. To reduce

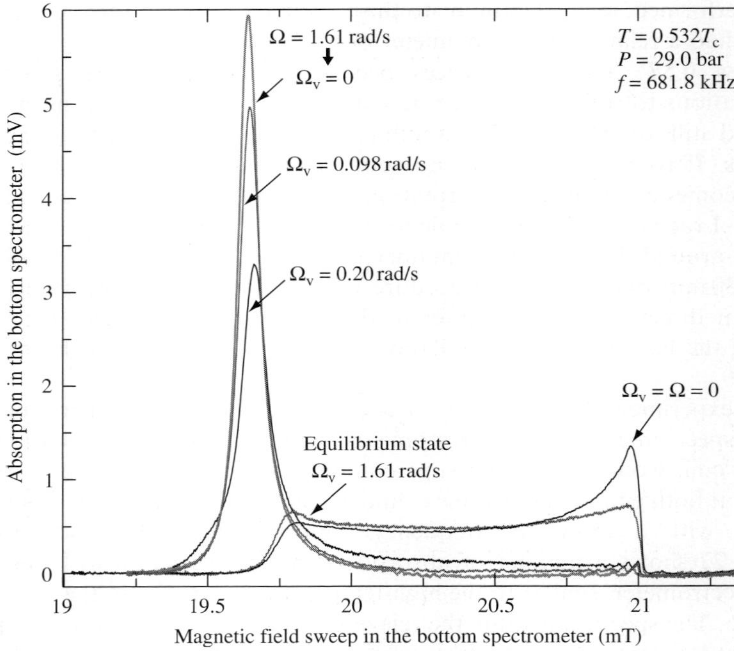

Fig. 22. NMR absorption spectra of ^{3}He–B in rotation at low temperatures. As in the top panel of fig. 6, changes in the line shape can be calibrated to give the number of vortex lines N. Here N is characterized by the rotation velocity $\Omega_v(N)$ at which a given number of lines N is in the equilibrium state. The different spectra have been measured with the rf excitation at constant frequency f, using a linear sweep of the axially oriented polarization field H. The spectra have been recorded at constant temperature and thus all have the same integrated total absorption. The sharp absorption maximum at low field is the counterflow peak (CF). Its shift from the Larmor field (at 21.02 mT) is used for temperature measurement. When a central cluster of rectilinear vortex lines is formed, the height of the CF peak is reduced. In the equilibrium vortex state ($\Omega = \Omega_v$), where the number of vortex lines reaches its maximum, the spectrum looks very different: it has appreciable absorption at high fields and borders prominently to the Larmor edge. This spectrum is more similar to that of the non-rotating state ($\Omega = 0$). As shown in fig. 23, when the vortex number is small, $\Omega_v \ll \Omega$, the reduction in the CF peak height can be calibrated to give Ω_v and thus N (from Finne et al. 2004a) Copyright 2004 Plenum Publishing Corporation; with kind permission from Springer Science and Business Media.

the dependence on drift and other irregularities it is normalized to the CF peak height $A_0(\Omega_{ref})$ of the vortex-free reference state. In contrast to the calibration in the right bottom of fig. 6, this procedure does not rely on single-vortex resolution. However like always, it does require that the order parameter texture is stable and reproducible and that new vortices are not formed during rotational acceleration.

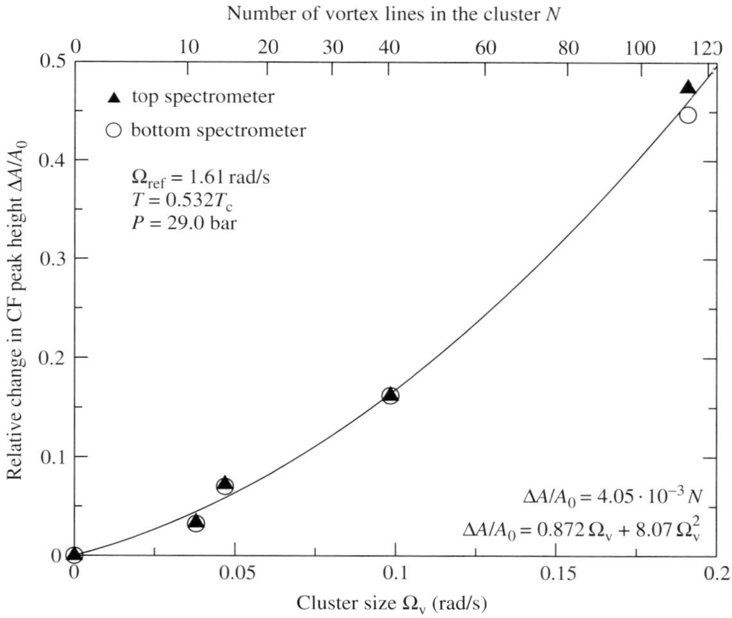

Fig. 23. Calibration of CF peak height A vs. vortex line number N. The CF peak height in the vortex-free state, $A_0(\Omega_{\rm ref})$, is compared to the peak height $A(\Omega_{\rm ref}, \Omega_{\rm v}(N))$, which is measured at the same rotation velocity $\Omega_{\rm ref}$ for a small vortex cluster of size $\Omega_{\rm v}$, which is prepared as described in the text. The quantity plotted on the vertical scale is the relative reduction in peak heights, $\Delta A/A_0 = [A_0(\Omega_{\rm ref}) - A(\Omega_{\rm ref}, \Omega_{\rm v}(N))]/A_0(\Omega_{\rm ref})$, measured in constant conditions. The solid line is a fit, given by the expression in the figure. The conversion from $\Omega_{\rm v}$ (bottom axis) to N (top axis) in the continuum picture is (Ruutu et al. 1998b) $N = N_0(1 - d_{\rm eq}/R)^2$, where $N_0 = \pi R^2(2\Omega_{\rm v}/\kappa)$ and $d_{\rm eq} = [\kappa/(8\pi\Omega_{\rm v})\ \ln(\kappa/2\pi\Omega_{\rm v} r_{\rm c}^2)]^{1/2}$. Here $r_{\rm c} \sim \xi(T, P)$ is the radius of the vortex core. Using this conversion it is found that $\Delta A/A_0$ is a linear function of N, similar to the bottom panel in fig. 6 (from Finne et al. 2004a) Copyright 2004 Plenum Publishing Corporation; with kind permission from Springer Science and Business Media.

The result in fig. 23 is a smooth parabola, which becomes a straight line if $\Delta A/A_0$ is plotted as a function of the number of rectilinear lines N. Any sample with an unknown number of lines, which is less than the maximum calibrated number, can then be measured under the same reference conditions ($\Omega_{\rm ref}$, T, and P) and compared to this plot, to determine N. In fig. 24 a measurement of vortex formation in neutron irradiation is shown for which the calibration was used. After each irradiation session at different bias flow velocity $v = \Omega R$ the rotation is changed to $\Omega_{\rm ref}$ and the NMR absorption spectrum is recorded. From this spectrum the reduction in CF peak height is determined, by comparing to the spectra of the vortex-free state which are measured regularly between neutron irradiation sessions.

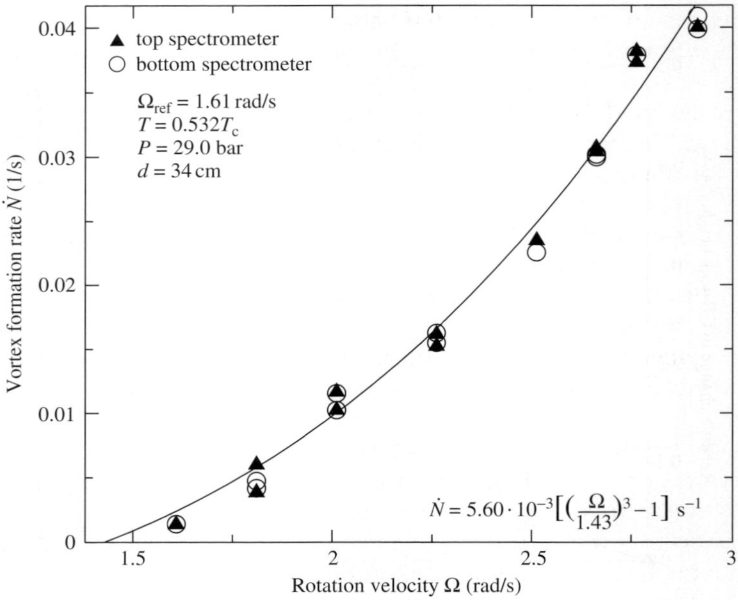

Fig. 24. Rate of vortex formation in neutron irradiation. The average number of rectilinear vortex lines created per unit time during the irradiation period is shown as a function of the rotation velocity Ω. The data are fitted with the expression $\dot{N} = 0.336[(\Omega/1.43)^3 - 1]\ \text{min}^{-1}$ (with Ω in rad/s). Depending on the rate $\dot{N}$, the irradiation time varies here from 30 min to 4.5 h, so that the number of accumulated vortices remains within the range of the calibration in fig. 23. This calibration is used to determine the number of vortices from the relative reduction in the CF peak height, $\Delta A/A_0$. The distance of the neutron source from the sample was $d = 34$ cm. The range of the bias flow is limited in these measurements between the critical velocity $\Omega_{cn} = 1.43$ rad/s and the upper limit $\Omega = 3.0$ rad/s, where the turbulent events start to occur (from Finne et al. 2004a) Copyright 2004 Plenum Publishing Corporation; with kind permission from Springer Science and Business Media.

2.13.2. *Measurement of vortex formation rate*

In fig. 24 the rate of vortex formation $\dot{N}$ is measured at $0.53T_c$ as a function of the bias velocity $v = \Omega R$, during neutron irradiation. The result supports the cubic rate equation (30). The threshold velocity $v_{cn} = \Omega_{cn} R = 4.3$ mm/s is in the same range as the data in fig. 13. This is plausible if the threshold velocity is determined only by the size of the neutron bubble which does not change appreciably with decreasing temperature, eq. (14). However, the rate factor γ is 36 times smaller than in the measurements above $0.80T_c$ and below 21.2 bar (when scaled to the same neutron flux). A second similar measurement at 10.2 bar and $0.57T_c$ gives a threshold velocity $v_{cn} = 3.6$ mm/s and a rate constant γ which is eight times larger than the measurement in fig. 24.

These results raise the question how do v_{cn} and γ vary over a wider temperature and pressure range. In particular, this becomes a problem at high pressures, where a wide range of stable A phase exists between T_c and the ambient B-phase bath temperature T_0, where the irradiation is performed. In contrast at 10 bar, A phase is stable in our NMR fields only over a range of some μK immediately below T_c (with reference to section 2.9.2). In both the 29.0 and 10.2 bar measurements the available range of bias velocities is limited (to approximately $v_{cn} < v \lesssim 2v_{cn}$), by the appearance of turbulent events at higher velocities. Both of these low-temperature measurements are consistent with the cubic form of the rate equation, but do not conclusively prove it owing to the limited velocity range (unlike the high-temperature result in fig. 19). Clearly more extensive measurements as a function of pressure and temperature are called for.

2.13.3. *Superfluid turbulence in neutron irradiation*

At temperatures below $0.60T_c$ neutron irradiation events may become turbulent, similar to vortex formation from other sources (Finne et al. 2003). This means that the vortex loops, which have been extracted from the neutron bubble and are injected into the bias flow, may start to interact and produce a localized network consisting of a large number of vortices. In this turbulent burst the number of individual vortices blows up locally and approaches that in the equilibrium vortex state. Ultimately after a period of transient evolution, the rotating sample settles down in the equilibrium vortex state with rectilinear lines. The sequence of events, which takes place during the transient epoch following the neutron absorption, is still under investigation. However, the fundamental observation is the sudden change in the yield of vortex lines from a neutron absorption event as a function of temperature.

The number of vortex loops, which is extracted from a neutron bubble and is injected in the bias flow, can be tuned to some extent by choosing the bias velocity. But as seen from fig. 19c this number is only a few (i.e. approximately from 1 to 6) in the limited range of the available rotation velocities. At high temperatures at large vortex damping the number of individual vortices is conserved when the loops develop to rectilinear lines in the rotating flow. At low temperatures one finds that such slow accumulation of vortices from successive neutron absorption events is suddenly terminated by a turbulent event which fills the sample with rectilinear vortex lines. This means that instead of a few vortices the turbulent event generates of order $\sim 10^3$ lines. Under these conditions the final equilibrium vortex state has little connection with the loop extraction process, since the properties of the neutron-induced vortex formation process are now hidden by the later turbulent proliferation of vortices. However, experimentally, neutron-induced turbulence becomes a handy means for creating suddenly the maximum possible number of vortices and for examining the dynamic configuration of vortices in which they propagate along the rotating column and replace vortex-free flow with the equilibrium vortex state.

The NMR signature of a turbulent event is dramatic. In fig. 25 the evolution of the CF peak height as a function of time is plotted while a turbulent neutron absorption event takesplace the NMR absorption

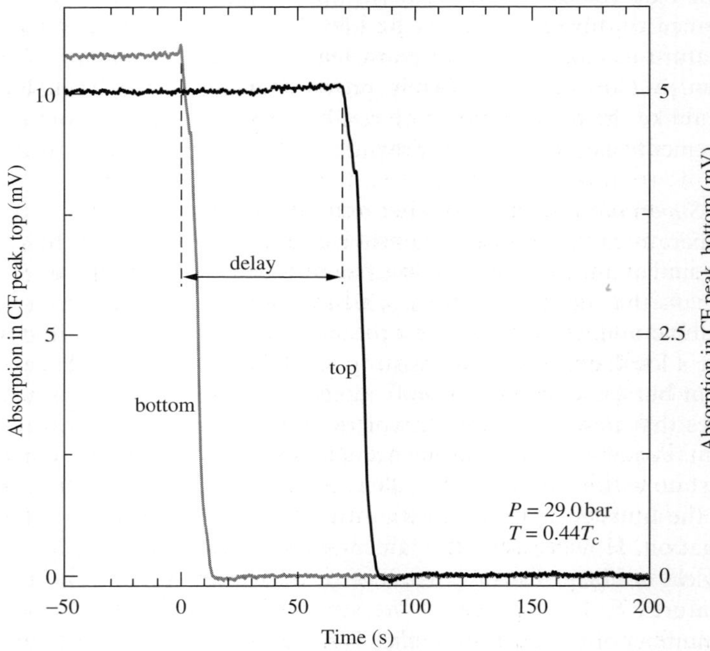

Fig. 25. NMR signature of turbulent vortex formation in neutron irradiation. The CF peak height (fig. 22) is monitored as a function of time in the two NMR detection coils. Here the turbulent burst occurs suddenly inside the bottom coil (at $t = 0$), when its CF peak rapidly starts to lose in absorption. After a delay of 70 s, which the front of vorticity needs to travel to the lower edge of the top coil 90 mm higher along the column, a similar collapse is recorded by the top spectrometer. The flight time $\tau = z/(\alpha\Omega R)$ corresponds to the situation where $z =$ 95 mm and turbulence is first formed in the middle of the bottom coil. This sequence of events is displayed here at $\Omega = 1.61$ rad/s. The sudden collapse of the CF peak means that the vorticity is rapidly polarized at these large rotating flow velocities and that the azimuthal global counterflow between the normal and superfluid components is thereby removed. Measurements of this type show that from the initial injection site, where the extracted vortex loops first start to intersect in the bias flow and produce a turbulent burst of new vortices, the vorticity expands by forming two fronts which move at constant velocity toward the top and bottom ends of the rotating column (from Finne et al. 2004a) Copyright 2004 Plenum Publishing Corporation; with kind permission from Springer Science and Business Media.

spectrum jumps (via a brief transitory period) from a line shape with a large CF peak to that of the rotating equilibrium vortex state of totally different form with no peak at all at the location of the former CF peak (fig. 22). To observe such a turbulent event, the rotation velocity has to exceed Ω_{c1} and the temperature has to be sufficiently low so that vortex motion is not heavily damped by mutual friction.

Low mutual friction damping is the most important requirement for turbulence (Finne et al. 2003), so that Kelvin wave excitations on existing vortices can readily expand in amplitude. This leads to loop formation and via self-reconnection of such loops or via inter-vortex reconnections of adjacent loops, the vortex number increases, starting a turbulent cascade of reconnections between intersecting loops. In the rotating flow the newly formed vortices are rapidly polarized, with their cores mainly oriented parallel to the rotation axis. In the first injection measurements, in which the onset of turbulence with decreasing mutual friction was discovered (Finne et al. 2003), it was found that the onset is Ω independent above $\sim$0.8 rad/s in the sample container of fig. 21. In these measurements an injection process was used where the size of the injected seed loops happens to be relatively large, their configuration rather reproducible from one injection to the next, one end of the loops always connects to the cylindrical wall, and the number of injected loops does not depend strongly on the bias flow velocity. In neutron irradiation the properties of the injection are not as reproducible and favorable. As a result turbulence starts at a slightly lower temperature and the probability to get a turbulent event depends on the bias flow velocity. With increasing bias velocity the number of loops extracted from the neutron bubble increases (fig. 19, bottom) and reconnections thereby become more likely. Close to the transition temperature of the turbulent regime $T \sim 0.6 T_c$ these processes are stochastic such that the vortex loops extracted from the neutron bubble only occasionally achieve the initial conditions in which turbulent loop extraction can start. However, at low temperature $T < 0.45 T_c$ even a singe loop starts a turbulent burst.

In fig. 25 the turbulent burst occurs inside the bottom coil. The collapse of the CF peak height is the fastest feature in the NMR absorption spectrum to signal the turbulent burst: the fast decay of the CF peak shows that vortices enter the coil, they are rapidly polarized and mimic on an average solid-body rotation so that the azimuthal flow from the vortex-free rotation drops to 0. Simultaneously the NMR absorption from the CF peak is transferred close to the Larmor edge of the spectrum where a complicated time-dependent evolution in the distribution of absorption intensity takes place. The interpretation of these NMR signals is still under consideration. However, one further feature emerges from fig. 25 which is more straightforward to understand, namely the timing between the two recorded signals.

In neutron irradiation the turbulent event can start randomly at any height of the long sample. In the example of fig. 25 it happens to start inside the bottom NMR coil. In this case the two signals monitoring the decay of the CF peak at both ends of the long sample do not overlap in time. Even the more slowly evolving responses in the Larmor region do not overlap in time when $T < 0.5T_c$ and the turbulent burst starts at one end of the sample tube. This means that the turbulent burst does not first fill the entire sample and then relax into the equilibrium vortex state. Instead, in fig. 25, by the time the NMR absorption in the top coil signals the first indication of the approaching change, the line shape in the bottom coil has already reached its stable equilibrium form. The NMR signals in the two coils have the same appearance as a function of time independently of at which height in the sample the turbulence starts, only the delay between the signals varies in this case. This means that there exists a stationary vortex structure which travels vertically along the rotating column in both directions away from the site of the neutron absorption event and the initial turbulent burst. This structure consists of a sharp front of vorticity which advances into the region with vortex-free flow and leaves behind a state which has little azimuthal flow and which continuously slowly transforms into the equilibrium vortex state.

The vortex front moves along the rotating column with constant axial velocity which is controlled by the dissipative mutual friction coefficient α, as defined in eq. (40). This velocity has the same value as that of a single short section of vortex filament, which is oriented perpendicular to the cylinder wall and drifts in the initial vortex-free bias flow along the wall: $v_z = \alpha \Omega R$. The result was concluded by measuring the velocity v_z (Finne et al. 2004b) and by comparing the extracted values of α to the original measurements of mutual friction by Bevan et al. (1997a). Knowing the axial velocity, we can now from the delay between the signals of the two detector coils in fig. 25 calculate the axial height z where the turbulent event started. This location has been plotted for the data in fig. 26 in the inset of this figure.

The probability of a neutron capture to trigger a turbulent process is studied in fig. 26. Two measurements are shown, at different temperatures and rotation velocities. The cumulative probability distribution has been plotted for the irradiation time needed to achieve a turbulent event. The sample is irradiated under constant conditions until the CF peak is observed to collapse suddenly. The irradiation time is plotted on the horizontal axis. On the vertical axis the number of turbulent events observed within this time is shown, normalized to the total number of events: 13 events in one case (at $0.53T_c$) and 9 in the second (at $0.45T_c$). As seen from the plot, the number of events is insufficient to produce smooth probability distributions, but the irradiation times are observed to be distributed over the same range in the

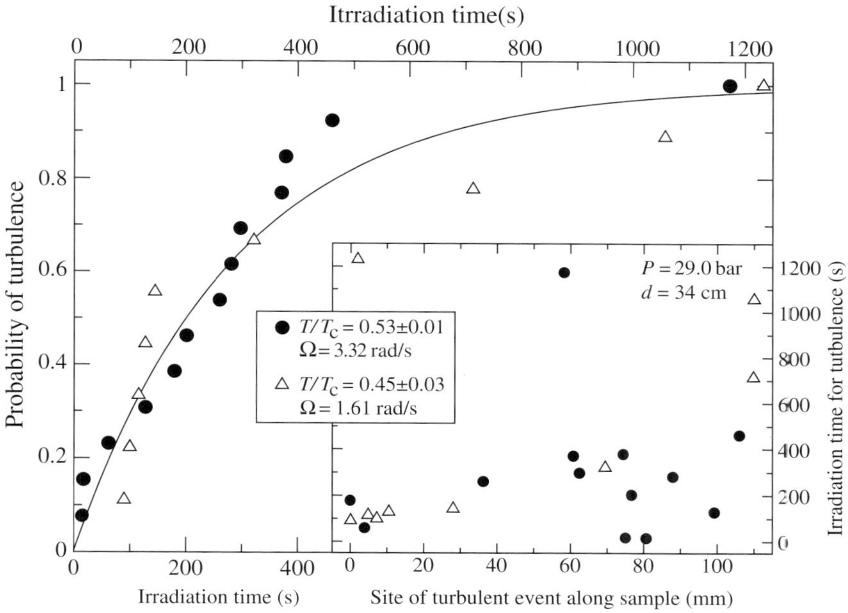

Fig. 26. Turbulent vortex formation in neutron irradiation at 29.0 bar pressure. The initially vortex-free sample is irradiated under constant conditions until a turbulent vortex expansion event takes place. The irradiation time required to achieve the turbulent event is measured. Results from measurements at two different constant conditions are shown in this plot in the form of cumulative probability distributions. The solid curve is a fit of the data measured at $0.53T_c$ to a distribution function of the form $P(t) = 1 - (1 - p_n)^{\dot{N}_e t}$, where the probability of a single neutron absorption event to start turbulence is $p_n = 0.092$. The rate $\dot{N}_e$ of the neutron-induced vortex injection events is taken as $0.55\dot{N}$ from fig. 24 as appropriate for $v/v_{cn} = 2.32$ (see fig. 19). There were no cases among the two sets of measurements where a turbulent event would not have been observed within an irradiation period of 20 min. The *inset* shows the axial location of the initial turbulent burst, the site of the neutron capture, measured from the orifice upward with a technique explained in fig. 25. The corresponding irradiation time needed to achieve the event is shown on the vertical axis. As expected, the sites are randomly distributed along the sample.

two cases, i.e. their distributions have similar average values and widths. This in spite of the fact that the measurements at $0.45T_c$ were performed at half the rotation of those at $0.53T_c$. Such a reduction in velocity results in a significant decrease in the yield of vortex loops from one neutron absorption event (fig. 19). The fact that the two distributions in fig. 26 do not differ significantly indicates that with decreasing temperature the transition to turbulence becomes more probable and less sensitive to the initial configuration of injected loops.

Let us consider the measurement at $0.53T_c$ in more detail. The state of the sample changes during the irradiation at constant Ω: (i) the initial state is vortex-free, (ii) during the irradiation rectilinear vortex lines are formed at a rate which can be extrapolated from fig. 24, (iii) until finally all vortex-free CF is completely terminated by one turbulent event. In the final step the state of the sample changes from one with only a small central vortex cluster to one with the equilibrium number of vortex lines ($N_{eq}{\sim}2600$). Note that to observe a new turbulent event the existing vortices have to be annihilated, by stopping rotation. Then the vortex-free state can be prepared again and a new irradiation session can be started. The longest irradiation time is here ~20 min. During this period the cluster grows at the rate $\dot{N} = 3.9$ vortices/ min, so that it contains about 80 vortices when the turbulent event finally starts. At this point the CF velocity at the sample boundary $v = v_n - v_s = \Omega R - \kappa N/(2\pi R)$ has been reduced by 2.7% from the initial state. Since the mean irradiation time of the measured distribution is only ~250 s, the reduction in CF velocity by vortices formed before a turbulent event is minor. We may thus view the result in fig. 26 as representative of these particular values of rotation and temperature.

One may wonder whether the initial turbulent burst results from a single neutron capture event or from the coincidence of two or more events. In the latter case the simultaneous events need to be sufficiently close not only in time, but also in space, so that expanding loops, which have been extracted from the two neutron bubbles, have a possibility to intersect. (The probability of the bubbles themselves to intersect is very small.) The intersection of the loops expanding from the two random positions along the height of the cylindrical sample is more likely to occur closer to the middle than at the top or bottom end. However, the events listed in the inset of fig. 26 occur randomly along the sample. In all cases the NMR signatures from the turbulent events are similar, there is no prominent variation in their appearance depending on where the event starts. Additionally, at $0.45T_c$, the turbulent events occur close above Ω_{cn}. Here successful neutron absorption events, which lead to the extraction of a vortex loop to the bulk, are rare but, nevertheless, turbulence becomes more probable. From this we presume that a single neutron capture event in injecting one single vortex ring must be able to start the turbulence.

To conclude, with decreasing temperature the onset of turbulence in neutron absorption events moves closer to the critical threshold v_{cn}. This means that the range of bias velocities, in which neutron-induced vortex formation can be studied, becomes further restricted from above and decreases. Ultimately below $0.45T_c$ a measurement of the bias dependence becomes obsolete. Nevertheless, here and at lower temperatures neutron absorption is a practical technique to start turbulence.

2.13.4. *Calorimetry of vortex network*

Another type of low-temperature neutron absorption experiment is the calorimetric measurement of Bäuerle et al. (1996, 1998a), which determines the amount of heat dumped into the liquid ^{3}He–B bath in each individual neutron absorption event. This method allows quantitative comparison to the KZ model. These measurements are performed in the very low-temperature limit of ballistic quasiparticle motion in a quiescent bath, with zero bias flow.

The measuring probe is a superconducting wire loop. It is oscillated in the liquid with a frequency of a few hundred Hz and a high Q value, by driving the loop at resonance with an ac current in a dc magnetic field oriented perpendicular to the wire. The damping of the wire oscillations measures the density of quasiparticle excitations in the liquid and can be calibrated to give the temperature. For this a second vibrating wire loop is used as a heater (fig. 27). The latter is excited with a known current pulse to heat up the liquid, by

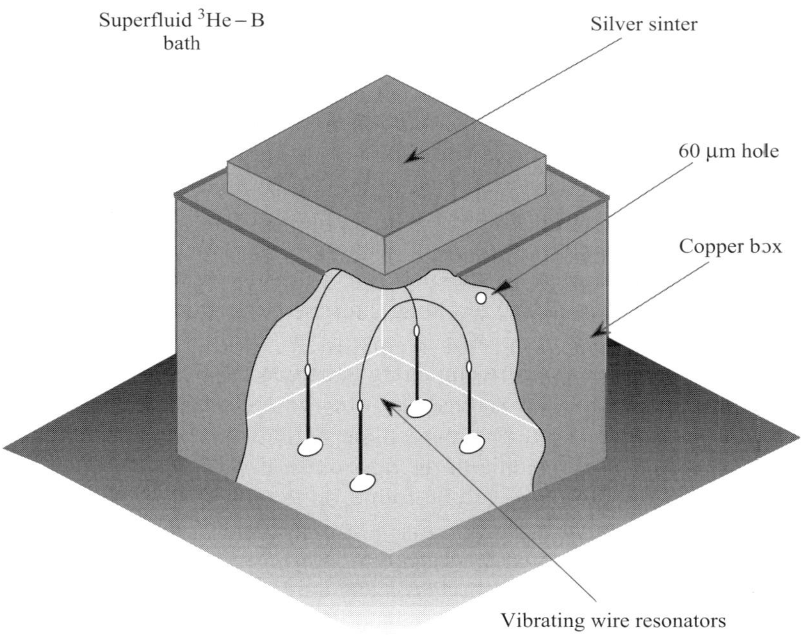

Fig. 27. Bolometer box, which is used to measure calorimetrically the energy balance of the neutron capture reaction in the low-temperature limit of ballistic quasiparticle motion. Two vibrating superconducting wire loops are included, one for measuring the temperature of the liquid within the box and the other as a heater to calibrate the calorimetric measurement (Reprinted figure with permission from Bäuerle et al. 1998b. Copyright (1998) by the American Physical Society).

driving the wire at supercritical velocities where the breakdown of Cooper pairs gives rise to a shower of quasiparticles. To contain the ballistically moving quasiparticles, both wire loops are placed inside a box from which the particles leak out through a pin hole at a well-known rate (Bäuerle et al. 1998b).

The thermal connection between the liquid in the calorimeter box and that in the surrounding ^{3}He–B bath is only via the small pin hole. A neutron absorption event inside the box heats up the liquid and a thermal pulse is recorded with the wire resonator. The rise time is determined by the resonator properties, while the trailing edge (with a time constant of about 1 min) monitors the much slower leakage of quasiparticle excitations from the box through the orifice. If energy is released into the box on a still longer timescale it is not recorded in the form of pulses.

It is found that neutron absorption events amount to roughly 100 keV smaller thermal pulses than the 764 keV, which a slow neutron is expected to yield for the absorption reaction with a ^{3}He nucleus. Since the various recombination channels of the ionized charge and the subsequent thermalization processes in liquid ^{3}He are poorly known, it is not quite clear how large an energy deficit one should expect. Obvious losses include the ultraviolet radiation absorbed in the walls of the bolometer box and the retarded relaxation of excited electronic states of helium atoms and molecular complexes. However, Bäuerle et al. (1996, 1998a) expect that these contributions are not strongly pressure-dependent and of the order of 7% of the reaction energy. They then ascribe the remaining energy deficit to the random vortex network which is created in the neutron bubble and which in the low-temperature limit should have a long lifetime, when mutual friction approaches 0 and no flow is applied.

In Table 1 the measured missing energy at three different liquid pressures has been compared to that estimated from the KZ model. The measured result $\Delta E_{\rm exp}$ is recorded on the topmost line while the calculated comparison proceeds stepwise from one line to the next in the downward direction of the table. The final result $\Delta E_{\rm theor}$ can be found on the lowermost line. The thermal diffusion constant, $D = 3k_{\rm T}/C_{\rm v} = v_{\rm F}^2\tau_{\rm T}$, is derived from the conductivity $k_{\rm T}$, which has been tabulated by Wheatley (1975), while all other liquid ^{3}He values are taken from Greywall (1986). The agreement between the top and bottom lines is within a factor of 2, which is surprising, if we remember the uncertainties which are built into this comparison. Table 1 has been included in this context to draw attention on the magnitudes of the different quantities.

The significance of the calorimetric measurement rests entirely on the quantitative analysis of the measured data – whether an energy deficit is present and can be ascribed to vortices. The quantitative analysis is hampered by three kinds of uncertainties:

TABLE 1

Comparison of measured and estimated total vortex line energies at different pressures in a random vortex network, as generated by a neutron absorption event according to the KZ model.

	P (bar)			Source
	0	6	19.4	
ΔE exp. (keV)	85	95	150	Bäuerle et al. (1996, 1998a)
T_c (mK)	0.93	1.56	2.22	Greywall (1986)
v_F (10^3 cm/s)	5.95	5.04	3.94	Greywall (1986)
C_v (10^3 erg/cm^3 K)	5.83	12.8	25.4	Greywall (1986)
ξ_0 (nm)	65	33	18	$\xi_0 = \sqrt{7\zeta(3)/(48\pi^2)}\hbar v_F/(k_B T_c)$
τ_0 (ns)	1.1	0.65	0.46	$\tau_0 = \xi_0/v_F$
R_b (μm)	27	17	12	$R_b = \sqrt{\frac{3}{2\pi e}}\left(\frac{E_0}{C_v(T_c - T_0)}\right)^{1/3}$
τ_T (μs)	0.59	0.17	0.12	Wheatley (1975)
D (cm^2/s)	21	4.3	0.94	$D = v_F^2 \tau_T$
τ_Q (μs)	0.16	0.31	0.69	$\tau_Q = (e/6)R_b^2/D$
ξ_v (μm)	0.23	0.15	0.11	$\xi_v = \xi_0(\tau_Q/\tau_0)^{1/4}$
$\tilde{D}_b$	210	200	190	$\tilde{D}_b = \left(\frac{4}{3}\pi R_b^3/\xi_v^3\right)^{1/3}$
L_v (cm)	79	43	28	$L_v = (4\pi/3a_l)R_b^3/\xi_v^2$
ρ_s (g/cm^3)	0.081	0.094	0.108	$\rho_s \approx \rho$ (Greywall 1986)
ΔE theor. (keV)	170	140	120	$\Delta E = (\rho_s \kappa^2/4\pi)L_v \ln(\xi_v/\xi_0)$

Note: The ambient temperature of the ^{3}He-B bath is taken to be $T_0 = 0.16$ mK. For the value of τ_Q we use the time it takes for the normal phase bubble to disappear. The coefficient $a_l = 2.1$, for estimating the total vortex-line length L_v, is taken from the simulation results in section 2.14.1, when the value of the dimensionless bubble diameter is $\tilde{D}_b = 200$.

(1) The proportions of the radiative and retarded contributions in the thermalization after neutron absorption are not known sufficiently well (Leggett 2002).

(2) To calibrate the bolometer box, the heater wire is vibrated at a velocity which exceeds the pair-breaking limit. It is now known that when the vibrating wire loop reaches this limit, then a beam of quasiparticles and vortex loops is created. The fraction of energy spent on generating vortices is temperature-dependent and appears to decrease rapidly with increasing temperature. With increasing drive level more and more vortices are formed at different points along the wire loop (Fisher et al. 2001, Bradley et al. 2004). The calibration of the bolometer box then becomes uncertain, if a sizeable fraction of the energy of the calibrating electrical pulse ends up in a long-lived turbulent vortex network.

(3) The decay time of vortex tangles is not well known in different situations at the lowest temperatures. The calorimetric measurement of the

energy deficit in the neutron absorption event requires a time constant of tens of seconds so that the decay of the tangle would not contribute to the measured thermal pulse. On the other hand, the measurements of Fisher et al. (2001) on vibrating-wire-generated tangles suggest that their time constant of decay is seconds. The simulation calculations of Barenghi and Samuels (2002) suggest even faster decay. However, the decay time is expected to depend on the topology and line density of the vortex tangle since it proceeds via reconnection into loops which fly away in the quiescent bath and finally annihilate on solid surfaces (where part of the energy may go directly to the wall, escaping the liquid). In this respect the random vortex network created in the neutron absorption event (section 2.14.1) is expected to be quite different from the vortex tangle generated by the vibrating wire: according to the measurements of Bradley et al. (2004), during stationary-state vortex generation and decay by a vibrating wire the line density reaches only a very low value of 20 mm^{-2} (an inter-vortex distance $\sim$0.2 mm). In comparison, in Table 1 the initial line density expected in the neutron bubble, as precipitated by the KZ process, is of order $10^7\ \mathrm{mm}^{-2}$.

The original interpretation of the calorimetric measurements supports the KZ model of vortex formation. However, to keep this statement valid in view of recent developments on vortex dynamics at ultralow temperatures, more work is required. In principle, a calorimetric measurement of the energy in the random vortex network, which is generated in a neutron absorption event, provides more direct proof of the KZ mechanism, than rotating measurements: here in the absence of applied flow, the discussion about the superflow instability at the neutron bubble boundary is irrelevant. Also the calorimetric measurement gives directly the domain size ξ_v of the inhomogeneity in the order-parameter distribution (eq. (3)) right after the quench, which can otherwise be inferred only indirectly from other types of measurement.

It is interesting to note that the understanding gained from the calorimetric work has turned ^{3}He–B into an attractive absorber material in the search for dark matter particles. In such a dark matter detector, the collision energy will be measured with an array of bolometer boxes equipped with micromechanically fabricated vibrating resonators (Bunkov 2003, Winkelmann et al. 2005). In the following sections we shall analyze further the initial loop formation in the neutron bubble and its later evolution, which we so far have omitted.

2.14. *Simulation of loop extraction*

In spite of the agreement, which we have listed so far between the measured characteristics of neutron-induced vortex formation and the KZ model, a deeper understanding of the processes involved would be important. Owing to its phenomenological content, the KZ model is based on general concepts and contains few media-dependent parameters (such as the superfluid coherence length ξ or the order-parameter relaxation time τ). In the case of superfluid ^{3}He, the predictions of the model for the initial state of the vortex network can be calculated. In the rotating experiments the initial state is connected with observable quantities only through the complex evolution of the network, governed by superfluid hydrodynamics. This means that a more rigorous comparison requires numerical simulation.

A number of numerical simulations exist on the evolution of a network of linear defects. These apply to cosmic strings, liquid crystals, and vortices in superfluid ^{4}He. Such results cannot be directly transferred to neutron-induced vortex formation in ^{3}He–B. The differences with the cosmic string and liquid crystal calculations arise from the different boundary conditions and the equations which govern the evolution of the network in the applied external bias fields. In the case of superfluid ^{4}He, studies of random vortex networks often concentrate on vortex flow driven by thermal counterflow between the normal and superfluid components in a stationary situation (Tough 1982). In contrast, the vortex network, which is produced in ^{3}He–B in a neutron absorption event, is in a state of rapid evolution. Nevertheless, standard techniques exist (see e.g. Schwarz 1978, 1985, 1988) and can be applied for solving also the transient problem in ^{3}He–B. Here we describe calculations which address the dependence of the vortex-formation rate on the normalized bias velocity $x = v/v_{\mathrm{cn}}$ (Ruutu et al. 1998a).

2.14.1. *Initial loop distribution*

Vachaspati and Vilenkin (1984) developed an approach for the simulation of the initial network of linear defects after a rapid phase transition, adapted to the case of cosmic strings. This technique of random phases has been used in most studies since then because of its simplicity in simulation calculations (see e.g. the reviews by Hindmarsh and Kibble 1995 and Bray 1994).

We approximate the neutron bubble with a cubic volume which is subdivided into smaller cubes, such that the size of these volume elements equals the length scale of the initial inhomogeneity in the order-parameter distribution at the moment of defect formation. This length is on the order of the coherence length and in the simulation it plays the role of a "unit" length: it is a parameter of the model, which has to be derived by other means. In the case of the superfluid transition this length is given by eq. (3).

An arbitrary phase between o and 2π is assigned to each vertex of the grid, to model the initial random inhomogeneity of the order parameter. It is usually assumed that the distribution of the phase in each segment of the grid between two vertices corresponds to the shortest path in the phase circle. Thus it is possible to determine whether a line defect pierces a given face. Then the centers of the corresponding faces are connected to form closed or open (depending upon specific boundary conditions) linear defects, strings in the cosmological case and vortices in the superfluid.

Vachaspati and Vilenkin assigned to each vertex a value of the phase from the following set: $\{0, 2\pi/3, 4\pi/3\}$ and then studied the statistical properties of the resulting network of strings. They found that most (70%) of the strings were in the form of open segments, which extended from one boundary of the system to another. For closed loops they found two scaling relations to hold:

$$n = CL^{-\beta} \tag{31}$$

and

$$\mathscr{D} = AL^{\delta}, \tag{32}$$

where $\beta \approx \frac{5}{2}$, $\delta \approx \frac{1}{2}$, n is the density of loops with a given length L, and $\mathscr{D}$ is the average spatial size of a loop. It is usually defined as an average of straight-line dimensions in x, y, and z directions. In this model both the characteristic inter-vortex distance and the radius of curvature are of the order of the length scale of the spatial inhomogeneity (i.e. the size of the small cubes, which here has been set equal to unity). Later, other variations of this model have been studied, including other types of grids and other sets of allowed phases. However, it has been found that the scaling relations (31) and (32) hold universally in each case.

The direct applicability of these results to the case of vortex formation in ^{3}He–B is not evident. First, the random phase approach with geodesic rule (Vachaspati and Vilenkin 1984) does not take into account all possible random vortex configurations. In addition to phase Φ with values in the interval from 0 to 2π, winding numbers must be assigned to each edge of the lattice, which describe the vortex degrees of freedom. (See e.g. the lattice model by Villain (1975).) Second, in cosmology the open lines are the most significant ones: only these strings may survive during the later evolution if no bias field exists to prevent closed loops from contracting and annihilating. In the case of a normal-liquid bubble within the bulk superfluid there exists an obvious boundary condition: the phase is fixed at the boundary and there are no open lines at all. Thus it is of interest to find out the influence of the boundary conditions on the scaling relations (31) and (32).

Another question, which arises in the case of superfluid vortices, concerns the interaction of the loops with the external bias field due to the normal–superfluid counterflow. The energy of a tangled vortex loop in the counterflow is proportional to its algebraic area $S = \oint y \, \mathrm{d}z$ in the direction x

of the counterflow velocity, eq. (21). Thus the dependence of S on the length of the loop is of interest as well.

The simulation (Ruutu et al. 1998a) is performed in a cubic lattice with a fixed (zero) phase at the outer boundary. The Vachaspati-Vilenkin random phase model is used, but phases on vertices can have any value between 0 and 2π. It is known that the set of allowed phases affect the number of open lines in cosmic-string simulations, which, however, may not be important for the network in the neutron bubble. Several vortices may pass through one cell in the grid. In this case the corresponding faces of the cell are connected at random. For calculating the length of a loop, it is assumed that both straight and curved segments of the loop inside one cell have unit length. The size of a loop in the direction of a specific coordinate axis is measured as the number of cells in the projection of the loop along this axis.

The size of the volume, which undergoes the superfluid transition in the neutron absorption event, is about 50 μm, while the characteristic inter-vortex distance in the initial network is of the order of 1 μm. Calculations have been performed for several "bubble" volumes: starting from $6 \times 6 \times 6$ up to $200 \times 200 \times 200$. For each bubble size the loop distributions obtained from a large number of (up to 1000) initial distributions of random phases are averaged. The resulting distributions of n, $\mathscr{D}$, and S are shown in fig. 28. One can see that despite differences in the boundary conditions the Vachaspati–Vilenkin relations (31) and (32) hold in the case of vortices in superfluid helium, but the exponents and prefactors are slightly different and depend on the size of the bubble.

For the algebraic area S of a loop as a function of the corresponding two-dimensional diameter $\mathscr{D}_2$ the additional scaling law is

$$|S| = B\mathscr{D}_2^{2-\zeta}, \tag{33}$$

where $\zeta \approx 0$. Here $\mathscr{D}_2$ is the average of the straight-line dimensions of a loop in y- and z-directions. Thus the oriented area of a tangled loop is proportional to the area of a circle of the same straight-line size. The scaling relation (33), as well as (31) and (32), are of the form expected for a Brownian particle, for which the square of the average displacement on the ith step is proportional to i and therefore the mean value of the square of the oriented area is given by

$$\begin{aligned}\langle S^2\rangle &= \left\langle \left(\sum_{1\leqslant i\leqslant L} y_i \Delta z_i \right)^2 \right\rangle = \sum_{1\leqslant i,j\leqslant L} \langle y_i \Delta z_i y_j \Delta z_j\rangle \\ &= \sum_{i,j} \langle y_i y_j\rangle \langle \Delta z_i \Delta z_j\rangle = \sum_i \langle y_i^2\rangle\langle \Delta z_i^2\rangle \propto \sum_i i \propto L^2.\end{aligned}$$

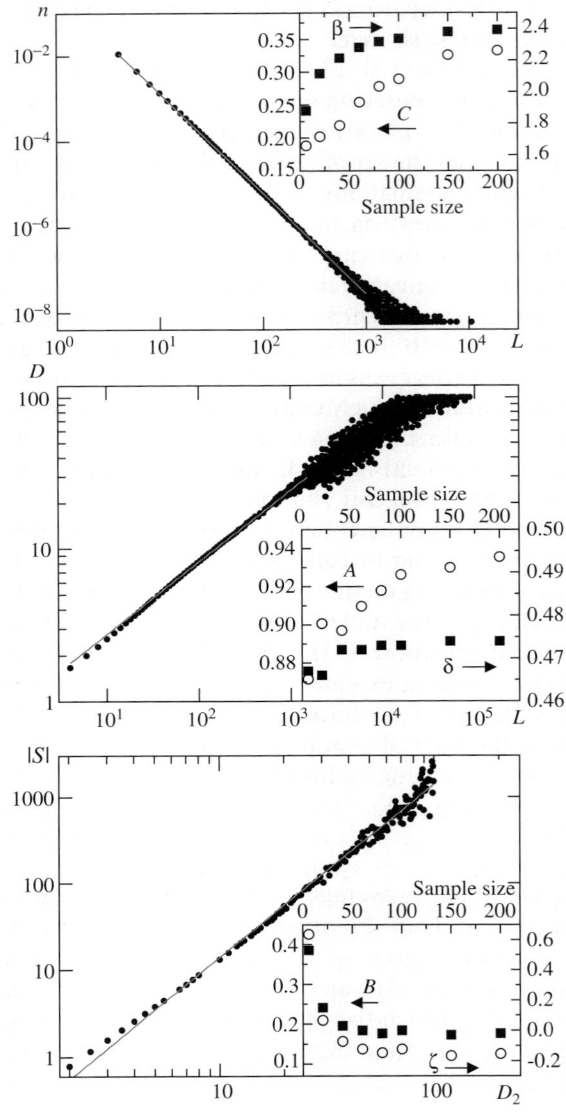

Fig. 28. Scaling properties of vortex loops in the initial network: density n (*top*) and average spatial size $\mathscr{D}$ (*middle*), plotted as a function of the length L of the loops. Below their algebraic area $|S|$ is shown (*bottom*), as a function of the two-dimensional spatial size $\mathscr{D}_2$ of the loop. The calculations are performed on a lattice of $100 \times 100 \times 100$ cells. The solid lines are fits to scaling laws obeying equations of the form of (31), (32), and (33), respectively. The inset in each plot shows the dependence of the scaling parameters on the size of the cubic neutron "bubble".

These scaling relations were used in the analytic model of vortex-loop extraction from the cooling neutron bubble in the bias flow (i.e. eqs. (22)–(24)).

2.14.2. *Network evolution under scaling assumptions*

After its formation, the vortex network evolves under the influence of the inter-vortex interactions and the normal–superfluid counterflow v. Its characteristic length scale $\tilde{\xi}(t)$ increases with time. Vortices start to escape from the bubble when the energy gain due to the external counterflow becomes larger than the energy of the superflow associated with the vortex itself. This corresponds to the critical value of $\tilde{\xi}$, expressed by eq. (11) or (26).

Tangled vortex flow in superfluid ^{4}He has been studied numerically by Schwarz (1978, 1985, 1988), Samuels (1992), Aarts and de Waele (1994), Nemirovskii and Fiszdon (1994), Barenghi et al. (1997), Tsubota and Yoneda (1995), and others. A large number of calculations has been devoted to the evolution of the initial network of cosmic strings (see reviews by Hindmarsh and Kibble 1995, Bray 1994) and also of linear defects in liquid crystals (Toyoki 1994, Zapotocky et al. 1995). In all three cases the initial state is quite similar, but the equations controlling the evolution are different. The common feature is that the interaction between the loops leads to reconnections when the loops cross each other. It has been shown by Kagan and Svistunov (1994) that the scaling relations remain valid in a random network if the vortices are allowed to reconnect when they cross each other, but all other interactions are neglected for simplicity. In most simulation work, both in the case of cosmic strings and liquid crystals, it has been found that the scaling relations are preserved during the evolution, even if inter-vortex interactions are included.

To calculate the escape rate from the network, two crude assumptions are made, which are essentially the same as in the analytic treatment in section 2.12.2: (1) we assume that the scaling laws remain valid until $\tilde{\xi}$ grows comparable in size to the critical value in eq. (26). (2) At this point the influence of the external counterflow becomes suddenly so significant that all sufficiently large loops immediately escape by expanding to a rectilinear vortex line. In the numerical simulations the state before escape is modeled by the same method as was used to construct the initial state. It is assumed that not only the scaling relations but also the statistical properties of the vortex tangle remain the same during the evolution as in the "initial" state with the characteristic length $\tilde{\xi}$.

At the moment of escape, $\tilde{\xi} \sim r_o \propto v_{cn}/v = 1/x$, and $\tilde{\xi}$ is used as the size of a cell in the Vachaspati–Vilenkin method. For integer values of x the bubble is represented by a grid with $x \times x \times x$ vertices and a random phase is assigned to each vertex. To satisfy the boundary condition, this grid is surrounded by a shell of vertices with fixed zero phase, representing the uniform bulk superfluid outside the heated bubble. Thus the whole grid contains $(x+1) \times (x+1) \times (x+1)$ cells. Such a correspondence between the size of the grid and the velocity is not too artificial even at small values of x: for example, at the critical velocity $v = v_{cn}$ one gets a grid with $(x+1)^3 = 8$ cells and $3 \times 3 \times 3$ vertices, but the phase can be non-zero only at one vertex in the middle of the grid. In this case no vortices can appear in agreement with the definition of v_{cn}.

By counting the vortex lines produced from a number of random-phase distributions it is possible to calculate the probability distribution for the number of loops escaping per absorption event. For these calculations it is assumed that each loop, which survives until the moment when $\xi \sim r_o$, will form an observable vortex line: in the case of a tangled loop the probability is high that at least one arc is oriented favorably with respect to the counterflow and will be extracted.

In fig. 29 the calculated probability distribution is plotted for observed neutron absorption events at different values of x ($x = 2, 3, 4$) expressed as the fraction of those neutron absorption events which produce a given number of vortices, normalized by the number of all events which give rise to at least one vortex. These results are in remarkably good agreement with the experimental data (without any fitting parameters). However, the agreement with the fraction of "zero" events, i.e. absorption events which produce no vortices at all, is poorer, underestimating their number, especially at $x = 2$. Experimentally the "zero" events can be extracted from the measured data by comparing the event rate at a given rotation velocity with the saturated event rate at the highest velocity.

One reason for this discrepancy is that at low R_b/r_o ratio (i.e. at low velocities) loops with a radius of curvature $\sim r_o$, which are capable of escape into the bulk superfluid, have a rather large probability to be oriented in such a way that they do not have sufficiently large segments oriented favorably with respect to the counterflow, owing to space constraints. Hence they contract and do not give a contribution to the observed signal. Taking this into account one could develop more elaborate techniques for counting vortices than the simple "count all" method described above. In this case better agreement with all experimental data could be achieved. Particularly, if only vortices with positive algebraic area with respect to the counterflow ($S = \oint y \, \mathrm{d}z$) are counted, then good agreement with the number of "zero" events would be obtained. However, such refinements do not change the results essentially, compared to uncertainties of the model.

2.14.3. *Direct simulation of network evolution*

There is no direct evidence for the validity of the assumptions, which were made in the previous section about the evolution of the network, i.e.

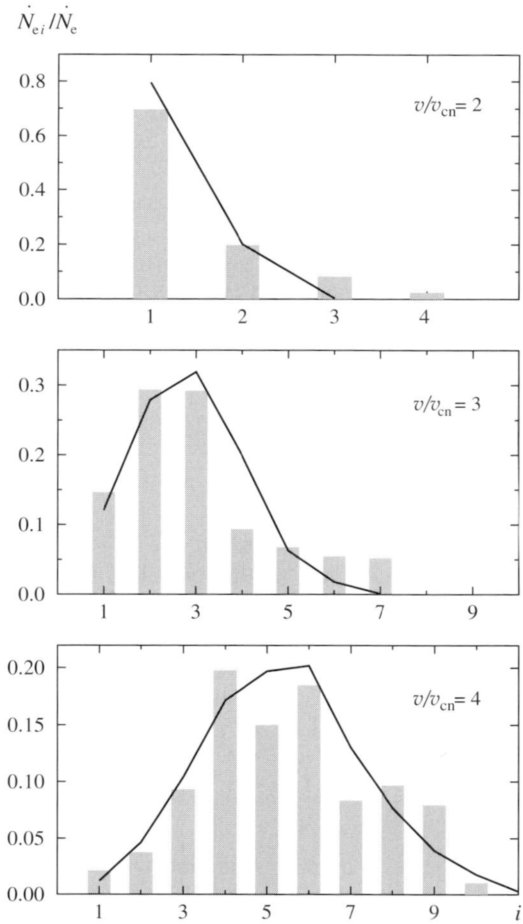

Fig. 29. Distribution of the number of loops escaping per absorption event, normalized to the total number of absorption events and given at three different values of the external bias field, v/v_{cn}. The solid lines represent the simulations while the bars denote the experiment.

whether the network remains self-similar (or scale-invariant) and the scaling laws, eqs. (31)–(33), can be applied during its later evolution. Preliminary calculations have been carried out using the techniques developed by Schwarz for the simulation of superfluid turbulence in ^{4}He. These methods give good numerical agreement with experimental data in ^{4}He (Schwarz 1978, 1985, 1988) and have been used extensively by Schwarz and many others (Samuels 1992, Aarts and de Waele 1994, Nemirovskii and Fiszdon 1994, Tsubota and Yoneda 1995, Barenghi et al. 1997).

In such calculations vortices are generally considered to be one-dimensional objects without internal structure. In ^{4}He the diameter of the core is much smaller than other characteristic lengths, foremost the average radius of curvature for the loops or the inter-vortex distance, and even $\ln(\tilde{\xi}/\xi)$ can be treated as a large parameter. This is not the case in ^{3}He–B due to the large coherence length, especially in the early stages of the evolution when the vortex density is largest. However, as mentioned in section 2.3, the characteristic inter-vortex distance and the radius of curvature may still be several times larger than the coherence length and the diameter of the vortex core. For now, we shall continue considering the vortices as linear objects.

There are several forces acting on a vortex line in a superfluid (Donnelly 1991). The Magnus force appears in the presence of superflow,

$$\mathbf{f}_\mathrm{M} = \rho_\mathrm{s}\kappa\nu\, \mathbf{s}' \times (\mathbf{v}_\mathrm{L} - \mathbf{v}_\mathrm{sl}), \tag{34}$$

where $\mathbf{s}$ is the radius vector of a point on the vortex line and the prime denotes the derivative with respect to the length of the line (i.e. $\mathbf{s}'$ is a unit vector tangent to the vortex line at $\mathbf{s}$), $\mathbf{v}_\mathrm{L} = \dot{\mathbf{s}}$ is the local velocity of the vortex line, and $\mathbf{v}_\mathrm{sl}$ is the local superfluid velocity at this point. For singular vortices in ^{3}He–B the number of circulation quanta is $\nu = 1$. The local superfluid velocity $\mathbf{v}_\mathrm{sl}$ is a sum of the superfluid velocity $\mathbf{v}_\mathrm{s}$ far from the network and the velocity induced by all the vortices in the tangle:

$$\mathbf{v}_\mathrm{sl}(\mathbf{r}) = \mathbf{v}_\mathrm{s} + \frac{\kappa}{4\pi} \int_\text{all loops} \frac{(\mathbf{s} - \mathbf{r}) \times \mathrm{d}\mathbf{s}}{|\mathbf{s} - \mathbf{r}|^3}. \tag{35}$$

The Iordanskii force arises from the Aharonov–Bohm scattering of quasiparticles from the velocity field of the vortex,

$$\mathbf{f}_\text{Iordanskii} = \rho_\mathrm{n}\kappa\nu\mathbf{s}' \times (\mathbf{v}_\mathrm{L} - \mathbf{v}_\mathrm{n}), \tag{36}$$

where $\mathbf{v}_\mathrm{n}$ is the velocity of the normal component (i.e. the heat bath of the fermionic quasiparticles). The Kopnin or spectral flow force has the same form, but originates from the spectral flow of the quasiparticle levels in the vortex core:

$$\mathbf{f}_\mathrm{sf} = m_3\mathscr{C}(T)\kappa\nu\mathbf{s}' \times (\mathbf{v}_\mathrm{L} - \mathbf{v}_\mathrm{n}), \tag{37}$$

where the temperature-dependent parameter $\mathscr{C}(T)$ determines the spectral flow in the core. All these three forces are of topological origin: they act in the transverse direction and are thus non-dissipative. They are discussed in more details in section 3.

In contrast, the non-topological friction force $\mathbf{f}_{\rm fr}$ acts in the longitudinal direction,

$$\mathbf{f}_{\rm fr} = -d_{\parallel}\rho_{\rm s}\kappa\nu\mathbf{s}' \times [\mathbf{s}' \times (\mathbf{v}_{\rm n} - \mathbf{v}_{\rm L})]. \tag{38}$$

Here the factor $\rho_{\rm s}\kappa\nu$ is the same as in the Magnus force.

Neglecting the vortex mass, we may write the force balance equation for the vortex element:

$$\mathbf{f}_{\rm M} + \mathbf{f}_{\rm Iordanskii} + \mathbf{f}_{\rm sf} + \mathbf{f}_{\rm fr} = 0. \tag{39}$$

It is convenient to rewrite the balance of forces in the following form:

$$\mathbf{s}' \times [(\mathbf{v}_{\rm L} - \mathbf{v}_{\rm sl}) - \alpha'(\mathbf{v}_{\rm n} - \mathbf{v}_{\rm sl})] = \alpha\mathbf{s}' \times [\mathbf{s}' \times (\mathbf{v}_{\rm n} - \mathbf{v}_{\rm sl})], \tag{40}$$

where α and α' are the dimensionless mutual friction coefficients. The α parameters are actually the experimentally determined mutual friction quantities (see e.g. Bevan et al. 1997b). The inverse coefficients are the d parameters:

$$\alpha + {\rm i}(1 - \alpha') = \frac{1}{d_{\parallel} - {\rm i}(1 - d_{\perp})}, \tag{41}$$

where ${\rm i} = \sqrt{-1}$. The transverse mutual friction parameter $d_{\perp}$ is expressed in terms of three temperature-dependent functions which determine the Magnus, Iordanskii and spectral-flow forces:

$$d_{\perp}(T) = (m_3\mathscr{C}(T) - \rho_{\rm n}(T))/\rho_{\rm s}(T). \tag{42}$$

Equation (40) is complicated because of the term (35) which contains the integral over the vortex network. To solve eq. (40), one may follow Schwarz and neglect the influence of all other vortex segments on $\mathbf{v}_{\rm sl}$, except the one containing the point of interest (the so-called local self-induced approximation). In this case

$$\mathbf{v}_{\rm sl} \approx \mathbf{v}_{\rm s} + \beta\mathbf{s}' \times \mathbf{s}'', \quad \text{where } \beta = \frac{\kappa}{4\pi}\ln\frac{\tilde{\xi}}{\xi}. \tag{43}$$

The leading correction to this simplification comes from the nearest neglected vortex segments and can be estimated to be of order $\tilde{\xi}/[b\ln(\tilde{\xi}/\xi)]$, where b is the average inter-vortex distance. For a network in ^{3}He–B, $\tilde{\xi}{\sim}b$ and $\ln(\tilde{\xi}/\xi){\sim}1$ in the early stages of the evolution and thus at this point the approximation is rather crude. However, the main effect in the initial stages arises from the reconnection of vortex lines which happen to cross each other and this effect is taken into account below.

A second simplification close to $T_{\rm c}$ comes from $\alpha' \ll \alpha$. We may rewrite the equation of vortex motion in the form

$$\mathbf{v}_{\rm L} = \mathbf{v}_{\rm s} + \beta\mathbf{s}' \times \mathbf{s}'' + \alpha\mathbf{s}' \times [(\mathbf{v}_{\rm n} - \mathbf{v}_{\rm s}) - \beta\mathbf{s}' \times \mathbf{s}'']. \tag{44}$$

In the rotating reference frame the normal component is at rest, $\mathbf{v}_n = 0$, and one has $\mathbf{v}_s = -\mathbf{v}$, where $\mathbf{v} = \mathbf{v}_n - \mathbf{v}_s$ is the counterflow velocity. We shall be comparing with experimental results measured at $T = 0.96T_c$ and $P = 18$ bar, where the most detailed data were collected. Under these conditions, $\alpha \approx 10$ (Bevan et al. 1995) and in the rotating frame eq. (44) can be simplified further to the form

$$\dot{\mathbf{s}} = \alpha\beta\mathbf{s}'' + \alpha\mathbf{s}' \times \mathbf{v}, \tag{45}$$

The first term on the right-hand side of eq. (45) causes a loop to shrink while the second represents growth or shrinking, depending upon the orientation of the loop with respect to the counterflow.

In the numerical calculation vortices are considered as lines in the same three-dimensional lattice as before. The temporal and spatial coordinates are discrete and therefore the network evolves in discrete steps: during each step in time a vertex of a loop can jump to one of the adjacent vertices of the lattice. The probability of a jump in any direction is proportional to the component of $\dot{\mathbf{s}}$ in eq. (45) in this direction, so that the average velocity equals $\dot{\mathbf{s}}$. Other more elaborate representations of vortex loops, based on splines for instance, have been used to model the dynamics of smooth vortex lines in the continuous space. However, for the preliminary study of the network evolution the simplest lattice representation was used. The simple case of one circular loop in the counterflow, where the result is known analytically, can be checked separately and it is found to give correct results.

It is well known that the interactions between neighboring vortex segments in the tangle play an important role in the evolution of a dense network. These interactions lead to the reconnection of loops which cross each other. The reconnections are the main source for the rarefication of the network at the initial stages of evolution, which process through formation and decay of small loops. The reconnection process has been studied numerically (Schwarz 1985, Koplik and Levine 1993), and was found to occur with a probability close to unity for vortex lines approaching each other. In the simulations the vertices are connected in such a way that closed loops are obtained after each time step during the evolution. If a cell is pierced by two vortices, the two segments coming in and the two segments going out are connected randomly. This corresponds to a reconnection probability of 0.5. It is reasonable to suppose that this convention leads to a slowing down in the evolution, in particular in the initial phase, but does not produce gross qualitative changes.

The vortex tangle is initially produced by the procedure described in section 2.14.2, in a lattice $40 \times 40 \times 40$, and its evolution is followed until all vortices have disappeared or loops have expanded and formed circular planar rings far from their location of origin. These large rings are counted, since it is fair to assume that each one of them will produce a detectable rectilinear vortex line in the rotating container. The calculation is repeated for a large number of random initial distributions of the phase at any given value of v. The results are averaged to obtain the average number of vortices N_b, which are produced from one simulated bubble as a function of v, and are compared to experimental data in fig. 30. It is also possible to study the validity of the scaling law (31) at different stages of the evolution in the course of these simulations. In the absence of counterflow ($v = 0$) it is found that the relation is valid for large loops, $L > 4\tilde{\xi}$, at least during much of the early evolution (in the late stages, when n approaches 0, the statistical noise exceeds n).

Figure 30 illustrates the simulation results. No vortices are obtained at low counterflow velocity $v < v_{cn}$, but when $v > v_{cn}$, their number N_b starts to increase rapidly with v. The value of the threshold velocity v_{cn} corresponds to the situation when the largest radius of curvature $r_o(v_{cn})$ becomes equal to the diameter of the initial volume of the vortex network (i.e. the diameter of the heated bubble). It then becomes possible for a loop to escape into the bulk superfluid if it consists of at least one arc with sufficiently large radius of curvature $\geqslant r_o$. The same calculations were also performed for a tangle with an initial volume of $20 \times 20 \times 20$, but no differences were found in the

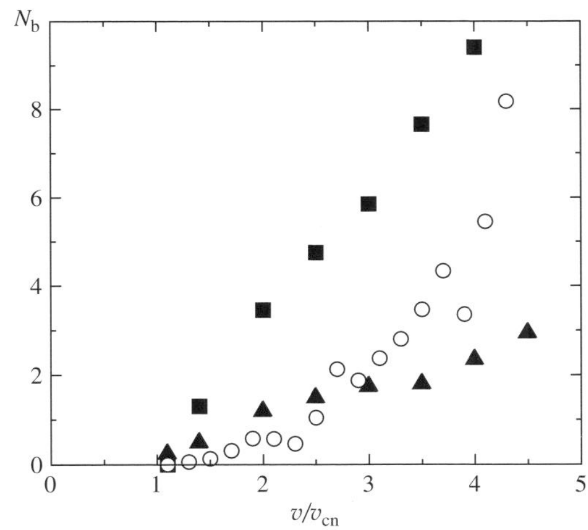

Fig. 30. The number of vortices N_b escaped from the bubble heated by one absorbed neutron as a function of v/v_{cn}: (■) simulation of the network evolution in the local self-induced approximation (eq. (43)) and (▲) by including in addition approximately the polarization of the tangle by the superflow (eq. (46)). Both calculations were performed on a $40 \times 40 \times 40$ lattice, but the results remain unchanged if the lattice is reduced in size to $20 \times 20 \times 20$. The experimental data (○) represent measurements at $P = 18$ bar and $T = 0.96T_c$ (Ruutu et al. 1998a).

dependence $N_{\mathrm{b}}(v/v_{\mathrm{cn}})$. This is additional circumstantial evidence for the fact that the scaling relations approximately survive during the evolution of the network also in the counterflow until the moment of escape and that the network can be approximated as scale-invariant.

The experimental data (denoted by ($\circ$) in fig. 30) fit the universal dependence $N_{\mathrm{b}}(v/v_{\mathrm{cn}}) = \tilde{\gamma}[(v/v_{\mathrm{cn}})^3 - 1]$, where $\tilde{\gamma}$ equals γ in eq. (8) divided by the neutron flux. The numerical results ($\blacksquare$) lie higher (note that no fitting parameters are involved) and do not display a cubic dependence in the experimental range $1 < v/v_{\mathrm{cn}} < 4$. The differences could be explained by the approximations in the expression for $\mathbf{v}_{\mathrm{sl}}$, which neglect significant contributions from the inter-vortex interactions. The polarization by the external counterflow causes the loops with unfavorable orientation to contract and loops with the proper winding direction and orientation to grow. However, as the polarization evolves, it also screens the vortex tangle from the external counterflow. To check whether the polarization has a significant effect on the results, the calculations were repeated by taking the screening approximately into account in the expression for $\mathbf{v}_{\mathrm{sl}}$:

$$\mathbf{v}_{\mathrm{sl}} = \mathbf{v}_{\mathrm{s}} + \beta \mathbf{s}' \times \mathbf{s}'' + \frac{\kappa}{4\pi} \int_{\text{other loops}} \frac{1}{V} \int \mathrm{d}\mathbf{r} \, \frac{(\mathbf{s} - \mathbf{r}) \times \mathrm{d}\mathbf{s}}{|\mathbf{s} - \mathbf{r}|^3}. \tag{46}$$

Here the contribution from each loop to the superflow is averaged over the volume. The results ($\blacktriangle$) show that this effect is significant and should be taken into account more accurately, to reproduce the network evolution correctly.

To summarize we note that the preliminary simulation work shows that it is possible to obtain numerical agreement between the KZ mechanism and the rotating experiments, if one assumes that the scaling relations of the initial vortex tangle are roughly obeyed also during its later evolution in the presence of the external superflow. This assumption should still be checked with sufficiently accurate simulation, with full account of the non-local inter-vortex interactions. Future simulation of transient networks should then answer the question how much information about the initial state of the vortex tangle can be retrieved from the rotating experiment, where only the final stationary state result is measured.

2.15. *Superfluid transition as a moving-phase front*

In any real laboratory system a rapid phase transition becomes inhomogeneous: the transition will be driven by a strong gradient in one or several of the externally controlled variables. In the case of the superfluid transition in the aftermath of a neutron absorption event it is a steep and rapidly relaxing thermal gradient. In this situation even a second-order transition becomes spatially inhomogeneous and is localized into a phase front. In the limit of very fast transitions the order-parameter relaxation slows down the propagation of the superfluid phase and the thermal front, where the temperature drops below T_c, may escape ahead. This means that the moving-phase boundary breaks down into a leading thermal front and a trailing superfluid interface. The width of the region between these two zones is determined by the relative velocities of thermal and superfluid relaxations. Thermal fluctuations in this region are amplified in the transition process and are carried over as order-parameter inhomogeneity into the new symmetry-broken phase. This is the central claim of the KZ model. Below we shall briefly discuss the influence of the thermal gradient on defect formation, as analyzed by Kibble and Volovik (1997) and Kopnin and Thuneberg (1999).

2.15.1. *Neutron absorption and heating*

The decay products from the neutron absorption reaction generate ionization tracks which can be estimated from a standard calculation of stopping power (Meyer and Sloan 1997). This leads to a cigar-shaped volume of ionized particles, with the largest concentration at the end points of the two tracks. The probabilities and relaxation times of the different recombination channels for the ionized charge are not well known in liquid ^{3}He. Also the thermalization of the reaction energy may not produce a heated region which preserves the shape of the original volume with the ionized charge.

Initially the recombination processes are expected to lead to particles with large kinetic energies in the eV range, which are well outside the thermal distribution and for which the recoil velocities become more and more randomly oriented. Energetic particles suffer collisions with their nearest neighbors and the mean free path increases only slowly for atoms participating in these collisions, until all particles are slowed down and become thermalized to the ambient conditions (Bunkov and Timofeevskaya 1998a, b). This means that the reaction energy remains initially localized. In the calorimetric experiments at the lowest temperatures (Bäuerle et al. 1996, 1998a) the thermal mean free path exceeds the container dimensions. Nevertheless, the energy is probably not immediately dispersed into the entire container volume, but remains centralized within a bubble of limited size during cooling through T_c, when the vortex network is formed. This is the conclusion to be drawn from the comparison between experiment and the KZ mechanism in Table 1, where it is assumed that the thermal diffusion mean free path is the same as in the normal fluid at T_c.

In an inhomogeneous initial state with large thermal gradients the second-order phase transition is turned into one where a normal-to-superfluid phase front with finite width sweeps through the heated bubble. If the velocity of the phase front, $v_T \sim R_b/\tau_Q \sim 6$ m/s, is sufficiently high, comparable to a critical value $v_{Tc} \sim v_F\ (\tau_0/\tau_Q)^{1/4}$, then the KZ mechanism is again expected to dominate, similarly as in the homogeneous case (Kibble and Volovik 1997).

In contrast, if the majority of the kinetic energy is assumed to be thermalized by quasiparticles with energies comparable to the high-energy tail of the thermal Maxwellian velocity distribution, then the mean free paths are long, the volume heated above the ambient becomes large, and its temperature distribution may even become non-monotonic like in a "Baked Alaska", as has been described by Leggett (1992). The Baked-Alaska scenario is also popular in high-energy physics, where the formation of the false vacuum with a chiral condensate after a hadron–hadron collision is considered (Bjorken 1997, Amelino-Camelia et al. 1997). In both cases, a rather thin shell of radiated high-energy particles expands, with the speed of light in a relativistic system and with the Fermi velocity v_F in ^{3}He, leaving behind a region at reduced temperature. Since this interior region is separated from the exterior vacuum by the hot shell, the cooldown into the broken-symmetry state in the center is not biased by the external-order-parameter state. The Baked-Alaska mechanism thus can solve the problem of the neutron-mediated formation of B phase from supercooled bulk A liquid (Leggett 1992), while in high-energy physics it opens the possibility for the formation of a bubble of chiral condensate in a high-energy collision (Bjorken 1997, Amelino-Camelia et al. 1997).

Under such conditions, when the quasiparticle mean free path exceeds or is comparable to the dimensions of the heated bubble, temperature is not a useful quantity for the description of its cooling. Most of the analysis of the previous sections is applicable only if we assume that the reaction energy remains reasonably well localized while the hot bubble cools through T_c. In this case there is no Baked-Alaska effect: no hot shell will be formed which would separate the interior from the exterior region. In this situation the exterior region could be imagined to fix the phase in the cooling bubble, while the phase front is moving inward, suppressing the formation of new order-parameter states, which are different from that in the bulk superfluid outside, and in the same manner suppressing the formation of vortices. However, it appears that there also exists another alternative: the influence of the exterior region may not be decisive if the phase transition front moves sufficiently rapidly. Which of these alternatives is realized in a particular situation is still very much a subject of discussion.

For the interpretation of the measurements in neutron irradiation, a sophisticated understanding of the shape and size of the constant temperature contours within the heated bubble is not vitally necessary. In the final results the bubble size does not enter, since the data can be normalized in terms of the measured threshold velocity v_{cn}. Its measurement provides an estimate of the circumference of the bubble, since the largest possible vortex ring has to be comparable in size to the neutron bubble. The diameter of this ring is 1–2 orders of magnitude larger than the expected average inter-vortex distance ξ_{v} in the initial random network which is created by the KZ mechanism.

2.15.2. *Thermal gradient and velocity of phase front*

For a rough understanding of the superfluid transition in a temperature gradient let us consider the TDGL model for a one-component order parameter $\Psi = \Delta/\Delta_0$:

$$\tau_0 \frac{\partial \Psi}{\partial t} = \left(1 - \frac{T(\mathbf{r}, t)}{T_{\mathrm{c}}}\right)\Psi - \Psi|\Psi|^2 + \xi_0^2 \nabla^2 \Psi, \tag{47}$$

where the notations for $\tau_0 \propto 1/\Delta_0$ and ξ_0 are the same as in section 2.3. This equation is the so-called over-damped limit of the more general TDGL equation which has a time derivative of second order. The over-damped limit has been used in numerical simulations (Laguna and Zurek 1997, Antunes et al. 1999, Aranson et al. 1999) and analytical estimations (Dziarmaga 1998, 1999) of the density of topological defects in a homogeneous quench. An extension of the above equation to superconductivity was used by Ibaceta and Calzetta (1999) to study the formation of defects after a homogeneous quench in a two-dimensional type II superconductor.

If the quench occurs homogeneously in the whole space $\mathbf{r}$, the temperature depends only on one parameter, the quench time τ_{Q}:

$$T(t) = \left(1 - \frac{t}{\tau_{\mathrm{Q}}}\right) T_{\mathrm{c}}. \tag{48}$$

In the presence of a temperature gradient, say, along x, a new parameter appears, which together with τ_{Q} characterizes the evolution of the temperature:

$$T(x - v_{\mathrm{T}} t) = \left(1 - \frac{t - x/v_{\mathrm{T}}}{\tau_{\mathrm{Q}}}\right) T_{\mathrm{c}}. \tag{49}$$

Here v_{T} is the velocity of the temperature front which is related to the temperature gradient (Kibble and Volovik 1997)

$$\nabla_x T = \frac{T_{\mathrm{c}}}{v_{\mathrm{T}} \tau_{\mathrm{Q}}}. \tag{50}$$

With this new parameter v_{T} we may classify transitions to belong to one of two limiting regimes: slow or fast propagation of the transition front. At slow velocities, $v_{\mathrm{T}} \to 0$, the order parameter moves in step with the temperature front and

$$|\Psi(x,t)|^2 = \left(1 - \frac{T(x - v_T t)}{T_c}\right)\Theta(1 - T(x - v_T t)/T_c). \tag{51}$$

Here Θ is the step function. In this case, phase coherence is preserved behind the transition front and no defect formation is possible.

The opposite limit of large velocities, $v_T \to \infty$, is more interesting. Here the phase transition front lags behind the temperature front (Kopnin and Thuneberg 1999). In the space between these two boundaries the temperature is already below T_c, but the phase transition did not yet happen, and the order parameter has not yet formed, $\Psi = 0$. This situation is unstable towards the formation of blobs of the new phase with $\Psi \neq 0$. This occurs independently in different regions of space, leading to vortex formation via the KZ mechanism. At a given point $\mathbf{r}$ the development of the instability can be found from the linearized TDGL equation, since during the initial growth of the order parameter Ψ from 0 the cubic term can be neglected:

$$\tau_0 \frac{\partial \Psi}{\partial t} = \frac{t}{\tau_Q} \Psi. \tag{52}$$

This gives an exponentially growing order parameter, which starts from some seed Ψ_{fluc}, caused by fluctuations:

$$\Psi(\mathbf{r},t) = \Psi_{\text{fluc}}(\mathbf{r}) \exp \frac{t^2}{2\tau_Q \tau_0}. \tag{53}$$

Because of exponential growth, even if the seed is small, the modulus of the order parameter reaches its equilibrium value $|\Psi_{\text{eq}}| = \sqrt{1 - T/T_c}$ soon after the Zurek time t_{Zurek}:

$$t_{\text{Zurek}} = \sqrt{\tau_Q \tau_0}. \tag{54}$$

This occurs independently in different regions of space and thus their order-parameter phases are not correlated. The spatial correlation is lost over distances exceeding ξ_v when the gradient term in eq. (47) becomes comparable to the other terms at $t = t_{\text{Zurek}}$. Equating the gradient term $\xi_0^2 \nabla^2 \Psi \sim (\xi_0^2/\xi_v^2)\Psi$ to, say, the term $\tau_0 \partial\Psi/\partial t|_{t_{\text{Zurek}}} = \sqrt{\tau_0/\tau_Q}\Psi$, one obtains the characteristic Zurek length which determines the initial length scale of defects,

$$\xi_v = \xi_0 \, (\tau_Q/\tau_0)^{1/4}, \tag{55}$$

which is the same as in eq. (3).

We can estimate the lower limit for the characteristic value of the fluctuations $\Psi_{\text{fluc}} = \Delta_{\text{fluc}}/\Delta_0$, which serve as a seed for vortex formation. If there are no other sources of fluctuations, caused by external noise for example, the initial seed is provided by thermal fluctuations of the order parameter in the volume ξ_v^3. The energy of such fluctuations is $\xi_v^3 \Delta_{\rm fluc}^2 N_F/E_F$, where $E_F \sim k_B T_F$ is the Fermi energy and N_F the fermionic density of states in normal Fermi liquid. Equating this energy to temperature $T \approx T_c$ one obtains the magnitude of the thermal fluctuations

$$\frac{|\Psi_{\rm fluc}|}{|\Psi_{\rm eq}|} \sim \left(\frac{\tau_0}{\tau_Q}\right)^{1/8} \frac{k_B T_c}{E_F}. \tag{56}$$

The same small parameter $k_B T_c/E_F \sim a/\xi_0 \sim 10^{-3} - 10^{-2}$ enters here, which is responsible for the extremely narrow temperature region of the critical fluctuations in ^{3}He near T_c. But in our case it only slightly increases the Zurek time by the factor $\sqrt{\ln[|\Psi_{\rm eq}|/|\Psi_{\rm fluc}|]}$ and thus does not influence vortex formation in a homogeneous quench or in an inhomogeneous quench at large velocities of the temperature front.

Clearly there must exist a characteristic velocity v_{T_c} of the propagating temperature front, which separates the fast and slow regimes, or correspondingly transitions with and without defect formation. The criterion for defect formation is that the Zurek time $t_{\rm Zurek} = \sqrt{\tau_Q \tau_0}$ should be shorter than the time $t_{\rm sw}$ in which the phase transition front sweeps through the space between the two boundaries. The latter time is $t_{\rm sw} = x_0/v_T$, where x_0 is the lag – the distance between the temperature front $T = T_c$ and the region where superfluid coherence starts (order-parameter front). If $t_{\rm Zurek} < t_{\rm sw}$, instabilities have time to develop. If $t_{\rm Zurek} > t_{\rm sw}$, both fronts move coherently and the phase is continuous. Let us consider the latter case, which we call "laminar" motion, and find how the lag x_0 depends on v_T. From the equation $t_{\rm Zurek} = x_0(v_{T_c})/v_{T_c}$ we find the criterion for the threshold where laminar motion becomes unstable and defect formation starts.

In steady laminar motion the order parameter depends on $x - v_T t$. Introducing a dimensionless variable z and a dimensionless parameter h,

$$z = (x - v_T t)(v_T \tau_Q \xi_0^2)^{-1/3}, \qquad h = \left(\frac{v_T \tau_0}{\xi_0}\right)^{4/3} \left(\frac{\tau_Q}{\tau_0}\right)^{1/3}, \tag{57}$$

the linearized TDGL equation becomes

$$\frac{d^2\Psi}{dz^2} + h\,\frac{d\Psi}{dz} - z\Psi = 0 \tag{58}$$

or

$$\Psi(z) = \text{const} \cdot e^{-hz/2} \chi(z), \qquad \frac{d^2\chi}{dz^2} - \left(z + \frac{h^2}{4}\right)\chi = 0. \tag{59}$$

This means that Ψ is an Airy function, $\chi(z - z_0)$, centered at $z = z_0 = -h^2/4$ and attenuated by the exponential factor $e^{-hz/2}$.

When $h \gg 1$, it follows from eq. (59) that $\Psi(z)$ quickly vanishes as z increases above $-h^2/4$. Thus there is a supercooled region $-h^2/4 < z < 0$, where $T < T_c$, but the order parameter is not yet formed: the solution is essentially $\Psi = 0$. The lag between the order parameter and temperature fronts is $|z_0| = h^2/4$ or in conventional units

$$x_0 = \frac{1}{4} \frac{v_T^3 \tau_Q \tau_0^2}{\xi_0^2}. \tag{60}$$

Setting $t_{\text{Zurek}} = x_0(v_{T_c})/v_{T_c}$ one can estimate the value for the velocity of the temperature front where laminar propagation becomes unstable:

$$v_{T_c} \sim \frac{\xi_0}{\tau_0} \left(\frac{\tau_0}{\tau_Q} \right)^{1/4}. \tag{61}$$

This result corresponds to $h \sim 1$ and is in agreement with that obtained from scaling arguments by Kibble and Volovik (1997). Here an exact numerical value cannot be offered for the threshold where laminar motion ends, but this can in principle be done using the complete TDGL equation (Kopnin and Thuneberg 1999). For the neutron bubble we might take $v_T \sim R_b/\tau_Q$, which gives $v_T \sim 10$ m/s. This value also provides an estimate for the limiting velocity v_{T_c}. These considerations suggest that the thermal gradient should be sufficiently steep in the neutron bubble such that defect formation is to be expected.

2.16. *Quench of infinite vortex tangle*

2.16.1. *Vorticity on microscopic and macroscopic scales*

Onsager (1949) was the first to interpret the λ-transition in liquid ^{4}He from the superfluid to the normal state in terms of quantized vortices: when the concentration of the thermally activated quantized vortices reaches the point where they form a connected tangle throughout the liquid (and their line tension vanishes), the liquid goes normal (fig. 31). This proliferation of vortices to an infinite network destroys superfluidity, since the phase slippage processes caused by the back reaction from the tangle to the superfluid current lead to the decay of this current.

In Ginzburg–Landau theory vortices are identified as lines of zeros in the scalar complex order parameter $\Psi = |\Psi| e^{i\Phi}$, around which the phase winding number v is non-zero. Above the second-order transition, in the symmetric phase, thermal fluctuations of the order parameter give rise to an

Fig. 31. Normal-to-superfluid transition, when described as a changeover from an infinite to a finite vortex network. (*Top*) The disordered phase ($T > T_c$) is an infinite network of defects. The coherence length ξ is the mean distance between the elements in this network. (*Bottom*) In the ordered phase ($T < T_c$) the defects form closed loops, but the coherence length is now the upper limit in the distribution of loop sizes. It is the existence of infinite strings as thermal fluctuations in the disordered phase that will lead to defect formation if the system is suddenly cooled through T_c (Buttencourt et al. 2000, Volovik 2003).

infinite network of zeros – to vortices which exist on the microscopic scale but are absent on the macroscopic scale (Kleinert 1989). This is another way of describing the fact that there is no long-range order in the symmetric phase. The properties of such microscopic vortices – topologically non-trivial zeros – have been followed across the thermodynamic phase transition in numerical investigations (Antunes and Bettencourt 1998, Antunes et al. 1998, Rajantie et al. 1998, Kajantie et al. 1998, Rajantie 1998). The renormalization group description of the phase transition, based on the Ginzburg–Landau free energy functional, also contains the microscopic quantized vortices, but in an implicit form of zeros (see e.g. Guillou and Zinn-Justin 1980, Albert 1982).

Some attempts have been made to reformulate the three-dimensional phase transition in terms of this vortex picture in the renormalization group approach (Chattopadhyay et al. 1993, Williams 1993a, b, 1999), in a manner similar to the two-dimensional Berezinskii–Kosterlitz–Thouless transition (Nelson and Kosterlitz 1977), i.e. by avoiding consideration of the Ginzburg–Landau free energy functional altogether. It is possible, though it has not been proven, that the Ginzburg–Landau model and the vortex model belong to the same universality class and thus give the same critical exponents for the heat capacity and superfluid density. Both ground and space-based measurements of the critical exponents (Goldner and Ahlers 1992, Lipa et al. 1996, 2003) have generally been consistent with renormalization-group estimations (Campostrini et al. 2001, Kleinert and Van den Bossche 2001, Strösser and Dohm 2003).

In an equilibrium phase transition the infinite network disappears below T_c, but in a non-equilibrium phase transition the tangle of microscopic vortices persists even into the ordered phase, due to critical slowing down. These vortices finally transform to conventional macroscopic vortices, when the latter become well defined. In this language the Kibble-Zurek mechanism corresponds to a quench of the infinite vortex network across the non-equilibrium transition from the normal to the superfluid phase (Yates and Zurek 1998). What is important for us here, is the scaling law for the distribution of vortices. According to numerical simulations (Antunes and Bettencourt 1998, Antunes et al. 1998) and the phenomenological vortex model (Chattopadhyay et al. 1993, Williams 1993a, b, 1999) the scaling exponent δ, which characterizes the distribution of vortex loops in eq. (32), is close to the value $\delta = 2/(d+2) = 0.4$ obtained using the concept of Flory calculations for self-avoiding polymers. This value, which is obtained in random vortex model, is different from the value given by the Vachaspati–Vilenkin random phase model.

2.16.2. *Scaling in equilibrium phase transitions*

Important differences exist between the phase transitions in the ^{4}He and ^{3}He liquids. These also are involved in a number of other phenomena. Let us therefore recall some of these differences. The temperature region of the critical fluctuations, where the simple Ginzburg–Landau theory does not work and one must introduce the thermal renormalization of the Ginzburg–Landau functional, can be derived from the following simplified considerations. Let us estimate the length L_T of the thermal vortex loop, i.e. a loop whose energy is comparable to temperature:

$$\rho_s(T)\frac{\kappa^2}{4\pi}L_T \ln\left(\frac{L_T}{\xi(T)}\right) = k_B T \approx k_B T_c. \tag{62}$$

In the broken-symmetry phase far below T_c, this length is less than the coherence length $\xi(T)$, which means that there are no real vortices with the energy of order $k_B T$, and real vortices with higher energy are exponentially suppressed. When the temperature increases, one approaches the point at which the length L_T becomes comparable to the coherence length ξ. This is the Ginzburg temperature T_{Gi}, determined by the condition $L_T(T = T_{Gi}) \sim \xi(T = T_{Gi})$:

$$\rho_s(T_{Gi})\xi(T_{Gi}) \sim \frac{k_B T_c}{\kappa^2} \tag{63}$$

or

$$\left(1 - \frac{T_{Gi}}{T_c}\right) \sim \left(\frac{k_B T_c}{\rho \xi_0 \kappa^2}\right)^2 \sim \left(\frac{T_c}{T_F}\right)^4. \tag{64}$$

Here as before $T_F \sim \hbar^2/ma^2 \sim 1$ K is the degeneracy temperature of the quantum fluid, with a being the inter-atomic spacing. The region of critical fluctuations – the Ginzburg region $T_{Gi} < T < T_c$ – is broad for ^{4}He, where $T_c \sim T_F$, and extremely small for ^{3}He, where $T_c < 10^{-2} T_F$.

In the region of Ginzburg fluctuations the scaling exponents for the thermodynamic quantities, such as $\xi(T)$ and $\rho_s(T)$, are different from those in the Ginzburg–Landau region, $T_c - T_{Gi} < T_c - T \ll T_c$:

$$\frac{\xi(T)}{\xi_0} \sim \left\{ \left(1 - \frac{T}{T_c}\right)^{-1/2}, \quad T_c - T_{Gi} < T_c - T \ll T_c, \right. \tag{65}$$

$$\frac{\xi(T)}{\xi_0} \sim \left\{ \left(1 - \frac{T_{Gi}}{T_c}\right)^{\nu - 1/2} \left(1 - \frac{T}{T_c}\right)^{-\nu}, \quad T_{Gi} < T < T_c, \right. \tag{66}$$

$$\frac{\rho_s(T)}{\rho} \sim \left\{ \left(1 - \frac{T}{T_c}\right), \quad T_c - T_{Gi} < T_c - T \ll T_c, \right. \tag{67}$$

$$\frac{\rho_s(T)}{\rho} \sim \left\{ \left(1 - \frac{T_{Gi}}{T_c}\right)^{1-\zeta} \left(1 - \frac{T}{T_c}\right)^{\zeta}, \quad T_{Gi} < T < T_c. \right. \tag{68}$$

There are two relations, the scaling hypotheses, which connect the critical exponents for $\xi(T)$ and $\rho_s(T)$ in the Ginzburg region, with the exponent of the heat capacity. In the Ginzburg region $T_{Gi} < T < T_c$, the coherence length is determined by thermal fluctuations, or which is the same thing, by thermal vortices. This gives for the relation between the coherence length $\xi(T)$ and superfluid density $\rho_s(T)$ in the Ginzburg region:

$$\rho_s(T)\xi(T) \sim \frac{k_B T_c}{\kappa^2} \sim \rho a \left(\frac{T_c}{T_F}\right), \quad T_{Gi} < T < T_c. \tag{69}$$

This equation gives the Josephson scaling hypothesis: $\nu = \zeta$.

Another relation between these exponents comes from a consideration of the free energy, which has the same scaling law as the kinetic energy of superflow:

$$F(T) \propto \left(1 - \frac{T}{T_c}\right)^{2-\alpha} \propto \rho_s v_s^2 \propto \frac{\rho_s(T)}{\xi^2(T)} \propto \frac{\rho}{\xi_0^2}\left(1 - \frac{T}{T_c}\right)^{\zeta+2\nu}. \quad (70)$$

Here α is the critical exponent for the heat capacity in the critical region, $C(T) = -T\partial_T^2 F \propto (1 - T/T_c)^{-\alpha}$. Equations (69) and (70) give $\nu = \zeta = (2 - \alpha)/3$.

2.16.3. *Non-equilibrium phase transitions*

The formation of vortices during a rapid transition into the broken-symmetry phase is the subject of dynamic scaling and a poorly known area in the field of critical phenomena. Dynamic scaling is characterized by an additional set of critical exponents, which depend not only on the symmetry and topology of the order parameter, but also on the interaction of the order parameter with the different dynamic modes of the normal liquid. The question first posed by Zurek was the following: what is the initial density ξ_v of macroscopic vortices at the moment when they become well defined? According to the general scaling hypothesis one has

$$\xi_v \sim \xi_0(\omega\tau_0)^{-\lambda} \quad \text{and} \quad \omega = \frac{1}{\tau_Q}, \quad (71)$$

where ω is the characteristic frequency of the dynamic process.

In the time-dependent Ginzburg–Landau model (eq. (47)) and also in its extension based on the renormalization group approach, the exponent λ is determined by the static exponents and by the exponent for the relaxation time $\tau = \tau_0(1 - T/T_c)^{-\mu}$:

$$\lambda = \frac{\nu}{1 + \mu}. \quad (72)$$

This follows from the following consideration. When we approach the critical temperature from the normal phase, at some moment t_{Zurek} the relaxation time $\tau(t)$ becomes comparable to the time t which is left until the transition takes place. At this moment

$$t_{\mathrm{Zurek}} = \tau_0\left(\frac{\tau_Q}{\tau_0}\right)^{\mu/(1+\mu)}, \quad (73)$$

the vortex network is frozen out. After the transition it becomes "unfrozen" when $\tau(t)$ again becomes smaller than t, the time after passing the transition. The initial distance between the vortices is determined as the coherence length $\xi = \xi_0(t/\tau_Q)^{-\nu}$ at $t = t_{\mathrm{Zurek}}$, which gives eq. (71) with λ as in eq. (72).

For the conventional time-dependent Ginzburg–Landau model in eq. (47) one has $\mu = 1$, $\nu = 1/2$, and thus $\lambda = 1/4$. In the Ginzburg regime, if one assumes that μ remains the same, while $\nu \approx 2/3$, one obtains $\lambda \approx 1/3$.

In numerical simulation of a quench in the time-dependent Ginzburg–Landau model, the main problem becomes how to resolve between the microscopic vortices (i.e. the zeros in the order parameter) and real macroscopic vortices. This requires some coarse-graining procedure, which is not well established. Also speculations exist (Rivers 2000) that if the quench is limited to within the region of the Ginzburg fluctuations, then the network of microscopic vortices might effectively screen out the real macroscopic vortices, so that the density of the real vortices after the quench is essentially less than its Zurek estimate. This could explain the negative result of the pressure-quench experiments in liquid ^{4}He (Dodd et al. 1998), where the final state after the decompression was well within the region of critical fluctuations. However, according to the simulations by Bettencourt et al. (2000), the Ginzburg regime does not suppress the density of defects. Thus in mechanical pressure-quench experiments vortices simply seem to decay before the observations start.

To summarize, we note that vortices play an important role, not only in the ordered state, but also in the physics of the broken-symmetry phase transitions. The proliferation of vortex loops with infinite size can be interpreted to destroy the superfluid long-range order above the phase transition. In a non-equilibrium phase transition from the symmetric normal phase to the superfluid state, the infinite vortex cluster from the normal state survives after the rapid quench and becomes the source for the remanent vorticity in the superfluid state. This is an alternative picture in which one may understand the formation of vortices in a rapid non-equilibrium transition.

2.17. *Implications of quench-cooled experiments*

2.17.1. *Topological-defect formation*

It is not obvious that a phenomenological model like the KZ mechanism, which is based on scaling arguments, should work at all: it describes a time-dependent phase transition in terms of quantities characterizing the equilibrium properties of the system. Numerical calculations on simple quantum systems, where one studies the fluctuations in the amplitude of the system wave function while it is quench cooled below a second-order phase transition, have provided much evidence for the KZ model and appear to agree with its qualitative features. Most attempts of experimental verification suffer from shortcomings. Measurements on superfluid ^{3}He are the first to test the KZ model more quantitatively. In this case, the experimental deviations from the ideal KZ setup include the presence of a strict boundary condition and a strong thermal gradient. At present time we can conclude that the experiments and the model are in reasonable harmony, assuming that open questions from section 2.14 can be answered satisfactorily, as seems likely.

However, even good agreement leaves us with an interesting question: what is the microscopic basis for the applicability of the KZ mechanism to such experiments? A rapid quench through the superfluid transition is more amenable to microscopic analysis in the case of liquid ^{3}He than in most other systems, since the freeze out of order-parameter domains can be demonstrated with physically acceptable calculations. The consequences from this are exciting and the prospects for a better understanding of non-equilibrium phase transitions look promising. Although detailed agreement has not yet been reached between experimental and theoretical work, nevertheless the effort by Aranson et al. (1999, 2001) illustrates that many aspects of the neutron absorption event in ^{3}He–B can be treated in realistic ways.

Are there implications from such work to cosmological large-scale structure formation? The combined existing evidence from condensed matter experiments supports the KZ mechanism at high transition velocities. This says that the KZ mechanism is an important process in rapid non-equilibrium phase transitions and cannot be neglected. Whether it was effective in the Early Universe and led to the formation of large-scale structure, rather than some other phenomenon like inflationary expansion, is a separate question. It can only be answered by measurements and investigations on the cosmological scale which directly search for the evidence on topological defect formation or its exclusion. The fact that no direct traces from topological defects have been identified, excepting the large-scale structure itself, is not yet sufficient proof that they can be discarded altogether. As measurements of the moving first-order AB interface in ^{3}He superfluids show, the interactions of the defects with the phase front and across it are complicated and no simple general rule exist on the final outcome.

2.17.2. *Phase transitions*

As discussed in sections 2.10.1 and 2.10.2, the KZ mechanism provides an attractive explanation for the radiation induced first-order transition from supercooled ^{3}He-A to ^{3}He–B. The nucleation of competing phases, with different symmetries and local minima of the energy functional, has been discussed both in superfluid ^{3}He (Volovik 2003) and in the cosmological context (Linde 1990).

Above the critical phase-transition temperature T_c, liquid ^{3}He is in its symmetric phase: it has all the symmetries allowed in non-relativistic condensed matter. The continuous symmetries, whose breaking are relevant for the topological classification of the defects in the non-symmetric phases of ^{3}He, form the symmetry group

$$\mathbf{G} = SO(3)_L \times SO(3)_S \times U(1)_N. \tag{74}$$

Here $SO(3)_L$ is the group of solid rotations of the coordinate space. The spin rotations of the group $SO(3)_S$ may be considered as a separate symmetry operation if one neglects the dipolar spin–orbit interaction. The magnetic dipole interaction between the nuclear spins is tiny in comparison with the energies characterizing the superfluid transition. The group $U(1)_N$ is the *global* symmetry group of gauge transformations, which stems from the conservation of the particle number N for the ^{3}He atoms in their ground states. $U(1)$ is an exact symmetry if one neglects extremely rare processes of excitations and ionization of the ^{3}He atoms, as well as the transformation of ^{3}He nuclei, in neutron radiation.

Below T_c the "unified GUT symmetry" $SO(3)_L \times SO(3)_S \times U(1)_N$ can be broken to the $U(1) \times U(1)$ symmetry of the A phase or to the $SO(3)$ symmetry of the B phase, with a small energy difference between these two states, but separated by an extremely high-energy barrier ($\sim 10^6 k_B T$) from each other. In this situation a thermally activated A → B transition becomes impossible. For the radiation-induced A → B transition two solutions have been suggested: (1) the Baked-Alaska configuration where the transition takes place inside a cool bubble isolated by a warm shell from the surrounding A-phase bath (Leggett 1984), and (2) the KZ mechanism, first proposed for cosmological domain wall formation by Kobzarev et al. (1974) and adopted ^{3}He by Volovik (1996) and Bunkov and Timofeevskaya (1998a, b).

This situation can be compared to that in the Early Universe. It is believed that the $SU(3) \times SU(2) \times U(1)$ symmetries of the strong, weak, and electromagnetic interactions (respectively) were united at high energies (or at high temperatures). The underlying grand unification (GUT) symmetry ($SU(5)$, $SO(10)$, or a larger group) was broken at an early stage during the cooling of the Universe. Even the simplest GUT symmetry $SU(5)$ can be broken in different ways: into the phase $SU(3) \times SU(2) \times U(1)$, which is our world, and into $SU(4) \times U(1)$, which apparently corresponds to a higher energy state. In supersymmetric models both phases represent local minima of almost equal depth, but are separated from each other by a high-energy barrier.

In the cosmological scenario, after the symmetry break of the $SU(5)$ GUT state, both new phases are created simultaneously with domain walls between them. The calculated probability for the creation of our world – the

$SU(3) \times SU(2) \times U(1)$ state – appears to be smaller than that of the false vacuum state of $SU(4) \times U(1)$ symmetry at higher energy. Thus initially the state corresponding to our world appears to have occupied only a fraction of the total volume. Later bubbles of this energetically preferred state grew at the expense of the false vacuum state and finally completely expelled it. However, before that interfaces between the two states were created in those places where blobs of the two phases met. Such an interface is an additional topologically stable defect, which is formed in the transition process and interacts with other topological defects.

3. Vortex dynamics and quantum field theory analogs

From the phenomenological point of view, vortex dynamics in ^{3}He–B is similar to ^{4}He-II, but the properties which govern the dynamics arise from the quite different microscopics of the p-wave Cooper-paired fermion superfluid. These properties turn out to have interesting analogies with various problems in quantum field theory (Volovik 2003). It is these connections which are in the focus of the discussion in this section.

The fundamental starting point is the notion that the bosonic and fermionic excitations in superfluid ^{3}He are in many respects similar to the excitations of the energetic physical vacuum of elementary particle physics – the modern ether. This similarity allows us to model, with concepts borrowed from ^{3}He physics, the interactions of elementary particles with the evolving strings and domain walls, which are formed in a rapid phase transition, for instance. Such processes become important after defect formation in the initial quench and give rise to the cosmological consequences which we are measuring today.

The quantum physical vacuum – the former empty space – is in reality a richly structured and asymmetric medium. Because the new quantum ether is such a complicated material with many degrees of freedom, one can learn to analyze it by studying other materials, i.e. condensed matter (Wilczek 1998). Fermi superfluids, especially ^{3}He-A, are the best examples, which provide opportunities for such modeling. The most pronounced property of ^{3}He-A is that, in addition to the numerous bosonic fields (collective modes of the order parameter which play the part of gauge fields in electromagnetic, weak, and strong interactions) it contains gapless fermionic quasiparticles, which are similar to the elementary excitations of the quantum physical vacuum (leptons and quarks).

It is important that the quantum physical vacuum belongs to the same class of fermionic condensed matter as ^{3}He-A: both contain topologically stable nodes in the energy spectrum of the fermionic excitations. As a result both of these fermionic systems display, for example, the gravitational and gauge fields as collective bosonic modes. Other fermionic systems belong either to a class, which is characterized by Fermi surfaces (such as normal metals and the normal ^{3}He liquid), or to a class with a gap in the fermionic spectrum (such as conventional superconductors and ^{3}He–B). High-temperature superconductors seem to belong to a marginal class, where the topological stability of lines of gap nodes is protected by symmetry. Such nodes become instable when the symmetry is violated. Recently the topological stability of different types of nodes has been discussed by Horava (2005). Thus ^{3}He-A (together with ^{3}He-A$_1$) is the condensed matter system in which the properties of the physical vacuum can, in principle, be probed with laboratory experiments.

3.1. *Three topological forces acting on a vortex and their analogs*

Here we consider the experimentally observed forces which act on a moving vortex in superfluid ^{3}He. As listed in sections 2.14.3, there are three different topological contributions to the total force. The more familiar Magnus force arises when the vortex moves with respect to the superfluid vacuum. In the case of a relativistic cosmic string this force is absent, since the corresponding superfluid density of the quantum physical vacuum is 0. However, the analog of this force appears if the cosmic string moves in a uniform background charge density (Davis and Shellard 1989, Lee 1994). The other two forces of topological origin – the Iordanskii force and spectral flow force – also have analogs in the case of a cosmic string (Volovik and Vachaspati 1996, Volovik 1998, Sonin 1997, Wexler 1997, Shelankov 1998).

We start with the Iordanskii force (Iordanskii 1964, 1965, Sonin 1975), which arises when the vortex moves with respect to the heat bath represented by the normal component of the liquid, or the quasiparticle excitations. The latter corresponds to the matter of particle physics. The interaction of quasiparticles with the velocity field of the vortex resembles the interaction of matter with the gravitational field induced by such a cosmic string, which has an angular momentum – the so-called spinning cosmic string (Mazur 1986). The spinning string induces a peculiar space-time metric, which leads to a difference in the time which a particle needs to orbit around the string at same speed, but in opposite directions (Harari and Polychronakos 1988). This gives rise to the quantum gravitational Aharonov–Bohm effect (Mazur 1986, 1987). We discuss how the same effect leads to an asymmetry in the scattering of particles from the spinning string and to the Iordanskii lifting force which acts on the spinning string or on the moving vortex.

The spectral flow force, which also arises when the vortex moves with respect to the heat bath, is a direct consequence from the chiral anomaly effect. The latter violates the conservation of fermionic charge. The anomalous generation by the moving vortex of fermionic charge or momentum (called "momentogenesis") leads to a net force acting on the vortex. The existence of this force was experimentally confirmed in the ^{3}He measurements of Bevan et al. (1997b). This phenomenon is based on the same physics as the anomalous generation of matter in particle physics and bears directly on the cosmological problem of baryonic asymmetry of our Universe: the question why the Universe contains so much more matter than antimatter ("baryogenesis").

The experimental observation of the opposite effect to momentogenesis has been reported by Krusius et al. (1998): the conversion of quasiparticle momentum into a non-trivial order parameter configuration or "texture". The corresponding process in a cosmological setting would be the creation of a primordial magnetic field due to changes in the matter content. Processes, in which magnetic fields are generated, are very relevant to cosmology since magnetic fields are ubiquitous now in the Universe. Our Milky Way and other galaxies, as well as clusters of galaxies, are observed to have a magnetic field whose generation is still not understood. One possible mechanism is that a seed field was amplified by the complex motions associated with galaxies and clusters of galaxies. The seed field itself is usually assumed to be of cosmological origin.

It has been noted that the two problems of cosmological genesis – baryo- and magnetogenesis – may be related to each other (Roberge 1989, Vachaspati 1994, Vachaspati and Field 1994, 1995). More recently an even stronger reason for a possible connection was proposed (Joyce and Shaposhnikov 1997, Giovannini and Shaposhnikov 1997). In this same way their analogs are related in ^{3}He, where the order parameter texture is the analog of magnetic field, while the normal component of the superfluid represents the matter: the moving vortex texture leads to an anomalous production of quasiparticles, while an excess of the quasiparticle momentum – the net quasiparticle current of the normal component – leads to the formation of textures. This mapping of cosmology to condensed matter is not simply a picture: the corresponding effects in the two systems are described by the same equations in the low-energy regime, by quantum field theory and the axial anomaly.

3.2. *Iordanskii force*

3.2.1. *Superfluid vortex vs. spinning cosmic string*

To clarify the analogy between the Iordanskii force and the Aharonov–Bohm effect, let us consider the simplest case of phonons propagating in the velocity field of the quantized vortex in the Bose superfluid ^{4}He-II. According to the Landau theory of superfluidity, the energy of a quasiparticle moving in the superfluid velocity field $\mathbf{v}_s(\mathbf{r})$ is Doppler shifted: $E(\mathbf{p}) = \varepsilon(\mathbf{p}) + \mathbf{p} \cdot \mathbf{v}_s(\mathbf{r})$. In the case of the phonons with the spectrum $\varepsilon(\mathbf{p}) = cp$, where c is the sound velocity, the energy–momentum relation is thus

$$(E - \mathbf{p} \cdot \mathbf{v}_s(\mathbf{r}))^2 = c^2 p^2. \tag{75}$$

Equation (75) can be written in the general Lorentzian form with $p_\mu = (-E, \mathbf{p})$:

$$g^{\mu\nu} p_\mu p_\nu = 0 \tag{76}$$

where the metric is

$$g^{00} = 1, \qquad g^{0i} = -v_s^i, \qquad g^{ik} = -c^2 \delta^{ik} + v_s^i v_s^k. \tag{77}$$

We use the convention to denote indices in the 0–3 range by Greek letters and indices in the 1–3 range by Latin letters. Thus the dynamics of phonons in the presence of the velocity field is the same as the dynamics of photons in the gravity field (Unruh 1976): both are described by the light-cone equation $\mathrm{d}s = 0$. The interval $\mathrm{d}s$ for phonons is given by the inverse metric $g_{\mu\nu}$ which determines the geometry of the effective space:

$$\mathrm{d}s^2 = g_{\mu\nu}\, \mathrm{d}x^\mu\, \mathrm{d}x^\nu, \tag{78}$$

where $x = (t, \mathbf{r})$ are physical (Galilean) coordinates in the laboratory frame.

A similar relativistic equation holds for the fermionic quasiparticles in superfluid ^{3}He-A in the linear approximation close to the gap nodes. In general, i.e. far from the gap nodes, the spectrum of quasiparticle in ^{3}He-A is not relativistic:

$$\varepsilon^2(\mathbf{p}) = v_F^2 (p - p_F)^2 + \frac{\Delta_A^2}{p_F^2} (\hat{\mathbf{l}} \times \mathbf{p})^2. \tag{79}$$

Here $v_F(p - p_F)$ is the quasiparticle energy in the normal Fermi liquid state above the transition, with p_F the Fermi momentum and $v_F = p_F/m^*$; m^* is the effective mass, which is of order of the mass m_3 of the ^{3}He atom; Δ_A is the so-called gap amplitude and the unit vector $\hat{\mathbf{l}}$ points in the direction of the gap nodes.

The energy in eq. (79) is 0 at two points $\mathbf{p} = e\mathbf{A}$ with $\mathbf{A} = p_F \hat{\mathbf{l}}$ and $e = \pm 1$. Close to the two zeros of the energy spectrum one can expand the equation $(E - \mathbf{p} \cdot \mathbf{v}_s(\mathbf{r}))^2 = \varepsilon^2(\mathbf{p})$ in $\mathbf{p} - e\mathbf{A}$ and write it in a form similar to the propagation equation for a massless relativistic particle in curved space-time in the presence of an electromagnetic vector potential:

$$g^{\mu\nu}(p_\mu - eA_\mu)(p_\nu - eA_\nu) = 0. \tag{80}$$

Here $A_0 = p_F(\hat{\mathbf{l}} \cdot \mathbf{v}_s)$ and the metric is anisotropic with the anisotropy axis along the $\hat{\mathbf{l}}$-vector:

$$\begin{aligned} g^{00} &= 1, \qquad g^{0i} = -v_s^i, \\ g^{ik} &= -c_\perp^2(\delta^{ik} - \hat{l}^i\hat{l}^k) - c_\parallel^2\hat{l}^i\hat{l}^k + v_s^i v_s^k, \\ c_\parallel &= v_F, \qquad c_\perp = \Delta_A/v_F. \end{aligned} \tag{81}$$

In the absence of superflow the quantities $c_\parallel$ and $c_\perp$ correspond to the speeds of light propagating along or transverse to $\hat{\mathbf{l}}$.

For simplicity, let us turn back to the case of phonons and vortices in ^{4}He-II which is described by eqs. (76) and (77). If the velocity field is generated by one vortex with ν quanta of circulation, $\mathbf{v}_s = \nu\kappa\hat{\boldsymbol{\phi}}/2\pi r$, then the interval (78) in the effective space, where the phonon is propagating along geodesic curves, becomes

$$ds^2 = \left(1 - \frac{v_s^2}{c^2}\right)\left(dt + \frac{\nu\kappa\, d\phi}{2\pi(c^2 - v_s^2)}\right)^2 - \frac{dr^2}{c^2} - \frac{dz^2}{c^2} - \frac{r^2\, d\phi^2}{c^2 - v_s^2}. \tag{82}$$

The origin of the Iordanskii force lies in the scattering of quasiparticles for small angles, so large distances from the vortex core are important. Far from the vortex v_s^2/c^2 is small and can be neglected, and one has

$$ds^2 = \left(dt + \frac{d\phi}{\omega}\right)^2 - \frac{1}{c^2}(dz^2 + dr^2 + r^2\, d\phi^2), \quad \omega = \frac{2\pi c^2}{\nu\kappa}, \tag{83}$$

The connection between time and the azimuthal angle ϕ in the interval suggests that there is a characteristic angular velocity ω. A similar metric with rotation was obtained for the so-called spinning cosmic string in 3 + 1 space-time, which has the rotational angular momentum J concentrated in the string core, and for the spinning particle in the 2 + 1 gravity (Staruszkievicz 1963, Deser et al. 1984, Mazur 1986, 1987):

$$ds^2 = \left(dt + \frac{d\phi}{\omega}\right)^2 - \frac{1}{c^2}(dz^2 + dr^2 + r^2\, d\phi^2), \quad \omega = \frac{1}{4JG}, \tag{84}$$

where G is the gravitational constant. This gives the following correspondence between the circulation $\nu\kappa$ around the vortex and the angular momentum J of the spinning string

$$\kappa\nu = 8\pi JG. \tag{85}$$

Although we here consider the analogy between the spinning string and vortices in ^{4}He-II, there exists a general statement that vortices in any superfluid have the properties of spinning cosmic strings (Davis and Shellard 1989). In particular, the spinning string generates a density of angular momentum in the vacuum outside the string (Jensen and Kuvcera 1993). The density of angular momentum in the superfluid vacuum outside the vortex is also non-zero and equals at $T = 0$

$$\mathbf{r} \times \rho \mathbf{v}_{\mathrm{s}} = \hbar \nu n_{\mathrm{B}} \hat{\mathbf{z}}, \tag{86}$$

where n_{B} is the density of elementary bosons in the superfluid vacuum: the density ρ/m_4 of ^{4}He atoms in superfluid ^{4}He-II or the density $\rho/2m_3$ of Cooper pairs in superfluid ^{3}He.

3.2.2. *Gravitational Aharonov–Bohm effect*

For the spinning string the gravitational Aharonov–Bohm effect is a peculiar topological phenomenon (Mazur 1986) which can be modeled in condensed matter. On the classical level the propagation of particles is described by the relativistic equation $\mathrm{d}s^2 = 0$. Outside the string the space metric, which enters the interval $\mathrm{d}s$, is flat, eq. (84). But there is a difference in the travel time for particles along closed paths around the spinning string in opposite directions. As can be seen from eq. (84), this time difference is (Harari and Polychronakos 1988)

$$2\tau = \frac{4\pi}{\omega}. \tag{87}$$

At large distances from the core the same equation is approximately valid for phonons propagating in the flow field of a vortex due to the equivalence of the metrics in eqs. (83) and (84). The asymmetry between the particles orbiting in different directions around the vortex implies that, in addition to the symmetric part of the cross section,

$$\sigma_{||} = \int_0^{2\pi} \mathrm{d}\theta (1 - \cos\theta)|a(\theta)|^2, \tag{88}$$

where $a(\theta)$ is a scattering amplitude, there should be an asymmetric part of the scattering cross section,

$$\sigma_{\perp} = \int_0^{2\pi} \mathrm{d}\theta \sin\theta |a(\theta)|^2. \tag{89}$$

The latter is the origin of the Iordanskii force acting on the vortex in the presence of a net momentum from the quasiparticles. Another consequence of eqs. (83) and (84) is displayed on the quantum level: the connection between the time variable t and the angular variable ϕ in eqs. (83) and (84) implies that the scattering cross sections of phonons (photons) from the vortex (string) should be periodic functions of energy, with the period equal to $\hbar\omega$.

Calculations which allow us to find both symmetric and asymmetric contributions to the scattering of quasiparticles in the velocity field of the vortex have been performed by Sonin (1997) for phonons and rotons in ^{4}He-II and by Cleary (1968) for the Bogoliubov–Nambu quasiparticles in conventional superconductors. In the case of phonons the propagation is described by the Lorentzian equation for the scalar field Φ: $g^{\mu\nu}\partial_\mu\partial_\nu\Phi = 0$, with $g^{\mu\nu}$ from eq. (77). We are interested in large distances from the vortex core. Thus the quadratic terms $\mathbf{v}_s^2/c^2$ can be neglected and the equation can be rewritten as (Sonin 1997)

$$E^2\Phi - c^2\left(-i\nabla + \frac{E}{c}\mathbf{v}_s(\mathbf{r})\right)^2\Phi = 0. \tag{90}$$

This equation maps the problem under discussion to the Aharonov–Bohm (AB) problem for the magnetic flux tube (Aharonov and Bohm 1959) with the effective vector potential $\mathbf{A} = \mathbf{v}_s$, where the electric charge e is substituted by the mass E/c^2 of the particle (Mazur 1987, Jensen and Kuvcera 1993, Gal'tsov and Letelier 1993). Actually $\mathbf{v}_s$ plays the part of the vector potential of the so-called gravimagnetic field (Volovik 2003). Because of the mapping between the electric charge and the mass of the particle, one obtains the AB expression (Aharonov and Bohm 1959) for the symmetric differential cross section

$$\frac{d\sigma_\parallel}{d\theta} = \frac{\hbar c}{2\pi E\sin^2(\theta/2)}\sin^2\frac{\pi E}{\hbar\omega}. \tag{91}$$

This expression was obtained for the scattering of particles with energy E in the background of a spinning string with zero mass (Mazur 1987) and represented the gravitational AB effect. Note the singularity at $\theta \to 0$.

Returning back to vortices, one finds that the analog with the spinning string is not exact. In more accurate calculations one should take into account that as distinct from the charged particles in the AB effect, the current in the case of phonons is not gauge-invariant. As a result, the scattering of the phonon with momentum p and energy E from the vortex is somewhat different (Sonin 1997):

$$\frac{d\sigma_\parallel}{d\theta} = \frac{\hbar c}{2\pi E}\cot^2\frac{\theta}{2}\sin^2\frac{\pi E}{\hbar\omega}. \tag{92}$$

The algebraic difference between the AB results in eqs. (91) and (92) is $(\hbar c/2\pi E)\sin^2(\pi E/\omega)$, which is independent of the scattering angle θ and thus is not important for the singularity at small scattering angles, which is present in eq. (92) as well. For small E the result in eq. (92) was obtained by

Fetter (1964). The generalization of the Fetter result for quasiparticles with arbitrary spectrum $\varepsilon(\mathbf{p})$ (rotons in ^{4}He-II and the Bogoliubov–Nambu fermions in superconductors) was recently suggested by Demircan et al. (1995): in our notations it is $(\nu\kappa^2 p/8\pi v_G^2)\cot^2(\theta/2)$, where $v_G = \mathrm{d}\varepsilon/\mathrm{d}p$ is the group velocity of a quasiparticle.

3.2.3. *Asymmetric cross section of scattering from a vortex*

The Lorentz force, which acts on the flux tube in the presence of an electric current, has its counterpart – the Iordanskii force, which acts on the vortex in the presence of a mass current of the normal component. The Lorentz-type Iordanskii force comes from the asymmetric contribution to the cross section (Sonin 1997, Shelankov 1998), which has the same origin as the singularity at small angles in the symmetric cross section and leads to a non-zero transverse cross section.

For the phonons in ^{4}He-II with the spectrum $E(\mathbf{p}) = cp$ the transverse cross section is (Sonin 1997)

$$\sigma_\perp = \frac{\hbar}{p}\sin\frac{2\pi E}{\hbar\omega}. \tag{93}$$

At low $E \ll \hbar\omega$ the result becomes classical: $\sigma_\perp = 2\pi c/\omega$ does not contain the Planck constant $\hbar$. This means that in the low-energy limit the asymmetric cross section can be obtained from the classical theory of scattering. In this case it can be generalized for an arbitrary spectrum $E(\mathbf{p})$ of scattering particles (Sonin 1997). Let us consider a particle with the spectrum $E(\mathbf{p})$ moving in the background of the velocity field $\mathbf{v}_s(\mathbf{r})$ around the vortex. The velocity field modifies the particle energy due to the Doppler shift, $E(\mathbf{p},\mathbf{r}) = E(\mathbf{p}) + \mathbf{p}\cdot\mathbf{v}_s(\mathbf{r})$. Far from the vortex, where the circulating velocity is small, the trajectory of the quasiparticle is almost a straight line parallel to, say, the axis y, with the distance from the vortex line being the impact parameter x. It moves along this line with almost constant momentum $p_y \approx p$ and almost constant group velocity $\mathrm{d}y/\mathrm{d}t = v_\mathrm{G} = \mathrm{d}\varepsilon/\mathrm{d}p$. The change in transverse momentum during this motion is determined by the Hamiltonian equation $\mathrm{d}p_x/\mathrm{d}t = -\partial E/\partial x = -p_y\partial v_{sy}/\partial x$, or $\mathrm{d}p_x/\mathrm{d}y = -(p/v_\mathrm{G})\partial v_{sy}/\partial x$. The transverse cross section is obtained by integration of $\Delta p_x/p$ over the impact parameter x:

$$\sigma_\perp = \int_{-\infty}^{+\infty}\frac{\mathrm{d}x}{v_\mathrm{G}}\int_{-\infty}^{+\infty}\mathrm{d}y\,\frac{\partial v_{sy}}{\partial x} = \frac{\nu\kappa}{v_\mathrm{G}}. \tag{94}$$

This result is purely classical: Planck's constant $\hbar$ drops out (it enters only via the quantized circulation $\nu\kappa$ which characterizes the vortex).

3.2.4. *Iordanskii force: quantized vortex and spinning string*

The asymmetric part of scattering, which describes the momentum transfer in the transverse direction, after integration over the distribution of excitations gives rise to the transverse force acting on the vortex if the vortex moves with respect to the normal component. This is the Iordanskii force:

$$\begin{aligned}\mathbf{f}_{\text{Iordanskii}} &= \int \frac{\mathrm{d}^3 p}{(2\pi)^3} \sigma_\perp(p) v_G n(\mathbf{p}) \mathbf{p} \times \hat{\mathbf{z}} \\ &= -\nu\kappa \hat{\mathbf{z}} \times \int \frac{\mathrm{d}^3 p}{(2\pi)^3} n(\mathbf{p}) \mathbf{p} = \nu\kappa \mathbf{P}_\mathrm{n} \times \hat{\mathbf{z}}. \end{aligned} \tag{95}$$

It is proportional to the density of mass current $\mathbf{P}_\mathrm{n}$ carried by excitations (matter). Of all the parameters describing the vortex, it depends only on the circulation $\nu\kappa$. This confirms the topological origin of this force. In the case of the equilibrium distribution of quasiparticles one has $\mathbf{P}_\mathrm{n} = \rho_\mathrm{n}\mathbf{v}_\mathrm{n}$, where ρ_n and $\mathbf{v}_\mathrm{n}$ are the density and velocity of the normal component of the liquid. Thus one obtains eq. (36). To avoid the conventional Magnus force in this derivation, we assumed that the asymptotic velocity of the superfluid component of the liquid is 0 in the frame of the vortex.

Equation (95) was obtained using the asymptotic behavior of the flow field $\mathbf{v}_\mathrm{s}$, which induces the same effective metric (83) as the metric around the spinning string (84). We can thus apply this result directly to the spinning string. The asymmetric scattering cross section of relativistic particles from the spinning string is given by eq. (93). This means that in the presence of momentum from matter the spinning cosmic string experiences a lifting force, which corresponds to the Iordanskii force in superfluids. This force can be obtained from a relativistic generalization of eq. (95). The momentum density $\mathbf{P}_\mathrm{n}$ of quasiparticles should be substituted by the component T^i_0 of the energy-momentum tensor. As a result, for $2+1$ space-time and for low-energy E, which corresponds to low-temperatures T for matter, the Iordanskii force on a spinning string moving with respect to the matter is

$$f^\alpha_{\text{Iordanskii}} = 8\pi J G \varepsilon^{\alpha\beta\gamma} u_\beta u_\mu T^\mu_\gamma. \tag{96}$$

Here u_α is the 3-velocity of the string and T^μ_γ is the asymptotic value of the energy-momentum tensor of the matter at the site of the string. Using the Einstein equations one can rewrite this as

$$f^\alpha_{\text{Iordanskii}} = J \varepsilon^{\alpha\beta\gamma} u_\beta u_\mu R^\mu_\gamma, \tag{97}$$

where R^μ_γ is the Riemannian curvature at the location of the string. This corresponds to the force which acts on a particle with spin J from the gravitational field owing to the interaction of the spin with the Riemann tensor (Thorne and Hartle 1985, Mino et al. 1997).

Note that previously we have shown how ideas first developed in the cosmological context (such as nucleation of bubbles of different broken-symmetry phases during a rapid phase transition) could be applied to superfluid helium. Here we have the opposite case when ideas originating in helium physics have found applications in other systems described by quantum field theory.

The Iordanskii force has been experimentally identified in rotating ^{3}He–B (Bevan et al. 1997a). Equation (42) describes the temperature dependence of the mutual friction coefficient $d_\perp$. Here the term $m_3\mathcal{C}(T)$ arises from the spectral flow force which is discussed in section 3.3.3 and vanishes at low temperatures. Thus at low temperatures $d_\perp \approx -\rho_n/\rho$ (Kopnin et al. 1995), where ρ is the total density of the liquid. The negative value of $d_\perp$ arises entirely from the existence of the Iordanskii force. This is in accordance with the experimental data, which show that $d_\perp$ does have negative values at low T (fig. 32). At higher T the spectral flow force dominates, which leads to the sign reversal of $d_\perp$. This fact can be interpreted to represent experimental verification of the analog of the gravitational Aharonov–Bohm effect for a spinning cosmic string.

3.3. *Spectral flow force and chiral anomaly*

3.3.1. *Chiral anomaly*

In the standard model of electroweak interactions there are certain quantities, like the baryon number Q_B, which are classically conserved but can be

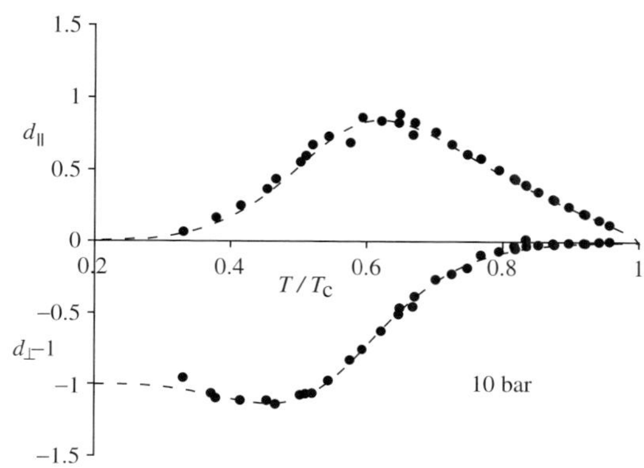

Fig. 32. Mutual friction parameters $d_{\parallel,\perp}$ as a function of temperature in ^{3}He–B at 10 bar. The negative sign of $d_\perp - 1$ at $T/T_c \approx 0.4$–0.5 constitutes experimental verification for the existence of the Iordanskii force. The curves are fits to measured data (from Bevan et al. 1997a) Copyright 1997 Plenum Publishing Corporation; with kind permission from Springer Science and Business Media.

violated by quantum mechanical effects known generically as "chiral anomalies". (Each of the quarks is assigned $Q_B = 1/3$ while the leptons (neutrinos and the electron) have $Q_B = 0$.) The process leading to particle creation is called "spectral flow", and can be pictured as a process in which fermions flow under an external perturbation from negative energy levels towards positive energy levels. Some fermions therefore cross zero energy and move from the Dirac sea into the observable positive energy world.

The origin for the axial anomaly can be seen from the behavior of a chiral particle in constant magnetic field, $\mathbf{A} = (1/2)\mathbf{B} \times \mathbf{r}$. We call a particle without mass but with spin $\vec{\sigma}$ [$= 1/2$] a chiral particle. The chiral particle can be classified as a right or left particle, depending on whether its spin is parallel or antiparallel to its momentum. The Hamiltonians for the right particle with the electric charge e_R and for the left particle with the electric charge e_L are

$$\mathcal{H} = c\vec{\sigma} \cdot (\mathbf{p} - e_R\mathbf{A}), \qquad \mathcal{H} = -c\vec{\sigma} \cdot (\mathbf{p} - e_L\mathbf{A}). \tag{98}$$

As usual, the motion of the particles in the plane perpendicular to $\mathbf{B}\|\hat{\mathbf{z}}$ is quantized in Landau levels. Thus the free motion is effectively reduced to one-dimensional motion along $\mathbf{B}$ with momentum p_z. Figure 33 shows the energy spectrum where the thick lines represent the occupied negative-energy states. The peculiar feature of the spectrum is that, because of the chirality of the particles, the lowest ($n = 0$) Landau level is asymmetric. It crosses zero only in one direction: $E = cp_z$ for the right particle and $E = -cp_z$ for the left. If we now apply an electric field $\mathbf{E}$ along z, particles are pushed from negative to positive energy levels according to the equation of motion $\dot{p}_z = e_R E_z$ ($\dot{p}_z = e_L E_z$) and the whole Dirac sea moves up (down) creating particles and electric charge from the vacuum. This motion of particles along the "anomalous" branch of the spectrum is called *spectral flow*. The rate of particle production is proportional to the density of states at the Landau level, which is

$$N_R(0) = \frac{|\, e_R\mathbf{B}\,|}{4\pi^2}, \qquad N_L(0) = \frac{|\, e_L\mathbf{B}\,|}{4\pi^2}. \tag{99}$$

The production rate of particle number $n = n_R + n_L$ and of charge $Q = n_R e_R + n_L e_L$ from vacuum is

$$\dot{n} = \frac{1}{4\pi^2}(e_R^2 - e_L^2)\mathbf{E} \cdot \mathbf{B}, \qquad \dot{Q} = \frac{1}{4\pi^2}(e_R^3 - e_L^3)\mathbf{E} \cdot \mathbf{B}. \tag{100}$$

This is an anomaly equation for the production of particles from vacuum of the type found by Adler (1969) and by Bell and Jackiw (1969) in the context of neutral pion decay. We see that for particle or charge creation, without creation of corresponding antiparticles, it is necessary to have an asymmetric branch in the dispersion relations $E(p)$, which crosses the axis

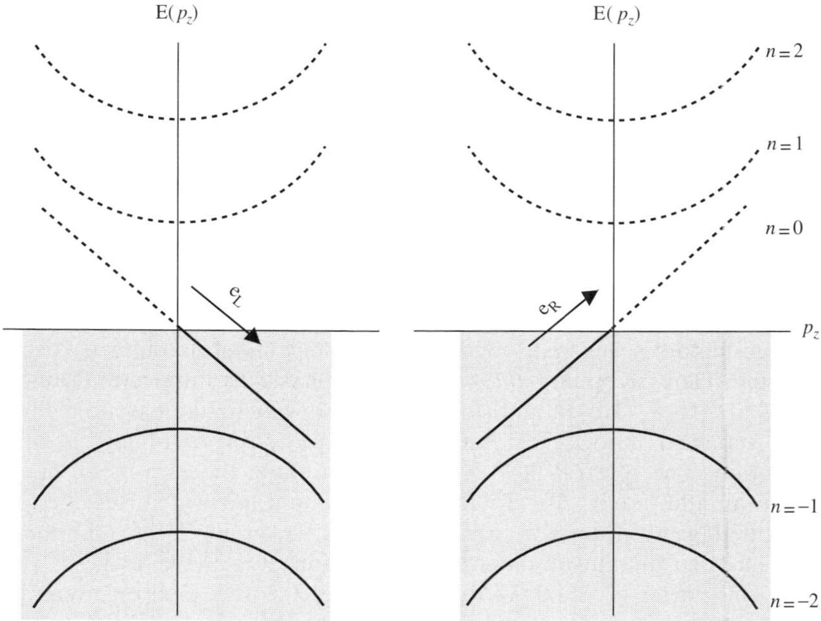

Fig. 33. The energy spectrum of a chiral particle in constant magnetic fields along $\hat{z}$ (Landau levels). The plots on left and right show spectra for left and right particles, respectively.

from negative to positive energy. Additionally, the symmetry between the left and right particles has to be violated: $e_{\mathrm{R}} \neq e_{\mathrm{L}}$ for charge creation and $e_{\mathrm{R}}^2 \neq e_{\mathrm{L}}^2$ for particle creation.

In the electroweak model there are two gauge fields whose "electric" and "magnetic" fields may become a source for baryoproduction: the hypercharge field $U(1)$ and the weak field $SU(2)$. Anomalous zero-mode branches exist in the core of a Z-string, where quarks, electrons, and neutrinos are all chiral particles with known hypercharge and weak charges. If we consider a process in which one electron, two u-quarks, and one d-quark are created, then lepton and baryon numbers are changed by 1 unit while electric charge is conserved (Bevan et al. 1997b). If we sum appropriate charges for all particles according to eq. (123) the rate of this process is

$$\dot{n}_{\mathrm{bar}} = \dot{n}_{\mathrm{lept}} = \frac{N_{\mathrm{F}}}{8\pi^2}(-\mathbf{B}_W^a \cdot \mathbf{E}_{aW} + \mathbf{B}_Y \cdot \mathbf{E}_Y), \tag{101}$$

where $N_{\mathrm{F}} = 3$ is the number of families (generations) of fermions, $\mathbf{B}_W^a$ and $\mathbf{E}_{aW}$ are the colored $SU(2)$ magnetic and electric fields, while $\mathbf{B}_Y$ and $\mathbf{E}_Y$ are the magnetic and electric fields of the $U(1)$ hypercharge. While a color and hypercharge magnetic flux is always present in the Z-string core, a color and hypercharge electric field can also be present along the string if the string is moving across a background electromagnetic field (Witten 1985) or in certain other processes such as the decoupling of two linked loops (Vachaspati and Field 1994, 1995, Garriga and Vachaspati 1995). Thus parallel electric and magnetic fields in the string change the baryonic charge and can lead to cosmological baryogenesis (Barriola 1995) and to the presence of antimatter in cosmic rays (Starkman 1996).

In superconductors and in superfluid ^{3}He an anomalous zero-mode branch exists for fermions in the core of quantized vortices. For electrons in superconductors it was first found by Caroli et al. (1964), and for vortices in superfluid ^{3}He by Kopnin (1993). One of the physically important fermionic charges in ^{3}He-A, ^{3}He–B, and superconductors which, like baryonic charge in the standard model, is not conserved due to the anomaly, is linear momentum. The spectral flow of momentum along the zero-mode branch leads to an additional "lift" force which acts on a moving vortex.

The analogy is clearest for the continuous vortex in ^{3}He-A (skyrmion), which has two quanta of superfluid circulation, $\nu = 2$ (Blaauwgeers et al. 2000). This vortex is similar to the continuous Z-vortex in electroweak theory: it is characterized by a continuous distribution of the order-parameter vector $\hat{\mathbf{l}}$, which denotes the direction of the angular momentum of the Cooper pairs (fig. 34). When multiplied by the Fermi wave number $k_F = p_F/\hbar$, this vector acts on the quasiparticles like an effective "electromagnetic" vector potential $\mathbf{A} = k_F\hat{\mathbf{l}}$. The quasiparticles in ^{3}He-A, which are close to the gap nodes, are chiral: they are either left or right handed (Volovik and Vachaspati 1996). As follows from the BCS theory of ^{3}He-A the

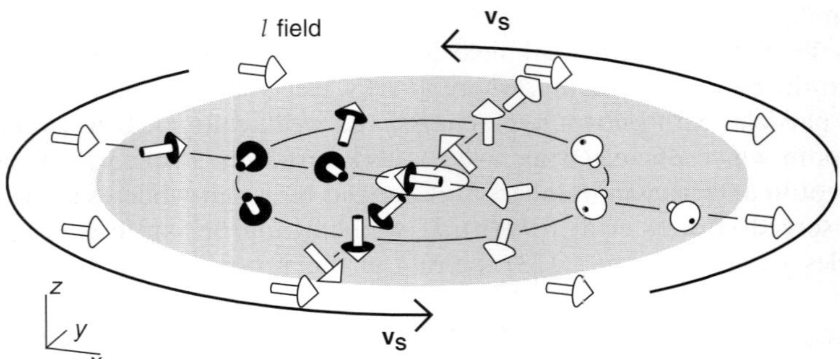

Fig. 34. The orbital order-parameter $\hat{\mathbf{l}}$ texture in the continuous soft core of the Anderson–Toulouse–Chechetkin vortex with skyrmion structure.

sign of the "electric" charge e, introduced in sections 3.2.1, simultaneously determines the chirality of the fermions. This is clearly seen from a simple isotropic example (with $c_{\parallel} = c_{\perp} = c$):

$$\mathscr{H} = -ec\vec{\sigma} \cdot (\mathbf{p} - e\mathbf{A}). \tag{102}$$

A particle with positive (negative) e at the north (south) pole is left-handed (right-handed). Here $\vec{\sigma}$ is a Bogoliubov spin.

For such gapless chiral fermions the Adler–Bell–Jackiw anomaly applies and the momentum ("chiral charge") of quasiparticles is not conserved in the presence of "electric" and "magnetic" fields, which are defined by

$$\mathbf{E} = k_{\mathrm{F}}\partial_t\hat{\mathbf{l}}, \qquad \mathbf{B} = k_{\mathrm{F}}\nabla \times \hat{\mathbf{l}}. \tag{103}$$

In ^{3}He-A each right-handed quasiparticle carries the momentum $\mathbf{p}_{\mathrm{R}} = p_{\mathrm{F}}\hat{\mathbf{l}}$ (we reverse the sign of momentum when it is used as fermionic charge), and a left-handed quasiparticle has $\mathbf{p}_{\mathrm{L}} = -\mathbf{p}_{\mathrm{R}}$. According to eq. (100) the production rates for right- and left-handed quasiparticles are (since $e_{\mathrm{R}}^2 = e_{\mathrm{L}}^2 = 1$ in this case)

$$\dot{n}_{\mathrm{R}} = -\dot{n}_{\mathrm{L}} = \frac{1}{4\pi}\mathbf{E} \cdot \mathbf{B}. \tag{104}$$

As a result there is a net creation of quasiparticle momentum $\mathbf{P}$ in a time-dependent texture:

$$\partial_t\mathbf{P} = \int \mathrm{d}^3r(\mathbf{p}_{\mathrm{R}}\dot{n}_{\mathrm{R}} + \mathbf{p}_{\mathrm{L}}\dot{n}_{\mathrm{L}}) = \frac{1}{2\pi^2}\int \mathrm{d}^3r p_{\mathrm{F}}\hat{\mathbf{l}}(\mathbf{E} \cdot \mathbf{B}). \tag{105}$$

What we know for sure is that the total linear momentum is conserved in nature. Then eq. (105) means that, in the presence of the time-dependent texture, the momentum is transferred from the superfluid motion of *vacuum* to *matter* (i.e. to the heat bath of quasiparticles which form the normal component).

3.3.2. *Anomalous force acting on a continuous vortex and baryogenesis from textures*

Let us take as an example the simplest model, namely the vortex with a *continuous soft core* structure in superfluid ^{3}He-A, the so-called *skyrmion*. The core has the following distribution of the unit vector $\hat{\mathbf{l}}(\mathbf{r})$, which points in the direction of the point-like gap nodes in the smooth core:

$$\hat{\mathbf{l}}(\mathbf{r}) = \hat{\mathbf{z}}\cos\eta(r) + \hat{\mathbf{r}}\sin\eta(r), \tag{106}$$

where z, r, ϕ are the cylindrical coordinates. Here $\eta(r)$ is a function which ensures that the $\hat{\mathbf{l}}$-vector orientation changes within the smooth core from $\hat{\mathbf{l}}(0) = -\hat{\mathbf{z}}$ to $\hat{\mathbf{l}}(\infty) = \hat{\mathbf{z}}$. The circulation of the superfluid velocity along a path far outside the soft core corresponds to $\nu = 2$: $\oint \mathrm{d}\mathbf{r} \cdot \mathbf{v}_{\mathrm{s}} = 2\kappa$. Such a continuous $\hat{\mathbf{l}}$ texture thus represents a doubly quantized vortex without singularities. In practice, the vortex measured in the NMR experiments is subject to an external magnetic field ($\mathbf{B}\|\hat{\mathbf{z}}$) and the $\hat{\mathbf{l}}$-vector at infinity is fixed to the (x, y) plane (see fig. 34 with $\hat{\mathbf{l}}(\infty) = \hat{\mathbf{x}}$), but this does not change the topological structure of the vortex and its circulation.

When the continuous vortex moves in ^{3}He-A with a velocity $\mathbf{v}_\mathrm{L}$ it generates a time-dependent $\hat{\mathbf{l}}$ texture $\hat{\mathbf{l}} = \hat{\mathbf{l}}(\mathbf{r} - \mathbf{v}_\mathrm{L}t)$. Hence both an "electric" and a "magnetic" field from eq. (103) exist and this leads to "momentogenesis". Integration of the anomalous momentum transfer in eq. (105) over the cross section of the soft core of the moving vortex gives an additional force which acts on the vortex due to spectral flow:

$$\mathbf{f}_\mathrm{sf} = \partial_t \mathbf{P} = \pi\hbar\nu\mathscr{C}_0\hat{\mathbf{z}} \times (\mathbf{v}_\mathrm{L} - \mathbf{v}_\mathrm{n}). \tag{107}$$

Here $\hat{\mathbf{z}}$ is the direction of the vortex, $\mathscr{C}_0 = k_\mathrm{F}^3/3\pi^2$, and $\mathbf{v}_\mathrm{n}$ is the heat bath velocity. Thus we have obtained eq. (37) with a temperature-independent parameter $\mathscr{C}(T) = \mathscr{C}_0$.

Measurements of the mutual friction coefficients in ^{3}He-A with the continuous vortex (Bevan et al. 1997a) provide experimental verification for the spectral flow force. According to eq. (42) it should be that $d_\perp = (m_3\mathscr{C}_0 - \rho_\mathrm{n})/\rho_\mathrm{s}$. The value of $m_3\mathscr{C}_0$ is the total mass density ρ in the normal phase. Its difference from ρ in the superfluid phase is thus determined by the effect of superfluidity on the particle density which is extremely small: $\rho - m_3\mathscr{C}_0 \sim \rho(\Delta_0/v_\mathrm{F}p_\mathrm{F})^2 = \rho(c_\perp/c_\|)^2 \ll \rho$. Thus one must have $d_\perp \approx 1$ for all practical temperatures. ^{3}He-A experiments at 29.3 bar and $T > 0.82\,T_\mathrm{c}$ are consistent with this conclusion within experimental uncertainty: it was found that $|1 - d_\perp| < 0.005$, as demonstrated in fig. 35.

3.3.3. *Anomalous force acting on a singular vortex and baryogenesis with strings*

The discussion of spectral flow in the previous sections and particularly eq. (100) cannot be directly applied to the singular vortex structures which are found in ^{3}He–B and superconductors. The reason is that the deviation in the magnitude of the order parameter from its equilibrium value in the cores of such vortices create a potential well for the core quasiparticles. In this well quasiparticles have discrete energy levels with some characteristic separation $\hbar\omega_0$ instead of continuous spectra (as a function of p_z), as was considered above. Thus the theory of spectral flow becomes more complicated, but can still be constructed (Kopnin et al. 1995, Kopnin 2002).

The basic idea is that discrete levels have some broadening $\hbar/\tau$, resulting from the scattering of core excitations by the free excitations in the heat bath outside the core (or by impurities in superconductors). At low temperatures, when the width of levels is much less than their separation, i.e. $\omega_0\tau \gg 1$,

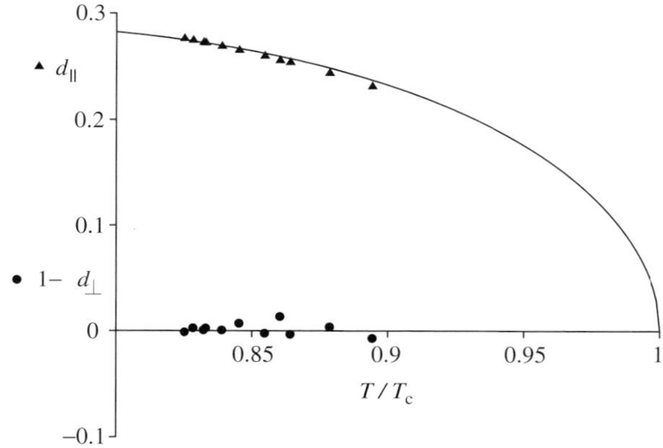

Fig. 35. Mutual friction parameters $d_{\parallel}$ and $1 - d_{\perp}$ in ^{3}He-A at 29.3 bar. The curve is a theoretical fit to the data points (from Bevan et al. 1997a) Copyright 1997 Plenum Publishing Corporation; with kind permission from Springer Science and Business Media.

spectral flow is essentially suppressed, $\mathscr{C}(T) = 0$. In the opposite case, $\omega_0\tau \ll 1$, the levels overlap and we have a situation similar to spectral flow in a continuous spectrum: $\mathscr{C}(T) \approx \mathscr{C}_0$. We may construct an interpolation formula between these two cases:

$$\mathscr{C}(T) \sim \frac{\mathscr{C}_0}{1 + \omega_0^2\tau^2}. \tag{108}$$

In fact, both the $d_{\parallel}$ and $d_{\perp}$ mutual friction coefficients are affected by this renormalization of the spectral flow force (Kopnin et al. 1995, Stone 1996):

$$d_{\parallel} - \mathrm{i}(1 - d_{\perp}) = \frac{\rho}{\rho_{\mathrm{s}}} \frac{\omega_0\tau}{1 + \mathrm{i}\omega_0\tau} \tanh\frac{\Delta(T)}{2k_{\mathrm{B}}T}. \tag{109}$$

Let us derive this equation using the Landau-type phenomenological description for fermions in the vortex core, as developed by Stone (1996). For simplicity we consider the axisymmetric vortex core; the general case of the asymmetric core is discussed by Kopnin and Volovik (1998) and by Volovik (2003). The low-energy spectrum of Caroli–de-Gennes–Matricon quasiparticles around a vortex contains an anomalous branch of fermion zero modes. In the case of the axisymmetric vortex, excitations on this branch are characterized by the angular momentum L_z,

$$E(L_z, p_z) = -\omega_0(p_z)L_z. \tag{110}$$

For superconductors with a coherence length ξ much larger than the inverse Fermi momentum, $p_{\mathrm{F}}\xi \gg 1$, the electron wavelength is short compared with the core size, and the quasiclassical approximation is relevant. The quasiclassical angular momentum L_z is a continuous variable; thus the anomalous branch crosses zero as a function of L_z at $L_z = 0$ and spectral flow can occur along this branch between the vacuum states with $E<0$ and the excited states with $E>0$. Such spectral flow occurs during the motion of the vortex with respect to the normal component, where it is caused by the interaction with impurities in superconductors or with the thermal scattering states in superfluids. In the quasiclassical approximation, the Doppler-shifted spectrum of the fermions in the moving vortex has the form

$$E(L_z, \mathbf{p}) = -\omega_0(p_z)L_z + (\mathbf{v}_\mathrm{s} - \mathbf{v}_\mathrm{L}) \cdot \mathbf{p}. \tag{111}$$

Here the momentum $\mathbf{p}$ is assumed to be at the Fermi surface: $\mathbf{p} = (p_\mathrm{F} \sin\theta \, \sin\varphi, p_\mathrm{F} \sin\theta \, \cos\varphi, p_\mathrm{F} \cos\theta)$. The azimuthal angle φ is canonically conjugated to the angular momentum L_z. This allows us to write the Boltzmann equation for the distribution function $n(L_z, \varphi)$ at fixed $p_z = p_\mathrm{F} \cos\theta$:

$$\partial_\mathrm{t} n - \omega_0 \partial_\varphi n - \partial_\varphi((\mathbf{v}_\mathrm{s} - \mathbf{v}_\mathrm{L}) \cdot \mathbf{p})\partial_{L_z} n = -\frac{n(L_z, \varphi) - n_\mathrm{eq}(L_z, \varphi)}{\tau}, \tag{112}$$

where the collision time τ characterizes the interaction of the bound state fermions with impurities or with the thermal fermions in the normal component outside the vortex core. The equilibrium distribution function is

$$n_\mathrm{eq}(L_z, \varphi) = f(E - (\mathbf{v}_\mathrm{n} - \mathbf{v}_\mathrm{L}) \cdot \mathbf{p}) = f(-\omega_0 L_z + (\mathbf{v}_\mathrm{s} - \mathbf{v}_\mathrm{n}) \cdot \mathbf{p}), \tag{113}$$

where $f(E) = (1 + \exp(E/T))^{-1}$ is the Fermi-function.

Introducing the shifted variable

$$l = L_z - (\mathbf{v}_\mathrm{s} - \mathbf{v}_\mathrm{n}) \cdot \mathbf{p}/\omega_0, \tag{114}$$

one obtains the equation for $n(l, \varphi)$

$$\partial_t n - \omega_0 \partial_\varphi n - \partial_\varphi((\mathbf{v}_\mathrm{n} - \mathbf{v}_\mathrm{L}) \cdot \mathbf{k})\partial_l n = -\frac{n(l, \varphi) - f(-\omega_0 l)}{\tau}, \tag{115}$$

which does not contain $\mathbf{v}_\mathrm{s}$. To find the force acting on the vortex from the heat bath environment, we are interested in the evolution of the total momentum of quasiparticles in the vortex core:

$$\mathbf{P} = \sum \mathbf{p} = \int_{-p_\mathrm{F}}^{p_\mathrm{F}} \frac{\mathrm{d}p_z}{2\pi} \mathbf{P}(p_z), \qquad \mathbf{P}(p_z) = \frac{1}{2} \int \mathrm{d}l \, \frac{\mathrm{d}\varphi}{2\pi} \, n(l, \varphi)\mathbf{p}. \tag{116}$$

It appears that the equation for $\mathbf{P}(p_z)$ can be written in closed form as

$$\partial_\mathrm{t} \mathbf{P}(p_z) - \omega_0 \hat{z} \times \mathbf{P}(p_z) + \frac{\mathbf{P}(p_z)}{\tau} = -\frac{1}{4} p_\mathrm{F}^2 \sin^2\theta \hat{z} \times (\mathbf{v}_\mathrm{n} - \mathbf{v}_\mathrm{L})(f(\Delta(T)) - f(-\Delta(T))). \tag{117}$$

Here we take into account that $\int \mathrm{d}l\partial_l n$ is limited by the bound states below the gap $\Delta(T)$ of bulk liquid: above the gap $\Delta(T)$ the spectrum of fermions is continuous, i.e. the inter-level distance $\omega_0 = 0$.

In steady state vortex motion one has $\partial_t \mathbf{P}(p_z) = 0$. Then, since $f(\Delta(T)) - f(-\Delta(T)) = \tanh(\Delta(T)/2T)$, one obtains the following contribution to the momentum from the heat bath to the core fermions due to the spectral flow of bound states below $\Delta(T)$

$$\mathbf{F}_{\mathrm{bsf}} = \int_{-p_{\mathrm{F}}}^{p_{\mathrm{F}}} \frac{\mathrm{d}p_z}{2\pi} \frac{\mathbf{P}(p_z)}{\tau} = \frac{\kappa}{4} \tanh \frac{\Delta(T)}{2T} \\ \times \int_{-p_{\mathrm{F}}}^{p_{\mathrm{F}}} \frac{\mathrm{d}p_z}{2\pi} \frac{p_{\mathrm{F}}^2 - p_z^2}{1 + \omega_0^2\tau^2} [(\mathbf{v}_{\mathrm{L}} - \mathbf{v}_{\mathrm{n}})\omega_0\tau + \hat{z} \times (\mathbf{v}_{\mathrm{L}} - \mathbf{v}_{\mathrm{n}})]. \tag{118}$$

The spectral flow of unbound states above $\Delta(T)$ is not suppressed, the corresponding $\omega_0\tau = 0$, since the distance between the levels in the continuous spectrum is $\omega_0 = 0$. This gives

$$\mathbf{F}_{\mathrm{usf}} = \frac{\kappa}{4} p_{\mathrm{F}}^2 \int_{-p_{\mathrm{F}}}^{p_{\mathrm{F}}} \frac{\mathrm{d}p_z}{2\pi} \sin^2\theta \left(1 - \tanh \frac{\Delta(T)}{2T}\right) \hat{z} \times (\mathbf{v}_{\mathrm{L}} - \mathbf{v}_{\mathrm{n}}). \tag{119}$$

Thus the total non-dissipative (transverse) and frictional (longitudinal) parts of the spectral-flow force are

$$\mathbf{F}_{\mathrm{sf}}^{\perp} = \frac{\kappa}{4} \int_{-p_{\mathrm{F}}}^{p_{\mathrm{F}}} \frac{\mathrm{d}p_z}{2\pi} (p_{\mathrm{F}}^2 - p_z^2) \times \left[1 - \tanh \frac{\Delta(T)}{2T} \frac{\omega_0^2\tau^2}{1 + \omega_0^2\tau^2}\right] \hat{z} \times (\mathbf{v}_{\mathrm{L}} - \mathbf{v}_{\mathrm{n}}), \tag{120}$$

$$\mathbf{F}_{\mathrm{sf}}^{\parallel} = (\mathbf{v}_{\mathrm{L}} - \mathbf{v}_{\mathrm{n}}) \frac{\kappa}{4} \tanh \frac{\Delta(T)}{2T} \times \int_{-p_{\mathrm{F}}}^{p_{\mathrm{F}}} \frac{\mathrm{d}p_z}{2\pi} (p_{\mathrm{F}}^2 - p_z^2) \frac{\omega_0\tau}{1 + \omega_0^2\tau^2} \tag{121}$$

In the limit $\omega_0\tau \ll 1$, the friction force disappears, while the transverse spectral flow force is maximal and coincides with that obtained for the continuous vortex in eq. (107) with the same value of $\mathscr{C}_0 = (1/4)\int_{-p_{\mathrm{F}}}^{p_{\mathrm{F}}} \mathrm{d}p_z(p_{\mathrm{F}}^2 - p_z^2) = p_{\mathrm{F}}^3/3\pi^2$. In the general case, the result in eqs. (120) and (121) is complicated since both ω_0 and τ depend on p_z. However, if one neglects this dependence and adds the Magnus and Iordanskii forces, one obtains eq. (109) for all T. If the temperature is not too high, so that $\tanh \Delta(T)/2T \approx 1$, one obtains eq. (108) for the renormalized spectral flow parameter.

Comparison of eq. (109) with experiment should take into account that neither ω_0 nor τ are known with good precision. However, the following combination of mutual friction coefficients does not depend on these parameters explicitly:

$$1-\alpha' = \frac{1-d_\perp}{d_\parallel^2+(1-d_\perp)^2} = \frac{\rho_s}{\rho}\left[\tanh\frac{\Delta(T)}{2k_B T}\right]^{-1}. \tag{122}$$

This equation (Kopnin et al. 1995) is compared to experimental data on mutual friction in ^{3}He–B (Bevan et al. 1997a) in fig. 36. The agreement is excellent in view of the approximations in the theory.

This shows that the chiral anomaly is relevant for the interactions of condensed matter vortices (analogs of strings) with fermionic quasiparticles (analogs of quarks and leptons). For vortices with continuous core structure in ^{3}He-A the spectral flow of fermions between the superfluid ground state (vacuum) and the heat bath of positive-energy particles forming the normal component (matter) dominates at any relevant temperature. For singular vortices in ^{3}He–B it is important at $T \sim T_c$ and vanishes at $T \ll T_c$.

It is interesting to note that in a homogeneous system the momentum (fermionic change) of the ground state (vacuum) and that of the excitations (matter) are conserved separately. Topological defects are thus the mediators for the transfer of momentum between these two subsystems. The motion of a vortex across the flow changes both the topological charge of the vacuum (say, the winding number of superflow in a torus geometry) and the fermionic charge (angular momentum in the same geometry). All these processes are similar to those involved in the cosmological production of baryons and can thus be investigated in detail.

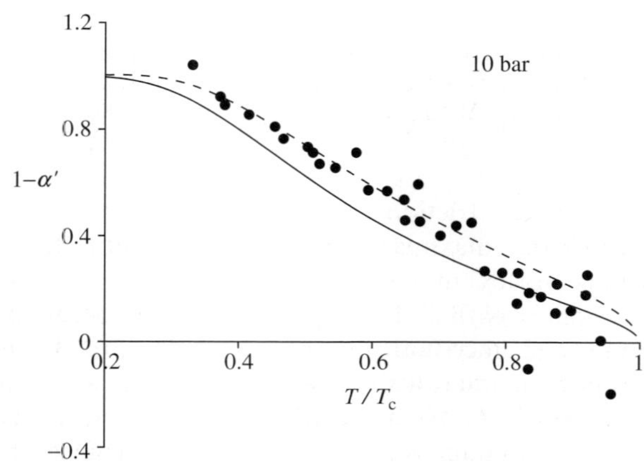

Fig. 36. Experimental values of $1-\alpha'$ in ^{3}He–B at 10 bar compared with the theoretical result in eq. (122). The full curve is for the theoretical value of $\Delta(T)$ and the broken curve is the fit with reduced Δ (from Bevan et al. 1997a) Copyright 1997 Plenum Publishing Corporation; with kind permission from Springer Science and Business Media.

However, it is important to note that although spectral flow leads to anomalous creation of fermionic charge from vacuum, the total fermionic charge of vacuum plus matter remains conserved. There is also no intrinsic bias in this process for the direction of charge transfer: from vacuum to matter or in the opposite direction. The necessary conditions for this mechanism to start operating are that the system is in a non-equilibrium state and under the influence of symmetry breaking which biases the direction of charge transfer. How such conditions were realized in the Early Universe is not clear at present – if we interpret the positive baryonic charge of matter in our Universe to arise from spectral flow. But it is commonly believed that broken P and CP invariances play crucial roles. In contrast in the ^{3}He case the situation is completely under control: the relevant symmetry breaking is achieved by applying rotation or a magnetic field, while the non-equilibrium state is generated by applying external superflow.

The experimental verification of momentogenesis in superfluid ^{3}He can thus be viewed to support these ideas of cosmological baryogenesis via spectral flow and points to a future where several cosmological problems are modeled and studied in light of the superfluid ^{3}He example (Volovik 2003). Now let us consider the opposite effect which leads to the analog of magnetogenesis.

3.4. *Analog of magnetogenesis: vortex textures generated in normal–superfluid counterflow*

From eq. (100) it follows that the magnetic field configuration can absorb fermionic charge. If this magnetic field has helicity, it acquires an excess of right-moving particles over left-moving particles:

$$(n_{\rm R} - n_{\rm L})_{\mathbf{A}} = \frac{1}{2\pi^2}\mathbf{A} \cdot (\nabla \times \mathbf{A}). \tag{123}$$

The right-hand side is the so-called Chern–Simons (or topological) charge of the magnetic field.

The transformation of particles into a magnetic field configuration opens the possibility to examine the cosmological origin of galactic magnetic fields from a system of fermions. This is the essential step in the scenario described by Joyce and Shaposhnikov (1997). In this model an initial excess of right-handed electrons, $e_{\rm R}$, was assumed to have been generated in the Early Universe. This excess would then have survived until the electroweak phase transition (at about 10^{-10} s after the Big Bang) at which point anomalous lepton (and baryon) number violating processes became efficient enough to erase the excess. However, it appears that well before the electroweak transition an instability developed, where the excess of the right electrons transformed into a hypermagnetic field. Then, when the electroweak transition took place, it transformed part of the hypermagnetic field into electromagnetic field, so that the universe was bestowed with a primordial (electro-) magnetic field.

We can now discuss the corresponding process in ^{3}He-A – the transformation of fermionic charge to magnetic field. In our case it is the quasiparticle momentum which plays the part of the relevant fermionic charge. The net quasiparticle momentum is generated by the relative flow of the normal and superfluid components. This fermionic charge is transformed via the chiral anomaly to order parameter texture which, as we have seen, plays the part of the magnetic field. The transformation occurs in the form of an instability which transfers the excess in momentum to the formation of $\hat{\mathbf{l}}$ textures. The process corresponds to the counterflow instability observed in ^{3}He-A, which has been discussed both experimentally and theoretically (Ruutu et al. 1997a). Thus the ^{3}He-A analogy closely follows the cosmological scenario described by Joyce and Shaposhnikov (1997).

Let us discuss this instability. In the presence of counterflow, $\mathbf{v} = \mathbf{v}_{\mathrm{n}} - \mathbf{v}_{\mathrm{s}}$, of the normal component of ^{3}He-A liquid with respect to the superfluid, the $\hat{\mathbf{l}}$-vector is oriented along the flow, $\hat{\mathbf{l}}_0 \| \mathbf{v}$. We are interested in the stability condition for such homogeneous counterflow with respect to the generation of inhomogeneous perturbations $\delta\hat{\mathbf{l}}$,

$$\hat{\mathbf{l}} = \hat{\mathbf{l}}_0 + \delta\hat{\mathbf{l}}(\mathbf{r}, t), \tag{124}$$

keeping in mind that the space and time dependences of $\delta\hat{\mathbf{l}}$ correspond to "hyperelectric field" $\mathbf{E} = k_{\mathrm{F}}\partial_t\delta\hat{\mathbf{l}}$ and "hypermagnetic field" $\mathbf{B} = k_{\mathrm{F}}\nabla \times \delta\hat{\mathbf{l}}$.

It is important for our consideration that the ^{3}He-A liquid is anisotropic in the same manner as a nematic liquid crystal. For the relativistic fermions this means that their motion is determined by the geometry of some effective space-time which in ^{3}He-A is described by the metric tensor in eq. (81). As we have already discussed above, in the presence of counterflow the energy of quasiparticles is Doppler shifted by the amount $\mathbf{p} \cdot \mathbf{v}$. Since the quasiparticles are concentrated near the gap nodes, this energy shift is constant and opposite for the two gap nodes: $\mathbf{p} \cdot \mathbf{v} \approx \pm p_{\mathrm{F}}(\hat{\mathbf{l}}_0 \cdot \mathbf{v})$. The counterflow therefore produces what would be an effective chemical potential in particle physics, which has opposite sign for the right- and left-handed particles:

$$\mu_{\mathrm{R}} = -\mu_{\mathrm{L}} = p_{\mathrm{F}}(\hat{\mathbf{l}}_0 \cdot \mathbf{v}). \tag{125}$$

The kinetic energy of the counterflow is

$$E_{\mathrm{kin}} = \tfrac{1}{2}\mathbf{v}\rho_{n\|}\mathbf{v}. \tag{126}$$

Here the density of the normal component is a tensor in the anisotropic ^{3}He-A liquid, and only the longitudinal component $\rho_{n\|}$ is involved.

Let us consider the low-temperature limit $T \ll T_c$, where $T_c \sim \Delta_0$ is the superfluid transition temperature. Then using the expression for the longitudinal density of the normal component of ^{3}He-A (Vollhardt and Wölfle 1990)

$$\rho_{n\|} = \pi^2 \rho \frac{m_3^*}{m_3} \left(\frac{k_B T}{\Delta_0} \right)^2 \tag{127}$$

and also the ^{3}He-A equivalent of the chemical potential (125) one obtains

$$E_{\text{kin}} \approx \frac{1}{6} m_3^* k_F^3 \frac{T^2}{\Delta_0^2} (\hat{\mathbf{l}}_0 \cdot \mathbf{v})^2 \equiv \frac{1}{6} \sqrt{-g} T^2 \mu_R^2. \tag{128}$$

In the last equality an overall constant appears to be the square root of the determinant of an effective metric in ^{3}He-A: $\sqrt{-g} = 1/c_\| c_\perp^2 = m_3^* k_F / \Delta_0^2$. In relativistic theories the rhs of eq. (128) is exactly the energy density of the massless right-handed electrons in the presence of the chemical potential μ_R. Thus the kinetic energy, stored in the counterflow, is exactly analogous to the energy stored in the right-handed electrons. The same analogy occurs between the net quasiparticle linear momentum, $\mathbf{P} = \rho_n \mathbf{v}$, along $\hat{\mathbf{l}}_0$ and the chiral charge of the right electrons,

$$n_R \equiv \frac{1}{p_F} \mathbf{P} \cdot \hat{\mathbf{l}}_0. \tag{129}$$

The inhomogeneity which absorbs the fermionic charge is represented by a magnetic field configuration in real physical vacuum and by a $\delta\hat{\mathbf{l}}$ texture in ^{3}He-A. However, eq. (123) applies in both cases, if in ^{3}He-A we use the standard identification $\mathbf{A} = k_F \delta\hat{\mathbf{l}}$.

Just as in the particle physics case, we now consider the instability towards the production of the "magnetic" texture due to the excess of chiral particles. This instability can be seen by considering the gradient energy of the inhomogeneous texture on the background of the superflow. In the geometry of the superflow, the textural contribution to the free energy of the $\delta\hat{\mathbf{l}}$-vector is completely equivalent to the conventional energy of the hypermagnetic field (Volovik 1992)

$$\begin{aligned} F_{\text{grad}} &= \ln\left(\frac{\Delta_0^2}{T^2}\right) \frac{p_F^2 v_F}{24\pi^2 \hbar} (\hat{\mathbf{l}}_0 \times (\nabla \times \delta\hat{\mathbf{l}}))^2 \\ &\equiv \frac{\sqrt{-g}}{4\pi e_{\text{eff}}^2} g^{ij} g^{kl} F_{ik} F_{jl} = F_{\text{magn}}. \end{aligned} \tag{130}$$

Here $F_{ik} = \nabla_i A_k - \nabla_k A_i$ and we again have included the effective anisotropic metric in eq. (81) appropriate for ^{3}He-A.

It is interesting that the logarithmic factor in the gradient energy plays the part of the running coupling $e_{\rm eff}^{-2} = (1/3\pi\hbar c)\ln(\Delta_0/T)$ in particle physics, where $e_{\rm eff}$ is the effective hyperelectric charge; while the gap amplitude Δ_0, the ultraviolet cutoff, plays the part of the Planck energy scale. Now if one has the counterflow in ^{3}He-A, or its equivalent – an excess of chiral charge produced by the chemical potential $\mu_{\rm R}$ – the anomaly gives rise to an additional effective term in the magnetic energy, corresponding to the interaction of the charge absorbed by the magnetic field with the chemical potential. This effective energy term is:

$$F_{\rm CS} = (n_{\rm R} - n_{\rm L})\mu_{\rm R} = \frac{1}{2\pi^2}\mu_{\rm R}\mathbf{A}\cdot(\nabla\times\mathbf{A})$$
$$= \frac{3\hbar}{2m}\rho(\hat{\mathbf{l}}_0\cdot\mathbf{v})(\delta\hat{\mathbf{l}}\cdot\nabla\times\delta\hat{\mathbf{l}}). \qquad (131)$$

The rhs corresponds to the well-known anomalous interaction of the counterflow with the $\hat{\mathbf{l}}$ texture in ^{3}He-A, where ρ is the mass density of ^{3}He (Volovik 1992) (the additional factor of 3/2 enters due to non-linear effects).

For us the most important property of this term is that it is linear in the derivatives of $\delta\hat{\mathbf{l}}$. Its sign thus can be negative, while its magnitude can exceed the positive quadratic term in eq. (130). This leads to the helical instability where the inhomogeneous $\delta\hat{\mathbf{l}}$ field is formed. During this instability the kinetic energy of the quasiparticles in the counterflow (analog of the energy stored in the fermionic degrees of freedom) is converted into the energy of inhomogeneity $\nabla\times\delta\hat{\mathbf{l}}$, which is the analog of the magnetic energy of the hypercharge field.

When the helical instability develops in ^{3}He-A, the final result is the formation of a $\hat{\mathbf{l}}$ texture, which represents a large jump from the vortex-free counterflow texture towards the global free energy minimum, the equilibrium vortex texture. The new texture contains a central part with a periodic $\hat{\mathbf{l}}$ texture, where the elementary lattice cell represents a so-called Anderson–Toulouse–Chechetkin (ATC) vortex with continuous structure which has the topology of a skyrmion (fig. 34). These vortices give rise to a characteristic satellite peak in the NMR absorption spectrum where their number is directly proportional to the height of the satellite peak (fig. 37) (Blaauwgeers et al. 2000).

Experimentally this is observed when the rotation velocity is slowly increased, starting from a state with vortex-free counterflow (i.e. with fermionic charge, but no hypermagnetic field). The helical instability is then observed as a discontinuity, when the vortex satellite is suddenly formed (fig. 38). Its peak height jumps from 0 to a magnitude which approaches that

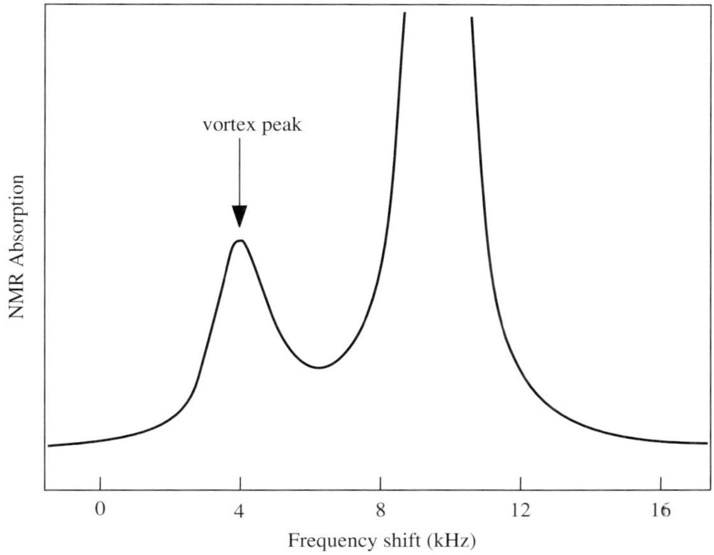

Fig. 37. NMR signal from an array of continuous ATC vortices, with the frequency shift from the Larmor value plotted on the horizontal axis. The large truncated peak on the right is generated by the bulk order-parameter texture, while the small satellite peak on the left is produced by the continuous soft-core vortex textures. The frequency shift of the satellite peak from the Larmor frequency indicates the type of vortex, while the intensity of the peak is proportional to the number of vortices of this particular type.

in the equilibrium vortex state, which means that the counterflow is essentially reduced and the number of vortex lines is close to that in equilibrium. Most of the counterflow (fermionic charge) thus becomes converted into the vortex texture (magnetic field).

Together with the results of Bevan et al. (1997b), the textural superflow instability shows that the chiral anomaly is an important mechanism in the interaction of vortex textures (the analog of the hypercharge magnetic fields and cosmic strings) with fermionic excitations (analog of quarks and leptons). These two experiments verified both processes which are induced by the anomaly: the nucleation of fermionic charge from vacuum, observed by Bevan et al. (1997b), and the inverse process of the nucleation of an effective magnetic field from the fermion current as observed by Ruutu et al. (1997a).

3.5. *Vortex mass: chiral fermions in strong magnetic field*

Until now we have assumed that the mass M_V of a vortex can be neglected in experimental vortex dynamics. How well justified is this approach? The

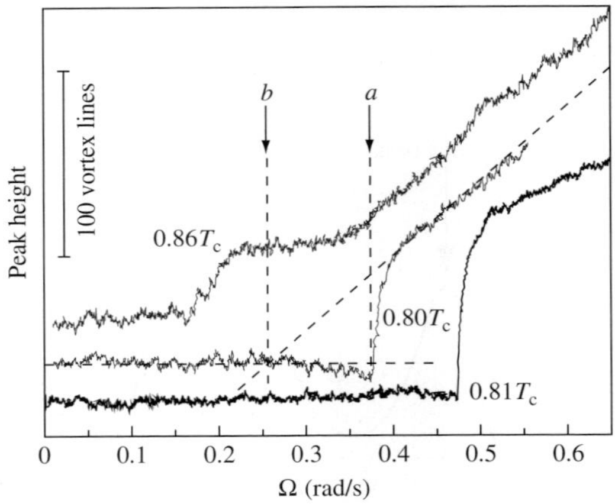

Fig. 38. Three measurements of the critical velocity of vortex formation in ^{3}He-A. In these examples from the same experimental setup (at 33.9 bar pressure) vortex formation starts with a phase transition in the global order-parameter texture while the rotation velocity Ω is slowly increased. On the vertical scale the peak height of the NMR satellite from continuous vortex lines is recorded. Initially there are no vortex lines (at low Ω on the left) and the peak height is 0. When the velocity of the counterflow $\mathbf{v}$ in the $\hat{\mathbf{l}}_0$ direction (corresponding to the chemical potential μ_R of the chiral electrons) exceeds some critical value (Ω value at arrow *a*), a transition in the global order-parameter texture occurs and the texture jumps into a configuration with a different, often lower critical velocity (corresponding to the extrapolated Ω value at arrow *b*). In the texture transition a large number of vortex lines is simultaneously formed. This means that the center of the sample is suddenly filled with the $\hat{\mathbf{l}}$ texture (or hypermagnetic field) formed by the central cluster of continuous vortex lines (based on Ruutu et al. 1996b).

term $M_V \partial_t \mathbf{v}_L$ in the force balance equation for the vortex contains the time derivative and thus at very low frequencies of vortex motion it can be neglected compared to other forces, which are directly proportional to $\mathbf{v}_L$. To be more quantitative, we estimate the vortex mass in the BCS superfluids and superconductors. There are several contributions to the vortex mass. It will also become clear that these are related to some most peculiar phenomena in quantum field theory. We start from the contribution, which is relevant for the Bose superfluid ^{4}He.

3.5.1. *"Relativistic" mass of the vortex*

In the hydrodynamic theory the mass of the vortex is non-zero owing to the compressibility of the liquid. Since sound propagation in fluids is similar to light propagation in vacuum, the hydrodynamic energy of the vortex, soliton or other extended object moving in the liquid is connected with its hydrodynamic mass per unit length by the "relativistic" equation $E = ms^2$, where the speed of sound s substitutes the speed of light (Davis 1992, Duan

1995, Kao and Lee 1995, Iengo and Jug 1995, Wexler and Thouless 1996). Thus the hydrodynamic mass of a vortex loop of length L at $T = 0$ is according to eq. (18),

$$M_{\text{compr}} = \frac{E_{\text{kin}}}{s^2} = \frac{\rho\kappa^2 L}{4\pi s^2} \ln\frac{L}{\xi}. \tag{132}$$

For Fermi superfluids s is of the order of the Fermi velocity $v_F \sim p_F/m$ (m is the mass of the electron in metals or of the ^{3}He atom), and the estimate for the hydrodynamic mass of a small vortex loop of length L is $M_{\text{compr}} \sim \rho a^2 L \ln L/\xi$, where a is the inter-atomic distance. For superfluid ^{4}He, where the core size $\xi \sim a$, we can roughly speaking associate the hydrodynamic vortex mass $\sim \rho a^2 L$ with the "mass of the liquid in the vortex core volume". However for ^{3}He superfluids and for superconductors, where $\xi \gg a$, the hydrodynamic vortex mass is much less than $\rho\xi^2 L$ and other contributions become more important.

3.5.2. *Contribution from bound states to the mass of a singular vortex*

It appears that the most important contribution to vortex mass originates from the quasiparticles occupying the bound states in the vortex core and thus forming the normal component concentrated in the core. For the singular-vortex structures in conventional superconductors and in ^{3}He–B this contribution to the vortex mass depends on $\omega_0\tau$ and in the clean-limit case it is proportional to the mass of the liquid in the vortex core, as was first found by Kopnin (1978) for superconductors and by Kopnin and Salomaa (1991) for superfluid ^{3}He–B: $M_{\text{Kopnin}} \sim \rho\xi^2 L$. This core mass is essentially larger than the logarithmically divergent contribution, which comes from the compressibility. In spite of the logarithmic divergence, the latter contains the speed of sound in the denominator and thus is smaller by the factor $(a/\xi)^2 \ll 1$, where a is the inter-atomic distance. The compressibility mass of the vortex dominates in Bose superfluids, where the core size is small, $\xi \sim a$.

According to Kopnin's theory the core mass comes from the fermions trapped in the vortex core (Kopnin 1978, Kopnin and Salomaa 1991, van Otterlo et al. 1995, Volovik 1997, Kopnin and Vinokur 1998). This Kopnin mass of the vortex can be derived using a phenomenological approach. Let us consider the limit of low T in the superclean regime $\omega_0\tau \gg 1$ (Volovik 1997). If the vortex moves with velocity $\mathbf{v}_L$ with respect to the superfluid component, the fermionic energy spectrum in the vortex frame is Doppler-shifted and has in eq. (111) the form $E = E_0(q) - \mathbf{p} \cdot \mathbf{v}_L$, where q represents the fermionic degrees of freedom in the stationary vortex. The summation over the fermionic degrees of freedom leads to the extra linear momentum of the vortex $\propto \mathbf{v}_{\mathrm{L}}$:

$$\mathbf{P} = \sum_q \mathbf{p}\theta(-E) = \sum_q \mathbf{p}(\mathbf{p}\cdot\mathbf{v}_{\mathrm{L}})\delta(E_0) = M_{\mathrm{Kopnin}}\mathbf{v}_{\mathrm{L}},$$

$$M_{\mathrm{Kopnin}} = \frac{1}{2}\sum_q \mathbf{p}_{\perp}^2\delta(E_0). \tag{133}$$

For the axisymmetric vortex, where $E_0 = -L_z\omega_0(p_z)$ and $\sum_q = \int \mathrm{d}L_z\, \mathrm{d}p_z\, \mathrm{d}z/2\pi$, one has

$$M_{\mathrm{Kopnin}} = L\int_{-p_{\mathrm{F}}}^{p_{\mathrm{F}}} \frac{\mathrm{d}p_z}{4\pi}\frac{p_{\mathrm{F}}^2 - p_z^2}{\omega_0(p_z)}. \tag{134}$$

Equation (134) can also be obtained from the time-dependent kinetic equation (112). It is the coefficient in the contribution to the longitudinal force, which is linear in external frequency ω: $\mathbf{F}_{\mathrm{sf}}^{\parallel} = -\mathrm{i}\omega M_{\mathrm{Kopnin}}\mathbf{v}_{\mathrm{L}}$.

Note that this vortex mass is determined in essentially the same way as the normal component density in the bulk system. The Kopnin vortex mass is non-zero if the density of fermionic states is finite in the vortex core, which is determined by the inter-level spacing ω_0 in the core: $N(0) \propto 1/\omega_0$. This gives for the Kopnin vortex mass an estimation, $M_{\mathrm{Kopnin}} \sim p_{\mathrm{F}}^3/\omega_0 L \sim \rho\xi^2 L$. A method to measure this contribution for the ^{3}He–B vortex with spontaneously broken axisymmetry of the vortex core, by driving the core into rapid rotation with an NMR excitation field, has been suggested by Kopnin and Volovik (1998). The much stronger connection to the normal component fraction in the core of a vortex with continuous structure, like the dominant vortex line in ^{3}He-A, will be examined in the next subsection.

3.5.3. *Kopnin vortex mass in the continuous-core model: connection to chiral fermions in magnetic field*

The continuous-core vortex in ^{3}He-A with skyrmion structure is the best example – a model case – which helps to understand the vortex core mass. The continuous-core model can also be applied to other Fermi superfluids and superconductors: the singular core can thereby be smoothed, so that the $1/r$ singularity of the superfluid velocity is removed, by introducing point nodes in the superfluid energy gap in the core region (fig. 39). As a result the superfluid/superconducting state in the vortex core acquires the properties of ^{3}He-A with its continuous vorticity and point gap nodes (Volovik and Mineev 1982, Salomaa and Volovik 1987). After that one can easily separate different contributions to the vortex mass. Actually this is not only a model: spontaneous smoothing of the velocity singularity occurs in the cores of both types of ^{3}He–B vortices (Salomaa and Volovik 1987), while in heavy

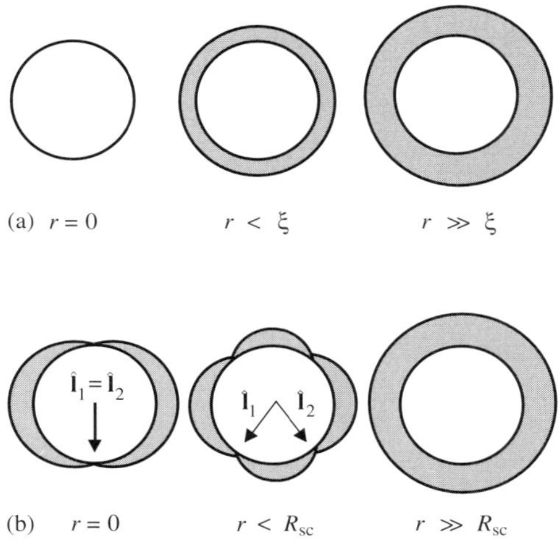

Fig. 39. Smoothing operation of the singular vortex core. (a) In the singular vortex the gap is continuously reduced and becomes 0 exactly on the vortex axis (at $r = 0$). (b) For some vortices it is energetically favorable to avoid the vanishing order parameter at $r = 0$. Instead, within the smooth core, $r < R_{sc}$, point-like gap nodes appear in the spectrum of fermions (Volovik and Mineev 1982). The unit vectors $\hat{\mathbf{l}}_1$ and $\hat{\mathbf{l}}_2$ show the directions to the nodes at different r. Close to the gap nodes the spectrum of the fermions is similar to that in ^{3}He-A (example at $r = 0$).

fermion and high-T_c superconductors such smoothing can occur due to the admixture of different pairing states in the vortex core.

For the smoothed singly quantized vortices of ^{3}He–B and superconductors one has two $\hat{\mathbf{l}}$ vectors: $\hat{\mathbf{l}}_1$ and $\hat{\mathbf{l}}_2$, each for one of the two spin projections. The simplest distribution of both fields is given by eq. (106) with such $\eta(r)$ that $\hat{\mathbf{l}}_1(0) = \hat{\mathbf{l}}_2(0) = -\hat{\mathbf{z}}$ and $\hat{\mathbf{l}}_1(\infty) = -\hat{\mathbf{l}}_2(\infty) = \hat{\mathbf{r}}$ (Salomaa and Volovik 1987). The region of radius R_{sc}, where the texture of $\hat{\mathbf{l}}$ vectors is concentrated, represents the smoothed *non-singular core* of the vortex.

In the continuous vortex, the normal component in the core can be considered to be a local quantity, determined at each point in the vortex core. This consideration is valid for a smooth core with radius $R_{sc} \gg \xi$, where the local classical description of the fermionic spectrum can be applied. The main contribution comes from the point-like gap nodes, where the classical spectrum has the form $E_0 = \sqrt{v_F^2(p - p_F)^2 + \Delta_0^2(\hat{\mathbf{p}} \times \hat{\mathbf{l}})^2}$ and Δ_0 is the gap amplitude. In the presence of a gradient in the $\hat{\mathbf{l}}$ field, which acts on the quasiparticles as an effective magnetic field, this gapless spectrum leads to the non-zero local DOS, discussed in section 3.3.1 for the relativistic chiral fermions. To apply the DOS in eq. (99) to the case of anisotropic ^{3}He-A, one must make the covariant generalization of the DOS, by introducing the general metric tensor and then substituting it by the effective ^{3}He-A metrics which describes the anisotropy of ^{3}He-A. The general form of the DOS of the chiral fermions in the curved space is

$$N(0) = \frac{|\,e\,|}{2\pi^2}\sqrt{-g}\sqrt{\frac{1}{2}g^{ij}g^{kl}F_{ik}F_{jl}}, \tag{135}$$

where F_{ik} is defined immediately following eq. (130). One now can apply this to ^{3}He-A where $|\,e\,|= 1$, and the metric tensor is given by eq. (81). Neglecting the dependence of the small velocity field v_s in the smooth core, one obtains the following local DOS for the fermions in the $\hat{\mathbf{l}}$ texture at $T = 0$:

$$N(0,\mathbf{r}) = \frac{p_F^2}{2\pi^2\Delta_A}\,|\hat{\mathbf{l}} \times (\nabla \times \hat{\mathbf{l}})|\ . \tag{136}$$

This DOS can be inserted to the expression for the local density of the normal component at $T = 0$ (see eq. (5.24) in the review by Volovik (1990)):

$$(\rho_n)_{ij}(\mathbf{r}) = \hat{l}_i\hat{l}_j p_F^2 N(0,\mathbf{r}). \tag{137}$$

For the axisymmetric continuous vortex (eq. (106)) one has

$$N(0,\mathbf{r}) = \frac{p_F^2}{2\pi^2\Delta_A}\sin\eta|\partial_r\eta|. \tag{138}$$

The integral of this normal density tensor over the cross section of the soft core gives the Kopnin mass of the vortex in the local density representation

$$\begin{aligned} M_{\text{Kopnin}} &= L\int d^2r(\hat{\mathbf{v}}_L\cdot\hat{\mathbf{l}})^2 p_F^2 N(0,\mathrm{r}) \\ &= \frac{p_F^2 L}{4\pi^2\Delta_A}\int d^2r\sin^3\eta(r)|\partial_r\eta|, \end{aligned} \tag{139}$$

where $\hat{\mathbf{v}}_L$ is the unit vector in the direction of the vortex velocity $\mathbf{v}_L$. The same equation for the mass can be obtained from eq. (134), using the exact expression for the inter-level distance $\omega_0(p_z)$ (Volovik 1997).

Since $v_F/\Delta_0{\sim}\xi$ one obtains that the Kopnin mass of the continuous vortex $\sim\rho\xi R_{sc}L$, i.e. it is linear in the dimension R_{sc} of the core (Kopnin 1995). Thus it follows that the area law for the vortex mass is valid only for vortices with a core size of order ξ (i.e. $R_{sc}{\sim}\xi$).

Note that the vortex mass discussed above comes from the normal component trapped in the vortex and which thus moves with the vortex velocity, $\mathbf{v}_n = \mathbf{v}_L$. In this consideration it is assumed that $\omega_0\tau \gg 1$. In this limit the normal component in the core and the normal component in the heat bath do not interact with each other and thus can move with different velocities. The local hydrodynamic energy density of the normal component trapped by the vortex is

$$F = \tfrac{1}{2}(\rho_{\rm n})_{ij}(\mathbf{r})(\mathbf{v}_{\rm L} - \mathbf{v}_{\rm s})_i(\mathbf{v}_{\rm L} - \mathbf{v}_{\rm s})_j. \tag{140}$$

This can be rewritten in a form, which is valid also for chiral fermions:

$$F = \frac{\mu_{\rm R}^2 + \mu_{\rm L}^2}{8\pi^2}\sqrt{-g}\sqrt{\frac{1}{2}g^{ij}g^{kl}F_{ik}F_{jl}}, \tag{141}$$

where, as before in eq. (125), the chemical potential of the left and right fermions in ^{3}He-A is expressed in terms of the counterflow: $\mu_{\rm R} = -\mu_{\rm L} = p_{\rm F}(\hat{\mathbf{l}}\cdot\mathbf{v})$. Equation (141) represents the magnetic energy of the system of the chiral particles with finite chemical potential in a strong magnetic field $B \gg \mu^2$.

3.5.4. *Associated hydrodynamic mass*

The problem of another source of vortex mass of hydrodynamic origin was raised by Sonin et al. (1998). It is the so-called backflow mass discussed by Baym and Chandler (1983), which also can be proportional to the core area. Here we compare these two contributions in the superclean regime and at low $T \ll T_{\rm c}$ using the model of a continuous core. The associated (or induced) mass appears when, say, an external body moves in the superfluid. This mass depends on the geometry of the body. For the moving cylinder of radius R it is the mass of the liquid displaced by the cylinder,

$$M_{\rm associated} = \pi R^2 L\rho, \tag{142}$$

which is to be added to the actual mass of the cylinder to obtain the total inertial mass of the body. In superfluids this part of the superfluid component moves with the external body and thus can be associated with the normal component. A similar mass is responsible for the normal component in porous materials, in aerogel for instance, where some part of the superfluid is hydrodynamically trapped by the pores. It is removed from the overall superfluid motion and thus becomes part of the normal component.

In the case when a vortex is trapped on a wire of radius $R_{\rm wc} \gg \xi$, such that the wire replaces the vortex core, eq. (142) gives the vortex mass due to the backflow around the moving core. This is the simplest realization of the backflow mass of the vortex discussed by Baym and Chandler (1983). For this vortex with a wire-core the Baym–Chandler mass is the dominant mass of the vortex. The Kopnin mass, which can result from normal excitations trapped near the surface of the wire, is essentially less.

Let us now consider the Baym–Chandler mass for the free vortex at $T = 0$, using again the continuous-core model. In the wire-core vortex this mass arises from the backflow caused by the inhomogeneity in $\rho_{\rm s}$:

$\rho_s(r > R_{wc}) = \rho$ and $\rho_s(r < R_{wc}) = 0$. Similar but less severe inhomogeneity of $\rho_s = \rho - \rho_n$ occurs in the continuous-core vortex due to the non-zero local normal density in eq. (137). Owing to the profile of the local superfluid density, the external flow is disturbed near the core according to the continuity equation

$$\nabla \cdot (\rho_s \mathbf{v}_s) = 0. \tag{143}$$

If the smooth core is large, $R_{sc} \gg \xi$, the deviation of the superfluid component inside the smooth core is small from its asymptotic value outside the core: $\delta\rho_s = \rho - \rho_s \sim (\xi/R_{sc})\rho \ll \rho$ and can be considered as a perturbation. Thus if the asymptotic value of the velocity of the superfluid component with respect of the core is $\mathbf{v}_{s0} = -\mathbf{v}_L$, the disturbance $\delta\mathbf{v}_s = \nabla\Phi$ of the superflow in the smooth core is given by:

$$\rho\nabla^2\Phi = v^i_{s0}\nabla^j(\rho_n)_{ij}. \tag{144}$$

The kinetic energy of the backflow gives the Baym–Chandler mass of the vortex

$$M_{BC} = \frac{\rho L}{v^2_{s0}} \int d^2r(\nabla\Phi)^2. \tag{145}$$

In the simple approximation that the normal component in eq. (137) is considered isotropic, one obtains

$$M_{BC} = \frac{L}{2\rho} \int d^2r\rho^2_n(r) \sim \rho\xi^2. \tag{146}$$

The Baym–Chandler mass does not depend on the core radius R_{sc}, since the large area R^2_{sc} of integration in eq. (146) is compensated by the small value of the normal component in the rarefied core, $\rho_n \sim \rho(\xi/R_{sc})$. That is why this mass is parametrically smaller than the Kopnin mass in eq. (139), if the smooth core is large: $R_{sc} \gg \xi$.

In conclusion, both contributions to the mass of the vortex result from the mass of the normal component trapped by the vortex. The difference between Kopnin mass and Baym–Chandler backflow mass is only in the origin of the normal component trapped by the vortex. The relative importance of the two masses depends on the vortex core structure: (1) for the free continuous vortex with a large core size $R_{sc} \gg \xi$, the Kopnin mass dominates: $M_{Kopnin} \sim \rho R_{sc}\xi L \gg M_{BC} \sim \rho\xi^2 L$. (2) For the circulation trapped around a wire of radius $R_{wc} \gg \xi$, the Baym–Chandler mass is proportional to the core area, $M_{BC} \sim \rho R^2_{wc} L$, and is parametrically larger than the Kopnin mass. (3) For the free vortex core with a core radius $R \sim \xi$ the situation is not clear since the continuous core approximation does not work any more. But extrapolation of the result in eq. (146) to $R{\sim}\xi$ suggests that the Baym–Chandler mass can be comparable with the Kopnin mass.

3.5.5. *Topology of the energy spectrum: gap nodes and their ramifications*

Zeros in the fermionic spectrum, such as Fermi surfaces (surfaces of zeros) and Fermi points (point zeroes) play an extremely important role in condensed matter and in analog models of the low-energy physics of the quantum vacuum (Volovik 2003, Horava 2005). In condensed matter, the gapless fermions interacting with the Bose fields of the order parameter lead to the anomalous behavior of superfluids and superconductors at low temperatures, $T \ll T_c$, such as spectral flow in vortex dynamics, non-analytic behavior of the current and gradient energy, non-linear and non-local Meissner effect, etc. The counterpart of this behavior in high-energy physics manifests itself in the axial anomaly, baryogenesis, zero-charge effect, running coupling constants, photon mass, etc. It is the zeros in the fermionic spectrum, through which the conversion of the vacuum degrees of freedom into that of matter takes place.

Similar zeros, but in real space, exist in the cores of topological defects, especially in quantized vortices (or cosmic strings in high-energy physics). Actually the real-space zeros and the momentum-space zeros are described by the same topology extended to the $8 = 4 + 4$-dimensional space. For example, from the topological point of view, the Fermi surface represents the vortex singularity of the Green's function in the $\omega, \mathbf{p}$ space, where ω is the Matsubara frequency. The Green's function $G(\omega,p) = 1/(\mathrm{i}\omega + v_F(p - p_F))$ displays a vortex in the ω, p plane with the winding number $\nu = 1$. This makes the Fermi surface topologically stable and robust under perturbations of the Fermi system. Even though the pole in the Green's function can disappear under some perturbations, the Fermi surface will survive in the marginal and Luttinger superfluids. The latter thus belong to the same class of Fermi systems as the Landau Fermi liquid.

In the same manner, superfluids and superconductors with a non-vanishing gap behave in the vicinity of the vortex core like superfluid ^{3}He-A in bulk – point-like gap nodes appear to be the common feature. Owing to the common topological origin of the point nodes, the fermions near the gap nodes in gapless Fermi superfluids and superconductors and the low-energy fermions localized in the cores of vortices in conventional gapped superconductors produce similar anomalous effects.

The splitting or annihilation of topological point nodes represent a new type of phase transition, a topological quantum phase transition, which may occur in superfluids, superconductors, ultracold fermionic atomic gases, and in the Standard Model (Klinkhamer and Volovik 2004). The formation of defects in such a quantum phase transition opens a new direction of study in the formation of topological defects (Dziarmaga et al. 2003).

4. Final remarks

We have examined a few cases in which superfluid ^{3}He has been used as an analog system, to model new concepts in quantum field theory. The focus in section 2 is on the question whether the observed vortex formation in a rapid second-order phase transition is described correctly by the Kibble-Zurek scaling theory in ^{3}He–B. In section 3 a number of other analog connections to vortex formation and dynamics are explored, this time also in ^{3}He-A. Vortex formation (Ruutu et al. 1997a) and dynamics (Eltsov et al. 2002) are very different in ^{3}He-A and -B, and so are the measuring techniques which are required to study these features. In fact, the rotating measurements on the KZ mechanism in section 2 cannot be repeated in ^{3}He-A, because vortex-free flow can be reached only up to $\lesssim 1$ mm/s while, as seen in fig. 13, the required threshold velocity $v_{\rm cn}$ is at least twice higher. No such fundamental limitations restrict calorimetric measurements on neutron absorption events in ^{3}He-A, but present measuring techniques are not sufficiently developed to tackle this question in an anisotropic orbital texture.

Other analog connections involving both ^{3}He-A and -B are described in the monograph by Volovik (2003). What is the role of analog models in physics in general? In examining the validity of a new concept we may assume that if a test in some system gives a positive result, then good grounds exist to claim that the principle has fundamental value as a physical model. However, if a particular concept has been shown to work in superfluid ^{3}He this does not imply universal applicability. Analog studies are not a replacement for cosmological observations or a means to spare investments in high-energy accelerators! From a condensed matter experiment no direct conclusions can be drawn on the presence or absence of the KZ mechanism and its consequences in the Early Universe. Although superfluid ^{3}He experiments in section 2 support the validity of the KZ mechanism, at present this scenario is not a popular explanation for the anisotropy of the cosmic background radiation, which was precipitated when the Early Universe had cooled to the point where it became transparent to light. Theories based on the presence of an epoch with exponentially accelerated inflationary expansion appear to provide better fits to the latest cosmological data. Nevertheless, defects are observed to form in non-equilibrium phase transitions in condensed matter physics and one is left wondering why cosmological transitions should be any different. Many intriguing questions wait for solid answers: what was the nature of the phase transitions of the Early

Universe, why defects would not have formed, or if indeed they did form, what was their role in shaping the Universe and how to discern their consequences today.

Irrespective of its cosmological origin, in condensed matter physics, the KZ mechanism has become an important new concept in rapid non-equilibrium phase transitions. It is one of the few intrinsic processes of the bulk material which accounts for defect formation in a wide spectrum of different systems in the absence of boundaries, impurities, or other non-ideal constraints. The strongest experimental support to date for the KZ mechanism comes from the neutron absorption measurements in ^{3}He–B where different types of defects are observed to form (usual mass-flow vortices, spin-mass vortices, AB interfaces). Here the conveniently controlled vortex formation in neutron irradiation has become a handy new tool for the experimentalist. In these measurements it is also directly seen that later phase transition fronts, which sweep through the system, interact with the existing defects. The final outcome depends on the relative velocity of the phase front with respect to the defects (Krusius et al. 1994). The amount of work performed on the KZ mechanism in the form of simulations, measurements, and analysis by far exceeds other cases of analog studies. It serves as a good example in general, on what can be achieved with analog studies in the unification of physical principles.

Acknowledgments

The first version of this review was finished in September 1998 (Eltsov et al. 1998). Although three revised editions were prepared since then over the years, the references were never updated to reflect properly the rapid progress of recent years in quantum analog models. We apologize for this shortcoming. We dedicate this work to Tom Kibble and Wojciech Zurek. We are indebted to numerous colleagues, Yu. Bunkov, A.P. Finne, A. Gill, H. Godfrin, H.E. Hall, J.R. Hook, T. Jacobson, N. Kopnin, A. Leggett, Yu. Makhlin, P. Mazur, B. Plaçais, J. Ruohio, V. Ruutu, E. Thuneberg, T. Vachaspati, G. Williams, Wen Xu, X. Zhang, and many others. Much of the joint work with our colleagues was made possible by the ULTI visitor programs under the EU Human Capital and Mobility program (contract no. CHGECT94-0069) and its continuation, the EU program Improving the Human Research Potential (no. HPRI-CT-1999-00050). This collaborative work has also been inspired by the European Science Foundation programs Cosmology in the Laboratory (COSLAB) and Vortex Matter in Superconductors at Extreme Scales and Conditions (VORTEX).

References

Aarts, R.G.K.M. and A.T.A.M. deWaele, 1994, Phys. Rev. B **50**, 10069.

Adams, J.S., S.R. Bandler, S.M. Brouer, R.E. Lanou, H.J. Maris, T. More and G.M. Seidel, 1995, Phys. Lett. B **341**, 431.

Adler, S., 1969, Phys. Rev. **177**, 2426.

Aharonov, Y. and D. Bohm, 1959, Phys. Rev. **115**, 485.

Albert, D.Z., 1982, Phys. Rev. B **25**, 4810.

Amelino-Camelia, G., J.D. Bjorken and S.E. Larsson, 1997, Phys. Rev. D **56**, 6942.

Andreev, A.F. and L.A. Melnikovsky, 2003, Pis'ma Zh. Exp. Teor. Fiz. **78**, 1063 [JETP Lett. 78, 574 (2003)].

Antunes, N.D. and L.M.A. Bettencourt, 1998, Phys. Rev. Lett. **81**, 3083.

Antunes, N.D., L.M.A. Bettencourt and M. Hindmarsh, 1998, Phys. Rev. Lett. **80**, 908.

Antunes, N.D., L.M.A. Bettencourt and W.H. Zurek, 1999, Phys. Rev. Lett. **82**, 2824.

Aranson, I.S., N.B. Kopnin and V.M. Vinokur, 1999, Phys. Rev. Lett. **83**, 2600.

Aranson, I.S., N.B. Kopnin and V.M. Vinokur, 2001, Phys. Rev. B **63**, 184501.

Awschalom, D.D. and K.W. Schwarz, 1984, Phys. Rev. Lett. **52**, 49.

Barenghi, C.F. and D.C. Samuels, 2002, Phys. Rev. Lett. **89**, 155302.

Barenghi, C.F., D.C. Samuels, G.H. Bauer and R.J. Donnelly, 1997, Phys. Fluids **9**, 2631.

Barriola, M., 1995, Phys. Rev. D **51**, 300.

Bäuerle, C., Yu.M. Bunkov, S.N. Fisher, H. Godfrin and G.R. Pickett, 1996, Nature **382**, 332.

Bäuerle, C., Yu.M. Bunkov, S.N. Fisher, H. Godfrin and G.R. Pickett, 1998a, J. Low Temp. Phys. **110**, 13.

Bäuerle, C., Yu.M. Bunkov, S.N. Fisher and H. Godfrin, 1998b, Phys. Rev. B **57**, 14381.

Baym, G. and E. Chandler, 1983, J. Low Temp. Phys. **50**, 57.

Beckurts, K.H. and K. Wirtz, 1964, Neutron Physics (Springer Verlag, Berlin, Germany).

Bell, J.S. and R. Jackiw, 1969, Nuovo Cim A **60**, 47.

Bennett, C.L., et al., 2003, Astrophys. J. Suppl. **148**, 1.

Bettencourt, L.M.A., N.D. Antunes and W.H. Zurek, 2000, Phys. Rev. D **62**, 065005.

Bevan, T.D.C., A.J. Manninen, J.B. Cook, A.J. Armstrong, J.R. Hook and H.E. Hall, 1995, Phys. Rev. Lett. **74**, 750.

Bevan, T.D.C., A.J. Manninen, J.B. Cook, H. Alles, J.R. Hook and H.E. Hall, 1997a, J. Low Temp. Phys. **109**, 423.

Bevan, T.D.C., A.J. Manninen, J.B. Cook, J.R. Hook, H.E. Hall, T. Vachaspati and G.E. Volovik, 1997b, Nature **386**, 689.

Bjorken, J.D., 1997, Acta Phys. Polon. B **28**, 2773.

Blaauwgeers, R., V.B. Eltsov, M. Krusius, J.J. Ruohio, R. Schanen and G.E. Volovik, 2000, Nature **404**, 471.

Blaauwgeers, R., V.B. Eltsov, G. Eska, A.P. Finne, R.P. Haley, M. Krusius, J.J. Ruohio, L. Skrbek and G.E. Volovik, 2002, Phys. Rev. Lett. **89**, 155301.

Bowick, M.J., L. Chandar, E.A. Schiff and A.M. Srivastava, 1994, Science **263**, 943.

Bradley, D.I., S.N. Fisher and W.M. Hayes, 1998, J. Low Temp. Phys. **113**, 687.

Bradley, D.I., S.N. Fisher, M.R. Lowe, A.M. Guénault, G.R. Pickett, A. Rahm and R.C.V. Whitehead, 2004, Phys. Rev. Lett. **93**, 235302.

Bray, A., 1994, Adv. Phys. **43**, 357.

Bunkov, Yu.M., 2003, Physica B **329**, 70.

Bunkov, Yu.M. and O.D. Timofeevskaya, 1998a, Phys. Rev. Lett. **80**, 4927.

Bunkov, Yu.M. and O.D. Timofeevskaya, 1998b, J. Low Temp. Phys. **110**, 45.

Campostrini, M., M. Hasenbusch, A. Pelissetto, P. Rossi and E. Vicari, 2001, Phys. Rev. B **63**, 214503.

Caroli, C., P.G. de Gennes and J. Matricon, 1964, Phys. Lett. **9**, 307.

Chattopadhyay, B., M.C. Mahato and S.R. Shenoy, 1993, Phys. Rev. B **47**, 15159.

Chuang, I., R. Durrer, N. Turok and B. Yurke, 1991, Science **251**, 1336.

Cleary, R.M., 1968, Phys. Rev. **175**, 587.

Davis, R.L., 1992, Physica B **178**, 76.

Davis, R.L. and E.P.S. Shellard, 1989, Phys. Rev. Lett. **63**, 2021.

Demircan, E., P. Ao and Q. Niu, 1995, Phys. Rev. B **52**, 476.

Deser, S., R. Jackiw and G. 't Hooft, 1984, Ann. Phys. **152**, 220.

Dodd, M.E., P.C. Hendry, N.S. Lawson, P.V.E. McClintock and C.D.H. Williams, 1998, Phys. Rev. Lett. **81**, 3703.

Donnelly, R.J., 1991, Quantized Vortices in Helium II (Cambridge University Press, Cambridge, UK).

Duan, J.M., 1995, Phys. Rev. Lett. **75**, 974.

Dziarmaga, J., 1998, Phys. Rev. Lett. **81**, 1551.

Dziarmaga, J., 1999, Phys. Rev. Lett. **82**, 4749.

Dziarmaga, J., A. Smerzi, W.H. Zurek and A.R. Bishop, 2003, H. Arodz, J. Dziarmaga, and W. Zurek, eds, Patterns of Symmetry Breaking, in NATO Science Series Vol. 127 (Kluwer Academic Publishers, Dordrecht) 313.

Eltsov, V.B., Krusius, M., Volovik, G.E., 1998, preprint http://arxiv.org/cond-mat/9809125v1.

Eltsov, V.B., T.W.B. Kibble, M. Krusius, V.M. Ruutu and G.E. Volovik, 2000, Phys. Rev. Lett. **85**, 4739.

Eltsov, V.B., R. Blaauwgeers, N.B. Kopnin, M. Krusius, J.J. Ruohio, R. Schanen and E.V. Thuneberg, 2002, Phys. Rev. Lett. **88**, 065301.

Fetter, A.L., 1964, Phys. Rev. 136A, 1488.

Finne, A.P., T. Araki, R. Blaauwgeers, V.B. Eltsov, N.B. Kopnin, M. Krusius, L. Skrbek, M. Tsubota and G.E. Volovik, 2003, Nature **424**, 1022.

Finne, A.P., S. Boldarev, V.B. Eltsov and M. Krusius, 2004a, J. Low Temp. Phys. **135**, 479.

Finne, A.P., V.B. Eltsov, R. Blaauwgeers, Z. Janu, M. Krusius and L. Skrbek, 2004b, J. Low Temp. Phys. **134**, 375.

Fisher, S.N., A.J. Hale, A.M. Guénault and G.R. Pickett, 2001, Phys. Rev. Lett. **86**, 244.

Gal'tsov, D.V. and P.S. Letelier, 1993, Phys. Rev. D **47**, 4273.

Garriga, J. and T. Vachaspati, 1995, Nucl. Phys. B **438**, 161.

Geller, M.J. and J.P. Huchra, 1989, Science **246**, 897.

Gill, A.J. and T.W. Kibble, 1996, J. Phys. A: Math. Gen. **29**, 4289.

Giovannini, M. and E.M. Shaposhnikov, 1997, Phys. Rev. Lett. **80**, 22.

Goldner, L.S. and G. Ahlers, 1992, Phys. Rev. B **45**, 13129.

Greywall, D.S., 1986, Phys. Rev. B **33**, 7520.

Guillou, J.C., Le and J. Zinn-Justin, 1980, Phys. Rev. B **21**, 3976.

Hakonen, P.J., M. Krusius, M.M. Salomaa and J.T. Simola, 1985, Phys. Rev. Lett. **54**, 245.

Harari, D. and A.P. Polychronakos, 1988, Phys. Rev. D **38**, 3320.

Hendry, P.C., N.S. Lawson, R.A.M. Lee, P.V.E. McClintock and C.D.H. Williams, 1994, Nature **368**, 315.

Hendry, P.C., N.S. Lawson and P.V.E. McClintock, 2000, J. Low Temp. Phys. **119**, 249.

Hindmarsh, M. and T. Kibble, 1995, Rep. Prog. Phys. **58**, 477.

Horava, P., 2005, Phys. Rev. Lett. **95**, 016405.

Ibaceta, D. and E. Calzetta, 1999, Phys. Rev. E **60**, 2999.

Iengo, R. and G. Jug, 1995, Phys. Rev. B **52**, 7537.
Iordanskii, S.V., 1964, Ann. Phys. **29**, 335.
Iordanskii, S.V., 1965, Zh. Exp. Teor. Fiz. **49**, 225 [Sov. Phys. JETP 22, 160 (1966)].
Jensen, B. and J. Kuvcera, 1993, J. Math. Phys. **34**, 4975.
Joyce, M. and M. Shaposhnikov, 1997, Phys. Rev. Lett. **79**, 1193.
Kafanov, S.G., A.Ya. Parshin and I.A. Todoshchenko, 2000, Zh. Exper. Teor. Fiz. **118**, 1143 [JETP 91, 991 (2000)].
Kagan, Yu. and B.V. Svistunov, 1994, Zh. Exper. Teor. Fiz. **105**, 353 [Sov. Phys. JETP 78, 187 (1994)].
Kajantie, K., M. Karjalainen, M. Laine, J. Peisa and A. Rajantie, 1998, Phys. Lett. B **428**, 334.
Kao, H.-C. and K. Lee, 1995, Phys. Rev. D **52**, 6050.
Karra, G. and R.J. Rivers, 1998, Phys. Rev. Lett. **81**, 3707.
Keto, J.W., F.J. Soley, M. Stockton and W.A. Fitzsimmons, 1974, Phys. Rev. A **10**, 872.
Kibble, T.W., 1976, J. Phys. A **9**, 1387.
Kibble, T.W.B., 2003, H. Arodz, J. Dziarmaga, and W. Zurek, eds, Patterns of Symmetry Breaking, in NATO Science Series Vol. 127 (Kluwer Academic Publishers, Dordrecht) 3.
Kibble, T.W.B., 2004, preprint http://arxiv.org/astro-ph/0410073v2.
Kibble, T.W.B. and A. Rajantie, 2003, Phys. Rev. B **68**, 174512.
Kibble, T.W. and G.E. Volovik, 1997, Pis'ma Zh. Exp. Teor. Fiz. **65**, 96 [JETP Lett. 65, 102 (1997)].
Kirtley, J.R., C.C. Tsui and F. Tafuri, 2003, Phys. Rev. Lett. **90**, 257001.
Kleinert, H., 1980, J. Low Temp. Phys. **39**, 451.
Kleinert, H., 1989, Gauge Fields in Condensed Matter (World Scientific, Singapore).
Kleinert, H. and B. Van den Bossche, 2001, Phys. Rev. E **63**, 056113.
Klinkhamer, F.R. and G.E. Volovik, 2004, JETP Lett **80**, 343.
Kobzarev, I.Yu., L.B. Okun and Ya.B. Zeldovich, 1974, Zh. Exp. Teor. Fiz. **67**, 3 [Sov. Phys. JETP **40**, 1 (1975)].
Kondo, Y., J.S. Korhonen, M. Krusius, V.V. Dmitriev, E.V. Thuneberg and G.E. Volovik, 1992, Phys. Rev. Lett. **68**, 3331.
Koplik, J. and H. Levine, 1993, Phys. Rev. Lett. **71**, 1375.
Kopnin, N.B., 1978, Pis'ma Zh. Exp. Teor. Fiz. **27**, 417 [JETP Lett. 27, 390 (1978)].
Kopnin, N.B., 1993, Phys. Rev. B **47**, 14354.
Kopnin, N.B., 1995, Physica B **210**, 267.
Kopnin, N.B., 2002, Rep. Prog. Phys. **65**, 1633.
Kopnin, N.B. and M.M. Salomaa, 1991, Phys. Rev. B **44**, 9667.
Kopnin, N.B. and E.V. Thuneberg, 1999, Phys. Rev. Lett. **83**, 116.
Kopnin, N.B. and V.M. Vinokur, 1998, Phys. Rev. Lett. **81**, 3952.
Kopnin, N.B. and G.E. Volovik, 1998, Phys. Rev. B **57**, 8526.
Kopnin, N.B., G.E. Volovik and Ü. Parts, 1995, Europhys. Lett. **32**, 651.
Kopu, J. and M. Krusius, 2001, Phys. Rev. B **64**, 094504.
Kopu, J., R. Schanen, R. Blaauwgeers, V.B. Eltsov, M. Krusius, J.J. Ruohio and E.V. Thuneberg, 2000, J. Low Temp. Phys. **120**, 213.
Korhonen, J.S., A.D. Gongadze, Z. Janu, Y. Kondo, M. Krusius, Yu.M. Mukharsky and E.V. Thuneberg, 1990, Phys. Rev. Lett. **65**, 1211.
Korhonen, J.S., Y. Kondo, M. Krusius, E.V. Thuneberg and G.E. Volovik, 1993, Phys. Rev. B **47**, 8868.
Krusius, M., E.V. Thuneberg and Ü. Parts, 1994, Physica B **197**, 376.
Krusius, M., Vachaspati, T., Volovik, G.E., 1998, preprint http://arxiv.org/cond-mat/9802005.
Laguna, P. and W.H. Zurek, 1997, Phys. Rev. Lett. **78**, 2519.
Lee, K.-M., 1994, Phys. Rev. D **49**, 4265.

Leggett, A.J., 1984, Phys. Rev. Lett. **53**, 1096.
Leggett, A.J., 1992, J. Low Temp. Phys. **87**, 571.
Leggett, A.J., 2002, J. Low Temp. Phys. **126**, 775.
Linde, A., 1990, Particle Physics and Inflationary Cosmology (Harwood Academic Publishers, Switzerland).
Lipa, J.A., J.A. Nissen, D.A. Stricker, D.R. Swanson and T.C.P. Chui, 2003, Phys. Rev. B **68**, 174518.
Lipa, J.A., D.R. Swanson, J.A. Nissen, T.C.P. Chui and U.E. Israelsson, 1996, Phys. Rev. Lett. **76**, 944.
Maniv, A., E. Polturak and G. Koren, 2003, Phys. Rev. Lett. **91**, 197001.
Mazur, P.O., 1986, Phys. Rev. Lett. **57**, 929.
Mazur, P.O., 1987, Phys. Rev. Lett. **59**, 2380.
Meyer, J.S. and T. Sloan, 1997, J. Low Temp. Phys. **108**, 345.
Mino, Y., Sasaki, M., Tanaka, T., 1997. Prog. Theor. Phys. Suppl. **128**, 373; preprint http://arxiv.org/gr-qc/9705073.
Nelson, D.R. and J.M. Kosterlitz, 1977, Phys. Rev. Lett. **39**, 1201.
Nemirovskii, S.K. and W.M. Fiszdon, 1994, Physica B **197**, 290.
Onsager, L., 1949, Nuovo Cimento **6**(Suppl. 2), 249.
Packard, R.E., 1998, Rev. Mod. Phys. **70**, 641 and references therein.
Parts, Ü., Y. Kondo, J.S. Korhonen, M. Krusius and E.V. Thuneberg, 1993, Phys. Rev. Lett. **71**, 2951.
Parts, Ü., V.M. Ruutu, J.H. Koivuniemi, Yu.M. Bunkov, V.V. Dmitriev, M. Fogelström, M. Huebner, Y. Kondo, N.B. Kopnin, J.S. Korhonen, M. Krusius, O.V. Lounasmaa, P.I. Soininen and G.E. Volovik, 1995, Europhys. Lett. **31**, 449.
Peacock, J.A., S. Cole, P. Norberg, C.M. Baugh, J. Bland-Hawthorn, T. Bridges, R.D. Cannon, M. Colless, C. Collins, W. Couch, G. Dalton, K. Deeley, R. de Propris, S.P. Driver, G. Efstathiou, R.S. Ellis, C.S. Frenk, K. Glazebrook, C. Jackson, O. Lahav, I. Lewis, S. Lumsden, S. Maddox, W.J. Percival, B.A. Peterson, I. Price, W. Sutherland and K Taylor, 2001, Nature **410**, 169.
Rajantie, A., 1998, Physica B **255**, 108.
Rajantie, A., 2001, J. Low Temp. Phys. **124**, 5.
Rajantie, A., Kajantie, K., Karjalainen, M., Laine, M., Peisa, J., 1998, preprint http://arxiv.org/hep-lat/9807042.
Rivers, R.J., 2000, Phys. Rev. Lett. **84**, 1248.
Roberge, A., 1989. Finite Density Effects in Gauge Theories. Ph.D. Thesis, University of British Columbia.
Ruutu, V.M., V.B. Eltsov, A.J. Gill, T.W. Kibble, M. Krusius, Yu.G. Makhlin, B. Plaçais, G.E. Volovik and Xu Wen, 1996a, Nature **382**, 334.
Ruutu, V.M., V.B. Eltsov, M. Krusius, Yu.G. Makhlin, B. Plaçais and G.E. Volovik, 1998a, Phys. Rev. Lett. **80**, 1465.
Ruutu, V.M.H., J. Kopu, M. Krusius, Ü. Parts, B. Plaçais, E.V. Thuneberg and Xu Wen, 1996b, Czechoslovak J. Phys. **46**(Suppl. S1), 7.
Ruutu, V.M.H., J. Kopu, M. Krusius, Ü. Parts, B. Plaçais, E.V. Thuneberg and Xu Wen, 1997a, Phys. Rev. Lett. **79**, 5058.
Ruutu, V.M., Ü. Parts, J.H. Koivuniemi, N.B. Kopnin and M. Krusius, 1997b, J. Low Temp. Phys. **107**, 93.
Ruutu, V.M., Ü. Parts, B. Plaçais, Xu Wen, G.E. Volovik and M. Krusius, 1996c, Czechoslovak J. Phys. **46**(Suppl. S1), 15.
Ruutu, V.M., J.J. Ruohio, M. Krusius, B. Plaçais and E.B. Sonin, 1998b, Physica B **255**, 27.
Salomaa, M.M. and G.E. Volovik, 1987, Rev. Mod. Phys. **59**, 533.

Samuels, D.C., 1992, Phys. Rev. B **46**, 11714.
Schiffer, P., M.T. O'Keefe, M.D. Hildreth, H. Fukuyma and D.D. Osheroff, 1992, Phys. Rev. Lett. **69**, 120.
Schiffer, P. and D.D. Osheroff, 1995, Rev. Mod. Phys. **67**, 491.
Schiffer, P.E., D.D. Osheroff and A.J. Leggett, 1995, W. Halperin, ed., Prog. Low Temp. Phys. Vol. XIV (Elsevier Science BV, Amsterdam) 159 and references therein.
Schwarz, K.W., 1978, Phys. Rev. B **18**, 245.
Schwarz, K.W., 1985, Phys. Rev. B **31**, 5782.
Schwarz, K.W., 1988, Phys. Rev. B **38**, 2398.
Shelankov, A., 1998, Europhys. Lett. **43**, 623 preprint http://arxiv.org/cond-mat/9802158.
Sonin, E.B., 1975, Zh. Exp. Teor. Fiz. **69**, 921 [Sov. Phys. JETP 42, 469 (1976)].
Sonin, E.B., 1997, Phys. Rev. B **55**, 485.
Sonin, E.B., V.H. Geshkenbein, A. van Otterlo and G. Blatter, 1998, Phys. Rev. B **57**, 575.
Spergel, D.N., L. Verde, H.V. Peiris, E. Komatsu, M.R. Nolta, C.L. Bennett, M. Halpern, G. Hinshaw, N. Jarosik, A. Kogut, M. Limon, S.S. Meyer, L. Page, G.S. Tucker, J.L. Weiland, E. Wollack and E.L. Wright, et al., 2003, Astrophys. J. Suppl. **148**, 175.
Starkman, G.D., 1996, Phys. Rev. D **53**, 6711.
Staruszkievicz, A., 1963, Acta Phys. Polon. **24**, 734.
Stockton, M., J.W. Keto and W.A. Fitzsimmons, 1971, Phys. Rev. A **5**, 372.
Stone, M., 1996, Phys. Rev. B **54**, 13222.
Strösser, M. and V. Dohm, 2003, Phys. Rev. E **67**, 056115.
Tang, Y.H., I. Hahn, H.M. Bozler and C.M. Gould, 1991, Phys. Rev. Lett. **67**, 1775.
Thorne, K.S. and J.B. Hartle, 1985, Phys. Rev. D **31**, 1815.
Tough, J.T., 1982, W. Halperin, ed., Prog. Low Temp. Phys. Vol. XIII (Elsevier Science BV, Amsterdam) 133 and references therein.
Toyoki, H., 1994, J. Phys. Soc. Jpn. **63**, 446.
Tsubota, M. and H. Yoneda, 1995, J. Low Temp. Phys. **101**, 815.
Unruh, W.G., 1976, Phys. Rev. D **14**, 870.
Vachaspati, T., 1994, in: Proc. NATO Workshop, Electroweak Physics and the Early Universe, Sintra, Portugal, Vol. 338, Series B: Physics (Plenum Press, New York).
Vachaspati, T. and G.B. Field, 1994, Phys. Rev. Lett. **73**, 373.
Vachaspati, T. and G.B. Field, 1995, Phys. Rev. Lett. **74**, 1258.
Vachaspati, T. and A. Vilenkin, 1984, Phys. Rev. D **30**, 2036.
van Otterlo, A., M.V. Feigel'man, V.B. Geshkenbein and G. Blatter, 1995, Phys. Rev. Lett. **75**, 3736.
Varoquaux, E., O. Avenel, P. Hakonen and Yu. Mukharsky, 1998, Physica B **255**, 55 and references therein.
Villain, J., 1975, J. Physique (Paris) **36**, 581.
Vollhardt, D., Wölfle, P., 1990, The Superfluid Phases of Helium 3 (Taylor & Francis, London).
Vollhardt, D., K. Maki and N. Schopohl, 1980, J. Low Temp. Phys. **39**, 79.
Volovik, G.E., 1990, in: Helium Three, ed W. Halperin, and L. Pitaevskii (Elsevier Science BV, Amsterdam) 27.
Volovik, G.E., 1992, Exotic Properties of Superfluid ^{3}He (World Scientific, Singapore).
Volovik, G.E., 1996, Czechoslovak J. Phys. **46**(Suppl. S6), 3048.
Volovik, G.E., 1997, Pis'ma Zh. Exp. Teor. Fiz. **65**, 201 [JETP Lett. 65, 217 (1997)].
Volovik, G.E., 1998, Physica B **255**, 86.
Volovik, G.E., 2003, The Universe in a Helium Droplet (Oxford Science Publications, Oxford).
Volovik, G.E. and V.P. Mineev, 1982, Zh. Exp. Teor. Fiz. **83**, 1025 [Sov. Phys. JETP 56, 579 (1982)].
Volovik, G.E. and T. Vachaspati, 1996, Int. J. Mod. Phys. B **10**, 471.

Wexler, C., 1997, Phys. Rev. Lett. **79**, 1321.
Wexler, C., Thouless, D.J., 1996, preprint http://arxiv.org/cond-mat/9612059
Wheatley, J.C., 1975, Rev. Mod. Phys. **47**, 415.
Wilczek, F., 1998, Phys. Today 51, N1 (January), 11.
Williams, G.A., 1993a, Phys. Rev. Lett. **71**, 392.
Williams, G.A., 1993b, J. Low Temp. Phys. **93**, 1079.
Williams, G.A., 1999, Phys. Rev. Lett. **82**, 1201.
Winkelmann, C.B., Yu.B. Bunkov and H. Godfrin, 2005, Gravitation & Cosmology **11**, 1.
Witten, E., 1985, Nucl. Phys. B **249**, 557.
Wen, Xu, B. Plaçais, V.M. Ruutu and M. Krusius, 1996, Czechoslovak J. Phys. **46**(Suppl. S1), 11.
Yates, A. and W.H. Zurek, 1998, Phys. Rev. Lett. **80**, 5477.
Zapotocky, M., P.M. Goldbart and N. Goldenfeld, 1995, Phys. Rev. E **51**, 1216.
Zurek, W.H., 1985, Nature **317**, 505.
Zurek, W.H., 1996, Phys. Rep. **276**, 177.

CHAPTER 2

ON THE HEAVY FERMION ROAD

BY

J. FLOUQUET

Institut de Physique de la Matière Condensée CEA – CNRS CNRS – BP 166X – 38042 Grenoble cedex France

Progress in Low Temperature Physics, Volume XV

ISSN: 0079-6417 DOI: 10.1016/S0079-6417(05)15002-1

Contents

1. Heavy Fermion instabilities

Essential points:

- Three singular pressures P_{KL}, P_C and P_V as $T \rightarrow 0$ K which are related respectively to 4f localisation, the magnetic instability, and valence (or orbital) fluctuation.
- The static description of the Doniach Kondo lattice.
- A first glance at the spin fluctuation approach and the non-Fermi liquid label.
- The possibility of a Kondo condensate. Over a Kondo coherence length $\ell_{KL} \sim m^*$, the spin memory may be preserved for long times, $\tau_{KL} \approx m^{*2}$.

1.1. *Introduction*

The heavy fermion compounds (HFC) belong to a large class of strongly correlated electronic systems (SCES) which also covers 3d intermetallic systems, organic conductors and the high-temperature oxide superconductors. They are also linked to the quantum matter of ^{3}He and systems, like manganite compounds, where the magnetic coupling affects the electronic conduction. Despite three decades of study, there are still some mysteries concerning the charge and spin dynamics. However, major results have been obtained on these specific materials with broad implications in condensed matter physics.

This chapter is more of our Grenoble laboratory's report on how to track the electronic excitations (charge and spin) and the nature of the ground states, than a review that covers all the published works. Furthermore, the approach is that of an experimentalist familiar with low-temperature physics, i.e. the main motivation is to clarify the complex nature of heavy fermions but with references to general basic questions. Most of the figures correspond to data obtained either by us or by collaborators. Of course, references are given to the original discoveries. Our approach was additionally motivated by the possibility to add new experimental insight.

For young researchers, a good introduction to the Kondo problem can be found in Anderson (1967), an excellent introduction to unconventional Cooper pairing is the review article on ^{3}He (Leggett 1975). Unconventional superconductivity is treated theoretically in Mineev and Samokin (1999) and Gorkov (1987). Reviews devoted to heavy-fermion systems can be found in Brandt and Moschalkov (1984), Grewe and Steglich (1991), Fulde et al. (1988), Springford (1997), Ott (1987) and Kuramoto and Kitaoka (2000).

A good summary can be found in Heffner and Norman (1996). Extensive discussions on unconventional superconductivity and magnetism have been recently given in the review of Thalmeier and Zwicknagl (2004) and Thalmeier et al. (2004). Recent lecture notes have been published (Aliev et al. 2001, Coleman 2002). A discussion on singular non-Fermi liquid can be found in Varma et al. (2002). Up to date points of view on Kondo problems can be found in the Kondo Festschrift edited recently (see Kondo 2005). Our favorite book on HFC is unfortunately up to now available only in Japanese (Ueda and Onuki 1998). Popular articles are Hess et al. (1993), Cox and Maple (1995), and Fisk et al. (1998). With these different reviews, the reader will see that the selection of material is often a question of personal taste. Thalmeier et al. (2004) have considered that the new ferromagnetic superconductors do not belong to the heavy fermion class, a subject extensively discussed in Section 5.

After discussing the link with the Kondo impurity problem, the relationship with the intermediate valence compounds (IVC), the relevance of spin fluctuations and the key issues in the Kondo lattice (Fermi surface, magnetic and valence instabilities), we will concentrate on the cerium heavy fermion normal phase properties. We will focus on the appearance of superconductivity in the vicinity of the antiferromagnetic (AF) instability (at a critical pressure P_C) for three-dimensional (3d) and quasi–2d compounds, notably the new 115 series discovered recently in Los Alamos.

The studies of transuranium compounds have been very successful for the understanding of unconventional superconductivity. The discovery of high T_C superconductivity in $PuCoIn_5$, again in Los Alamos, illustrates the interplay between the electronic bandwidth and magnetic fluctuations in order to optimize the superconducting critical temperature T_C. The recent observation of the coexistence of superconductivity and ferromagnetism (F) in UGe_2 has led to a discovery of new materials with these tendencies, as well as to a revival of theoretical interest in ferromagnetic superconductors. The data on the four archetypal heavy-fermion uranium superconductors UPt_3, UPd_2Al_3, URu_2Si_2 and UBe_{13} will be examined with special focus on determination of the superconducting order parameters (UPt_3, UPd_2Al_3) on the low-temperature excitation characteristic of unconventional pairing (UPt_3) and on the respective temperature (T), magnetic field (H) and pressure (P) phase diagrams.

1.2. *Localization, valence and magnetism*

The cerium HFC are just at the frontier of classical rare-earth intermetallic systems (here the localized 4f electrons and light itinerant electron form two different baths) (Elliot 1972) and of IVC (Wachter 1993, Newns and Read 1987), where strong mixing occurs at the Fermi level between the two types of electrons. In the IVC case, the occupancy n_f of the electron in the 4f shell is less than unity and its valence on a given Ce site, $v = 4 - n_f$, is clearly intermediate. In HFC, it is the weak departure from $n_f = 1$ which leads to strong low-energy magnetic fluctuations at the Fermi level and also to the memory of local 4f character.

We will first discuss the case of Ce metal where a discontinuity in n_f occurs in pressure and temperature. Our idea is that this discontinuity may be less dramatic and even smooth in HFC but as $T \rightarrow 0\,K$, it will correspond to a pressure P_V where the 4f electron loses its local sensitivity to the environment and notably the lifting of its angular momentum degeneracy by the crystal field. Another pressure, $P_{KL} < P_V$, will correspond to the critical pressure above which the 4f electron is included in the volume of the Fermi surface. The duality between the localized and itinerant part of the f electron is the core of the heavy fermion problem. In the duality model introduced by Miyake and Kuramoto (1990) and Kuramoto and Miyake (1990), the aim is to use the known results for the single site and to add the renormalization flow for the low-temperature regime (see later).

The cerium metal (T, P) phase diagram shows occurrence of a high-temperature trivalent γ phase ($n_f = 1$) and of a low-temperature α phase ($n_f \sim 0.9$). A first-order isostructural line $T_{\gamma\alpha}$ ends up at a critical point around $T_{cr} = 600\,K$, $P_{cr} \sim 22\,kbar$ (Jayaraman 1965). As the intercept $T_{\gamma\alpha}$ $\sim 100\,K$ at $P = 0$ (fig. 1) is high by comparison to any hypothetical magnetic ordering temperature T_N, the magnetism was treated crudely (Lavagna et al. 1982, Allen and Martin 1982). Recent theoretical developments can be found in Held et al. (2001) and in recent publications using a new approach for the band structure (see later).

However, if the volume can be expanded to negative pressure the linear extrapolation of $T_{\gamma\alpha}(P)$ to zero occurs at a low negative pressure $P_{KL} \sim -0.3\,GPa$. Of course, at $T = 0\,K$, the initial slope $dT/dP = \Delta V/\Delta S$ of the first-order transition (ΔS and ΔV respectively entropy and volume discontinuity) must be vertical, according to the Clausius–Clapeyron relation. At $T = 0\,K$ no entropy discontinuity can exist and just the volume jump remains. Near P_{KL}, an important feature is the possibility of long-range magnetic ordering (antiferromagnetic (AF) or ferromagnetic (FM)) for $P < P_C$. P_C can coincide or not with P_{KL}. In the paramagnetic phase (PM) the 4f electron is found to be itinerant ($P_{KL} < P_C$).

In usual HFC, both P_{KL} and P_C are positive, and the interplay between magnetism and the localization of the f electron is central. In contrast to the case of Ce metal, there are only a few cases of a first-order transition between γ and α phases in HFC. That may be due to the presence of other

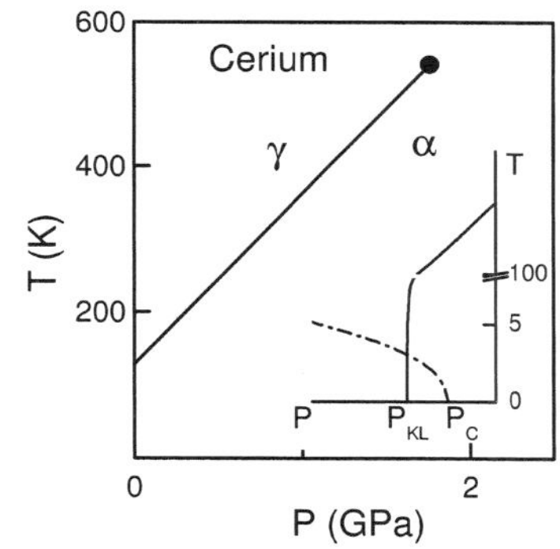

Fig. 1. The simplified (T, P) phase diagram of Ce metal full line: the first-order γ–α transition. When it is extended to negative pressure, in the inset, it ends up at P_{KL} for $T \rightarrow 0$ K. The dashed line represents the hypothetical variation of its magnetic ordering temperature which will collapse at P_C. In HFC, $T_{\gamma\alpha}$ ($T \rightarrow 0$) will become positive. T_{cr} will collapse to 0 K. The memory of P_{cr} will appear in P_V. The cascade of instabilities may be $P_{KL} < P_C < P_V$.

ligand ions in the lattice and consequently to complex electronic structures with a large number of bands. The P_{KL} hypothesis gives the possibility to discuss the localization of the f electron, notably its contribution to the Fermi surface. The pressure P_C marks the disappearance of long-range magnetism. The usual consensus is that P_C is a second-order transition at a quantum critical point (QCP). Evidence will be given that a first-order transition may occur at P_C. Furthermore, if P_{KL} and P_C are two first-order singularities, phase separation may appear between magnetic and paramagnetic phases (see $CeIn_3$).

In HFC, the temperature T_{cr} has dropped and often vanishes to a negative value and the pressure P_{cr} may be associated with large valence or orbital fluctuations (see $CeCu_2Si_2$, $CeCu_2Ge_2$). Our physical intuition is that it collapses with the pressure P_V where the f electron loses its sensitivity to the crystal field environment since above P_V its angular momentum becomes quenched by the electronic Kondo coupling. In the case of cerium ions at low temperatures, below P_V, the effective spin of the 4f moment is 1/2 while above P_V the full degeneracy $J = 5/2$ must be taken into account. That will wash out the intersite magnetic coupling and restores a situation of strong mixing between the electrons without magnetic correlation. This intermediate valence regime corresponds to a Kondo temperature $T_K \geqslant 100$ K and $n_f < 0.9$ (see below).

Historically, the research on IVC was very active three decades ago. As it involves rather high-energy scales ($T_K > 100$ K), the magnetism was very often ignored. The field of heavy fermion systems began with the discovery

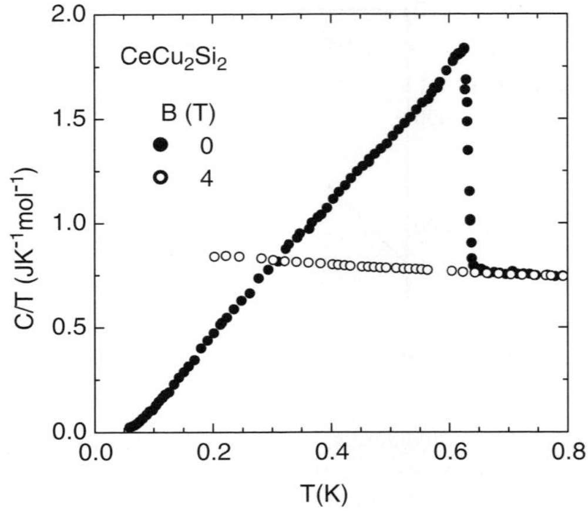

Fig. 2. Thermal variation of C/T of $CeCu_2Si_2$ at (•) $H = 0$ and (○) $H = 4\,T$ above $H_{c_2}(0)$ (Helfrich 1995, Steglich et al. 1996) on a superconducting crystal without A phase component. The complicated interplay between A and S phases is discussed in Gegenwart et al. (1998).

of the huge value of the linear temperature coefficient $\gamma \sim 1500\,\mathrm{mJ\,mol^{-1}\,K^{-2}}$ of the specific heat $C = \gamma T$ in $CeAl_3$ (Andres et al. 1975). The discovery of the first superconducting heavy fermion superconductor (HFS) ($CeCu_2Si_2$) was reported in 1979 (fig. 2) (Steglich et al. 1979). The importance of this observation was boosted by the successive reports of superconductivity in uranium compounds UBe_{13} (Ott et al. 1983), URu_2Si_2 (Schlabitz et al. 1986) and UPt_3 (Stewart et al. 1984). The possible link of superconductivity with the magnetic instability as $T \rightarrow 0\,\mathrm{K}$, i.e. to the critical density or pressure (P_C) was clear in the pioneering pressure experiments on $CeCu_2Ge_2$ (Jaccard et al. 1992) and reinforced by the observation of superconductivity in $CePd_2Si_2$ and $CeIn_3$ (Mathur et al. 1998). Direct evidence of heavy quasiparticles was realized in UPt_3 (Taillefer and Lonzarich 1988); effective masses m^* up to $100m_o$ were detected (m_o is the free electronic mass). It was a major breakthrough as it really demonstrated that heavy fermion particles move on Fermi surface orbits.

1.3. *From Kondo impurity to Kondo lattice*

HFC are complex systems where a large effective mass m^* appears due to the slow motion of the f electron by hybridization with the light electron. This magnification is reminiscent of the Kondo effect observed for a single magnetic impurity in a non-magnetic normal host. Below a characteristic temperature T_K, a strong coupling occurs leading locally to the disappearance of the magnetism. For $n_f \sim 1$ the coupling of the localized spin S and the spin of the conduction electron s can be reduced to an exchange term (see Blandin 1973):

$$H = -\Gamma \vec{S} \cdot \vec{s}.$$

T_K is related to Γ and to the density of states of the light electrons $N(E_F)$ at the Fermi level:

$$T_K = \frac{1}{N(E_F)} \exp\left(-\frac{1}{\Gamma N(E_F)}\right).$$

The negative value of Γ is related to the position E_0 of the virtual f level relative to the Fermi level and its width Δ_0 according to the relation

$$\Gamma N(E_F) = \frac{\Delta_0}{E_0}$$

when the on-site Coulomb potential tends to infinity.

At low temperature, Fermi liquid behavior replaces the high-temperature Curie like paramagnetic behavior. As $T \to 0$ K, the specific heat, the susceptibility and the resistivity vary as

$$\gamma = \frac{C}{T} = \frac{1}{T_K},$$

$$\chi \approx \frac{1}{T_K},$$

$$\rho = \rho_0 \left(1 - a\left(\frac{T}{T_K}\right)^2\right).$$

The resistivity reaches a maximum at $T = 0$ which corresponds to the unitary limit (~200 μΩ cm per impurity for a metal with one carrier per mole) (see Anderson 1967). The beauty of the single Kondo impurity problem is that, due to the coupling with the Fermi sea of the electronic bath, the entropy can collapse as $T \to 0$ K without further coupling with other impurities.

A supplementary effect is that the local interaction, mediated by the polarization of the Kondo singlet, leads to the famous enhancement (the Wilson ratio $R = 2$ for $S = 1/2$) of the susceptibility χ over C/T with respect to the free electron value R_0 (see Nozières 1974; Hewson 1992):

$$R = 2R_0 \quad \text{for } S = 1/2 \quad \text{with } R_0 = \frac{3(g\mu_B)^2}{4\pi^2 k_B^2}.$$

For 4f electrons, the spin–orbit interaction $\lambda_{SO}\vec{L}\cdot\vec{S}$ between the $L = 3$ orbital momentum and the spin S governs the formation of the magnetic momentum J carried by the particle. For less than a half-filled 4f shell, λ_{SO} is positive and thus L and S are antiparallel. For the cerium ($4f^1$) case, the $J = 5/2$ level lies roughly 0.3 eV above the $J = 7/2$ excited level. For the trivalent Yb ions ($4f^{13}$), the spin–orbit interaction is negative, the ground state has an angular momentum $J = 7/2$. Due to further coupling with the environment, the effective degeneracy of the level changes for $P < P_V$; often an effective doublet will be the crystal field ground state. The Coqblin and Schrieffer (1969) model was developed to take into account an orbitally degenerate site. It was used extensively to discuss the competition between Kondo effect (energy $k_B T_K$) and the crystal field splitting Δ_{CF} (Cornut and Coqblin 1972). If $k_B T_K < \Delta_{CF}$, the low-temperature Kondo impurity problem may correspond to the ideal case of a doublet. The proximity to the nonmagnetic unfilled f level of La explains the importance of hybridization and thus of a Kondo mechanism. When $k_B T_K > \Delta_{CF}$, only the full $2J + 1 = 6$ degeneracy (N_f) of the 4f level of the trivalent configuration must be considered over the entire temperature range. The magnetic interaction drops drastically compared to $k_B T_K$ (Ramakrishnan and Sur 1982). The change from $k_B T_K < \Delta_{CF}$ to $k_B T_K > \Delta_{CF}$ is induced at $P = P_V$ under pressure; as we will show, T_K increases under pressure while the crystal field splitting is weakly pressure-dependent (Schilling 1979, Thompson and Lawrence 1994). The relative strength of the Kondo temperature with respect to other energy scales such as the hyperfine coupling (Flouquet 1978), the intersite coupling (Doniach 1977), the pair interaction (Jones et al. 1988) or the crystal field splitting is the key parameter to define the ground state properties. By comparison with the cerium impurity, the problem of the 3d Kondo impurity like Mn, Fe or Co is far more complex, as the nature of the magnetism far above T_K involves already the difficult unsolved question of orbital quenching. The experimental irony is that the study of the rare-earth Kondo impurity was undertaken much later than that of the 3d elements and almost by the time when the Kondo problem was solved theoretically (Nozières 1974, Wilson 1975, Yamada 1975, see also Kondo et al. 2005).

Let us stress the feedback of T_K with the valence. The Kondo phenomena is linked with the release of $1-n_f$ from 4f shell to the Fermi sea. In the so-called $1/N_f$ expansion, the Kondo energy has been expressed as a function of n_f and Δ_0 (Hewson 1992):

$$k_B T_K = \frac{1-n_f}{n_f}\Delta_0 N_f.$$

A large T_K corresponds to a low 4f occupancy. Under pressure, n_f will decrease and thus T_K will increase. To discuss the strong pressure dependence of T_K, the pressure variation of Δ_0 must be known. This is not so obvious from first principles. In the Anderson Hamiltonian, the width Δ_0 is connected to the mixing potential V_{df} between the 4f and the d light electron and to $N(E_F)$ by the relation

$$\Delta_0 = \pi V_{df}^2 N(E_F).$$

From high-energy spectroscopy (Malterre et al. 1996), Δ_0 increases under pressure as the Fermi energy $k_B T_F$ does. So let us assume the proportionality $\Delta_0 = 10^{-2} k_B T_F$. The strong pressure response of T_K is due to the weakness of $\Delta / k_B T_F \sim 10^{-2}$ by comparison to $E_o / k_B T_F \sim 0.2$. Neglecting the pressure shift of E_o toward the Fermi level, the Kondo formula with $\Delta = 10^{-2} k_B T_F$ leads to a Kondo–Grüneisen parameter:

$$\Omega_{T_K} = -\frac{\partial \mathrm{Log}\, T_K}{\partial \mathrm{Log}\, V} = 10 \Omega_{T_F}.$$

If the shift of E_o with pressure is taken into account, Ω_{T_K} will be again enhanced. Neglecting the degeneracy dependence, the physical insight is that the large value of Ω_{T_K} is linked to the quasi-trivalence of the Ce ion ($n_f = 1$). A weak relative variation of n_f magnifies the relative increase in $\partial T_K / T_K$ by $(1-n_f)^{-1}$. Assuming that n_f varies linearly in the volume according to Vegard's law with a volume difference of 50% between the Ce^{3+} and Ce^{4+} configuration, Ω_{T_K} reaches 20 for $n_f = 0.9$. For cerium intermetallic compounds n_f does not drop below 0.8–0.85 (Malterre et al. 1996). As the $4f^0$ and $4f^1$ configurations of the cerium atoms are separated by 2 eV, it will cost too much kinetic energy to drop further n_f. For the Sm, Tm or Yb cases, the separation between the two valence states (2+ or 3+) are far less (100 meV) and thus the valence can vary by 1.

Now for a regular array of Ce ions, the Kondo lattice, an extra temperature scale T_{KL} may appear below T_K as the electronic reservoir (carrier number n_e) cannot be regarded as an independent infinite bath and furthermore intersite magnetic interactions must be considered. For example, neglecting extra sources of light conduction electrons than the release or absorption of an electron on the 4f electron, the valence equilibrium of Ce^{3+} or Yb^{3+},

$$Ce^{3+} \rightleftharpoons Ce^{4+} + 5d,$$

$$Yb^{2} \rightleftharpoons Yb^{3+} + 5d,$$

leads to a quite different Fermi temperature (T_F) for the 5d electrons (resp. proportional to $(1-n_f)^{2/3}$ and $n_f^{2/3}$). Consequently, there are strong differences in T_{KL} for the trivalent configuration if an extrapolation $k_B T_{KL} \sim (1-n_f) D(n_f)$ with $D(n_f) \sim 10^{-2} k_B T_F$ is made for the lattice; $D(n_f)$

represents an unrenormalized 4f band proportional to the Fermi temperature of the 5d electrons (Flouquet et al. 2005a). In this naïve frame, the traveling electron seems to be the 5d electron. It gets its effective mass $(1-n_f)^{-1}$ by jumping from the 4f shell; its Fermi temperature goes as $(1-n_f)^{+2/3}$ or $n_f^{+2/3}$ depending on the valence equilibrium. The image is similar to the motion of x atoms of He^3 ($x<0.05$) in dilute solution in liquid ^4He (see Lounasmaa 1974).

A full understanding of the low-energy spectrum of the Kondo lattice even in the PM state has not been given. The difficult controversial point (see Bergmann 1991) is the so-called exhaustion principle (Nozières 1985) which points out the impossibility to conserve a rigid picture of the Kondo screening on each site as the required number of available itinerant electrons $N(E_F)xk_BT_K$ is far lower than 1. Within a large N_f approach, in the framework of the slave boson technique, analytical calculations show that $T_{KL}{\sim}T_K n_e^{1/3}$ at low carrier content n_e (Burdin et al. 2000). The ratio T_{KL}/T_K does not depend on the T_K strength. That is not so surprising since magnetic correlations are treated roughly, i.e. the coherence length is restricted to atomic distances. For experimentalists, the message is that the relation between γ and the number of 4f sites and carrier is not trivial when $n_e \ll 1$. For an intermediate value of $n_e(0.5)$, the differentiation between T_{KL} and T_K becomes difficult. Discussions on band-filling effects on a Kondo lattice within a mean-field approximation can be found in Coqblin et al. (2003), with references to other approaches to Kondo lattice and the periodic Anderson model.

In contrast, it was proposed (Nozières 1998) from considerations of phase memory that $T_{KL}{\sim}T_K^2$ is independent of n_e. This relation can be found by assuming the motion of the quasiparticles on a finite path ℓ_{KL} extending far above atomic distances, i.e. basically in the vicinity of a QCP. To take into account the motion of the heavy quasiparticle (m^*) and their strong correlation, the simple step is to introduce an extra correlation length ℓ_{KL}. A physical image may be that the quasiparticle of effective mass $m^*{\sim}T_K^{-1}$ circulated along a Kondo loop of length $\ell_{KL}{\sim}m^*$ leading to a lifetime $\tau_{KL}{\sim}m^{*2}{\sim}T_K^{-2}$ (in agreement with Nozières) and even to a small magnetic moment $M_o=(m_0/m^*)\mu_B$ on each visited site (m_0 being the bare electronic mass). For a classical magnetically ordered rare-earth compound $\left(\frac{T_K}{T_N}\to 0\right)$, the length of the magnetic correlation at $T=0\,K$ corresponds to atomic distances and the information is carried from site to site by fast light electrons with $m^*=m_0$. Thus a 10^4 or 10^6 longer time constant for HFC ($\tau_{KL}{\sim}m^{*2}$) will be the result of a slow motion on large distances.

1.4. *The "Doniach model"*

A first discussion of the pressure collapse of long range magnetism in HFC was given by Doniach (1977). In the popular Doniach picture, the interplay is between a local Kondo fluctuation given by $k_B T_K$ and the indirect Ruderman–Kittel–Kasuya–Yosida oscillating interaction (RKKY) E_{ij} between two paramagnetic sites i and j mediated by the conduction electrons:

$$E_{ij} \sim \Gamma^2 N(E_F).$$

As T_K has a strong exponential dependence on Γ and E_{ij} a smoother parabolic dependence, the first idea is that above a critical value of Γ_C, long-range magnetism, either ferromagnetic (FM at T_{Curie}) or antiferromagnetic (AF at T_N), will collapse at P_C (fig. 3). If the collapse is continuous through a second-order transition, then P_C corresponds to a quantum critical point. Such a scenario has been discussed (see later) by Hertz (1976) and revisited in the case of spin fluctuation theory (Moriya 1985, 2003a and Moriya and Takimoto 1995), renormalization group theory (Millis 1993) or in the framework of universality in phase transitions (Continentino 2001).

From a purely phenomenological point of view, the Doniach approach can be related to the discussion of induced magnetism for an array of initially singlet paramagnetic ions characterized by a deep minimum of the energy E_i at zero local magnetization (m_i) ($E_i = am_i^2$) plus a further interaction term $bm_i m_j$, taking into account first- and second-neighbor coupling (Benoit et al. 1978, 1979). Of course, magnetism disappears for a critical value of a/b, i.e. T_K as shown in fig. 3. The problem is quite similar to that

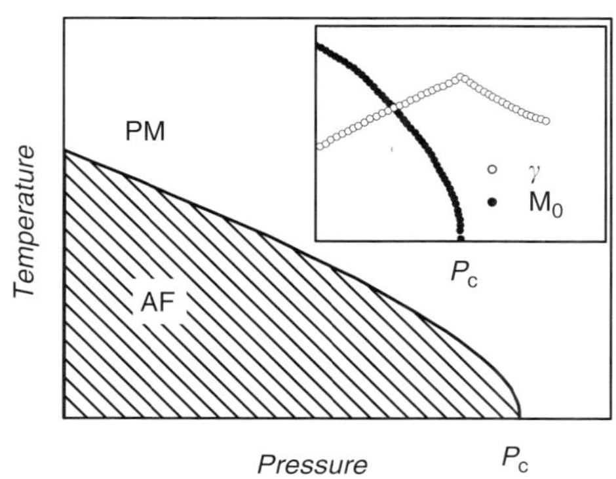

Fig. 3. Phase diagram (T, P at $H = 0$) (Benoit et al. 1979). In this model, the magnetic transition as $T \to 0$ K is first-order with $T < T_K$. The full line gives the dependence of T_N of M_o (●) and of γ(○), as a function of pressure.

for the appearance of magnetism for a singlet crystal field ground state (Wang and Cooper 1968, 1969).

Assuming that, on the Fermi sea, the electronic properties can be derived by a Lorentzian density of states:

$$\gamma = \frac{k_B T_K}{(k_B T_K)^2 + (g\mu_B H_m)^2},$$

where H_m is the molecular field, g the g-factor and μ_B the Bohr magneton a maxima of γ i.e. of the effective mass, will occur at P_C adjusting the relative variation of a and b with the pressure dependence of T_K and E_{ij} (Benoit et al. 1979). The extension of the Doniach model to the magnetic field gives an excellent description of the experiments realized on the magnetic Kondo lattice of $CeAl_2$ (Steglich 1977, Bredl et al. 1978) at $P = 0$.

Let us look at some magnetic Kondo lattices. Figure 4 represents the variation of C/T in the normalized scale T/T_N for three cubic magnetically ordered Kondo lattices $CePb_3$, $CeAl_2$ and $CeIn_3$ (Steglich et al. 1978, Peyrard 1980, Lin et al. 1985, Pietri and Andraka 2000). The compound $CePb_3$ (Vettier et al. 1986) as well as $CeAl_2$ (Barbara et al. 1979) exhibit incommensurate magnetic Bragg reflections which are stable down to 60 mK. The critical exponent associated with the order parameter M (the sublattice magnetization) is the Wilson value 0.3 for the 3d Heisenberg antiferromagnetic whereas in $CeIn_3$ it is near 0.5, the mean field limit. The trend is that, when the ratio T_K/T_N increases by applying pressure, the magnetic Bragg reflection is commensurate with the same propagation vector (1/2, 1/2, 1/2)

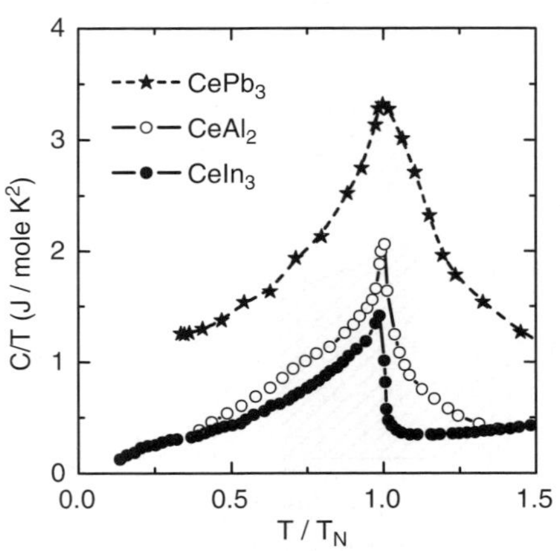

Fig. 4. Specific heat of three magnetically ordered compounds: $CeIn_3$, $CeAl_2$ and $CePb_3$ as a function of T/T_N (Bredl et al. 1978, Peyrard 1980, Pietri and Andraka 2000).

(Morin et al. 1988). The refinement of the magnetic structure of $CeAl_2$ on a single crystal (Forgan et al. 1990) shows that it is not a modulated structure with a single k_0 component but a double k structure involving the composition of two helicoidal modes. The Kondo coupling is invoked for the pressure collapse of M_o. The understanding of the magnetic structure of this system is an interesting topic; an overview of the different facets of exotic structures completely determined by neutron scattering can be found in the review article of Rossat Mignod (1986).

In the two cases of $CeIn_3$ and $CeAl_2$ the estimations γ_{LT} and γ_{HT} of C/T far below T_N and just above T_N are quite similar while for $CePb_3 \gamma_{LT} \gg \gamma_{HT}$ (see the table below):

	T_N (K)	T_K (K)	γ_{LT} (mJ mol^{-1} K^{-2})	γ_{HT} (mJ mol^{-1} K^{-2})	M_0^{exp} (μ_B mol^{-1})	M_0^{th} (μ_B mol^{-1})
$CeAl_2$	3.8	10	120	150	0.53	0.89
$CeIn_3$	10	10	140	140	0.55	0.48
$CePb_3$	1.2	2	1300	200	0.50	0.6

At least in the two first cases, the establishment of the AF phase seems decoupled from the existence of heavy quasiparticles ($\gamma_{LT} = \gamma_{HT}$). Furthermore, the decoupling looks efficient as an evaluation of M_0^{th} in a classical model with localized moments (susceptibility at T_N equal to $\chi(T_N)$). This gives a rather good estimation of the measured value M_0^{exp} according to the relation (Marcenat et al. 1988) $M_0^2 = 2\chi(T_N) \int_0^{T_N} C(T)\,dT$ even for $CePb_3$.

For $CePb_3$, there is a drastic change in C/T through T_N. Two components seem to exist: the ordinary magnetic one and the heavy fermion one. Coherence (crossing through T_N) leads to a rapid approach to the low-temperature limit of C/T ($\gamma_{LT} > \gamma_{HT}$). In paramagnetic HFC above P_C, the increase of C/T in cooling from γ_{HT} to γ_{LT} will occur through a continuous slow process over a large temperature range. One may think that $CePb_3$ at $P = 0$ is near a QCP and thus a slight increase in pressure will drive the system right to the QCP. That seems to be supported by an initial pressure decrease of T_N. But the reality is different. At $P^* = 5$ kbar, T_N increases again. The magnetic structure becomes commensurate. The pressure P_C is pushed above 3 GPa (Morin et al. 1988, Welp et al. 1987, Welp 1988). HFC are rather subtle toys with a number of different possibilities. Notice that even far below $P_C > 3$ GPa at $P = 0$, in its AF phase, $CePb_3$ has a γ term near 1000 mJ mol^{-1} K^{-2}.

1.5. *Spin fluctuations and the non-Fermi properties*

In the almost opposite framework where the f electrons are considered to be completely itinerant ($P_{KL} < P_C$), i.e. characterized by an effective Fermi temperature $T_F \sim T_K$ it is worthwhile to refer to the results of spin fluctuation theory, developed to understand the magnetism of 3d elements (Moriya 1985, Lonzarich and Taillefer 1985) and recently revisited in connection with heavy fermion, organic and high T_C oxide compounds.

For simplicity, let us look at the case of the ferromagnetic instability (fig. 5) (Moriya 1985; Nozières 1986). For $P < P_C$ the ground state is ferromagnetic. Slowly, on approaching P_C, the effective mass m^* will be dressed by spin fluctuations. At P_C, m^* will diverge. In the Hubbard scheme, this will occur when the product $UN(E_F) = I$ of the on-site coulomb repulsion U by the density of states $N(E_F)$ at the Fermi level, approaches 1 ($m^* = \log(1-I)$) (Moriya 1985). Far below P_C, undamped spin waves can be detected. On approaching P_C, they become overdamped. The regime where Fermi liquid properties can be observed will be pushed to the low-temperature T_I which collapses at P_C. Above T_{Curie}, there is a large regime (III) where the paramagnetism has strong variations with temperature. The uniform susceptibility χ_0 follows a $T^{-4/3}$ law. For the singular pressure P_C, the collective singlet never enters into a low-temperature Fermi liquid regime. This leads to the non-Fermi liquid (NFL) label (see recent review of Stewart 2001).

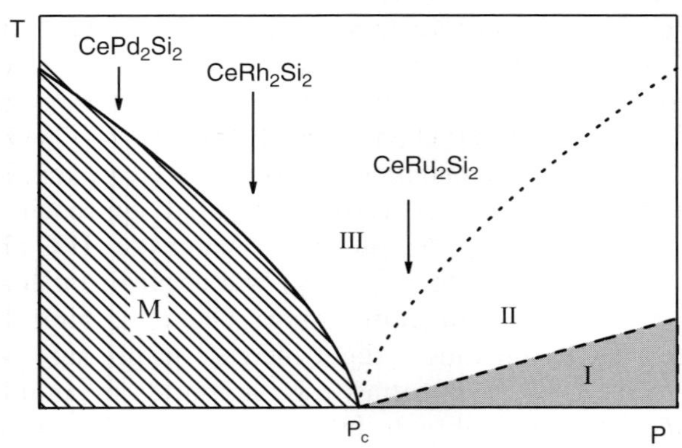

Fig. 5. Magnetic phase diagram predicted for spin fluctuation. In the domain I, Fermi liquid properties will be achieved. In domains II and III, non-Fermi liquid behavior will be found. The location of three HFC described in the text is shown at $P = 0$. Depending on the nature of the interactions (AF or F), the contour lines I, II, III change. Here the contour is drawn for antiferromagnetism. In case of ferromagnetism, the T_I line starts as $(P - P_C)^{3/2}$.

For $P > P_C$, the FL regime is reached only at very low temperature below $T_I \sim T_F(1-I)^{3/2}$. Between $T_I < T < T_{II}$, a large crossover regime appears ($T_{II} \sim T_F(1-I)^{3/4}$) before recovering on warming into domain III. Since the ratio of T_{II}/T_I diverges as $(1-I)^{-3/4}$ when $I \to 1$, the crossover regime II covers a relatively wide temperature range. Near T_I, for three-dimensional (3d) nearly ferromagnetic metals, C/T varies as log T in region II; for a three-dimensional, near-antiferromagnet, the leading term will go as $\sqrt{T}$. The interesting concept is that the proximity of quantum criticality will be felt by the electron already at high temperature ($T \gg T_I$) via its non-Fermi liquid behavior i.e. already at $T \gg T_I$ (fig. 5). The electron "knows" its destiny to be or not to be near $P \sim P_C$. This statement must be true for simple elements such as cerium metal or later ε-Fe. The Ce and Yb HFC will give a nice illustration of this in Sections 2 and 3.

Table 1 (Moriya 1985–2003a) indicates, for 3d ferromagnetic or antiferromagnetic systems, the respective pressure dependence of T_I, T_{II}, T_{II}/T_I, the ordering temperature T_{Curie} or T_N and their relation with the extrapolated sublattice magnetization M_o as $T \to 0$ K assuming $(1-I) \sim (P-P_C)$.

For 3d systems the extrapolated value of γ, the susceptibility χ_Q at the ordering wave vector Q and the average amplitude A of the T^2 inelastic term of the resistivity ρ, as P decreases to P_C are given in Table 2 (Moriya 2003a):

The predictions of their quantum critical behaviors with temperature (plus the nuclear relaxation time T_1) for the 3d and 2d case (Moriya 2003a) are given in Table 3.

The phenomenological model developed by Moriya (self-consistent renormalization theory: SCR) allows one to classify the different systems with four parameters: two dimensionless parameters Y_0 and Y_1 (Moriya and Takimoto 1995), two characteristic temperatures T_0 and T_A. Y_0 is directly proportional to the inverse of the staggered susceptibility χ_Q^{-1} as $T \to 0$; $Y_1 = J_k/\Delta J_k$ is the ratio of the exchange at $k = Q$ and its dispersion ΔJ_k in k (wavevector) space. T_0 is related to the frequency response and $T_A \sim \Delta J_k$ is linked to the exchange wavevector dispersion. Figure 6 (Kambe et al. 1997) shows the comparison between the experiments and the spin fluctuation (SF) fitting in a reduced temperature scale $t = T/T_0$ for the temperature variation of C/T for three typical HFC: $CeCu_6$, $CeNi_2Ge_2$ and $CeRu_2Si_2$ as

TABLE 1
Pressure dependence of the characteristic temperatures T_I, T_{II}, T_{Curie} or T_N. Last column link between T_{Curie} or T_N and sublattice magnetization M_0.

	T_I	T_{II}	T_{II}/T_I	T_{Curie} or T_N	$T_{Curie}(m_0)$ or (T_N)
F	$(P-P_C)^{3/2}$	$(P-P_C)^{3/4}$	$(P-P_C)^{-3/4}$	$(P_C-P)^{3/4}$	$M_0^{3/2}$
AF	$(P-P_C)$	$(P-P_C)^{2/3}$	$(P-P_C)^{-1/3}$	$(P_C-P)^{2/3}$	$M_0^{4/3}$

TABLE 2
Pressure dependence of γ, χ_Q and A at QCP.

	γ	χ_Q	A
F	$\log(P-P_C)$	$(P-P_C)^{-1}$	$(P-P_C)^{-1}$
AF	γ_0–const $\sqrt{(P-P_C)}$	$(P-P_C)^{-1}$	$(P-P_C)^{-1/2}$

TABLE 3
Temperature variation of C/T, χ_Q, ρ ant T_1 in the NFL regime.

		C/T	χ_Q^{-1}	$\rho \sim T^n$	T_1^{-1}
F	3d	$-\ln T$	$T^{4/3} \rightarrow$ CW	$T^{5/3}$	$T\chi$
	2d	$T^{-1/3}$	$-T \ln T$	$T^{4/3}$	$T\chi^{3/2}$
AF	3d	$T^{1/2}$	$T^{3/2} \rightarrow$ CW	$T^{3/2}$	$T\chi_Q^{1/2}$
	2d	$-\ln T$	$-T/\ln T$	T	$T\chi_Q$

well as two cases where P_C is approached by doping $CeCu_{5.9}Au_{0.1}$ and $Ce_{0.925}La_{0.075}Ru_2Si_2$. The corresponding parameters are given in Table 4.

The coefficient Y_1 is far smaller for $CeRu_2Si_2$ (1.6) than in $CeCu_6$ (10) and $CeNi_2Ge_2$ (7): a more pronounced q structure in $\chi''(q, \omega)$ is observed for $CeRu_2Si_2$. Rather similar values of $T_0(K)/T_A(K)$ are found for $CeCu_6$ and $CeRu_2Si_2$. The low temperatures (T_0, T_A) of $CeCu_6$ combined with the large Y_1 parameter explains why $CeCu_6$ may correspond to a different situation than $CeRu_2Si_2$. For $CeNi_2Ge_2$, $T_A(K)/T_0(K) \sim 3$ differing greatly from 1. We will see later that two characteristic energies (4 and 0.6 meV) characterize the magnetic fluctuations of $CeNi_2Ge_2$.

Let us stress that, by comparison to other strongly correlated electron systems (SCES), the particular interest of heavy fermions is that the Néel temperature and the Fermi temperature are comparable and already low (a few Kelvin) far below P_C. Furthermore small shifts in pressure of the bare parameters (E_0, Γ, $N(E_F)$) will be magnified by large shifts of the effective temperatures (T_N, T_F, T_I,...). The corresponding pressure derivatives $\partial T_N/\partial P, \partial T_I/\partial P, \partial T_K/\partial P$, are enhanced by two orders of magnitude (huge Grüneisen parameter). This leads to huge effects on thermal expansion, sound velocity and magnetostriction according to Maxwell relations. It gives the unique opportunity to scan through P_C by moderate pressure variations (a few GPa) or to sweep through the anomalies of the density of states with a magnetic field H of a few tesla. Such conditions are almost unique in condensed matter.

Qualitative agreement with spin fluctuation (SF) theory is found when other quantities like the dynamical susceptibility or the resistivity are

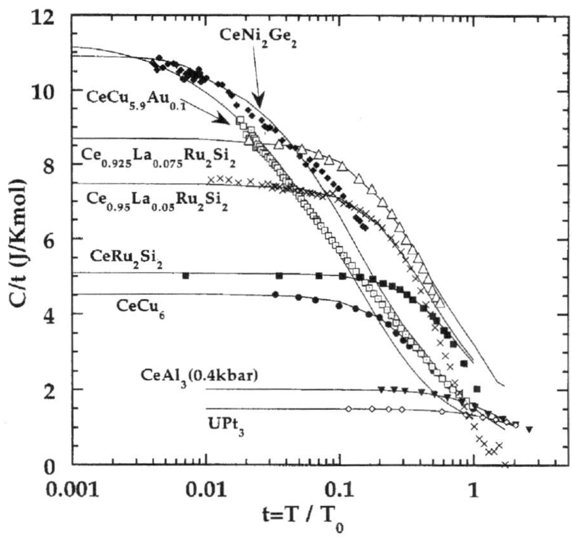

Fig. 6. Comparison between C/t data (symbols) and the SF model (solid lines) in various heavy electron compounds. Only some experimental points are presented for clarity and a normalized temperature scale $t = T/T_0$ is adopted (T_0 for some compounds is presented in Table 4). (Kambe et al. 1996).

TABLE 4

HFC parameters derived through a Moriya-Takimoto scheme (1995) (see Kambe et al. 1997).

	Y_0	Y_1	T_0 (K)	T_A (K)
$CeRu_2Si_2$	0.31	1.6	14.1	16
$Ce_{0.925}La_{0.075}Ru_2Si_2$	0.05	0.77	14.7	14
$CeCu_6$	0.4	10	3	5.5
$CeCu_{5.9}Au_{0.1}$	0.003	16.7	3.4	6.7
$CeNi_2Ge_2$	0.007	7	29.7	94

calculated in this framework. The diversity of the HFC initial conditions leads to quite different values of Y_1 which is related to the problem of the localization of the interaction. In a first approximation, spin fluctuation theory gives a good description of the low-energy excitations. Quantitatively, there are discrepancies like the $C/T = \log T$ crossover law observed over a large temperature range (von Löhneysen et al. 1994–2000) which is not predicted for 3d itinerant antiferromagnets. Different routes have been proposed: reduction of the dimensionality (3d→2d), occurrence of distributions for the Kondo temperature linked to disorder (Miranda et al. 1997, Castro Neto and Jones 2000) or valence fluctuation or drastic change of the Fermi surface's volume (Coleman 1999). For example for $CeCu_{6-x}Au_x$ it was proposed that three-dimensional conduction electrons are coupled to two-dimensional critical ferromagnetic fluctuations near the quantum critical point $x_c = 0.1$ (Rosch et al. 1997). A picture of dynamical heterogeneities (Bernhoeft 2001) invokes different distributions of space and time motion. Of course, broad time and space responses can cover a large diversity of phenomena. This approach can only be a first step to visualize the occurrence of different lengths and times before entering the Fermi liquid regime.

Let us mention that a simple form has been proposed to describe the temperature variation of the specific heat and the susceptibility of HFC. The high-temperature Kondo gas of non-interacting Kondo impurities condenses slowly into a heavy electron Kondo liquid of Wilson ratio = 2 with a fraction f. The analysis on the cerium 115 series shows that f increases linearly with T on cooling before saturating at a fixed value near 0.9 for $P = P_C$. The resistivity is dominated by the fraction $1-f$ of isolated Kondo impurities. For $P \ll P_C$, deep inside the AF state $f = 1$ (Nakatsuji et al. 2003; see application to NMR analysis by Curro et al. 2004). This continuous two-component description assumes that the critical point (T_{cr}, P_{cr}) will never be achieved ($T_{cr} < 0$). It was stressed that both components have protected behavior according to the "classification" made by Laughlin and Pines (2000) in their "theory of everything."

1.6. *Quantum phase transition*

In the Doniach or SF pictures, a second-order QCP will mark the transition from long-range magnetic ordering to the paramagnetic phase at $T = 0\,K$. Since it must correspond to simultaneous collapsing of the specific heat and thermal expansion anomalies for $T_N \to 0$, the initial slope $\partial T_N/\partial P$ can have any finite value. The magnetic coherence length ξ_m will diverge at T_N. For example, in AF–SF theory (Hasegawa and Moriya 1974; Makoshi and Moriya 1975),

$$\xi_m(T) = \xi_m^o \left[1 - \left(\frac{T}{T_N}\right)^{3/2}\right]^{-1/2}$$

and the magnetic coherence length ξ_m^o diverges on both sides of P_C at $T = 0\,K$.

On the AF site, SF theory predicts

$$\xi_{\mathrm{m}}^{\mathrm{o}} = a\left(\frac{T_0}{T_{\mathrm{N}}}\right)^{-3/4}.$$

The jump of the specific heat anomaly collapses with T_{N} (Zülicke and Millis 1995). In many HFC, the initial slope $\partial T_{\mathrm{N}}/\partial P$ is steep. We will see that the occurrence of a second-order QCP is certainly not a general rule. When the transition becomes first order, there will no longer be a divergence of $\xi_{\mathrm{m}}(T_{\mathrm{N}})$ but a jump appears instead.

The field of quantum phase transitions has attracted the interest of both experimentalists and theorists since even if the transition occurs at $T = 0\,\mathrm{K}$, it can govern the physical properties over a wide range in the (T, P, H) phase diagram. A quantum phase transition corresponds to competing ground states which may be changed by external variables such as pressure or magnetic field. If the crossing corresponds to a second-order transition at a quantum critical point (P_{c} or H_{c}), universal behavior will appear depending on the initial spin and charge dimensionality.

Quantum fluctuations destroy the long-range order at $T = 0\,\mathrm{K}$ in a different way than thermal fluctuations since now statics and dynamics are mixed. Within the renormalization group approach (Hertz 1976; Millis 1993) the effective dimension of the system $d_{\mathrm{eff}} = d + z$ couples the geometrical dimension d and the exponent z characteristic of the dynamics (resp. 3 or 2 for ferromagnetic and antiferromagnetic dimension). Since the $d_{\mathrm{eff}} = 5$ for 3d itinerant antiferromagnets is higher than 4, the upper critical dimension, a molecular field treatment (as done in previous SF approaches) may describe the experiments. Two important points must be verified in a microscopic neutron scattering experiment:

- the singular temperature dependence of the static susceptibility $\chi'(k_0,T)$ at the ordered wavevector k_0 by comparison to the uniform one,
- the absence of scaling in ω/T of the dynamical susceptibility $\chi''(k,\omega)$ since the data must follow a ω/T^{β} law with $\beta>1$ due to coupling among paramagnons.

As we will see in the section on for $\mathrm{CeCu}_{6-x}\mathrm{Au}_{x,}$ these SF predictions seem to fail indicating that previous arguments should be reconsidered (Coleman 1999). It was suggested that a scenario called local quantum criticality might be relevant. The term local emphasizes that surprisingly $\chi'(0,T)$ and $\chi'(\mathrm{k}_0,T)$ have the same temperature dependence. This possibility was supported by Si and co-workers (Si et al. 2001, 2003a, b) in the treatment of the Kondo lattice for the specific case of 2D quantum magnetic fluctuations. However the introduction of an extra tridimensional fluctuation leads to the recovery of the classical spin fluctuation picture. Numerical identification of a locally quantum critical point is also found for Ising anisotropy and again 2d magnetic correlations. (Grempel and Si 2003, Zhu et al. 2003a). As the phenomena involve all wave vectors, it was proposed that it may be associated with a Fermi surface (FS) melting with the image that the FS is small on the AF magnetic side and large on the PM side with a divergence of the effective mass (Coleman 1999). Quite different conclusions were given by Pankov et al. (2004) and Sun and Kotliar (2003). At least, there is yet no consensus on the relevance of quantum local criticality. In the standard SF framework, FS is assumed to be large and conserved over the entire pressure range ($P \gtrless_; P_c$, i.e. $P_{\mathrm{KL}} \to 0$).

There is a possibility that the main fluctuations responsible for the non-Fermi liquid behavior come from destruction of the large Fermi surface of the PM state. This was supported by using insights from the theory of deconfined quantum criticality of insulating antiferromagnets (Senthil et al. 2004a, b). In our classification, it corresponds to $P_{\mathrm{KL}} = P_{\mathrm{C}}$. The experimental implifications will be weak moment magnetism, due to low-energy instability of the small Fermi surface, and spinon excitations of the AF state. Additional spinon excitations must enhance the thermal transport by comparison to the electric one leading to a violation of the Wiedeman-Franz law.

Testing these recent theoretical proposals will be an experimental challenge for the future. In our view, an important point is the proximity of P_{V} to P_{C} which will favor a first-order transition. It is even amazing to remark than in the quantum Monte Carlo approach with extended dynamical mean field theory, the finite temperature magnetic transition is first order, while the extrapolated zero temperature magnetic transition on the other hand is continuous and locally critical (Zhu et al. 2003a). In many cases the critical point is weakly first order with the consequence of large fluctuations but with finite values of the low-energy excitations and coherence length at P_{C}. As we will see, often the QCP does not appear to be governed strictly by the spin dynamics but is also associated with a change in electronic structure.

Any first-order transition, whatever the discontinuity in volume at $T = 0\,\mathrm{K}$, may have drastic effects. The universality may be lost since the volume jump (or drop) leads to a non-perturbative mechanical work $P\Delta V$ which implies a temperature change. An interesting issue is the large pressure spreading which may occur between the first-order transition P_{C} and its disappearance at $P_{+\mathrm{C}}$. This type of problem was addressed in the lecture notes of Levanyuk (2001) for a structural transformation with the prediction that $P_{+\mathrm{C}}$ is the pressure predicted in the frame of Landau–Ginzburg theory. But, as he emphasized, spin matter will offer a large diversity. Depending on microscopic parameters (defects, mismatch between them) a large pressure range may be required before the usual behavior after a second-order phase transition will be recovered. New magnetic matter may appear as a Griffiths phase or glass state or phase separation with a continuous variation in the mixing between antiferromagnetism and paramagnetism: $f = 1$ at P_C and 0 at P_{+C}. Recent discussions on the quantum phase transition can be found in the review of Vojta (2003) and in the recent survey by Continentino (2004a). A general reference is Sachdev (1999).

It is worthwhile pointing out that the discontinuity ΔV of the volume gives rise to significant mechanical work (W). A ppm or 100-ppm volume variation for a molar volume $V = 40\,\mathrm{cm^3\,mol^{-1}}$ leads respectively to $W = 4 \times 10^2$ and 4×10^4 erg. If this work is absorbed by thermally isolated heavy quasiparticles of $\gamma = J\,\mathrm{mol^{-1}\,K^{-2}}$, it will cause warming to $T = 10$ and 100 mK, respectively, from $T = 0$ K. Of course, for an isolated system, as the entropy discontinuity between the two phases drops at $T \to 0$ K, the warm up will proceed until a significant entropy drop is recovered between the two phases. It happens that W on He^3 on its melting curve at $P_o \sim 34$ bar and $T \to 0$ K can be quite comparable to that of HFC at its first-order transition at P_C. For a HFC case of $\Delta V/V \sim 2.5 \times 10^{-5}$, $V = 48\,\mathrm{cm^3\,mol^{-1}}$ and $P_C = 30$ kbar, W is the same for of He^3 matter where $\Delta V/V = 5\%$, $V = 24\,\mathrm{cm^3\,mol^{-1}}$ and $P_o = 34$ bar (see Lounasmaa 1988).

Experimental evidence of first-order transitions in HFC comes from the finite value of the characteristic fluctuation at $T = 0$ K for $P \sim P_C$ (Section 2); strong departure of the specific anomalies at T_N ($T_N \to 0$ K) from the molecular field prediction, as well as the observation of a coexisting pressure range of PM and AF phases (Section 4). Often, the transition appears as a weakly first-order transition. So the expected relative discontinuity in volume at P_C may be near or below 10^{-7}. To our knowledge, no HFC with a second-order ferromagnetic QCP has been found. Theoretical arguments for a first-order transition at the ferromagnetic instability can be found in Belitz et al. (1999) and Chubukov et al. (2003). The first evidence for an inhomogeneous medium, and thus a phase separation below the ordering temperature, in related HFC was given in the NMR work of Thessieu et al. (1998, 1999) on MnSi. Such a possibility was rediscovered recently by Doiraud et al. (2003), Pfleiderer et al. (2004) and Yu et al. (2004). The occurrence of inhomogeneous intrinsic behavior in HFC may belong to a class of phenomena like the geometric order of stripes proposed for high T_C materials (see Zaanen 2001) or the formation of droplets.

1.7. *Fermi surface/mass enhancement*

In a crystal without Galilean invariance, the FS cannot grow without feeling the Brillouin zone: different bands will occur. Taking into account all possible mobile electrons, the band picture is often that of a spaghetti plate containing more than 10 bands. A large number of orbits need to be detected to discuss the full FS issue. Due to the large number of orbits a distinction between small ($P < P_{KL}$) and large Fermi surfaces is partly misleading. Furthermore the magnetic order often creates new Brillouin zones.

Generally whatever localization the f electron has, the FS is not small with respect to the volume of the Brillouin zone. Far below P_C, one must recover the situation of a normal rare-earth compound with the 4f electron localized and the Fermi sea given by the content of the other electrons. For a well-ordered compound like $CeAl_2$ or $CeRu_2Ge_2$ ($P \ll P_C$) their Fermi surface is that calculated for the isostructural lanthanium host, i.e. $LaAl_2$, $LaRu_2Ge_2$ (a small FS in a single band model) (Lonzarich 1988). There is evidence in a system like $CePd_2Si_2$, with $T_K \sim T_N$ at $P = 0$, that the Fermi surface has some itinerant character (Sheikin et al. 2003).

On the paramagnetic side, dHvA experimental data are well understood assuming the 4f electron is itinerant (Sections 2 and 4). This observation follows either arguments based on the 1d Kondo lattice (Tsunetsugu et al. 1988) or the invariance of the Fermi surface volume with the interactions. When Coulomb repulsion is switched on in the Anderson Hamiltonian (Fazekas 1999), the Fermi surface is predicted to be large. With the new conjectures of local criticality or deconfinement, a new generation of quantum oscillations experiments are underway. To know the present status, the reader can refer to the work in Osaka by the group of Onuki (Onuki et al. 2003).

On warming one may reach a regime where the 4f electron loses its itinerant character before recovering its single impurity T_K behavior. Above T_K the 4f electron plays the role of a paramagnetic Kondo center for the light electronic band of the lanthanium isostructural compound ($LaRu_2Si_2$ for $CeRu_2Si_2$). Quantum oscillation and *high-energy spectroscopy* experiments on $CeRu_2Si_2$ ($P > P_{KL}$) show the crossing through T_K at $P = 0$. As $T \to 0$ K close to P_C, the Fermi surface of $CeRu_2Si_2$, measured by quantum oscillation and derived in band calculations, corresponds to an itinerant 4f electron. However on warming (*high-energy spectroscopy* response), the 4f electrons are localized (see Section 2).

A method which takes into account the renormalization was successfully applied to HFC with phenomenological adjustable parameters (Zwicknagl 1992). The FS topology is not affected by strong local correlations; its contour is already well defined in band calculations with weakly correlated electronic bands. The strong correlations lead to a large effective mass and corresponding anisotropy. The renormalized band approach was successfully applied to $CeRu_2Si_2$.

Recent progress has been made in the treatment of the correlation (including even the feedback to the choice on the crystal field arrangement) using the so-called dynamical mean field theory (DMFT) (Georges et al. 1996). The combination of DFMT with electronic structure methods is very promising (see Georges 2004). Already application has been made to the Ce phase diagram (Amadon et al. 2005, Sakai et al. 2005). For example, in this last work it has been found that at $P = 0$ in the γ phase, the crystal field splitting due to the hybridization is 250 K.

A particular situation must appear when the number of carriers n_e is just equal to the number of magnetic sites n_m or when the balance between two valence configurations releases a light electron. That leads to the prediction of the Kondo insulator (Jullien et al. 1979). However in many cases of HFC no insulating phase appears when the magic ratio seems to be achieved as in CeB_6 or CeTe. Insulating phases exist in the Sm or Tm chalcogenides in the intermediate valence regime but collapse when the valence approaches three. Kondo insulators found in intermetallic compounds seem to correspond to specific conditions, as found in ordinary metals with an even number of electrons (Takabatake et al. 1998, Aeppli and Fisk 1992). We will focus mainly on metallic material with $(n_m + n_e)$ non-integer.

We will take a pedestrian view in considering the mass enhancement. A common point with the previous view is that the magnetic intersite interaction in HFC does not modify drastically the mass enhancement originating from Kondo local fluctuations. Different mechanisms are responsible for the dressing of the effective mass m^* relative to m_o of the bare electron mass. Schematically, one can invoke the renormalization m_K/m_o by strong local fluctuations (T_K or T_0 in the Kondo lattice or spin fluctuation approach) and the further renormalization m^*/m_K due to spin fluctuation or Kondo lattice enhancement. Studies under pressure through P_C and P_V (Brodale et al. 1985), under magnetic field (through metamagnetic transition) (Flouquet et al. 2002a) and also the analysis of the upper critical field $H_{c2}(T)$ of the superconducting state (simultaneous fit of the effective mass and the strong coupling constant λ, ($m^*/m_K = 1 + \lambda$) can give an evaluation of the respective weights of m^*/m_K and m_K/m_o. For HFC cases where AF fluctuations dominate, m^*/m_K seems to be near 2 even at P_C while m_K/m_o may reach 100. Of course, when FM fluctuations play an important role, m^*/m_K can diverge at P_C. Later we will come back to the recent claim of the divergence of the effective mass induced by the magnetic field in $YbRh_2Si_2$ (Custers et al. 2003).

1.8. *Comparison with ^{3}He*

The comparison of heavy fermion systems with ^{3}He (see Benoit et al. 1978, 1981, Beal Monod and Lawrence 1980, Leggett 1987) is interesting as it involves similar considerations: the localization of the ^{3}He particle (difference between the solid and liquid phases), the link with magnetism (AF of the solid phase, paramagnetism of the liquid phase), proximity of the liquid phase to a magnetic instability, tunneling character of the exchange and multiparticle exchange interaction in the solid and unconventional nature of the p wave superfluidity (see Leggett 1975, Vollhardt and Wölfle 1990). The quantum He^3 community likes to represent their phase diagram taking pressure as the y-axis and temperature as the x-axis. The HFC community turns the representation by 90° choosing respectively T and P for the y- and x-axes. As experimentalists, we recommend that the HFC physicists turn their heads by 90°. Then they extend the territory to negative pressure where somewhere a first-order phase transition will appear at $T = 0\,\mathrm{K}$ with a horizontal line on the (P, T) frame but vertical line on (T, P) frame. The UGe_2, URu_2Si_2, $CeRh_2Si_2$, $CeIn_3$ and $CeRhIn_5$ phase diagrams will be like the localization phase diagram of He^3 and Ce metal. This observation highlights the association of two mechanisms: delocalization of the particle and drastic change in the spin dynamics. The comparison between HFC and quantum ^{3}He matter was our first motivation to search for the magnetic structure of solid ^{3}He on the melting curve by neutron diffraction (Benoit et al. 1985). Attempts are actually made to improve our data which have confirmed the assignment by NMR (Osheroff et al. 1980) of the up-up-down-down nuclear spin structure of ^{3}He. Results and analysis of excitations in the liquid ^{3}He phase can be found in Fak et al. (1998) and Glyde et al. (2000).

The strong point of the quantum liquid and solid phases of He^3 is first its purity. Secondly the bare parameters such as the bare mass (m_{He^3} nuclear mass), the magnetic moment of the carrier (μ_n nuclear magnetism) and the experimental condition on its density studies are very well controlled. For the liquid phase, knowledge of the bare parameters (m_{He^3}, μ_n) allows one to derive Landau parameters. Notably F_1 and Z_0 are connected with the mass enhancement and the reinforcement of the Pauli susceptibility by the spin fluctuation mechanism. As already pointed out, in HFC, different mechanisms are involved in the mass enhancement and even the choice of bare magnetic parameters for the carrier is ambiguous. Basically for He^3 there is one type of bare carrier; in HFC it is a complex two-band system.

From the weak pressure variation of the Landau parameter Z_0, the low value of the Grüneisen parameter ($\Omega^* = -2$), and the analysis of the neutron scattering experiments, it is well established that liquid ^{3}He is far from a ferromagnetic instability (Anderson and Brinkmann 1975). The role of ferromagnetic spin fluctuations is however crucial in order to explain the stability of the A superfluid phase with respect to the B phase on heating. So HFC are ideal to study the magnetic instabilities. In ^{3}He, the transition from the liquid phase to the solid phase corresponds to a huge volume contraction at $T \rightarrow 0\,K$ at the melting pressure $P = 34\,\text{bar}$. As there is no hysteresis at this first-order transition, the ^3He melting curve can be used for very low-temperature thermometry.

Of course, in HFC, the carrier is not neutral and the motion and pairing of the charge and the spin has its own interest with potential applications. Pumping of itinerant electrons can be induced by pressure or magnetic field (P_{KL}, magnetostriction in UBe_{13}). In comparison with high T_C oxide superconductors, the great advantage is that this low-temperature physics deals only with electrons, the normal phase can be studied down to very low temperature without being masked by another superconducting phase transition. Nice scans in (P, T, H) will allow one to explore and understand a large diversity of phenomena. Concerning unconventional superconductivity, the low value of T_C allows restoration of the normal phase with moderate magnetic field, as the upper critical field H_{c_2} generally does not exceed 10 T.

1.9. *Experiments*

1.9.1. *Material measurements*

For an experimentalist, the study of complex materials is rich as it requires the handling of different aspects which require collaborations. The interplay of the different ingredients (T_F, T_{CF}, T_K, T_{KL}) and their magnetic field and pressure dependence leads to a large variety of situations with critical temperatures (T_N, T_{Curie}, T_C), crossover temperatures (T_{KL}, and T_I), magnetic field (H_a, H_c, H_M , H_{c_2}) (I) or pressure instabilities (P_{KL}, P_C, P_V and P_{-S}, P_S, the two pressures between which superconductivity occurs). The discovery of new materials can be a major breakthrough.

The appearance of new compounds may open a completely new perspective as happened for the high T_C superconductors, or decisive possibilities to clarify basic issues (see Ce 115 and Yb 122 compounds) and also effects which are magnified by the interplay between parameters. For scientists outside the field, HFC may appear as a labyrinth with no exit. They must realize that this physio-chemical spadework is essential to select a clear situation with the possibility to tune later through different phases by application of pressure, uniaxial strain or magnetic field.

Here we focus on extreme regions of the (T, P, H) phase diagram of HFC. Progress has been made at the frontier of instrumentation (see Salce et al. 2000) to successfully make simultaneous specific heat, resistivity and susceptibility experiments with low-temperature tuning of the hydrostatic pressure and also with the increasingly routine pressure experiments developed more than a decade ago at the high pressure Institute of Troitsk (Eremets

1996). However, often a major development becomes possible only if new detail has been resolved. Special attention must be taken to the contact quality in order to be sure that the output thermal power is directly transmitted to the electronic bath. In Grenoble, the boost was given by our Japanese visitor, Y. Okayama (1996), who taught us how to realize tiny beautiful electrical contacts with a microwelding machine on almost any Ce or U intermetallic compound.

Concerning experimental methods, focus is given on the microscopic probe of neutron scattering done in Grenoble and on NMR performed in Osaka. Inelastic neutron scattering is still the only way to probe the spin dynamics in a wide range of frequency and wavevector. NMR is a "light" probe, which gives access to the ultimate low-frequency limit at the expense of wavevector integration. It has now been performed with good accuracy up to P_C in different HFC (Sections 4 and 5).

Thermodynamic measurements such as the specific heat C and the magnetization are fundamental quantities, which can be measured with great accuracy, within 1% or better. They are a severe test of any theoretical proposal near the quantum critical point (P or H). Their associated pressure derivatives (thermal expansion and magnetostriction) or field derivatives (magnetocaloric effect) are very powerful. They reveal the importance of deformation (compression and shear mode), i.e. of density fluctuations. Theoretical calculations at constant volume will miss the main issue. The full detection of dHvA oscillations in UPt_3 (Lonzarich and Taillefer 1985) show that the FS topology is rather conventional (given roughly by any band calculation) but with huge renormalized effective masses that establishes the validity of the heavy fermion quasiparticle. Few complete FS have been drawn experimentally apart from $CeRu_2Si_2$ and UPt_3. A large number of orbits have been determined in the materials addressed in this review (see Sections 2, 4 and 6).

As it was underlined, resistivity is certainly the best and fastest method to detect major breakthroughs in the discovery of superconductivity. Thermal conductivity is a very powerful technique to study unconventional superconductivity: an illustration will be the UPt_3 study. The Hall effect is complex in this multiband material. The possibility of a large effect at P_{KL} (at any FS reconstruction) now gives new experimental fever (Coleman et al. 2001), see, for example, the recent applications to V(Cr) (Yeh et al. 2002) and to $YbRh_2Si_2$ (Paschen et al. 2004) and for the theoretical comments (Norman et al. 2003).

If the physicist enjoys measuring a signal which can change its signal amplitude in T, H or P he will certainly select the thermoelectric power S_{TEP} in these narrow renormalized bands (see Jaccard and Flouquet 1985) as well as the Nernst effect. These possibilities were extensively exploited in high T_C

materials for the pseudogap issue (Ong and Wang 2004, Wang et al. 2002). As with thermal expansion, large variations are expected near P_{KL} or P_C. New experiments are underway notably in Paris (Bel et al. 2004a). As the interplay between FS and magnetic instabilities are still obscure, TEP is an excellent probe to detect subtle effects (see Abrikosov 1988).

The readers can find reviews on other specific techniques on HFC, for the powerful ultrasound probe (Thalmeier and Lüthi 1991) for electrodynamic response of HFC (Degiorgi 1999) for high-energy spectroscopy (Malterre et al. 1996) and for muon spectroscopy (Amato et al. 1997). Few experiments have been performed using the tunneling technique (see UPd_2Al_3) but point contact spectroscopy was extensively used (Naidyuk and Yanson 1998). Most of the figures describe results obtained in Grenoble, Geneva (D. Jaccard) and Osaka (Y. Kitaoka). Using the internet, readers can recover the important results obtained in the groups of Los Alamos (J. Thompson, J. Sarrao), Tallahassee (Z. Fisk), Zurich (H. R. Ott), Dresden (F. Steglich), Karlsruhe (H. von Löhneysen), Bristol (S. Hayden), London (G. Aeppli), Ames (P. Canfield), Sherbrooke (L. Taillefer), Berkeley (N. Phillips), Wien (E. Bauer), Cambridge (G. Lonzarich), Toronto (S. Julian), Osaka (Onuki), Tokyo (Sakibara, H. Sato), and so on.

Finally the applied side has not been discussed despite a large possibility for entropy changes in magnetic field, in pressure and of resistivity and thermoelectric power variations in pressure, temperature and magnetic field. At least HFC has close analogies to other physical systems such as the equivalent of the Pomeranchuk effect or fast decompression at constant field in quantum ^{3}He, the efficiency of adiabatic demagnetization or of Peltier cooling. Our proposal made in March 2004 was applied later by Continentino and Ferreira (2004) and by Continentino et al. (2005) to $YbInCu_4$. It is also possible to play with a machine where the vapor or fuel will be the spin. The problem is particularly interesting if there is a (P, H, T) range where phase separation occurs.

1.9.2. *From transport measurements to heavy Fermion properties*
Resistivity is often the only measurement performed under pressure since it is a sensitive technique (detection of voltage down to 10^{-3} nV) with a low dissipative power and the possibility to be coupled thermally with the cold source by excellent electronic contacts. Let us state the related specific problems due to the interplay between charge carriers and scattering.

Since the material is not perfect, residual impurities lead to a residual elastic term ρ_0 which is decoupled from inelastic quasiparticle collision at very low temperature. The simple Drude expression for a single type of carrier: $\rho_0 = k_F/n_e^2\ell$ shows that ρ_0 depends only on the carrier's number ($k_F \sim n_e^{1/3}$) and on the electronic mean free path ℓ which does not depend directly on the effective mass but may be enhanced near P_C and P_V ($\rho_0 \approx 1/n_e^{2/3}\ell$). Often an increase of the impurity scattering is detected near a QCP (Flouquet et al. 1988, Thessieu et al. 1995, Wilhelm et al. 2001). The ρ_0 enhancement, i.e. a decrease of ℓ due to quantum critical fluctuations, was calculated by Miyake and Narikiyo (2000) for the spin and Miyake and Maebashi (2002) for the valence. The effect is quite strong near a ferromagnetic (F) QCP and less for an AF QCP. This increase will tend to suppress superconductivity just near P_C for F systems. Critical valence fluctuations near $P = P_V$ also produce a sharp peak of ρ_0. Of course, if there is a FS reconstruction (for example at P_{KL}), ρ_0 will be a basic probe to detect a change of the carrier. Finally, as the impurity sites may correspond to a so-called Kondo hole, i.e. a deformed Cerium site near a lattice imperfection (intersticial, dislocation, stacking fault, etc.), they may also give a temperature-dependent term ρ_{imp} (T) to the resistivity. Before claiming that any T variation of ρ is an intrinsic property, proof must be given that it does not depend on ρ_0.

The low-energy excitations of the dressed particles appear in the temperature contribution of the resistivity. In SF, below T_I, the AT^2 law is one of the signatures of the Fermi liquid regime. The behaviors reported in Table 3 for the SF predictions of $\rho(T)$ at the QCP assume an average scattering over all the Fermi surface, i.e. basically one type of carrier but different scattering processes. Using the parameters (Y_0, Y_1, T_0, T_A) extracted from the specific heat and inelastic neutron scattering data in the $CeRu_2Si_2$, the calculated SF contribution (ρ_{th}) with the hypothesis of an average on relaxation times is greater than that measured, $\rho < \rho_{th}$. An extra source of electronic conductivity, i.e. from different types of carriers may occur. It was first assumed that a source of the new conduction channel might come from an impurity band (Kambe and Flouquet 1997).

Another hypothesis is to assume that on the Fermi surface, the hot spots (corresponding to a momentum $k = k_0$ transfer on the Fermi surface) and the cold spots (insensitive to AF SF) will give two different channels of conduction. Such a model was developed by Rosch (1999) for 3d HFC. Right at the AF QCP, the electronic conductivity σ is written as

$$\sigma = \sigma_{hot} + \sigma_{cold} = \frac{\sqrt{t}}{x_{imp} + t} + \frac{1 - \sqrt{t}}{x_{imp} + t^2}.$$

The impurities are represented by a reduced parameter x_{imp} inversely proportional to ℓ^{-1} $x_{imp}^{-1} \sim \ell$, and the temperature by its normalized value $t = T/T_I$. The fraction of the FS considered as hot spots is $\sqrt{t}$; its spin fluctuation rate is linear in T. The second term describes the cold regions where FL quasiparticle scattering proportional to T^2 occurs. In the clean limit $x_{imp}<t$, the cold quasiparticle term dominates and $\Delta\rho{\sim}T^2$, in the dirty limit $x_{imp}>t$ and $\Delta\rho{\sim}T^{3/2}$. For a value of $x_{imp}=0.01$, a linear T term appears in an intermediate range of temperature.

As shown by Rosch (2000), a departure from P_C (i.e. the appearance of a finite coherence length) restores the observed situation with a minimum $n=3/2$ for the inelastic SF term ($\rho{\sim}T^n$) at P_C. In the specific case of HFC, the k structure of the AF correlation emerges on top of a large continuous signature of the local fluctuations. The recovery of the T^2 term may be faster than in the SF model where all the weight of the fluctuation is at $k{\sim}k_0$. The effect of dimensionality can modify the present scenario: 2d fluctuations lead to $n=1$ at the QCP as well as valence fluctuations (see $CeCu_2Si_2$ Section 4). In some HFC, different AF wavevectors (k_0, k_1, k_3 for $CeRu_2Si_2$) may have different instability points. However in many cases, n reaches a minimum at P_C, which may be a simple test to check for a critical pressure.

A scaling of A with γ^2 contrary to SF predictions, the so-called Kadowaki–Woods (1986) relation ($A/\gamma^2=10^{-5}\,\mu\Omega\,\mathrm{cm\,mol^2\,K^2\,mJ^{-2}}$) is often observed in HFC due to the broad wavevector response in contrast to the strong frequency dependence. Theoretical discussions on the validity of this assumption can be found in Miyake et al. (1989) and Takimoto and Moriya (1996). A recent discussion on the Kadowaki–Woods relation is given in Tsuji et al. (2003), with an analysis for its deviation in Yb-based intermediate valence systems.

It was stressed that another interesting ratio is the Seebeck coefficient (S_{TEP}/T extrapolated to $T\to 0\,\mathrm{K}$) by the corresponding term γ of the specific heat ($C=\gamma T$) (Sakurai 1994, 2001, Behnia et al. 2004). A log–log plot of S_{TEP}/T versus γ for different SCES going from cuprate, organic conductor, HFC and even simple metals (Behnia et al. 2004) gives results aligned mainly on the same line with the ratio

$$S_{TEP}/T\gamma{\sim}(N_{av}e)^{-1}$$

where N_{av} is Avogrado's number. Deviations may indicate:

- a large difference between the number of itinerant carriers and the heat carrier generally assigned to the f electron (one per formula unit in the case of Ce),
- different channels for scattering (case of IVC),
- specific cancellation of S_{TEP}/T in compensated metals where hole and electron-like bands can give a difference in the sign of the thermoelectric power.

However it is remarkable for the Ce HFC that, whatever is T_K at very low temperature and also, whatever is the band structure and notably the degree of compensation between holes and electrons, S_{TEP} is positive. The current flow from a free electron gas state to a heavy fermion phase gives the remarkable result of $S_{TEP} \sim C$ as $T \rightarrow 0\,K$ as pointed out by Sakurai (1994, 2001). This very low-temperature behavior is quite different from that observed at intermediate temperature ($T \sim T_K$), where S_{TEP} seems to be sensitive to the energy derivative of the density of states (see Jaccard and Flouquet 1985). Experimental references on the thermoelectric power of cerium and ytterbium intermetallics can be found in Zlatic et al. (2003). Calculation of the TEP, specific heat and susceptibility of mixed valence systems or of high T_K using $1/N_f$ expansion can be found in Bickers et al. (1985) and Houghton et al. (1987). At the limit of large degeneracy the proportionality $S_{TEP}/T\gamma \sim (N_{av}e)^{-1}$ is recovered. It was recently shown that the quasi-universal ratio of the Seebeck coefficient to the specific heat term γ can be understood on the basis of the Fermi liquid description of strongly correlated metals (Miyake and Kohno 2005). The Kadowaki–Woods rule for resistivity and its equivalent for the thermoelectric power are not golden rules. At least, they emphasize that the strong renormalization is due to the strong frequency dependence of the response.

Finally it is interesting to read old papers published seven decades ago on electrical conductivity and thermoelectricity to elucidate the properties of transition metals taking into account two bands with light and heavy carriers (Mott 1935; Baber 1937; Wilson 1938).

2. Cerium normal phase properties

Essential points:

- HFC magnetism is furtive: its ground state cannot be predicted from first principles.
- The $CeRu_2Si_2$ Kondo lattice ($P_C < P_V$) shows that the collapse of long range magnetism may be not associated with a divergence of the coherence length. The tiny moment is an intrinsic property which can be tuned by pressure or field. The transition from a weakly polarized paramagnetic phase to a strongly polarized paramagnetic phase is driven by the Kondo collapse at the pseudo-metamagnetic transition,
- Experiments on $CeCu_6$ were the first indication of local quantum criticality.
- Just above P_C, $CeNi_2Ge_2$ looks like a nearly antiferromagnetic metal … but a complete study has not yet been achieved.
- The hole Kondo lattice of Yb-HFC appears different from the electron analog of Ce-HFC.

2.1. *Magnetic furtivity of $CeAl_3$*

To illustrate the difficulty to predict *ab initio* the ground state of HFC, the story of $CeAl_3$ is briefly summarized. In the Ce, Al series, the cerium system appears as a "normal" rare-earth ion sensitive to the crystal field splitting down to low temperature (10 K). However, tiny differences (huge volume and anisotropy sensitivity) will lead to drastic changes in the ground state. Figure 7 represents the magnetic entropy of three Ce, Al compounds: $CeAl_2$, $CeAl_3$ and Ce_3Al_{11} (Steglich et al. 1977, Bredl et al. 1978, Peyrard 1980, Flouquet et al. 1982). Already at $T = 10\,K$, the magnetic entropy of a free doublet ($R \log 2$) is recovered. Thus down to 10 K, the magnetic properties of these compounds appear similar. On this basis it would be very difficult to predict their low-temperature destinies. The drop in entropy for $CeAl_2$ at $T_N = 3.8\,K$ and for Ce_3Al_{11} at $T_{Curie} = 6.2\,K$ and $T_N = 3.2\,K$ mark the entrance to magnetic phases. As for $CeAl_3$ (Andres et al. 1975), there is no trace of magnetic ordering; it was realized, after more than a decade, that $CeAl_3$ ends up as a Pauli paramagnet (PM). The maxima of C/T for $T{\sim}350\,mK$ was taken as evidence of the entrance to a low-temperature correlated Fermi liquid regime. Furthermore, the combination of thermal expansion and specific heat shows that the huge C/T value and the concomitant large negative thermal expansion ($\partial V/\partial T = -\partial S/\partial P$) (as for liquid ^{3}He) is not a continuation of the single impurity Kondo picture. As T_K increases under pressure, the classical dilatation of a metallic solid on

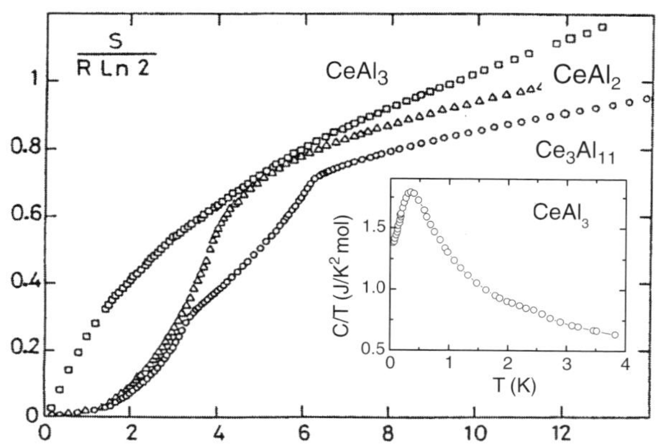

Fig. 7. Entropy of $CeAl_3$ (□), $CeAl_2$ (△), and Ce_3Al_{11} (○) measured on polycrystal (Flouquet et al. 1982). Inset: temperature variation of C/T of $CeAl_3$.

heating is expected and thus a positive thermal expansion for a single Kondo impurity. At least, it was obvious that a characteristic temperature other than T_K occurs (Ribault et al. 1979, Flouquet et al. 1982).

Experimentally, the main difficulty with $CeAl_3$ is that the compound is not a "line" compound, i.e. it is formed in a peritectic solid reaction. All data were taken on polycrystals. The growth of tiny single crystals of this hexagonal lattice shows clearly the occurrence of AF ordering with pronounced AF anomalies in specific heat and resistivity (Jaccard et al. 1987, 1988, Lapertot et al. 1993). The data on polycrystals appear to be a result of a broad distribution of Néel temperatures due to its unusual high sensitivity to pressure and uniaxial stress. At $P = 0$, $CeAl_3$ may be on the AF boundary with $P_C \sim 2$ kbar. The negative value of the thermal expansion can be well explained as being on the verge of AF approaching P_C. The Kondo like picture and the spin fluctuation approach predict an increase of γ with pressure through the Maxwell relation ($\partial V/\partial T = -\partial S/\partial P$) that will lead to a negative thermal expansion.

In this situation close to P_C at ambient pressure, the difference in sample preparations is directly linked to the fact that the compound $CeAl_3$ appears as a solid phase only below 1135 °C, i.e. its peritectoid formation. Starting with the magic composition 1–3 of $CeAl_3$, the first solid created phases will be $CeAl_2$ and Ce_3Al_{11} on cooling from $T \sim 1500$ °C. The realization of nice polycrystals with a single $CeAl_3$ phase is achieved by a slow interdiffusion of $CeAl_2$ and Ce_3Al_{11} at $T = 850$ °C during a week; residual resistivity down to μΩ cm was achieved. Using a highly non-equilibrium process with a start on the Ce_3Al_{11} side, Lapertot et al. (1993) succeed to produce large separate millimetric grains of $CeAl_3$ and Ce_3Al_{11}. In Geneva, the growth of single crystals was successful by a mineralization just below the temperature 1135 °C of the peritectoid landing (Jaccard et al. 1987, 1988). In both cases, the $CeAl_3$ crystals are extracted from a mixture of Ce_3Al_{11} or $CeAl_2$ aggregates. Whatever is the process, single crystals are characterized by residual resistivity one order of magnitude higher than that of the polycrystal. Obviously, that leads to enhancement of the disorder with respect to the polycrystal line material with the consequence of the appearance of clear magnetic transitions. Weak perturbations (not observable by relative changes in the lattice parameters down to 10^{-4}) cause the material to select either a pure AF ordering or a dominant PM phase. Whatever the sample elaboration (polycrystal–single crystal), an extra pressure of $P = 0.2$ GPa or a magnetic field of $H = 2\,T$ pushes the system into an ordinary FL–PM state (Flouquet et al. 1988, Cibin 1990).

2.2. *The Kondo lattice $CeRu_2Si_2$: P, T phase diagram*

$CeRu_2Si_2$ is a key system for the understanding of HFC (Besnus et al. 1985) as the tetragonal lattice has an axial symmetry and the local χ_L magnetic susceptibility shows an Ising behavior (Flouquet et al. 2002). The crystal field ground state of $CeRu_2Si_2$ is almost a pure $\pm 5/2$ doublet with g-factors along the c- and a-axes being $g_{//} = 5g_J$ and $g_\perp = 0$, respectively. The growth of large excellent crystals has allowed a large diversity of microscopic studies in different laboratories, notably extensive high energy spectroscopy, neutron scattering, NMR and quantum oscillation experiments. The possibility to achieve high electronic mean free paths of $\ell \sim 1000$ Å gives the opportunity of a full determination of the Fermi surface in de Haas van Alphen measurements. The high quality of the shiny surface is a guarantee of excellent robust metallic connections, which allow reliable thermal and pressure studies.

At $P = 0$, the pure compound $CeRu_2Si_2$ can be located a few tenths of a GPa above $P_C = -0.3$ GPa. Positive pressure experiments move the system away from P_C. Negative pressure experiments have been realized artificially, expanding the lattice by doping. A dilution of the Ce ions by lanthanium ions with $x_c = 7.5\%$ drives the system to $P_C \sim -0.3$ GPa. For simplicity, a unique variable pressure will be used. The negative pressure was calibrated knowing the change of the lattice parameters and the compressibility.

The NFL behavior of $CeRu_2Si_2$ can be recognized by a slow increase of C/T on cooling before reaching the Fermi liquid regime (fig. 8) (Fisher et al. 1991). Neutron scattering experiments were successfully explained in the framework of spin fluctuation theory (Raymond 1999a). New data of Kadowaki et al. (2004) confirm the validity of SF. A scan in wavevector at constant energy transfer (fig. 9) shows a large local response, i.e. invariant in wavevector and superimposed AF correlations characterized by different vectors k_0, k_1, k_2 (Rossat Mignot et al. 1988, Regnault et al. 1988). For a negative pressure, below P_C, long-range magnetism appears at the wavevector k_0.

A new generation of inelastic neutron scattering experiments (Raymond et al. 2001, Knafo et al. 2004) was performed recently; notably on $Ce_{0.87}La_{0.13}Ru_2Si_2$ (i.e. 0.3 GPa below P_C at $P = 0$) by pressure tuning through P_C and on $Ce_{0.925}La_{0.075}Ru_2Si_2$ at the critical concentration x_c but only at $P = 0$. In this last study, the imaginary part of the susceptibility χ'' was almost continuously analyzed as a function of frequency ω, at different T, for the AF wavevector k_o and for k_s, a wavevector characteristic of the local response.

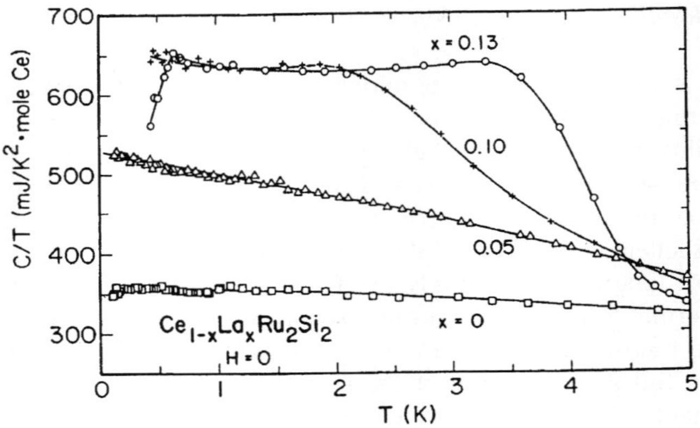

Fig. 8. Variation of C/T as a function of T for $Ce_{1-x}La_xRu_2Si_2$ for $x = 0$, 0.05, 0.10 and 0.13 (Fisher et al. 1991).

At a phase transition, the frequency enters in the scaling function

$$\chi'' \sim T^{-\alpha} f\left(\frac{\omega}{T^{\beta}}\right).$$

As pointed out previously, when dangerous irrelevant interactions must be considered, $\omega \xi_m^{z} = \omega / T^{\beta}$ with an exponent $\beta > 1$ (Continentino 2001, Sachdev 1999). If the fit is made over a large temperature window from 1 to 100 K, the respective value of α and β are found for the AF wavevector k_o ($\alpha = 1$, $\beta = 0.8$) and for k_s a wavevector far from magnetic instabilities ($\alpha = 1$, $\beta = 0.60$) at least down to the temperature where saturation occurs. In any temperature range, the inelastic response is well described by a Lorentzian form

$$\chi'' = \frac{A}{\Gamma(1 + (\omega/\Gamma)^2)}$$

but the linewidth Γ_{k_o} at k_o stays finite ($\sim$0.2 meV) below $T_{k_o} = 3$ K while the linewidth Γ_{k_s} ($\sim$1.5 meV) at k_s saturates below $T_{k_s} = 17$ K (fig. 10). So, contrary to the SF prediction ($\Gamma_{k_o} \sim Y_0$, $Y_0 \to 0$ at P_C), Γ_{k_o} does not collapse. The so-called magnetic QCP may not exist. A finite coherence length seems to occur at P_C (fig. 10). This residual value may be linked to the observation that a tiny sublattice magnetization (M_o) is observed on cooling (Raymond et al. 1997) at k_o ($M_o = 0.02\mu_B$, $T_N \sim 2$ K). Since a concentration gradient might occur in the crystal, it was suggested that it may lead to "residual" long-range magnetic order which will disappear under pressure. However the quasi-coincidence in temperature ($T \sim 2$ K), where the inelastic linewidth

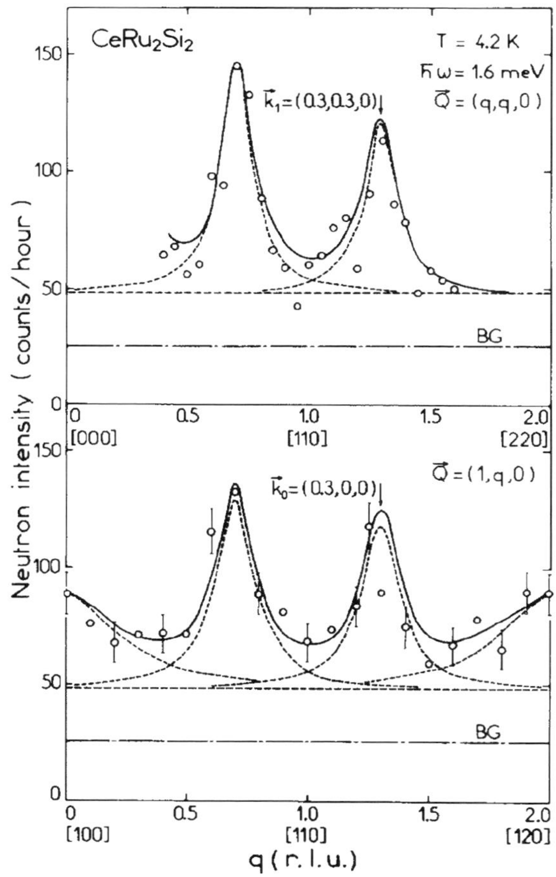

Fig. 9. q scans performed at a finite energy transfer $\hbar\omega = 1.6\,\text{meV}$ along the directions [110] at $T = 4.2\,\text{K}$ for $CeRu_2Si_2$, showing the incommensurate wavevector $k_0 = (0.3, 0, 0)$ and $k_1 = (0.3, 0.3, 0)$ (Rossat-Mignod et al. 1988).

Γ_{k_s} saturates and the signal for elastic magnetic diffraction appears, favors a homogeneous picture. This is reinforced by the fact that similar saturation of Γ_{k_0} right below P_C was found at low temperature in the pressure measurements on $Ce_{0.87}La_{0.13}Ru_2Si_2$ (Raymond et al. 2001).

Small moment antiferromagnetism (SMAF) (tiny M_o) may be the signature of a new type of heavy fermion matter (Kondo condensate), a fact that seems to be supported in excellent single crystals of $CeRu_2Si_2$, where tiny ordered moments $M_o{\sim}10^{-3}\,\mu_B$ have been detected (fig. 11) below 2 K on top of dynamic spin fluctuations. Two components may be involved in this

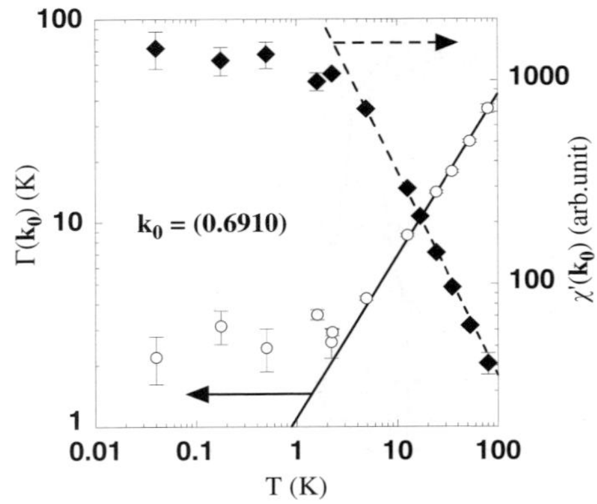

Fig. 10. Temperature variation of the quasi-elastic energy width Γ_{k_0} and of the static susceptibility $\chi'(k_0)$. The wavevector k_0 is characteristic of AF correlations. The dashed and full lines are fits to T^{-1} and $T^{0.8}$ (Knafo et al. 2004).

heavy fermion matter (Amato et al. 1993). Migration of one electron from one constituent to the other produces a thermoelectric power S_{TEP} (inset of fig. 11) (Amato et al. 1989). As explained by D.O. Edwards for the mechanism of a ^3He–^4He dilution refrigerator (Edwards 1970), the migration of an electron from one component to the other can produce a Peltier cooling effect.

Even if relatively good agreement has been obtained with the SF approach, there are still uncertainties at P_C: order of the transition, role of disorder, Fermi surface formation. On warming above T_0 or T_K, the derived β exponent <1 is not what is predicted for a quantum phase transition. This behavior is due to the fact that T_K ($\sim$20 K) occurs just in the temperature window of the fit (3–100 K). Taking into account that one order of magnitude of temperature lower than T_K or T_{KL} is necessary to establish the simple regime ($T_F/10$ for a free electron gas), it is not so surprising that the T^β scaling reflects single impurity dynamics and the slow formation of the Fermi surface.

The analysis of $CeRu_2Si_2$ neutron scattering data may require introduction at low temperature of an inelasticity (ω_0) for the frequency at least to explain consistently the dynamical and static response. As Γ increases strongly on warming about 20 K, the inelasticity is smeared out. Macroscopic evidence of a pseudogap i.e. a dip in the renormalized density of states near the Fermi level appears clearly from the occurrence of a susceptibility maximum χ_M at T_M identified often as the Kondo lattice temperature (fig. 12) and concomitant metamagnetic phenomena. For the magnetically ordered compound ($P<P_C$), T_M is just above T_N. When T_N vanishes at P_C, the maximum χ_M

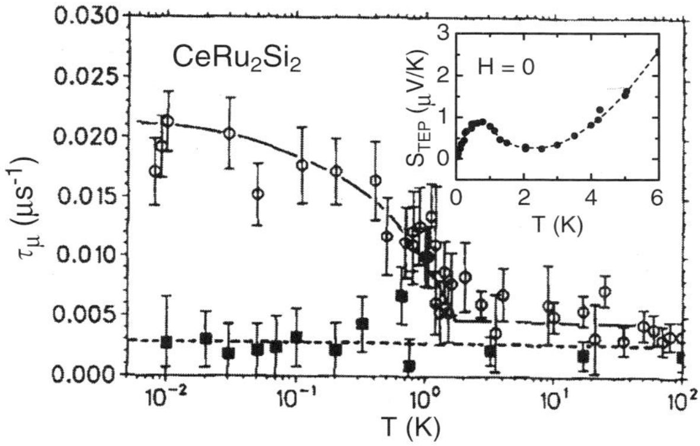

Fig. 11. Evidence for static ultrasmall moment magnetism in $CeRu_2Si_2$ (Amato et al. 1993). Relaxation rate $1/\tau_\mu$ of the $\mu+$ polarization measured in zero-field (0) and longitudinal applied field (■), $H_{ext} = 6\,kOe$). The initial $\mu+$ polarization is along the $\hat{c}$-axis. The lines are a guide to the eye. In the inset, the temperature dependence of TEP is measured in the basal plane at $H = 0$ (Amato et al. 1989).

subsists as a signature of the interplay between AF exchange coupling and Kondo like fluctuations. The behavior is opposite to that of a spin 1/2 Kondo impurity where the susceptibility continuously increases on cooling. Translated into the density of states a pseudogap is required. In good agreement with this picture, the Zeeman splitting of the spin up and spin down band can induce pseudo-metamagnetism with a continuous jump of the magnetization.

To demonstrate the complexity of this heavy fermion matter, the simple visualization is via old fashioned thermodynamic language (Grüneisen 1912, Peyrard 1980, Benoit et al. 1981, Takke et al. 1981). In the same spirit as the Clapeyron (or Ehrenfest relations) for the pressure and temperature dependence on discontinuities in the entropy (or specific heat) and volume (or thermal expansion) for first (or second order) transitions, the Grüneisen parameter is defined as the ratio of α over C at each temperature:

$$\Omega^*(T) = \frac{\alpha}{C}\frac{V_0}{\kappa}$$

(where α, V_0 and κ are respectively the volume thermal expansion, the molar volume and the isothermal compressibility). This parameter is a excellent probe of single component scaling. It will be reduced to a constant $\Omega^*(0)$ independent of the temperature only if the free energy F can be expressed by

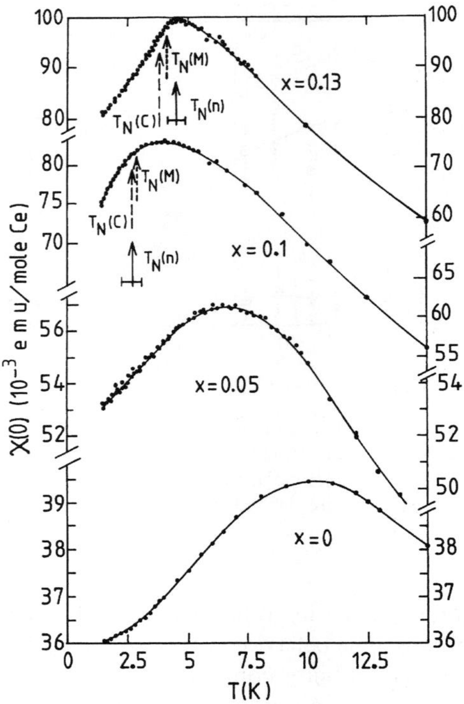

Fig. 12. Susceptibility of $Ce_{1-x}La_xRu_2Si_2$ for $x = 0.13$, 0.10 on the AF side and $x = 0.05$, 0 on the Pa side. In terms of pressure with reference to $CeRu_2Si_2$ at $P = 0$, $P_C = -0.3$ GPa, $x_c = 0.075$. T_N (M, n, C) indicate the Néel temperature determined by magnetization, neutron scattering and specific heat (Fisher et al. 1991).

a single parameter T^*, i.e.

$$F(T) = T\Phi\left(\frac{T}{T^*}\right) \quad \text{with} \quad \Omega^*(0) = -\frac{\partial \log T^*}{\partial \log V}$$

Figure 13 represents the temperature variation of the Grüneisen parameter of $CeRu_2Si_2$ at $P = 0$, i.e. ~3 kbar above the critical pressure $P_C = -3$ kbar. The two singular points are: (i) the huge extrapolated value of $\Omega^*(0) = +190$ and (ii) the slow entrance into a simple regime (T~1 K), where α and C become proportional. At low pressure close to P_C, $\Omega^*(P, T = 0\,\text{K})$ decreases strongly with pressure and then reaches a plateau $\Omega^*(P_V)$~80 at P_V~3.5 GPa before decreasing again above 5 GPa (Payer et al. 1993, Flouquet et al. 2004).

Extensive studies of $CeRu_2Si_2$ were performed by dHvA experiments (Aoki. et al. 1993, 1995, Julian et al. 1994). Quantitatively below the

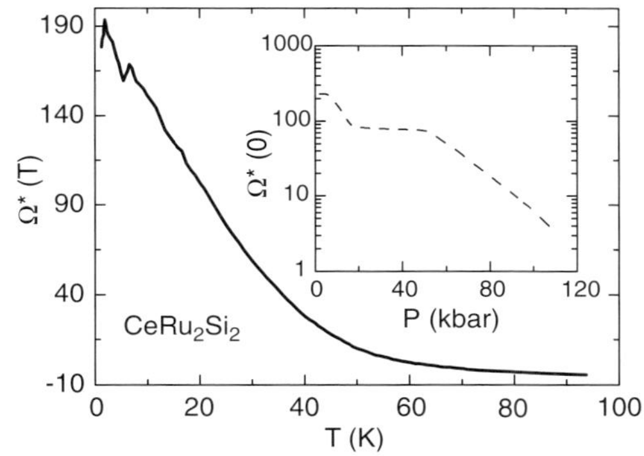

Fig. 13. Heavy fermion Grüneisen parameter ($\Omega^*(T)$) of $CeRu_2Si_2$, the phonon contribution has been subtracted (Lacerda et al. 1989). Inset: pressure dependence of $\Omega^*(0)$ (Flouquet et al. 2004).

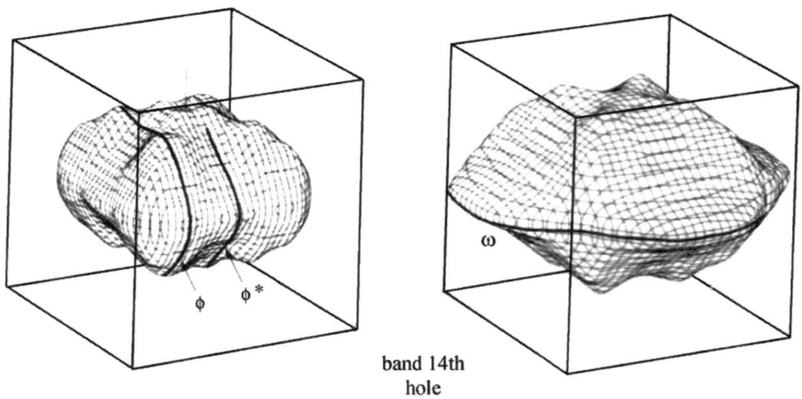

Fig. 14. Hole Fermi surface of $CeRu_2Si_2$ on both side of H_M (Aoki et al. 1995).

metamagnetic field H_M, i.e. in the PM phase, the data are well understood by assuming itinerant 4f electrons. For the main hole orbit ψ detected below H_M (fig. 14), the dHvA signal can be only observed for H close to the basal plane (100) axis. Since its effective mass reaches $120m_o$ very large magnetic fields and very low temperature are needed for its detection. As $H_M \rightarrow \infty$ in the basal plane, there is no field limitation to observe the low-field PM phase. On warming above T_M, photoemission spectroscopy (Denlinger et al. 2000) is well explained assuming that the 4f electron is localized, i.e. excluded from the FS which corresponds to the $LaRu_2Si_2$ Fermi surface. Qualitative arguments can be found in Fulde (1994).

2.3. *The Kondo lattice $CeRu_2Si_2$: (H, T) phase diagram*

The application of a magnetic field along the easy *c*-axis leads to switching (through a drastic crossover at $H = H_M$) from a low-field paramagnetic phase (PM), dominated by AF correlations and large local fluctuations, to a highly polarized state (PP), dominated by the low wavevector q (ferromagnetic) excitation and the surviving local fluctuations (Haen et al. 1987) (figs. 15a and b). Neutron scattering experiments show the continuous spread of the AF response with the same characteristic energy $\omega_{AF}{\sim}1.6\,\text{meV}$ up to H_M (Raymond et al. 1999a) and the emergence, just in the vicinity of H_M, of an inelastic ferromagnetic signal at far lower energy transfer $\omega_F{\sim}0.4\,\text{meV}$ than ω_{AF} (Flouquet et al. 2004, see also Sato et al. 2004).

A particular mechanism, characteristic of the Kondo lattice $CeRu_2Si_2$, controls the dominant magnetic interaction. Quite remarkably, the field and temperature response is well reproduced by a simple form of the thermodynamic quantities such as the entropy:

$$S = S\left(\frac{T}{T^*}, \frac{H}{H_M}\right)$$

with equal Grüneisen parameters $\Omega_{T^*} = \Omega_{H_M} = +190$ (Lacerda et al. 1989). With this hypothesis one can predict the field variation of γ_H knowing the field variation of the linear thermal expansion $\alpha_v = a_H T$. A satisfactory agreement is found between this scaling derivation of γ_H and the bare determination either from specific heat or from magnetization (Paulsen et al. 1990). Using the values of a_H/γ_H, the corresponding field variation of Ω_H shows a positive and negative divergence through H_M (fig. 16) (Holtmeier 1994). It has been recently pointed out that a hyperbolic divergence of $\Omega_H = 1/H - H_M$ is associated with a field-dependent quantum phase transition (Zhu et al. 2003b). The phenomenon is not driven by a sole change of the AF phase transition but involves a concomitant transfer to ferromagnetic fluctuations. The driving mechanism is the field shift of the Fermi level in the pseudogap.

The Ising spin character of the bare local magnetic ion plays a crucial role in the sharpness of this electronic substructure and thus for the pseudo-metamagnetism. The pseudogap shape of the density of states of $CeRu_2Si_2$ is the manifestation of the strong anisotropic hybridization induced by the $|\pm 5/2\rangle$ Ising crystal field ground state (Hanzawa et al. 1987) and by the periodicity of the lattice (Evans 1992). In a Fermi liquid theory, based on the periodic Anderson model (Ikeda 1997, Ikeda and Miyake 1997), the density of states of the quasiparticle bands has a singularity in $\sqrt{\varepsilon}$ of the energy, ε, in agreement with the observed T and H dependence of the specific heat (Aoki et al. 1998). Satoh and Ohkawa (2001) introduced a pseudogap as an input

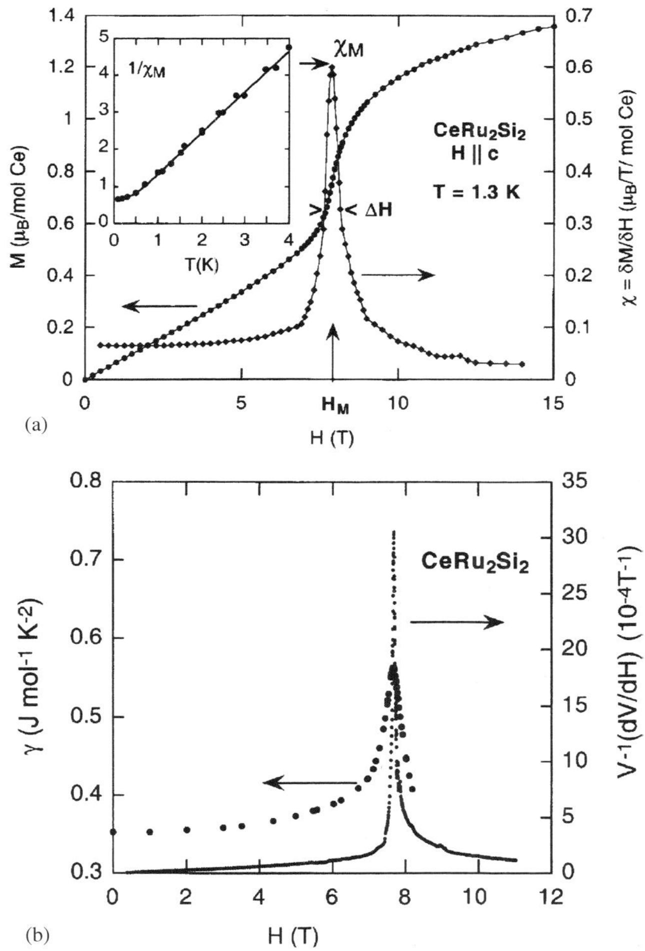

Fig. 15. (a) Low-temperature magnetization $M(H)$ of $CeRu_2Si_2$; Inset: temperature dependence of the differential susceptibility χ_M at H_M. (b) Field variation of $\gamma = (C/T)$, in the limit $T \to 0$, and of the derivative of the magnetostriction (1dV/VdH) (Flouquet et al. 2002).

parameter in the Anderson lattice. The magnetic exchange interaction $J(k)$ between the quasiparticles, caused by the virtual exchange of pair excitations of quasiparticles, depends on the structure of the density of states. The field sweep in the pseudogap produces a change in sign of $J(k)$ at H_M. As the volume dependence of J_k (M) mimics that of T_K, J_k (M, x) scales with T_K

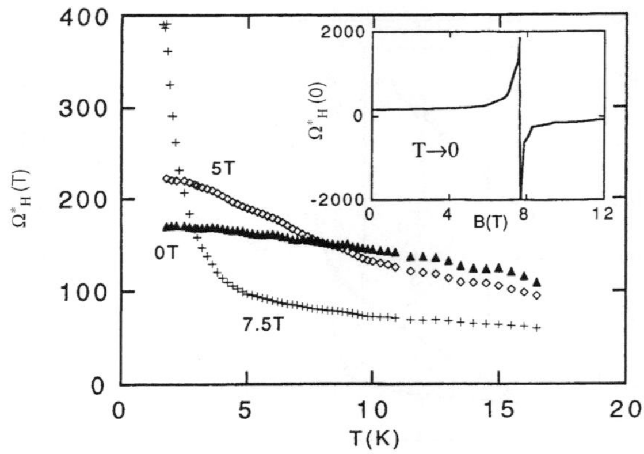

Fig. 16. Ω^*_H (T) Grüneisen parameter at constant magnetic field of $CeRu_2Si_2$ for $H < H_M$. In inset, the extrapolation Ω^*_H (0) for $T \rightarrow 0$ K, Ω^*_H through H_M (Holtmeier 1994).

whatever its sign. Despite the H switch from AF to F, a scaling can occur through H_M (the real mechanism comes from the Kondo lattice).

Reminiscent of the $\alpha-\gamma$ collapse of cerium metal, the associated spectacular lattice softening (~30%) (Kouroudis et al. 1987) and the large volume expansion ($\sim 10^{-3}$) illustrate the interplay (see below) between magnetic, electronic and lattice instabilities in the vicinity of this critical end point. By the sensitive technique of thermal expansion, it was possible to match the crossover line T_α between PM, PP and the uncorrelated paramagnetic state (fig. 17). Below T_α, where $\alpha_v \sim a_H T$, T_α(I) defines the singular contour where this low-temperature electronic property is recovered. Up to $H = H_M$, the thermoelectric power (TEP) shows a maximum at $T^{\mathrm{Max}}_{\mathrm{TEP}} = 0.50\,\mathrm{K}$ with roughly the same initial positive slope: $(\partial S_{\mathrm{TEP}}/\partial T)$ (Amato et al. 1989). Above H_M , $T^{\mathrm{Max}}_{\mathrm{TEP}}$ increases rapidly with the field. The quasi-invariance of $T^{\mathrm{Max}}_{\mathrm{TEP}}$ below H_M coincides also with the weak H variation of the temperature $T_A \sim 0.3$ K below which AT^2 is obeyed in the resistivity despite changes in the amplitude of A by 80% (Kambe et al. 1995). Future experiments may demonstrate if M_o collapses above the metamagnetic transition, i.e. entering in the homogeneous polarized paramagnetic phase. The invariance of T_A and $T^{\mathrm{Max}}_{\mathrm{TEP}}$ for $H < H_M$ may be an indication of the occurrence of weak antiferromagnetism.

At $P = 0$, i.e. for $P = P_C + \varepsilon$, the thermal expansion is zero at H_M. The effective mass cannot diverge but reaches a pronounced maxima. For AF/SF as well as for ferromagnetic fluctuations in finite magnetic field, no divergence of the effective mass is expected. At least with the spin dynamics, only a drastic decrease in m^* will appear above H_M. In the Doniach Kondo

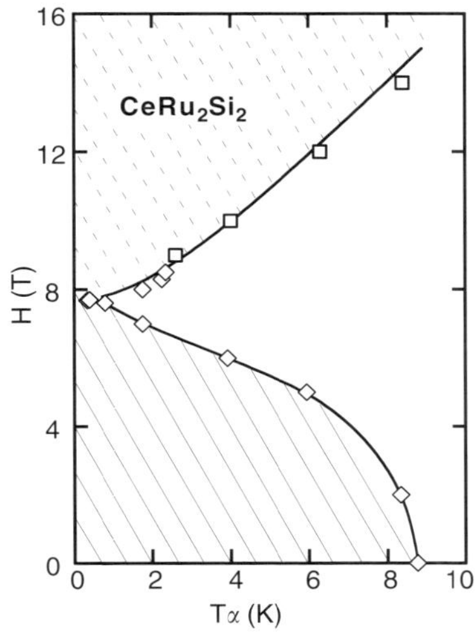

Fig. 17. The crossover phase pseudodiagram $T_\alpha(H)$ derived from the thermal expansion measurements (◇). The high-field data (□) are the temperature of the C/T maxima (Lacerda et al. 1989).

picture, no divergency will be expected as it will occur for $T_K \rightarrow 0$ but here classical magnetism will lead to $m^*/m_o \rightarrow 1$.

A volume expansion occurs with $x = 0.20$ La substitution, corresponding a 6 kbar negative pressure by comparison to $CeRu_2Si_2$. Then the AF ordered phase at $P = 0$ shows two successive first-order metamagnetic transitions at H_a and H_c (fig. 18). Their weak initial pressure dependence corresponds to the observation that the magnetization jump occurs at the critical values M_a and M_c independent of the pressure (Haen et al. 1987, 1996). The pseudometamagnetism at H_M emerges on warming above H_c. As P approaches $P_C \sim +3$ kbar, its differentiation from H_c $(T = 0)$ becomes less pronounced. In the ordered phase, the initial field variation of the magnetization, i.e. here the Pauli susceptibility, appears quasi-independent of the pressure. Constant M_a and M_c corresponds to fixed values of H_a and H_c. In the PM state, the Pauli susceptibility χ_0 becomes strongly pressure-dependent and correlates with H_M since the product $\chi_0 H_M$ is invariant. The discontinuous disappearance of H_C and H_a at P_C has not been observed under ideal conditions as the lanthanum doping introduces disorder. The basic idea is that H_C reaches its critical point at P_C and H_M is the crossover continuation of the metamagnetic field H_C above P_C. Figure 19 is a schematic picture of the pressure variation of H_C and H_M in $CeRu_2Si_2$. It will be emphasized later that, for $CeNi_2Ge_2$ and also for $YbRh_2Si_2$, loss of the strong Ising character

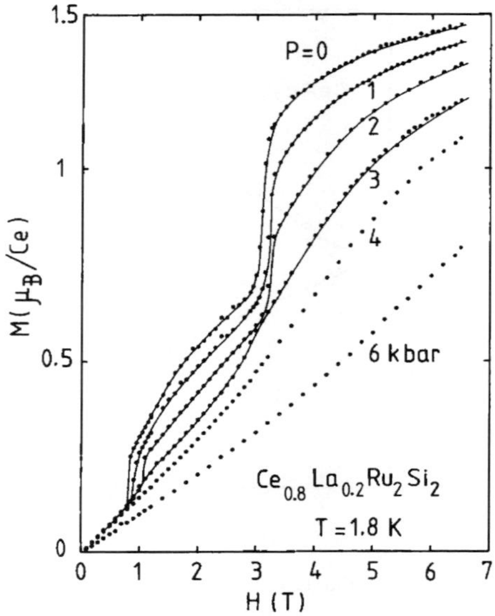

Fig. 18. Magnetization vs. H (//c) of $Ce_{0.8}La_{0.2}Ru_2Si_2$ at 1.8 K at different pressures indicated in kbar (0.1 GPa). At P_C, H_M will be the extension of H_C which ends up at the critical field H_M at P_C. T_N decreases strongly with the pressure (Haen et al. 1996).

leads to a quite different situation with evidence of a decoupling Kondo field H_K. In $CeRu_2Si_2$, $T_A \sim T_O$, thus $H_K \sim H_M$. Of course, the metamagnetic field H_M changes with temperature. Its collapse on warming near 40 K corresponds to the disappearance of an inflection point in the magnetization, of the maximum of the magnetoresistivity and of the temperature window where AF correlations are no longer visible in neutron scattering.

An intriguing question is the FS evolution (fig. 14). Above H_M there is no trace of the heavy (Ψ) orbit ($m^* \sim 120 m_o$), a new orbit (w) is detected now for H close to the (0, 0, 1) axis with rather moderate effective mass. As such, an orbit is predicted in the $LaRu_2Si_2$ band calculation, it was first claimed that the magnetic field leads through H_M to a localization of the 4f electron (Aoki 1993, 1995). However important orbits are still missing in dHvA experiments as the measured FS is too small to explain the remaining large contribution of the electronic specific heat. Attempts to detect a carrier change at H_M by Hall effect and transverse magnetoresistivity failed to detect any variation (Kambe et al. 1996). It may happen that the FS volume does not change but drastic modification occurs in the Kondo lattice. At least, from the analysis of the dHvA amplitude, the effective masses of spin-up and spin-down electrons are different above H_M. A simple physical idea

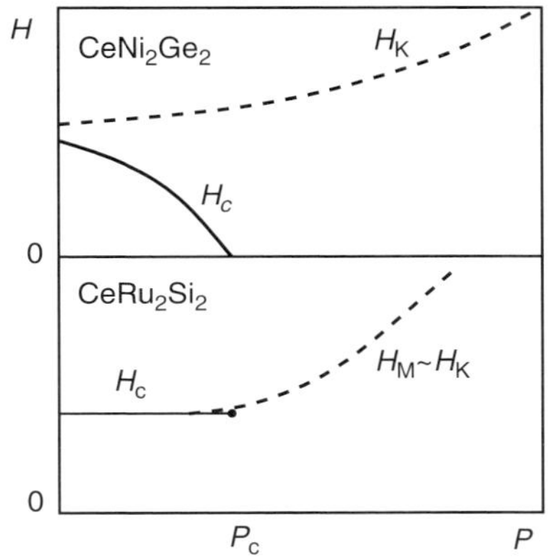

Fig. 19. Schematic pressure dependence of the critical field H_C and the crossover field H_M or H_K for the two cases with H_C ending at a critical point P_C ($CeRu_2Si_2$) or H_C collapsing with T_N ($CeNi_2Ge_2$, $YbRh_2Si_2$).

is that the minority spin bands get a high mass as, on travelling, they feel the repulsion of the majority spin electrons. Below H_M, the spin-up $n_\uparrow$ and spin-down $n_\downarrow$ carriers can be regarded as indistinguishable and their effective masses are equal, $m^*_\uparrow = m^*_\downarrow$. In the polarized frame, drastic changes will occur. Then the current flow is not given by a single type of quasiparticle but by two types. It will be the sum of the current of each spin entity, running with their respective masses which change with magnetic field. For $H \to \infty$, the spin-down carrier must become immobile $m^*_\downarrow \to \infty$, the spin-up carrier becomes completely undressed, $m^*_\uparrow = m_0$.

It is worthwhile to compare the FS of the polarized state of $CeRu_2Si_2$ versus that of the ferromagnetic $CeRu_2Ge_2$ (King and Lonzarich 1991, Ikezawa et al. 1997). $CeRu_2Ge_2$ represents the situation where the lattice of $CeRu_2Si_2$ is expanded by a virtual negative pressure near 8 GPa. For a positive pressure of 8 GPa, the electronic properties of $CeRu_2Ge_2$ reproduce the same behavior as $CeRu_2Si_2$ at $P = 0$. As shown in the (T, P) phase diagram (fig. 20) (Wilhelm and Jaccard 2004), at $P = 0$, $CeRu_2Ge_2$ presents two successive magnetic states on cooling (Raymond et al. 1999b). The first AF phase has the same incommensurate propagation vector k_0 as that found previously while the low-temperature phase is ferromagnetic. This last phase disappears rapidly under pressure at $P = P^*$. The Curie temperature T_{Curie} does not collapse to zero but stays at a finite temperature of 1.6 K. Above 3 GPa, only the AF order survives. In the low-temperature F phase, FS measurements show that at $P = 0$, the f-electrons appear localized, i.e. the

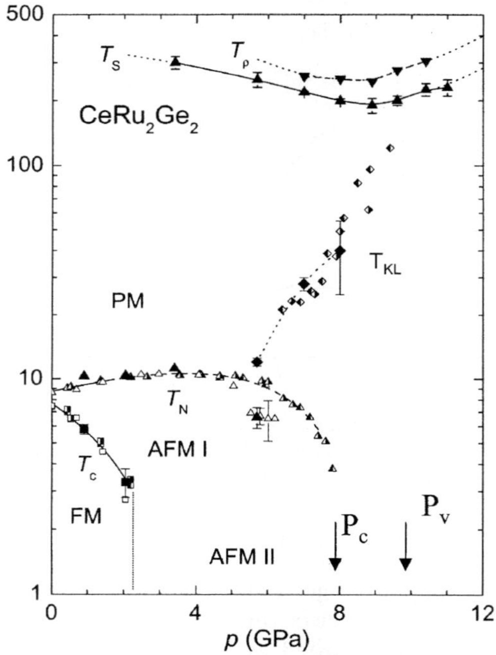

Fig. 20. (T, P) phase diagram of $CeRu_2Ge_2$ obtained from electrical resistivity (half open symbols) calorimetric (open symbols) and the combined $\rho(T)$ and S_{TEP} (bold symbols) measurements. At low pressure a paramagnetic (PM) to antiferromagnetic (AFM I) phase transition occurs at T_N and a subsequent transition into a ferromagnetic phase (FM) takes place at T_C. The FM ground state is suppressed at 2.3 GPa.The combination of all data suggests that long-range magnetic order is suppressed at a critical pressure $P_C \sim 8.0$ GPa. T_ρ and T_S are maxima in $\rho(T)$. The half-filled diamonds indicate $T_{KL} \propto 1/\sqrt{A}$. (Wilhelm and Jaccard 2004).

orbits are those found in band calculations for $LaRu_2Si_2$. Thus, the transition of F to AF ground states at P^* appears discontinuous. It may coincide with a discontinuous changes in FS in contrast with the previous case of $CeRu_2Si_2$ through H_M at $P = 0$. Clarifying the situation above H_M is still an experimental challenge; progress needs to be made in band calculations taking the magnetic field into account since the powerful method of quantum oscillation requires strong magnetic fields, which leads to finite polarization ($M > 0.1\mu_B$). One can see on the phase diagram that T_K will reach the maxima of the resistivity T_{Max} for $P = P_V \sim 10$ GPa, i.e. 2 GPa above P_C. This observation agrees with the conclusion reached for $CeRu_2Si_2$ where it was also found that $P_C < P_V$.

The absence of unconventional superconductivity in $CeRu_2Si_2$, associated with AF fluctuation, can be due to the Ising character of the magnetism, which precludes favorable transverse fluctuations (Monthoux and Lonzarich 2001) (see Section 3). As we will see for $CeRh_2Si_2$, the superconducting dome may be restricted in a narrow pressure range at P_C. Already, at $P = 0$, superconductivity might be lost. One might speculate whether superconductivity will occur for $P \sim P_V$ since valence fluctuations may be an efficient mechanism (see Section 4). The superconductivity of $CeRu_2Si_2$ (or $CeRu_2Ge_2$) is still an open experimental problem as the results were obtained only in a Bridgman cell with a weakly hydrostatic solid medium for pressure transmission. A new generation of measurements must be performed.

If superconductivity occurs at P_C , the interesting situation is if the upper critical field H_{C_2} (0) becomes larger than H_M. The switch in the magnetic interactions may induce change in the Cooper pairing and in the nature of the different transitions. Quite generally, an interplay of a previous (H, T) phase diagram as observed in $CeRu_2Si_2$ and other phase diagrams, such as superconductivity or long-range magnetism, may lead to novel phases. Such a case may happen for the superconductor $CeCoIn_5$ and the hidden order phase of URu_2Si_2 (Sections 4 and 6). In the next section let us compare $CeRu_2Si_2$ with three examples of highly documented NFL behavior: $CeCu_6$, $CeNi_2Ge_2$ and $YbRh_2Si_2$.

2.4. *$CeCu_6$, $CeNi_2Ge_2$: local criticality versus spin fluctuations*

The $CeCu_6$ family (doping with Au, Ag) (von Löhneysen et al. 1994, 1996, 2000) was extensively and carefully studied in the past since it was the first large HFC ($\gamma = 1500\,\text{mJ mole}^{-1}\,\text{K}^{-2}$) (Onuki and Komatsubara 1987, Amato et al. 1987) where large single crystals could be obtained in contrast to $CeAl_3$. By comparison with $CeRu_2Si_2$, its orthorhombic crystal structure is far less symmetric and its residual resistivity is higher ($\rho_0 \sim 10\,\mu\Omega\,\text{cm}$). At low temperature, $CeCu_6$ becomes even less symmetric, that is to say monoclinic. The anisotropies of the susceptibility along the a-, b- and c-axes are $\chi_c/\chi_a \sim 5$, $\chi_a/\chi_b = 2$, while in $CeRu_2Si_2$, $\chi_c/\chi_a = 15$. FS measurements are still incomplete with the detection of only a few orbits (Reinders et al. 1987). In $CeCu_6$, the neutron inelastic spectrum shows less pronounced peaks in the wavevector response than in $CeRu_2Si_2$ (Aeppli et al. 1986, Rossat Mignot et al. 1988). Special focus was given to the critical doping $x_c = 0.1$ of the $CeCu_{6-x}Au_x$ series. Careful thermodynamic measurements (von Löhneysen et al. 1994) point out NFL laws with unexpected temperature variation for 3d itinerant AF. For example, the specific heat follows a $T \log T$ law over two decades of temperature; the uniform static susceptibility has a strong increase on cooling, $T^{-0.75}$. From neutron scattering data, it was claimed that the main effect is due to 2d magnetic fluctuations (as maxima of intensity at a given frequency occur on wavevector rods (Stockert et al. 1998)). Ignoring the spreadout of the rods, their existence is given as justification for the validity of a 2d treatment of the magnetic fluctuations (see Si et al. 2001). In contrast to the previous case of $Ce_{1-x}La_xRu_2Si_2$, a unique ω/T scaling describes $\chi''(q, \omega)$ with $\alpha = 0.75$, $\beta = 1$ whatever the wavevector (Schröder et al. 1998). The same $T^{+0.75}$ variation of the inverse susceptibility $\chi^{-1}(q)$ is found at all q (fig. 21) with no sign of a saturation in temperature. As such invariance of α contradicts a SF description, a new picture called a Fermi-liquid-destroying, spin-localizing transition, i.e. with $P_{KL} = P_C$ has been suggested (Coleman 1999) with the intuition that m^* may diverge at a single point, P_C. The stimulating remark was made that, at P_C, Fermi surface stability must be reconsidered. As we discussed in Section 1, this scenario seems supported by theoretical developments (Si et al. 2001, 2003).

Comparing different HFC is not a easy task as the different ingredients (Δ_{CF}, T_K, T_{KL}, n_e, P_C,...) may give quite different temperature ranges for the observation of non-Fermi liquid behaviors (domains I–III of fig. 5). Furthermore, the other ions (the ligands) play a role in the band structure. Typical parameters of the pure lattice for $CeCu_6$ and $CeRu_2Si_2$ are shown in Table 5. Both of them are located a few tenths of a GPa above P_C respectively

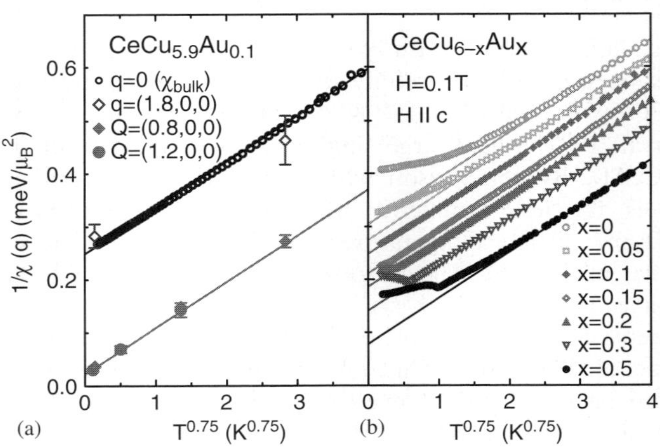

(a) (b)

Fig. 21. Results on $CeCu_{6-x}Au_x$. In (a), the inverse of the static susceptibility χ_q measured at different wavevectors in $CeCu_{5.9}Au_{0.1}$ as a function of $T^{0.75}$ (Schröder et al. 2000). In (b), the inverse of the uniform susceptibility $\chi_q = 0$ is shown for different x as a function of $T^{0.75}$. From the proximity of $x_c = 0.1$ near a QCP there is a large T range where NFL behavior in $T^{0.75}$ is observed.

TABLE 5
Comparison of $CeCu_6$ and $CeRu_2Si_2$ characteristic temperatures.

	γ (mJ mol^{-1} K^{-2})	T_A (K)	T_M (K)	$T^{(a)}_{corr}$ (K)	$T^{(b)}_{corr}$ (K)
$CeCu_6$	1500	0.1	0.8	5	0.15
$CeRu_2Si_2$	360	0.5	10	40	70

near −0.4 and −0.3 GPa. In Table 5, γ is the extrapolated value of C/T as $T \to 0$ K, T_A is the temperature below which the AT^2 law in resistivity is observed; T_M is the temperature where the susceptibility reaches its maximum; $T^{(a)}_{corr}$ the temperature characteristic of the magnetic correlations (Jacoud 1989); and $T^{(b)}_{corr}$ the temperature below which the usual positive magnetoresistivity of metals appears (values can be found in the references). A very low temperature is required for $CeCu_6$ in order to enter the FL regime. The drastic contrast between $CeCu_6$ and $CeRu_2Si_2$ is in $T^{(a)}_{corr}$ and $T^{(b)}_{corr}$ which respectively involve the onset of magnetic correlations and the formation of the Fermi surface . The low $T^{(b)}_{corr}$ value of $CeCu_6$ suggests a slow construction of the Fermi surface for the $CeCu_{6-x}Au_x$ series. It is an open problem if the very low-temperature regime is different from that of $CeRu_2Si_2$.

As indicated in Section 1, the specific heat and resistivity data indicate that $CeNi_2Ge_2$ may be located close to P_C ($Y_0 = 0.007$). For $CeNi_2Ge_2$, in contrast to $CeRu_2Si_2$, the ratio χ_c/χ_a is relatively weak ($\chi_c/\chi_a \sim 1.4$ at $T = 4.2$ K) (Fukuhara et al. 1996). $CeNi_2Ge_2$ may appear ideal to study a 3d QCP as it is very close to the QCP ($Y_0 = 0.007$) at $P = 0$. Neutron scattering experiments have been performed recently with mono-isotopic Ni in order to avoid the large incoherent elastic scattering of natural Ni (Kadowaki et al. 2003). This allows one to detect the low-energy excitations at $\omega \sim 0.6$ meV around the AF vectors (1/2, 1/2, 0) and (0, 0, 3/4). This energy range is lower than the previous excitations at 4 meV centered at the incommensurate wavevector (0.23, 0.23, 0.5) (Fak et al. 2000). The magnetic fluctuation becomes independent of the temperature below 2 K. A crude fit of the specific heat can be obtained with this new low energy of 0.6 meV in the SF framework. But here again the low characteristic energy is finite despite the fact that Y_0 is considered near zero. This may support the previous statement on the $CeRu_2Si_2$ series of a non-divergence of the coherence length at P_C or reduction of QCP in $CeNi_2Ge_2$ to a lower negative pressure than that predicted.

Recently new thermodynamic measurements (notably coupled specific heat and thermal expansion) were realized on $CeNi_2Ge_2$ down to 50 mK by Küchler et al. (2003) (fig. 22). Now in agreement with AF spin fluctuation theory at a QCP, the Grüneisen parameter diverges at very low temperature

($\Omega^* = 70$ at $T = 1\,\text{K}$ goes up to 1000 at $T = 0.1\,\text{K}$). The observed temperature variation of C and α is predicted by 3d AF spin fluctuation theory ($C/T = \gamma - \alpha\sqrt{T}$, $\gamma(Y_0) = \gamma_{0-a}\sqrt{Y_0}$, $A(Y_0) \sim Y_0^{1/2}$ and $\Omega(Y_0) \sim Y_0^{-1}$ neglecting the pressure dependence of γ_0) (Moriya and Takimoto 1995). Thus thermodynamic measurements confirm the location of $CeNi_2Ge_2$ almost right at a QCP. The finite value of Γ_0 is still an enigma. The metallurgy of $CeNi_2Ge_2$ is quite sensitive to the chemical composition (see Cichorek et al. 2003). Appearance of superconductivity at P_C or P_V is still unclear (Grosche et al. 2001, Braithwaite et al. 2000). Differences may exist between large and small crystals as shown from studies on $CePd_2Si_2$. Thus, the location right at P_C may be not so obvious. A complete understanding of $CeNi_2Ge_2$ has still not been achieved.

Thermal conductivity experiments (Kambe et al. 1999) have been performed on $CeNi_2Ge_2$ to test if the Wiedeman–Franz law $\rho K/T = L_0$ is well obeyed close to $T \to 0\,\text{K}$, at least close to P_C on the PM side. Excellent agreement was found. Of course, the effect of the proximity to the magnetic instability appears in the thermal response of both ρ and K quantities. At least in this study, the charge carriers are also the heat carriers: the Wiedeman–Franz law is verified. Now, the remaining question is if the Wiedeman–Franz law also will be obeyed on approaching P_C from the AF side.

Pseudo-metamagnetism in $CeNi_2Ge_2$ has been found at $H_K = 42\,\text{T}$ but the microscopic origin of the cross-over field is different than that of $CeRu_2Si_2$ where an equal balance exists between the intersite and Kondo

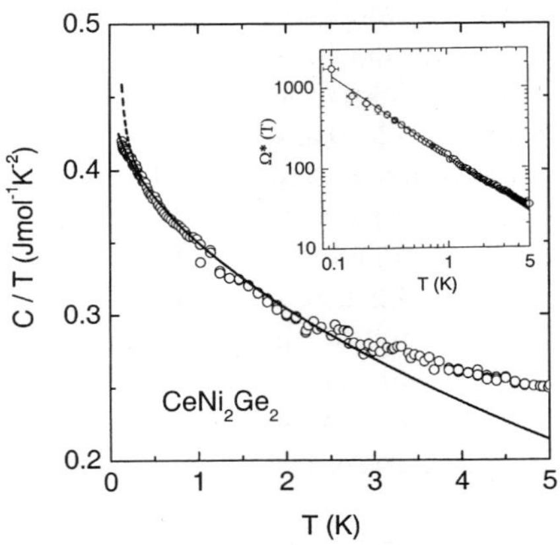

Fig. 22. Temperature variation of C/T and of Ω^* of $CeNi_2Ge_2$ (Küchler et al. 2003). From the raw data (dashed line at low T) a nuclear contribution has been subtracted giving the low T (open circles). The solid line is a fit with SCR-SF, i.e. assuming C/T varies as $C/T = \gamma_0 - C\sqrt{T}$.

coupling ($T_0 \sim T_A$ table Section 1) (Fukuhara et al. 1996). In $CeNi_2Ge_2$, T_0 and T_A are quite different, weak magnetic fields rapidly restore Fermi liquid behavior with a concomitant strong H decrease in γ (Gegenwart et al. 1999). Thus bare AF correlations may strongly collapse with the magnetic field or interfere rapidly with ferromagnetic fluctuations. However, high magnetic fields at $H > H_K$ are required to wipe out the Kondo lattice gap, i.e. for the breaking of the local Kondo fluctuations.

To summarize for the $CeNi_2Ge_2$ series, we propose that H_C is quite decoupled from H_K. For $P \rightarrow P_C$, $H_C \rightarrow 0$ while H_K is finite. The same phenomenon seems to occur for $YbRh_2Si_2$ (Tokiwa et al. 2004). Consequently, there is evidence that the SF approach and its excitation spectrum describes experiments to a first approximation as reported here for $CeRu_2Si_2$ or $CeNi_2Ge_2$ above P_C. However a zoom, for a given ω or T window, shows the need for theoretical and experimental improvement. The focus just at P_C or just on its PM side ignores the AF boundary as $P \rightarrow P_C$. In the reports on $CeCu_{6-x}Au_x$ (von Löhneysen 1994, 1996, 2000), $Ce_{1-x}La_xRu_2Si_2$ (Fisher 1991) or $Ce_{1-x}Pd_xNiGe_2$ (Knebel et al. 1999), a maximum of γ at P_C rarely exists. The dogma of a unique singularity at P_C must be challenged. We prefer to break the popular consensus of a QCP with a universal second-order transition.

2.5. *On the electron symmetry between Ce and Yb Kondo lattice: $YbRh_2Si_2$*

The recent fashionable material is $YbRh_2Si_2$, considered to be the hole Kondo lattice analog of $CeRh_2Si_2$. A strong similarity is expected between Yb and Ce intermetallic compounds with the difference being that, in Ce HFC, the pressure induces a transition from AF to PM while, in Yb HFC, the pressure induces a reverse effect with a transition from PM to AF. The stable trivalent state corresponds to low pressure for the Ce center and high pressure for the Yb center. In the trivalent Yb^{3+} configuration, the Kondo effect is produced by an absorption of an extra electron on the 4f shell leading to its full saturation (14 electrons on the 4f shell), while for Ce^{3+} the Kondo effect is created by the release of the electron from the 4f shell leaving an empty 4f shell.

Excellent crystals of $YbRh_2Si_2$ have been obtained by the flux technique but in the form of thin tablets. The novelty in $YbRh_2Si_2$ is that nice specific heat and resistivity anomalies (fig. 23) occur at $T_N = 70\,\text{mK}$ for $P = 0$ even for a tiny ordered moment ($M_o = 10^{-3}\,\mu_B$) (Trovarelli et al. 2000, Ishida et al. 2003). In Ce compounds, the corresponding anomalies become highly difficult to follow below $M_o \sim 0.1\,\mu_B$ i.e. $x < 0.2$ in the inset of fig. 23 (see von Löhneysen et al. 1996). This strongly suggests that the ytterbium case is not

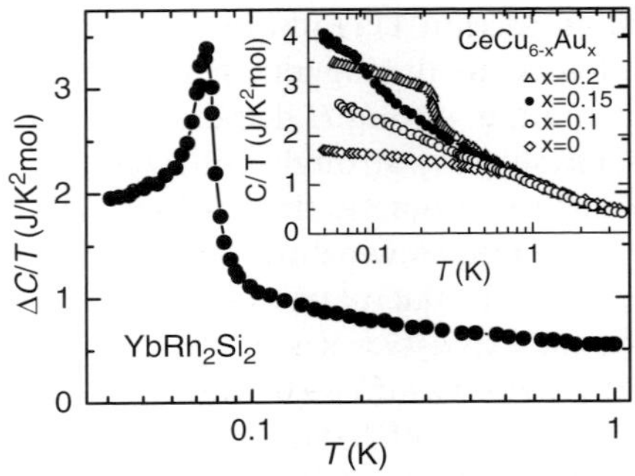

Fig. 23. Specific heat as $\Delta C/T$ vs. T (on a logarithmic scale) for $YbRh_2Si_2$. (from Trovarelli et al. 2000, Gegenwart et al. 2002). Comparison with the specific C/T of $CeCu_{1-x}Au_x$ in the border of the QCP ($x_c = 0.1$).

equivalent to the cerium case at least concerning their Kondo lattices close to P_C. Evidence for this is given by the observation of an electron spin resonance at low temperature (Sichelschmidt et al. 2003). Obviously, internal structures stabilize the existence of Yb^{3+} moments below T_K suspected to be near 25 K and even favor a large magnetic anisotropy. In the Yb Kondo lattice, the spin coherence appears to be preserved during a long period and thus the electronic spin may be sensitive to any fine structure. Even the bare hyperfine interaction A between the nuclear spin I and the localized spin S is not a small perturbation (AIS~1 K for some Yb nuclei). Four different stable isotopes exist for Yb. Two, ^{171}Yb and ^{173}Yb with the isotopic abundance of 14 and 16%, have nuclear spin of $I = 1/2$ and $I = 5/2$ (see Flouquet 1978) A large variety of phenomena must be considered with different spin and orbital channels. The hyperfine coupling AIS may play the role of a cut off, which must be compared with the characteristic energies $k_B T_K$ for the single Kondo impurity (see Frossati et al. 1976), or $k_B T_{KL}$ or $k_B T_N$ for the Kondo lattice. It is amazing that when $T_N = 20$ mK in the studies of $YbRh_2Si_2$ doped with 5% of Ge (Custers et al. 2003), the critical field H_C~200 Oe for restoring the PM state is roughly the field where the electronic and nuclear spin of ^{171}Yb and ^{173}Yb will become decoupled. In order to test if the hyperfine coupling plays a role, a crystal was measured with monoisotopic ^{174}Yb which, has no nuclear moment. As for the natural Yb case, the specific heat shows a very pronounced peak at $T_N = 80$ mK and also the linear temperature dependence of the resistivity just above T_N persists. Thus the particular effects of this Yb compound is not due to a fancy hyperfine dynamic but must be linked to the microscopic description of the 4f ytterbium electrons.

It is our opinion that the electron–hole symmetry between Ce and Yb may be not so significant as compared with the fact that their respective couplings with d electrons are not equivalent. Of course, there is similarity between $YbRh_2Si_2$ and $CeRh_2Si_2$. For the same strength of T_K there are similarities in the amplitude of the maxima of the resistivity, the T-linearity of ρ just above T_N, and the shape of the resistivity anomaly at T_N. But there are also differences such as the sharpness of the 4f density of states in $YbRh_2Si_2$ by comparison to $CeRh_2Si_2$ found in band calculations with in the local density approximation (Harima 2004). The 4f orbital of Yb is far more localized (0.25 Å) than that of Ce (0.37 Å) (Waler and Cromer 1965). Thus the degree of hybridisation must be higher in Ce HFC than in Yb HFC. In the picture of virtual bound states, the width Δ_{Yb} will be an order of magnitude smaller than that of Ce in the same non-magnetic lattice. Thus, for a given T_K^{3+}, the departure $(1 - n_f)$ of the occupation number n_f from unity will be 10 times higher for Yb than for Ce the case. This may lead to drastic differences in the appearance of long-range magnetism. Let us point out that a supplementary factor to preserve the local character of the Yb atoms is that the spin-orbit coupling λ'_{SO} of each 4f electron in the $j = \ell - s$ and $j = \ell + s$ individual configuration (see Abragam and Bleaney 1970) is one order of magnitude higher than in the cerium case, i.e. in the Yb case, $\lambda'_{SO} \gg \Delta \sim C_{CF}$; in the Ce case $\lambda'_{SO} \sim \Delta > C_{CF}$. The hidden problem is the role of the strength of the hybridization (Δ, T_K and n_f) on the crystal field and thus the relation between C_{CF} and T_K, which will govern Ising, planar or Heisenberg spin dynamics.

A manifestation of the difference between $YbRh_2Si_2$ and $CeRh_2Si_2$ (see Section 4, fig. 29) is the pressure dependence of their Néel temperatures. In $CeRh_2Si_2$ basically a collapse of T_N occurs at $P_C = 1\,GPa$ with a corresponding broadening and collapse of the specific heat anomaly. In $YbRh_2Si_2$, (Plessel et al. 2003, Knebel et al. 2005, Dionicio et al. 2005 – fig. 24), increasing the pressure leads to a first regime where T_N increases at the rate of $\partial T_N/\partial P \sim 0.4\,K/GPa$ up to $P_1 = 2\,GPa$; then T_N reaches a smooth pressure variation with a maximum of $T_N \sim 1\,K$ between 2 and 7 GPa. Suddenly, above $P^* \sim 9\,GPa$, T_N strongly increases up to 12 GPa with a slope $\partial T_N/\partial P \sim 0.8\,K/GPa$. Finally above 12 GPa, T_N seems to saturate to $T_N \sim 8\,K$ with a sublattice magnetization $M_o \sim 2\,\mu_B$ (Plessel et al. 2003).

Our feeling is that due to the already reported differences in valence mixing between Ce and Yb, the Yb^{3+} configuration can live longer than the Ce^{3+} one. Similar phenomena have been recently observed for intermediate valence compounds of Sm where the Kondo effect on Sm^{3+} also corresponds to an absorption of a 5d electron in its 4f shell (Barla et al. 2004, 2005). Even if the d electrons from the Rh ions form a narrow band, locally

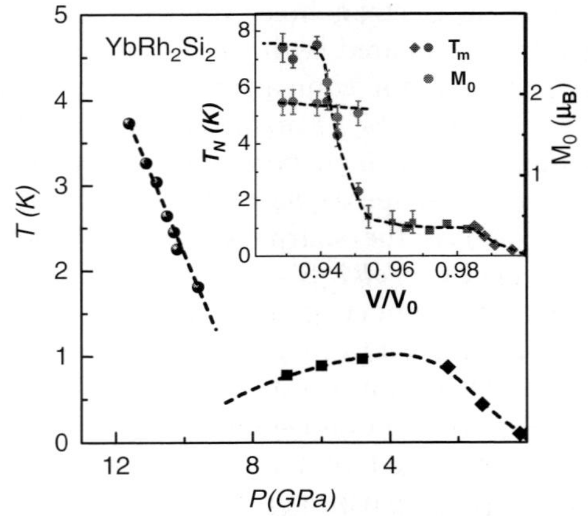

Fig. 24. The inset shows the pressure–temperature magnetic phase diagram for $YbRh_2Si_2$ based on the experimental data for the ordering temperature T_N determined by Mössbauer effect (●) and electrical resistance (■) and (◆) (Plessel et al. 2003). The inner part show the recent result obtained with resistivity and microcalorimetric experiment under pressure (Knebel 2005).

they will be trapped on the Yb site. Increasing the pressure will delocalize strongly the d electron, i.e. decrease the f–d correlation and thus lead to recovering a situation rather antisymmetric from the Ce case. Following the idea that for the same T_K^{3+}, $(1 - n_{Yb}) \sim 10\ (1 - n_{Ce})$, the new phenomena in the Yb case is that AF can occur for a relative large departure of n_f from unity. In this domain, the magnetic interaction will not be given by the simple RKKY dependence of E_{ij} in $\Gamma^2 N(E_F)$ but will contain a strong dependence on n_f. The competition betsween T_K and E_{ij} (n_f) can lead to different pressure regimes with the recovery of the Doniach picture only when $n_f \to 1$, i.e. for $P > P^*$.

The unusual temperature variations of the specific heat and of the resistivity just above T_N and of their field dependences in $YbRh_2Si_2$ and $YbRh_2Si_{2-x}Ge_x$ at $P = 0$ was a supplementary boost toward the possibility of local criticality. In the spirit of the breakdown of the heavy Fermi surface, proposed for $CeCu_6$, a divergence of m^* at the transition field H_C from AF ($T_N = 80\,mK$, $H_C = 600\,Oe$) was recently suggested (Gegenwart et al. 2002). The claims of the divergence of m^* by complementary studies on $YbRh_2Si_{1.95}G_{0.05}$ ($T_N \sim 20\,mK$, $H_c = 200\,Oe$) leads to the statement that: "all ballistic motion of electron vanishes at the magnetic quantum critical point H_C forming a new class of conductor in which electrons decay into collective current carrying motions of the electron fluid" (Custers et al. 2003).

In $YbRh_2Si_2$ $(Si_{0.95}Ge_{0.05})_2$, $\gamma(H)$ has mainly a log H decrease above $H^* = 10H_C$. Below $10H_C$, in the large field region $10H_C > H > H_C$, $\gamma(H)$ seems to vary like $((H-H_C))^{-0.33}$; however there is no convincing evidence that m^* diverges at H_C since for the closest value reported of H to H_C, $\gamma(H_C + \varepsilon)$ is quite similar to the zero field value $\gamma_0 = 1.7\,\mathrm{J\,mol^{-1}\,K^{-2}}$ measured for the pure lattice of $YbRh_2Si_2$. The large field quantum regime $\Delta H_C \sim 2000$ Oe, where $\gamma(H)$ increases strongly on approaching $\gamma(H_C)$, is quite comparable to that found in $CeRu_2Si_2$ at H_M. As in this field window, AF and F compete (Ishida 2002) it is not surprising that the Kadowaki–Woods ratio is not observed and that A/γ^2 increases as $H \to H_C$. The relevance of 2d critically with strong local fluctuations (Si 2001, 2003, a, b) may be favored by the large anisotropy between the susceptibility χ_a and χ_c, respectively ⊥ and //, to the c-axis ($\chi_a/\chi_c = 200$). Thus the local ion has a clear planar anisotropy. Preliminary dHvA experiments with the detection of only a few orbits does not show a 2d character, which will reinforce the hypothesis of 2d fluctuations. The results on $YbRh_2Si_2$ continue to favor new theoretical developments such as fractionalization of the Fermi surface (Pépin 2004) and the underscreened Kondo model (Coleman and Pépin 2003).

To summarize, $YbRh_2Si_2$, as for $CeRu_2Si_2$, is a clean system with a simple axial symmetry. At $P = 0$, its position right on the AF side of P_C opens a view which is quite complementary to that achieved in $CeRu_2Si_2$ or $CeNi_2Ge_2$, where studies concern the PM phase at $P = 0$. Furthermore, the planar local anisotropy of the spin in $YbRh_2Si_2$ is quite different from the respective Ising and Heisenberg characters found in $CeRu_2Si_2$ and $CeNi_2Ge_2$. More systematic measurements on Yb HFC need to be realized to specify its microscopic description. This includes low-energy experiments as well as high-energy spectroscopy.

3. Unconventional superconductivity

Essential points:

- Conventional superconductivity: only gauge symmetry is broken.
- Unconventional superconductivity: additional symmetry is broken.
- In HFS, interplay between the orbital and Pauli limit of the superconducting upper critical field H_{C_2}.
- In conventional superconductivity, antiferromagnetism and superconductivity may coexist while ferromagnetism usually destroys superconductivity.
- Unconventional superconductivity can be found in F and AF spin fluctuation approaches. Magnetic and electronic anisotropies are favorable factors to increase the pairing strength.

- For a Kondo lattice, an efficient Cooper pairing mechanism is also density fluctuations.

3.1. *Generalities*

Usual conventional superconductors are well described by the Bardeen Cooper Schrieffer theory (BCS) first established for a s-wave singlet state with pairing of electrons with opposite spin and zero angular momentum. The order parameter $\Delta(k)$ is mainly isotropic. When the superconducting transition occurs in conventional superconductors, gauge symmetry is the only symmetry broken at the superconducting transition. Due to strong coulomb repulsion among the f electrons, the existence of conventional s-wave Cooper pairs with finite amplitude on a given site is precluded. This prohibition can be overcome with an anisotropic pairing like the triplet p-wave (case of ^{3}He) or the spin singlet d-wave channel. We will see that ferromagnetic fluctuations or antiferromagnetic fluctuations can lead to these two situations.

In unconventional superconductivity, another symmetry is broken: point group, odd parity or time reversal. The latter occurs when the superconducting state has orbital, spin moments or odd frequency pairing. $\Delta(k)$ can be written for even- and odd-parity pairing respectively as

$$\Delta(k) = \Psi(k)\mathrm{i}\sigma_\mathrm{y}$$

$$\Delta(k) = (\sigma \cdot \hat{d}(k))\mathrm{i}\sigma_y$$

where $\Psi(k)$ and $\hat{d}(\mathrm{k})$ are, respectively, even scalar and odd vector functions of the momentum k; σ is the Pauli spin matrix. Often due to the additional broken symmetry, $\Delta(k)$ vanishes on point nodes or lines of nodes on the FS. The occurrence of zeros will allow low-energy excitations and produce temperature power-law dependences of transport and thermodynamic quantities. This contrasts with the exponential collapse of the number of low-energy thermal excitations for s-wave superconductors. As gapless superconductors can also have power law temperature dependence, the proof of unconventional superconductivity requires careful studies as a function of the purity (i.e. of residual resistivity, for example). Important support can be given by the discovery of multiple superconducting phases (case of UPt_3), triplet spin pairing (via NMR or other related techniques), of anisotropy in the thermal conductivity, in the penetration depth and in ultrasound or of a violation of time-reversal symmetry. Classification of the different order parameters based on group theory arguments has been given in Gorkov (1987). We will later discuss the case of UPt_3 and UPd_2Al_3 in more detail.

Non-magnetic impurities are effective pair breakers in unconventional superconductors since impurity scattering destroys the anisotropic Cooper pair wavefunction. In conventional superconductors, only magnetic impurities are efficient pair breakers as explained by Abrikosov and Gorkov (1961). Furthermore, it was stressed by Pethik and Pines (1986) and Schmidt-Rink et al. (1986) that in order to explain the results of ultrasound or thermal conductivity in UBe_{13} or in UPt_3, a large phase shift $\delta = \pi/2$ (in the strong scattering unitary limit) must be assumed. This assumption seems to be a "rule" that is now applied to all strongly correlated electronic systems. For unconventional superconductivity, it is generally admitted that the clean limit must be achieved, i.e. an electronic mean free path $\ell \gg \xi_0$ where ξ_0 is the superconducting coherence length. Generally as $\ell \sim \rho_0^{-1}$ increases, T_C increases with increases in the values of the upper critical field. We will discuss later how the impurity and pressure gradient can modify the (P, T) contour of superconductivity. For s-wave dirty superconductors, the addition of non-magnetic centers is a well-known process that increases $H_{C_2}(T)$ without drastically changing T_C. In unconventional superconductors (HFS), any impurity will be pair-breaking, thus both $H_{C2}(T)$ and T_C depend strongly on ℓ. Improving the mean-free path leads to obtain the optimal values of T_C and $H_{C2}(0)$.

Before focusing on the mechanism of unconventional superconductivity driven by spin fluctuations, let us stress the particular situation of heavy fermion systems with regard to the field restoration of the normal phase at $H_{C2}(T)$. The consequence of a large effective mass m^* is that the orbital limitation of $H_{C2}(T)$ is given by

$$H_{C2}^{orb}(T) = \frac{\Phi_0}{2\pi\xi^2(T)} \quad \text{with} \quad H_{C2}^{orb}(T=0) \approx m^{*^2} T_C^2,$$

which is large since the coherence length goes as $(m^*)^{-1}$,

$$\xi_0 = 0.18 \frac{\hbar k_F}{m^* k_B T_C}$$

(k_F and k_B being, respectively, the Fermi wavevector and the Boltzman constant).

At $T \sim T_C$, the dominance of the orbital limitation leads to an initial linear temperature variation of H_{C2}. As the orbital contribution can be large, the breaking of the Cooper pair by the Zeeman effect (the so-called Pauli limit H_P) can be effective at low temperatures. For s-wave superconductors, as $T \to 0\,K$, $H_P(0)$ is equal to

$$H_P(0) = \frac{\sqrt{2}}{g\mu_B} \Delta_0 = 1.85 T_C \ \ (T)$$

assuming $g = 2$ for the g factor of the conduction electrons. It will be effective for unconventional superconductors when the spin susceptibility decreases below T_C. For triplet pairing, when H is perpendicular to d (direction where the spin component S_z is zero), no Pauli limit occurs. For s-wave superconductors, the reference article for $H_{C2}(T)$ is Wertharmer et al. (1966). Scharnberg and Klemm (1988) derived $H_{C2}(T)$ for a p-wave triplet superconductor assuming no effective mass anisotropy.

When the Pauli limit dominates, for s-wave superconductors in the case of the clean limit (generally required), Fulde and Ferrel (1964) and Larkin and Ovchinnikov (1965) (FFLO) predicted that the entrance in the superconducting state below $T_{FFLO} \sim 0.56 T_C$ is not the isotropic state but a spatially modulated structure. When the orbital limitation becomes comparable to the Pauli limit, T_{FFLO} is lower and the occurrence of the FFLO state is governed by the strength of the Maki parameter defined by

$$\alpha = \sqrt{2}\frac{H_{C2}^{orb}(0)}{H_P(0)} = 0.27g\left(\frac{\partial H_{C2}}{\partial T}\right)_{T_C}$$

Favorable conditions seems to exist in $CeCoIn_5$, UPd_2Al_3 and URu_2Si_2 ($H//c$) (Brison et al. 1997). When the orbital limit H^{orb} (0)$<1.27H_P$ (0), the FFLO state disappears. The analysis of the superconducting properties of UBe_{13} will lead us to discuss the interplay between different mechanisms as well as new effects due to strong coupling. Evidence of a FFLO state has been recently invoked in the new HFC $CeCoIn_5$ (see Section 4). It was proposed for UPd_2Al_3 (Gloos et al. 1993) and then rejected (Norman 1993) and suggested in UBe_{13} (see Section 6) but without confirmation.

As in heavy fermion systems, T_C may be comparable to the effective Fermi temperature $T_F^* \sim T_K$, and strong coupling can be considered. For $T_C \sim T_K$, the thermal disorder leads to a decrease in T_C; however on cooling, the superconducting properties will be reinforced. For s-wave superconductors in an Einstein phonon model, it can be shown that $\Delta(0) \sim \sqrt{\lambda} k_B T_C > 1.76\ k_B T_C$ for $\lambda \gg 1$. The extreme strong coupling phenomena are rare in HFC. Examples may be found in UBe_{13} and $CeCoIn_5$ according to the criterion based on the specific heat jump (ΔC) at T_c since for strong coupling superconductors $\Delta C/C$ can surpass the value (1.43) predicted for a BCS superconductor ($T_C < T_F$). However, this is an artifact in that, at T_C, the Kondo lattice may not have achieved its zero temperature value thus leading to a larger $\Delta C/\gamma T_C$ ratio if γ is identified with C/T at T_C.

3.2. *Magnetism and conventional superconductivity*

The pair-breaking of s-wave superconductivity by paramagnetic impurities was formulated by Abrikosov and Gorkov (1961). Experimental and theoretical discussions can be found in the volume *V* of the series Magnetism (Maple 1973, Fischer and Peter 1973, and Müller-Hartmann). The theory was extended to Kondo impurities (Müller-Hartmann and Zittartz 1971). We will focus here on the interplay between long-range ordered magnetism and superconductivity.

The coexistence of antiferromagnetism and superconductivity is well established in conventional s-wave superconductivity. Basically on the scale of the superconducting coherence length, the Cooper pairs feel zero exchange interaction. The field was very active after the discovery of the ternary compounds of rare-earth (RE) elements and molybdenum sulfide (RE Mo_6S_8) in 1975 and of a series of rhodium boride alloys (RE Rh_4B_4) in 1977 (see Fischer 1978, 1990). It has been revived with the appearance of superconductivity in the new borocarbide family (RE Ni_2B_2C) in 1994. The major improvement brought by this family was the possibility to grow large crystals of high quality, which allow a large diversity of experiments (Canfield et al. 1998). The interesting features are the initial value of T_C (16 K for $LuNi_2B_2C$) and the persistence of superconductivity even when T_N becomes larger than T_C ($T_N \sim 10$ K, $T_C \sim 5$ K for $DyNi_2B_2C$). For further reading on the interplay between antiferromagnetism and superconductivity in borocarbide systems, the review article by Müller (2001) and by Canfeild et al. (1998) as well as the recent review of Thalmeier and Zwicknagl (2004a) are recommended.

The problem of the interplay between s-wave superconductivity and ferromagnetism is more intriguing. It was continuously discussed in the past decades starting with the first paper by Ginzburg (1957), who points out that "the probability of finding superconductivity of ferromagnets in ordinary measurements is as small as that of finding non-ferromagnetic superconductors placed in an external field with a magnetization of several thousand oersted". At this time, before the discovery of type II superconductivity, the first consideration was on the internal magnetic induction $B_0 = 4\pi M_0$ created by the magnetization density. Typical values are reported in Table 6.

Ferromagnetism places stringent limits on the existence of superconductivity. It can easily break the Cooper pair. In the experiments on erbium rhodium boride ($ErRh_4B_4$) and holmium molybdenum sulfide ($HoMoS_8$), it is now well established that superconductivity is destroyed by the onset of a first-order ferromagnetic phase transition. For example, $ErRh_4B_4$ is a superconductor below 8.7 K. When it is cooled to a temperature $T_m \sim 1$ K a

TABLE 6
Ferromagnetic material and value of the internal magnetic induction B_0 in a single domain.

Material	B_0 (Gauss)
Fe	22,000
Co	18,500
Ni	6,400
Gd	24,800
UGe_2	2400
$ZrZn_2$	400

modulated magnetic structure appears rather than ferromagnetic ordering. Neighboring magnetic moments are aligned in the same direction but a magnetic modulation occurs with a period $d < \xi_0$. The material is not ferromagnetic but presents domain like structures of period $d/2$ which have been detected by neutron diffraction ($d \sim 80$ Å, $\xi_0 = 200$ Å) (see Fischer 1978, 1990). From the point of view of superconductivity, this magnetic structure is like an antiferromagnetic one but almost ferromagnetic on atomic scales. In these compounds, the energy gained by the atoms through the magnetic ordering $T_N \sim \Gamma^2/T_F$ exceeds the gain related to the superconducting transition as the Cooper pair modifies the electronic spectrum in a very small energy window $k_B T_C \ll k_B T_F$. As the number of Cooper pairs is small ($T_C/T_F \ll 1$), the energy gain per atom due to the Cooper condensation is low, $k_B\, T_C^2/T_F < k_B T_N$. At least for static magnetic centers, the magnetism is a much more robust phenomenon than superconductivity. Superconductivity cannot prevent the magnetic transition and is able only to modify it slightly. (In SCES this argument must be revisited as the magnetic ordering is driven by $k_B T_K$ and the pairing energy per atom becomes comparable to $k_B T_K$ if $\Delta \sim k_B T_K$ since $T_F \sim T_K$: superfluidity of fermions may look like boson superfluidity). However such a phase is not stable at the lowest temperatures as the creation of microdomains costs energy. Cooling to $T_{Curie} = 0.8$ K brings $ErRh_4B_4$ via a first-order transition into a ferromagnetic phase with the disappearance of superconductivity (full restoration of the resistivity). The ferromagnetic phase destroys superconductivity as the exchange field, acting on the light conduction electrons, exceeds T_C (see Flouquet and Buzdin 2002).

Superconductivity and weak ferromagnetism were found to coexist in $ErNi_2B_2C$ below $T_{Curie} = 2.3$ K (Gammel et al. 2000). The ferromagnetic component ($0.33\,\mu_B$ /Er atom) is weak as only one of the 20 Er atoms contributes to the ferromagnetic order (periodicity near 35 Å) (Choi et al. 2001, Kawano-Furukawa 2002, Deflets et al. 2003). Up to now, a spontaneous flux lattice has not been observed at $H = 0$ as suggested by Ng and Varma (1997) but enhanced critical currents characterize the entrance into the weakly ferromagnetic state (Gammel et al. 2000). Ferromagnetism and superconductivity have been claimed to coexist in $Eu_{1.5}Ce_{0.5}RuSr_2Cu_2O_{10}$ (Felner et al. 1997) and $RuSr_2GdCu_2O_8$ (Bernhard et al. 1999). In these cases, the two states might occur in different structural layers. An exotic case is that of nuclear ferromagnetism ($T_{Curie} = 37\,\mu K$) on the superconductivity of $AuIn_2$ ($T_C = 207\,mK$, $H_C = 1.45\,mT$) (Rehmann et al. 1997). However it was suggested that the contact hyperfine exchange interaction may also lead to a domain structure as described for $HoMo_6S_8$ and $ErRh_4B_4$ (Kulic et al. 1997).

In these cases, different electrons are involved in magnetic and Cooper pairing. Switching to the heavy fermion case we will now assume that the same electrons are involved in the magnetism and superconductivity. Of course, unconventional superconductivity and ferromagnetism can occur as emphasized previously for triplet superconductivity and notably for equal spin pairing (EPS) between parallel (up–up or down–down) spins. In such a case, the exchange field cannot break the Cooper pair via the Pauli limit and the upper critical field limitation will be determined only by the orbital limit, which can be very high if the effective mass is huge. The discovery of superconductivity in the ferromagnet UGe_2 with a Curie temperature $T_{Curie}\sim30\,K$ far higher than $T_C\sim0.7\,K$ at its optimum $P_{opt} = 1.3\,GPa$ opens new perspectives (Section 5).

3.3. *Spin fluctuations and superconductivity*

The relevance of nearly ferromagnetic spin fluctuations for anisotropic BCS states was illustrated by the p-wave superfluidity of liquid ^3He (Anderson and Brinkman 1973, Nakajma 1973). The p-wave superconducting transition temperature for paramagnon-induced pairing in nearly ferromagnetic itinerant systems was first calculated by Layzer and Fay in 1971. The vanishing of T_C at P_C is correlated with the vanishing of T_I, i.e. the divergence of the effective mass as the pairing is a strong function of frequency. In the paramagnetic state, T_C is the same for parallel and antiparallel spin pairs. Of course, a different situation (fig. 25) will occur in the ferromagnetic state calculated by Fay and Appel (1980). The ESP interaction between $\uparrow\uparrow$ or $\downarrow\downarrow$ component of the triplet with an angular momentum's transfer q is related to the non-interacting Lindhard response of the spin susceptibility $\chi_0^{\uparrow}$ or $\chi_0^{\downarrow}$ and the onsite Coulomb repulsion U by the relation

$$V^{\uparrow\uparrow} = \left(\frac{U^2\chi_0^{\downarrow\downarrow}}{1 - U^2\chi_0^{\uparrow\uparrow}(q)\chi_0^{\downarrow\downarrow}(q)}\right).$$

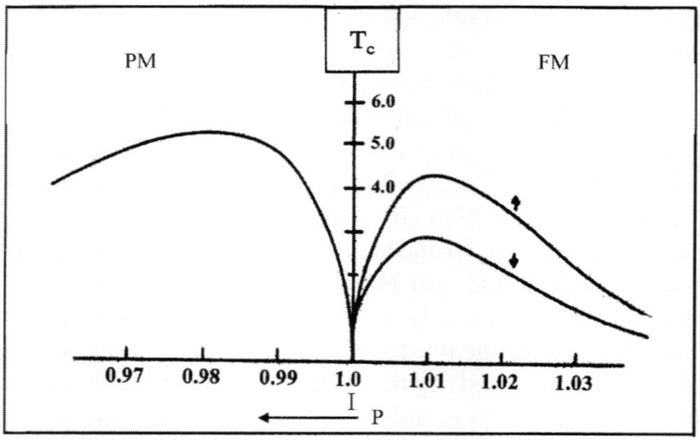

Fig. 25. First prediction for superconducting transition T_C of a ferromagnet under pressure P (Fay and Appel 1980).

Due to spin conservation, the exchange of spin waves or transverse fluctuations does not contribute to $V^{\uparrow\uparrow}$. As for the PM phase, T_C reaches a maximum before collapsing at P_C. Figure 25 shows the variation of T_C versus I or $P-P_C$. In the ferromagnetic domain, T_C differs for the majority (↑) and minority (↓) spin.

A more complete treatement (Roussev and Millis 2001) shows that T_C does not vanish at P_C but reaches a minimum in the vicinity of P_C. Furthermore, for small values of $(P-P_C)$, the T_C behavior is predicted to be universal. The coexistence of ferromagnetism and superconductivity was revisited recently by Kirkpatrick and Belitz (2003). We have already pointed out that an enhanced impurity scattering occurs at P_C (Miyake and Narikiyo 2002).

The discussion of superconductivity induced by antiferromagnetic fluctuations has been boosted by the discovery of superconductivity in high T_C oxides and the rapid demonstration that the spin dynamics plays an important role in the normal phase properties and reacts with the onset of superconductivity (see Rossat Mignod et al. 1992). Early considerations on superconductivity mediated by antiferromagnetic fluctuations can be found in Emery (1983), Hirsch (1985), Miyake et al. (1986). Reviews have been written on the AF spin fluctuation model and d-wave superconductivity ($d_{x^2-y^2}$ for high T_C) (Moriya and Ueda 2000, 2003b, Chubukov et al. 2002) with discussions of the relevance to the different SCES. It was stressed that a key point is the occurrence of the so-called hot spots in the Fermi surface, i.e. peaked magnetic coupling at the wavevector k_o, and that d-wave pairing is favored more by AF spin fluctuation than magnon like excitations. Two distinct frequencies have strong impact on the pairing: the characteristic spin fluctuation energy $\hbar\omega_{sf}\sim k_B T_1$ and the energy $\hbar\tilde{\omega}_0\sim k_B T_0$ related to the effective Fermi temperature.

The d-wave spin pairing in nearly AF systems is generally stronger than triplet pairing in the nearly F case as both longitudinal and transverse fluctuations can mediate superconductivity (Monthoux and Lonzarich 2001). For comparable parameters, the strength of the pairing increases with the magnetic, α_m, and electronic, α_t, anisotropy (Monthoux and Lonzarich 2003) independent of the type of the magnetic instability (ferromagnetic or antiferromagnetic). Few theoretical studies address the respective boundary of P_C and the pressures P_{-S}, P_{+S} where superconductivity emerges and disappears ($T_C = 0\,\mathrm{K}$ at P_{-S} and P_S). A recent discussion can be found in (Takimoto and Moriya 2002) where different cascades of ground states are pointed out: for example a first-order transition from commensurate spin wave (SDW) to superconductivity (*S*), a cascade of SDW to incommensurate (*I*) SDW followed by the coexistence of ISDW and *S*, then of *S* and PM and finally PM phases. Due to the occurrence of soft modes near electronic, magnetic or structural instabilities, very often T_C reaches its maximum at the instability point in agreement with the famous MacMillan optimization between coupling and energy cutoff. A good example is given by the disappearance of the charge density wave of α uranium at P_{CDW} where T_C has its maximum (see Smith and Fisher 1973). Let us indicate the new approach where the orbital splitting energy ε plays a key role of control parameter for the AF and *S* quantum phase transition in HFC (Takimoto et al. 2003). As ε increases, i.e. for example C_{CF}/k_B versus T_K, the system transits from PM, and AF with a superconducting dome right at P_C (see also Hotta and Ueda 2003).

3.4. *Atomic motion and retarded effect*

An important aspect of HFC is their huge Grüneisen parameters, which points out huge anharmonicity or strong mixing between different modes. Neglecting the electron, in a harmonic crystal, the coefficient of the thermal expansion (α) must be zero as the pressure required to maintain a given volume does not vary with temperature. Anharmonic terms play a big role and are responsible for the phonon thermal expansion $\alpha\sim T^3$. A typical phonon Grüneisen parameter is $\Omega(\theta_D)\sim 2$ for the associated phonon contribution which is characterized by its Debye temperature θ_D (Aschcroft and Mermin 1976). The electronic contribution for a normal metal such as copper will lead to a linear temperature thermal expansion ($\alpha\sim T$) as its Fermi temperature varies as $V^{-2/3}$ and effective mass is near the free electron mass. Again the corresponding Grüneisen parameter is weak $\Omega(T_F) = +0.66$. In HFC, as emphasized before, Ω^* can reach 100 or 1000. The electronic thermal expansion $\partial V/\partial T$ goes basically as m^{*2} (i.e. $\partial V/\partial T \sim \Omega^* m^*$ and $\Omega^* \sim m^*$). An amplification of four orders of magnitude of $\partial V/\partial T$ can be achieved by comparison to a normal metal at low temperature. Thus the displacement of the atom can reach a value at 1 K only achieved in an ordinary metal above 100 K through the volume variation of the Fermi temperature or through the phonon anharmonicity. The density fluctuation of HFC is very large. In conventional superconductors, although the direct electrostatic interaction is repulsive, the ion motion overscreens the coulomb interaction and leads to attraction (see Aschcroft and Mermin 1976). As the interaction spreads over an energy interval $k_B\theta_D$, it operates over a finite interval.

In HFC the two ingredients for pairing, such as for the electron–phonon coupling, may exist: atomic motion linked to large density fluctuation and retarded effect linked to the long lifetime τ_{KL} of the Kondo cloud or the spin fluctuation near P_C. Thus the picture of its superconductivity may be rather similar to that developed for the electron–phonon interaction. In the early times of discovery of $CeCu_2Si_2$ it was proposed that the electron phonon mechanism can explain the superconductivity of $CeCu_2Si_2$ (see Razafimanchy et al. 1984, Tachiki and Maekawa 1984, Ambrumenil and Fulde, 1985,). As pointed out in Section 4, the Kondo coupling favors longitudinal fluctuations and thus the difference in pairing between AF and F may be not so high. An interesting idea is the possible difference due to large retarded effects close to P_C (long lifetime of excitations) and instantaneous coupling far from P_C (see Fuseya et al. 2003). Near P_C, the possible achievement of p-wave spin singlet superconductivity with a gap function $\Delta(k, i\omega)$ odd in momentum and frequency has been found to be more likely than d-wave singlet superconductivity. This phase will have no gap in the quasiparticle spectrum anywhere on the FS due to the odd frequency. It will exist on both sides of P_C if P_C is a second-order QCP. The conditions for this stabilization in a narrow pressure range around P_C are that (1) The FS is not nested, i.e. basically the origin of AF is the exchange interaction between the local spin components, (2) strongly retarded effects which occur at $P \sim P_C$ where the effective interaction is strongly frequency-dependent. A crucial theoretical point is to clarify the Meissner effect since at a simple level of calculation a negative Meissner effect has been found in odd frequency pairing (Abrahams et al. 1995). Evidence of gapless superconductivity close to P_C in HFC has been found in the $CeCu_2(Si_{1-x}Ge_x)_2$ series (Kitaoka et al. 2001) and will be reported here for $CeRhIn_5$ (Kawasaki et al. 2003b, Knebel et al. 2004). In this last example the associated diamagnetic response is broad and shifted to temperatures higher than T_C. The tiny specific heat or NMR superconducting anomalies at T_C might be due to extrinsic properties. The debate on intrinsic gapless properties is still open.

4. Superconductivity and antiferromagnetic instability in cerium compounds

Essential points:

- A universal second-order singularity at P_C may not be the correct principle.
- Observation of a magnetic phase separation or first-order transition in $CeIn_3$, $CeRh_2Si_2$ and $CeRhIn_5$.
- Two distinct superconducting phases in $CeCu_2Si_2$: two mechanisms?
- The new 115 series: importance of quasi-bidimensional fluctuations.
- A new field-induced superconducting phase in $CeCoIn_5$.
- T_C dependence on the anisotropy ratio c/a between c and a lattice parameters; the record with $PuGaIn_5$.
- Recent exotic superconductors: $CePt_3Si/PrOs_4Sb_{12}$.

4.1. *Superconductivity near a magnetic quantum critical point $CeIn_3$, $CePd_2Si_2$ and $CeRh_2Si_2$*

For clarity, we will not use the chronological order of the discovery of superconductivity near P_C but select successive systems where the magnetic instability can be tuned under pressure. In the chosen examples of $CeIn_3$, $CePd_2Si_2$ and $CeRh_2Si_2$, P_C is respectively 2.6, 2.8 and 1.0 GPa and the superconductivity domain is centered around P_C. In the cases of $CeCu_2Si_2$ and $CeCu_2Ge_2$ superconductivity appears just below P_C but its temperature maximum T_C^{max} is shifted to higher pressures. We will discuss this situation later. The main difference seems to be that for the first three examples, P_C and P_V coincide or at least are very near to each other while in the $CeCu_2(Si_{1-x}\,Ge_x)_2$ family, P_C and P_V are quite distinct.

4.1.1. *$CeIn_3$: phase separation*

At $P = 0$, $CeIn_3$ is a AF HFC with $T_N = 10\,K$, $M_o = 0.5\,\mu_B$ and (1/2, 1/2, 1/2) propagation vector (Lawrence and Shapiro 1980, Benoit et al. 1980). The spin dynamics at zero pressure are now very well documented by neutron scattering experiments with the determination of the crystal field splitting ($C_{CF} = 10\,meV$), the observation of a quasi-elastic line and damped spin waves (Knafo et al. 2003). Neutron diffraction experiments under pressure have already suggested that AF in the cubic lattice of $CeIn_3$ will collapse near $P_C = 2.6\,GPa$ (Morin et al. 1988).

P_C is marked by a deep minimum of the exponent n of the temperature dependence of the resistivity $\rho = A_n T^n$. Fermi liquid properties are recovered on both sides of P_C; n reaches 2 for $P > 3.7\,GPa$ above P_C and a value $n > 2$ for $P = 2.4\,GPa$ below P_C as the spin wave contribution will add an extra scattering contribution to ρ (fig. 26). It is worthwhile to reemphasize that P_C can be more easily defined via the deep minimum of n than by its resistivity anomaly at T_N which becomes difficult to detect in the vicinity of P_C. A strong broadening of the resistivity anomaly at T_N has already been pointed out 1 GPa below P_C (Knebel et al. 2002). The analysis of the resistivity shows that P_C and P_V coincide. dHvA experiments under pressure have just been performed through P_C (Settai et al. 2003). In the experiments, only a slight change occurs for the spherical FS (referred to as d, d′) with a hump along (1,1,1). At least the effective mass increases smoothly on approaching P_C (factor 1.3 and 2, respectively, for d and d′). The recent publication of Ebihara et al. (2004), Endo et al. (2004) and Settai et al. (2005) report new features in magnetic field and pressure not discussed in the previous works of Settai et al. (2003).

Experiments on a high-quality crystal in Cambridge ($\rho_0 = 1\,\mu\Omega\,cm$) (Mathur et al. 1998, Grosche et al. 2000) show that superconductivity occurs

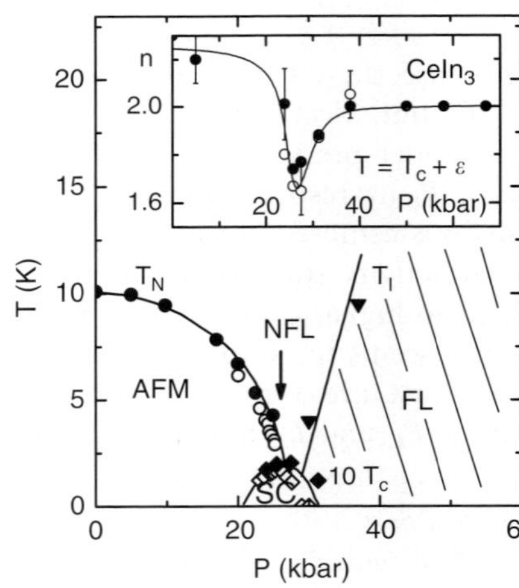

Fig. 26. Phase diagram of $CeIn_3$. T_N indicates the Néel temperature, and T_I the crossover temperature to the Fermi liquid regime. The superconducting transition temperature T_C is scaled by a factor 10 (◇) after Mathur et al. (1998). The pressure dependence of the resistivity exponent n is shown in the inset. The minimum of the exponent n in the temperature dependence T^n of the resistivity corresponds to the critical pressure P_C (Knebel 2002).

($T_C^{max} = 400\,mK$) in a narrow pressure range around P_C . The large initial slope of the upper critical field $H_{C2}(T) \approx (m^*)^2$ demonstrates that the heavy quasiparticles themselves condense in Cooper pairs. A full superconducting resistive transition was confirmed by further experiments made in Grenoble (Knebel et al. 2002) and Osaka (Kobayashi et al. 2001).

The upper critical field can be analyzed in terms of a single band superconducting model assuming a g factor equal to 1.4 and a strong coupling parameter $\lambda = 1.3$. This points out that the mass enhancement, coming from non-local fluctuations, m^*, is not large by comparison to the band mass renormalization m_K driven by local Kondo fluctuations: $m^*/m_K = \lambda + 1 = 2.3$. Nuclear quadrupolar resonance (NQR) on the In site was very successful to study the spin dynamics notably in the (AF and S) coexistence regime (Kohori et al. 2000a, Kawasaki et al. 2001). The second-order nature of the magnetic collapse at P_C must be questioned as two NQR signals (AF and Pa) appear at P_C - ε (fig. 27). The coexistence of both phases points to a phase separation in a pressure interval $P_{KL} - P_C \sim 0.3\,GPa$ (Kawasaki et al. 2004). Evidence for the unconventional nature of the superconductivity in both phases is mainly given by the temperature variation of the nuclear relaxation time T_1 which follows the $1/T_1 \sim T^3$ law reported in many unconventional exotic superconductors. It is taken as indication of a line node (Asayama 2002, Kohori et al. 2000a). The superconductivity of $CeIn_3$ has been studied theoretically on the basis of a three-dimensional Hubbard model (Fukazawa and Yamada 2003). The suggested d-wave pairing is induced by AF spin fluctuations. In agreement with other theoretical studies, T_C is lower by one order of magnitude compared to that for the 2d case.

4.1.2. *$CePd_2Si_2$: questions on the range of the coexistence*

At $P = 0$, $CePd_2Si_2$ is a AF with $T_N \sim 10\,K$ and a (1/2, 1/2, 0) propagation vector. As for $CeIn_3$, the $P = 0$ situation is very well documented by macroscopic and microscopic measurements. By contrast to the cubic $CeIn_3$ compound, in this tetragonal crystal (1, 2, 2), a strong axial anisotropy occurs. The easy direction of the magnetization is found in the basal plane. Below T_N both spin waves and quasi-elastic excitations can be detected (van Dijk et al. 2000). Above T_N, the magnetic fluctuations are dominated by Kondo fluctuations (Fak et al. 2004). The exchange constants extracted from the spin wave excitations have only a weak anisotropy. The Fermi surface was first investigated by the dHvA torque technique on tiny crystals. At least parts of the Fermi surface show partial 4f itineracy (Sheikin et al. 2003) ($P_{KL} < P_V$). As in $CeIn_3$, a minimum in the resistivity exponent n appears at P_C with a lower value 1.3 instead of 1.6 for $CeIn_3$. Reasons for this difference can be disorder, dimensionality or proximity to a valence

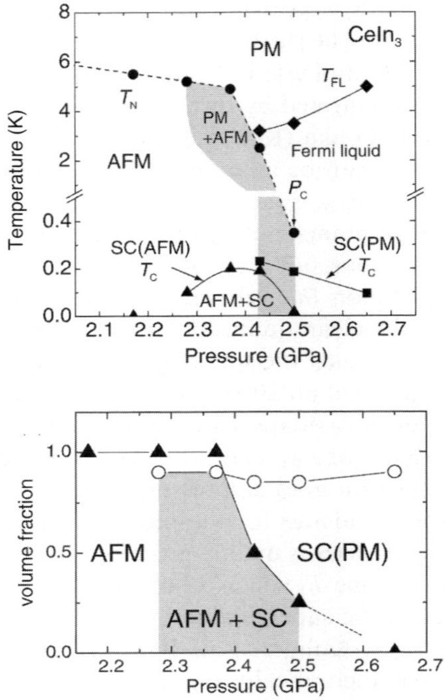

Fig. 27. NMR data of Kawasaki et al.: T_N, T_C, T_I of $CeIn_3$ and domain of phase separation (gray area). Volume fraction of AF and S states as a function of P (Kawasaki et al. 2004).

transition as we will see for $CeCu_2Si_2$ or $CeCu_2Ge_2$. The first one is very unlikely as $n = 1.3$ appears to be a robust limit whatever is the residual resistivity ρ_0 (Demuer 2000; Demuer et al. 2002a). The temperature analysis of the resistivity also shows that $P_C \sim P_V$. However no P experiments, either by resistivity, specific heat, or neutron scattering (Demuer et al. 2001, 2002a, Kernavanois et al. 2003) succeed in following the magnetic ordering close to P_C and thus cannot answer the question of the order of the magnetic transition at P_C. At least, in recent neutron scattering experiments up to $P = 2.45$ GPa, the magnetic transition as a function of temperature appears second order and any first-order transition at P_C will be weak (Kernavanois et al. 2005).

Our aim was to clarify the process of the magnetic collapse at the so-called QCP by simultaneous ac calorimetry and resistivity experiments with a zoom on their respective anomalies at T_N (Demuer 2000). To avoid any parasitic effect due to non-hydrostaticity, the measurements were realized with a diamond anvil cell and with helium as the transmitting medium. In the orthodox vision of QCP, for $T \to 0\,\mathrm{K}$, the molecular field description must become more and more valid for a second-order phase transition (Zülicke and Millis 1995). Contrary to these classical statements, both normalized anomalies are gradually broadened. For example, for $P = 2.3\,\mathrm{GPa}$, $T_N = 3\,\mathrm{K}$, the specific heat broadening reaches 15% (we will see that similar effects are observed for $CeRh_2Si_2$ and $CeRhIn_5$). However, explanation of the gradual broadening of the specific heat anomaly at T_N as $T_N \to 0$ may come from the weakness of a first-order transition with large fluctuations.

The discovery of superconductivity in $CePd_2Si_2$ was also made in Cambridge (Mathur et al. 1998). Further experiments have been completed in Geneva (Raymond and Jaccard 2000, Demuer et al. 2002b) and Grenoble (Sheikin et al. 2001a, Demuer 2002a). The analysis of $H_{C2}(T)$ requires an anisotropy in the g factor for the weight of the Pauli limit in qualitative agreement with the magnetic basal plane anisotropy and again a moderate $\lambda = 1.5$ coefficient (Sheikin et al. 2001b). The new experimental feature is the observation of the superconducting specific heat anomaly at 2.7 GPa (Demuer et al. 2002b). However, this superconducting calorimetric signature is observed only very close to P_C. Bulk homogeneous gapped superconductivity may occur in a narrower region than that claimed from resistivity measurements. This statement is reinforced by the difficulty to reach zero resistivity on each side of P_C. The realization of the clean limit is crucial for unconventional superconductivity ($\ell > \xi_0$). When T_C collapses with change in pressure, ξ_0 increases as T_C^{-1}, so the impurities are a severe cut-off.

Simultaneous pressure measurements, on two crystals oriented with c-axis parallel or perpendicular to the force load direction in a Bridgman anvil cell with a quasi-hydrostatic steatite pressure medium, show a large shift in the pressure phase diagram as indicated, fig. 28. Good correspondence is demonstrated between the magnetic instability, the position of P_V and the optimum conditions for superconductivity (Demuer et al. 2002b). In this axial crystal, the application of a strain along the c- and a-axes leads to opposite shifts of T_N (van Dijk et al. 2000). Thus the high magnetic sensitivity of the $CePd_2Si_2$ tetragonal crystal to a non-uniform pressure (even a weak percent of P_C) was suspected. The nice result is that a non-ideal pressure set-up but a carefully designed experiment proves that AF and S boundaries remain bounded at P_C.

The fast disappearance of the superconducting specific heat anomaly when P is not near P_C seems to be a consequence of the fact that the intrinsic superconducting dome $T_C^0(P)$ is associated with a high sensitivity to impurities as in the pair breaking mechanism represented by the well known Abrikosov–Gorkov formula (Abrikosov–Gorkov 1961). The important parameter is $x = \xi_0/\ell$. As $T_C^0(P)$ decreases, ξ_0 will increase and thus x. This

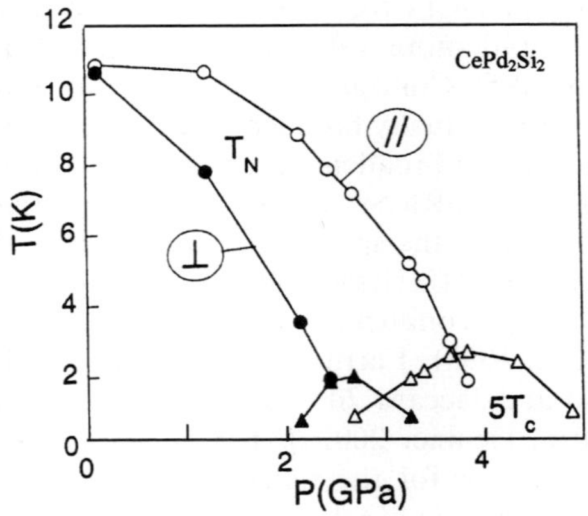

Fig. 28. Phase diagram of the two samples (filled and open symbols for samples ⊥ and // respectively) pressurized as described Demuer et al. (2002b). (●, ▲: T_N; π, ρ: T_C).

leads to a supplementary decrease of the pressure range P_{-S}–P_S where superconductivity can be detected. Indeed, numerical simulation (Brison 2004) shows that the specific heat anomaly at T_C is rapidly smeared out. These considerations can be applied to any quantum phase transition at P_C or P_S, where a steep pressure collapse of the critical temperature may occur. Our message is that the determination of the contour of S and AF extrapolated to ideal conditions is a "tour de force" (see $CeRhIn_5$).

The additional experimental pressure anisotropy will lead to minor effects for a cubic material like $CeIn_3$. It becomes a major perturbation in anisotropic materials like the 122 cerium family (like $CePd_2Si_2$) and even more the 115 series. Even at $P = 0$ (see later $CeIrIn_5$) pressure gradients of 0.1 GPa exist near imperfections: dislocations, stacking faults. These may produce superconducting nanostructures inside the material. Superconductivity can also occur by proximity near these objects. (For an array of stacking faults. Abrikosov and Buzdin (1988) have even proposed the possibility of a tiny splitting for the superconductivity transition of conventional superconductors.)

4.1.3. *$CeRh_2Si_2$: first order and superconductivity*

In contrast with the first two examples where, at $P = 0$, T_K is around 10 K, in $CeRh_2Si_2$ the Kondo temperature is higher, $T_K = 30$ K. However the magnetic exchange among the Ce ions is strong enough for AF to occur at $T_N = 36$ K the Rh ions may play an important role in the strength of T_K and E_{ij}. Magnetism disappears at $P_C \sim 1.0$ GPa (Kawarazaki et al. 2000). For $P = 0$, just above T_N, the anisotropy of the magnetic susceptibility between the c- and the a-axes is near 4, but it drops to an almost isotropic situation for the Pauli susceptibility at $T \ll T_N$(Mori et al. 1999). Just below P_C, the propagation vector k_0 of AF is again (1/2, 1/2, 0) and the magnetic moment is transverse with respect to k_0 and oriented along the (0, 0, 1) direction. Evidence that the transition may be first order at P_C comes from the steep pressure variation of T_N on approaching P_C, from a rapid wipe-out of the specific heat anomaly on approaching P_C (Haga 2005) (figs. 29 and 30), and also from drastic modifications of the FS as detected by the change of dHvA frequencies which can be analyzed with a localized 4f picture below P_C and an itinerant one above P_C (Araki et al. 2002) ($P_{KL} \sim P_C$). Another indirect evidence of a discontinuity (i.e. volume) is given by the persistence of a T^2 resistivity law on each side of P_C and the absence of NFL behavior down to very low temperature as $P \rightarrow P_C$ (Araki et al. 2002). A recent confirmation can be found in Ohashi et al. (2003).

Kawarazaki et al. (2000) have carefully followed the pressure evolution of $T_N(P)$ and $M_o(P)$. Up to 0.85 GPa, M_o is proportional to T_N. The magnetic moment is a longitudinal variable. The reason lies in the fact that it is created through an induced magnetism in accordance with the Doniach picture. In the model of Benoit et al. (1979), $M_o \sim T_N$. As in the case for linear polarized light, which is the sum of equal right and left circular polarizations, the induced electronic moment is the combination of the electron and hole components. The proportionality $T_N \sim M_o$ (quite general in magnetically ordered HFC) deserves special attention. It is neither predicted by SF theory, where $T_N \sim M_o^{4/3}$, nor for local Heisenberg magnetism $T_N \sim M_o^2$

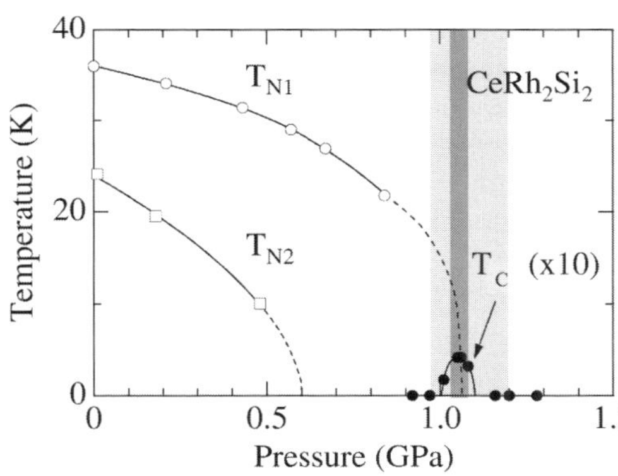

Fig. 29. $CeRh_2Si_2$ (T, P) phase diagram drawn by resistivity (○, ●) ac calorimetry (■) and neutron scattering (□) (see Araki et al. 2000, Kawarazaki et al. 2000, Haga 2005). The superconducting border (× 10) in dark corresponds to the full resistivity drop and in gray to the onset of the resistivity (Araki et al. 2002). We will not discuss the low-pressure AF phase at T_{N2}.

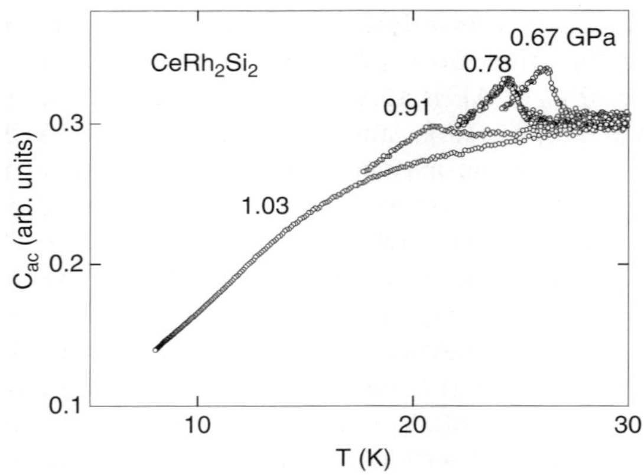

Fig. 30. Ac specific heat of $CeRh_2Si_2$ with the broadening and collapse of the AF anomaly on approaching P_C. A quite common phenomena in HFC (Haga et al. 2004).

with a pressure-invariant exchange. A linear relation of T_N with M_o has also been observed for the antiferromagnetism in chromium alloys (Koehler et al. 1966). This has been explained for spin density wave structures by a feedback effect on the Fermi surface due to nesting. With the dual character of the magnetism here, one may expect rather similar phenomena even for commensurate structures with missing Fermi surface related to AF gap (see Fuseya et al. 2003).

The occurrence of two time scales is manifest by the discrepancy between M_o detected by neutrons ($M_o \sim 1.38\,\mu_B$ Ce) and by NMR ($M_o \sim 0.3\,\mu_B$) (Kawasaki et al. 2002). "There may be a longitudinal fluctuation of the f electron moment which has a lifetime longer than the characteristic time of observations for neutrons and shorter than for NMR" (Kawarazaki et al. 2000). This switch between two frequencies may be the result of the slow motion of the heavy fermion condensate in the ordered magnetic medium. In $CeRh_2Si_2$ at P_C, the localized moment becomes delocalized and some itinerant magnetic picture occurs above P_C. The specific heat anomaly at T_N has almost collapsed at 0.9 GPa (fig. 30). Above $P_C \sim 1$ GPa, a drastic difference appears between the pressure evolution of T_N and M_o. A slight pressure variation of T_N contrasts with a continuous drop of M_o. This exotic SMAF signature was pointed out for $CeRu_2Si_2$ and will appear later for UPt_3 and URu_2Si_2.

Superconductivity in $CeRh_2Si_2$ was first reported by Movshovich et al. (1996). Recently it was stressed even for a high-quality crystal ($\rho_0 \sim 0.8\,\mu\Omega\,cm$) that the achievement of zero resistivity or a sharp resistive transition in the superconducting phase is only realized close to P_C in a narrow pressure region (1.04–1.07 GPa) (Araki et al. 2002). Precursor effects of superconductivity (resistivity onset) will give a wider pressure window (~0.2 GPa) (fig. 29). The situation is rather similar to that found in $CePd_2Si_2$. Again in $CeRh_2Si_2$ the thermodynamic boundary of superconductivity and long-range magnetism is not yet defined.

4.2. *$CeCu_2Si_2$ and $CeCu_2Ge_2$: spin and valence pairing*

For $CeCu_2Si_2$ and $CeCu_2Ge_2$ the new phenomenon is that the maximum of T_C is located far above P_C (a few GPa). For $CeCu_2Si_2$, the precise location for P_C was unknown until recent successful neutron scattering experiments (Stockert et al. 2004). We will not enter into the details of all systematic studies realized on this material but roughly P_C, may be a few kbar below $P = 0$. Tiny differences in composition can induce AF. Recently, the $CeCu_2(Si_{1-x}Ge_x)_2$ series was explored intensively by elastic neutron scattering experiments. The same incommensurate wavevector was observed for all concentrations suggesting a spin density wave instability given by a nesting property of the Fermi surface (Stockert et al. 2004). The difficulty to grow large crystals of $CeCu_2Si_2$ has precluded up to now collecting inelastic information, however, extensive NMR work has been realized (see Kawasaki et al. 1998, Ishida et al. 1999). Detailed discussion on the superconductivity of $CeCu_2Si_2$ can be found in the recent article of Thalmeier et al. (2004b) with the following new highlights that:

(1) Below $T = T_A$ in the mysterious A phase, which was found to envelop superconductivity in the (H, T) phase diagram, the reported spin density wave is in good agreement with the nesting wavevector predicted from the heavy fermion Fermi surface sheets (Zwicknagl 1992, Zwicknagl and Pulst 1993). When T_C and T_A are close, superconductivity and long-range magnetism seem to repel; similar effects are observed in $CeRhIn_5$ for $P \sim P_C$.

(2) For a proper doping of Ge the large pressure stability of superconductivity can be broken in two domes associated with magnetic and valence transitions (Yuan et al. 2003). This last observation confirms a series of pressure experiments realized by Jaccard and co-workers on $CeCu_2Si_2$ and $CeCu_2Ge_2$.

$CeCu_2Ge_2$ was the first case (Jaccard et al. 1992), where it was found that superconductivity occurs near $P_C = 6\,\mathrm{GPa}$ at first with a rather flat pressure variation of $T_C = 0.6\,\mathrm{K}$ followed by a bump to $T_C^{\max} = 2\,\mathrm{K}$ at $P = 17\,\mathrm{GPa}$. The relevance of such a structure of T_C was reinforced by similar observations already made in $CeCu_2Si_2$ (Bellarbi et al. 1984, Jaccard et al. 1985), where P_C is assumed to be shifted almost to zero. It is amazing to notice that this key result, obtained two decades ago, has had a real impact only in recent years. Simultaneous resistivity and ac specific heat measurements under pressure on $CeCu_2Si_2$ over 6 GPa (Holmes et al. 2004a) clarify the correlation between the collapse of the two maxima T_1^{max} and T_2^{max} of the temperature variation of the resistivity on warming at P_V and the optimum T_C (fig. 31). In agreement with the previous cases, a first superconducting domain will be centered on P_C and a second one at P_V suggesting that new superconductivity is mediated by valence fluctuations. The particular pressure variation of T_C was already interpreted via two contributions (Thomas et al. 1996): a smooth one assumed to be due to the pressure increase of T_K and sharper additional features reflecting topological changes in the renormalized heavy bands. However the monotonic scaling of $H_{C_2}(0)$ by $m^{*2}T$ in reduced temperature T/T_C leaves questionable two different sources of pairing (Vargoz et al. 1998).

Valence fluctuations near P_V (Onishi and Miyake 2000, Holmes et al. 2004a) seem to be a favorable factor for the increase of T_C. The departure from P_C will decrease the spin pairing potential. So a new attractive potential is needed. The importance of valence fluctuations is stressed by the fact that near P_C the usual NFL 3d AF behavior is recovered ($\rho \sim T^n$ with $\sim 3/2$) while near P_V a linear temperature crossover in the resistivity is observed. This dependence is well explained by soft valence fluctuations, which

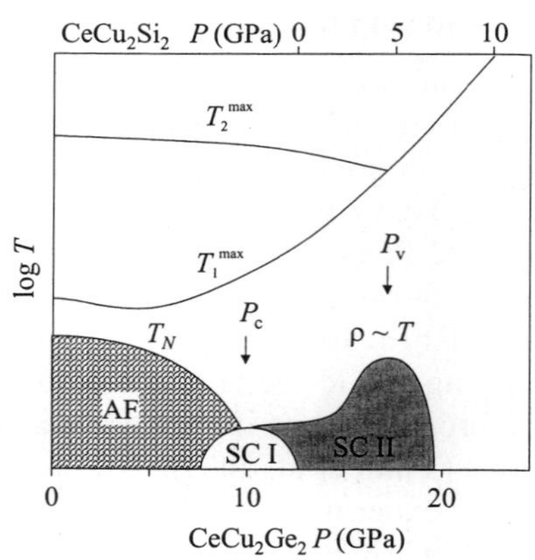

Fig. 31. Schematic P–T phase diagram for $CeCu_2(Si/Ge)_2$ showing the two critical pressures P_C and P_V. At P_C, where the antiferromagnetic ordering temperature $T_N \rightarrow 0$, superconductivity in region SC is governed by antiferromagnetic spin fluctuations. Around P_V, in the region SC II, valence fluctuations provide the pairing mechanism and the resistivity is linear in temperature. The temperatures T_1^{MAX}, and T_2^{MAX} of the maxima of the resistivity merge at a pressure coinciding with P_V (Holmes et al. 2004a).

will produce large angle scattering of the quasiparticles. They are efficient in a wide region of the Brillouin zone and are strongly coupled to Umklapp process of the quasiparticle scattering. At least, the shift of P_V from P_C seems a convincing explanation for the large shift of T_C^{max} from P_C. The direct mark of a valence change in $CeCu_2Si_2$ comes from L_{III}-X-ray absorption experiments (Roehler et al. 1988). Other indirect evidence is the decrease of the Kadowaki–Wood ratio, a small hump in the γ term and strong enhancement of the residual resistivity. In $CeIn_3$, $CePd_2Si_2$, and $CeRh_2Si_2$ P_C and P_V cannot be separated.

4.3. *From 3d to Quasi-2d systems: the new 115 family: CeRhIn$_5$ and CeCoIn$_5$*

The link between superconductivity and magnetism was recently boosted with the Los Alamos discovery of superconductivity (see Thompson 2001) in the so-called 115 cerium compounds like $CeRhIn_5$, $CeIrIn_5$ or $CeCoIn_5$. A planar anisotropy is induced by inserting a single layer of MIn_2 into $CeIn_3$. Single layers of $CeIn_3$ are stacked sequentially along the c-axis according to the relation $Ce_nM_mIn_{3n+2m}$ ($n = 1, m = 1$) in 115 series. For $n = 2$, there will be two adjacent layers of $CeIn_3$ separated by a single layer $m = 1$ of MIn_2 (M, transition metal). A planar anisotropy is produced. The Fermi surface is dominated by a slightly warped cylindrical sheet even in $LaRhIn_5$. This contrasts with the previous cases of 3d complex Fermi surfaces.

The gold "mine" that these compounds represent is that they offer three examples (where the crystal growth by flux is easy) that cover all possibilities. $CeRhIn_5$ (AF at $P = 0$ with $T_N = 3.8\,K$) has $P_C{\sim}2\,GPa$. The two others ($CeIrIn_5$ and $CeCoIn_5$) are already on the PM side of the magnetic instability and superconductors at $T_C = 0.4\,K$ and $T_C = 2.3\,K$ at $P = 0$ (Petrovic et al. 2001, Thompson et al. 2001) (fig. 32). As for the other cerium heavy fermion compounds, in $CeRhIn_5$, superconductivity emerges near P_C. There is already a large diversity of studies on the 115 series notably for the FS determinations of the normal phases at $P = 0$. In $CeRhIn_5$, the 4f electrons are localized and itinerant in $CeIrIn_5$ and $CeCoIn_5$ (Shishido et al. 2002). A new generation of pressure studies through P_C for $CeRhIn_5$ (Shishido et al. 2005) show that the FS changes and the 4f electrons become itinerant above $P{\sim}2.4\,GPa$. This is slightly higher than P_C but magnetostriction corrections must be made. Indeed, the (H, T) domain of the AF boundary must be made precise close to P_C in order to extrapolate the data of quantum oscillations at finite H to determine the FS topology at $H = 0$.

Extensive NQR experiments on the In site demonstrate the 3d behavior for the spin fluctuations of $CeRhIn_5$ above T_N (Mito et al. 2001). The quasi-2d behavior of the PM phase compound in $CeCoIn_5$ has been studied on In and Co sites (Kohori et al. 2002, Kawasaki et al. 2003a). Unconventional singlet d-wave superconductivity has been tested by a large panoply of techniques: for $CeCoIn_5$, NMR (Kohori et al. 2000b, 2001, 2002) with the proof of d-wave pairing (from the drop of the Knight shift below T_C) and of a line of nodes (with the T^3 decrease of T_1^{-1}), the observation of power laws in the T dependence of the specific heat, thermal conductivity (Movshovich et al. 2001), and microwave conductivity (Ormeno et al. 2002). Anisotropy has been detected in the angular variation of the thermal conductivity (Izawa et al. 2001) and of the specific heat (Aoki et al. 2004) in a magnetic field.

4.3.1. *$CeRhIn_5$: Coexistence and exclusion*

We will focus here on the coexistence of superconductivity and magnetism near P_C in $CeRhIn_5$ which can be studied under pressure but also by alloying. Extensive work can be found for the $CeRh_{1-x}Co_xIn_5$ and $CeRh_{1-x}Ir_xIn_5$ (Zapf et al. 2001, Pagliuso et al. 2001). In the first resistivity report on $CeRhIn_5$ under pressure (Hegger et al. 2000), magnetism and superconductivity seem to repel each other, i.e. a drastic decrease of T_N coincides with the sudden appearance of superconductivity at $P_C \sim 1.5$ GPa. Careful specific heat measurements in a quasi-hydrostatic solid pressure medium indicate nice magnetic specific heat anomalies below P_C and superconducting jumps above P_C (Fisher et al. 2002). Just at P_C, the interpretation is complex as a maximum appears in the temperature variation of C/T above P_C, i.e. in its PM state. The differentiation between the PM and AF states, i.e. identification of P_C, has been made through the location of

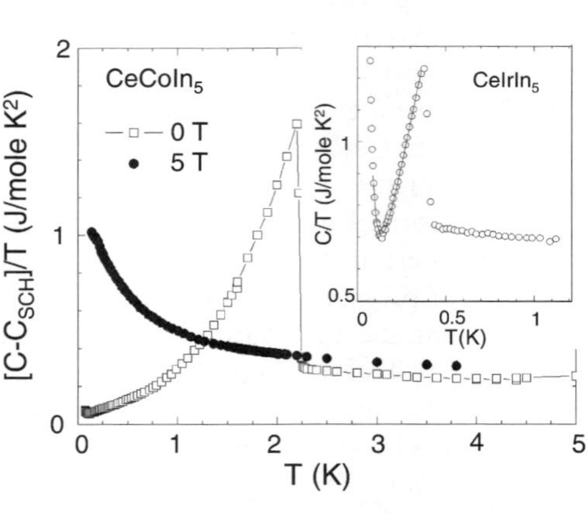

Fig. 32. Superconducting specific heat anomaly of $CeCoIn_5$ and $CeIrIn_5$ (from Petrovic et al. 2001, and Movshovich et al. 2001). The nuclear Schottky contribution of In has been subtracted for $CeCoIn_5$ but not for $CeRhIn_5$.

the maximum of γ. With such a procedure, superconductivity appears only on the PM side.

By improving the sample quality and the accuracy in the resistivity measurements (Llobet et al. 2003), it was found that a complete resistive superconducting transition at $T_C(\rho)$ occurs far below P_C ($T_C = 0.7\,K$ at $P\sim1.6\,GPa$). Simultaneous measurements of ac susceptibility, NQR spectrum, and nuclear spin relaxation time (T_1) (Kawasaki et al. 2003b) show that the temperature derivative of χ_{ac} has its maximum at T_C lower than the previous determination of $T_C(\rho)$ by resistivity. Furthermore a strong broadening of the diamagnetic signal occurs below P_C, which is located near 1.8 Gpa for this case.

To clarify the situation, a recent attempt has been made (Knebel et al. 2004) by ac calorimetry to detect T_N and T_C under more hydrostatic conditions (with argon as a pressure-transmitting medium) than in the previous calorimetric experiment of Fisher et al. (2002). The critical pressure as shown in fig. 33 seems to be located near 1.9 GPa. Qualitatively the important feature is that there are clear AF specific heat anomalies observed below P_C and there are clear superconducting specific heat signatures just above P_C. Tiny AF or superconducting (S) anomalies are detected in the vicinity of P_C but it can be due to residual internal stress (fig. 34). The domain of homogeneous coexistence of AF and gapped superconducting phases may not exist. Furthermore, ac susceptibility experiments on a sample coming from the same batch show only a broadened diamagnetism below at higher temperatures than the superconducting specific heat anomaly. A sharp diamagnetic transition occurs only for $P > P_C$, now in good

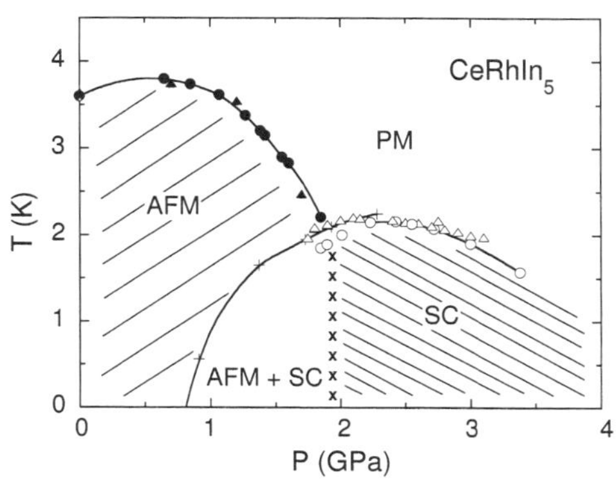

Fig. 33. (T, P) phase diagram of $CeRhIn_5$ (●, ○) specific heat anomalies at T_N, T_C and corresponding susceptibility (ρ). The AF anomaly disappears suddenly at $P_C\sim1.9\,GPa$. Gapped superconductivity is observed above P_C. Inhomogeneous ungapped superconductivity may occur below P_C (white domain) (Knebel et al. 2004).

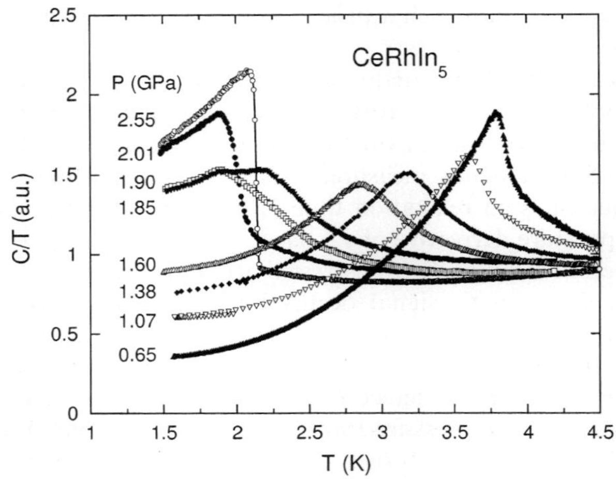

Fig. 34. Specific heat anomalies of $CeRhIn_5$ at different pressure. At 1.9 GPa, there is a superposition of tiny superconducting and magnetic anomalies (Knebel et al. 2004).

agreement with the specific anomaly. It was proposed that the observation of the inhomogeneity may not be a parasitic effect but an intrinsic property of a new gapless superconducting phase of odd-parity pairing directly linked with strong retarded effects which may occur near P_C (see Section 3). An interesting observation is that only the cascade AF→S has been observed on cooling (see also $CeIn_3$) but, to our knowledge there is no other example of a HFC, where a AF or F phase appears after the entrance into a superconducting state at high temperature. This strongly suggests a clear demarcation between AF and PM phases i.e. a first-order transition. In conventional magnetic superconductors such as Chevrel phases, usually $T_C > T_N$.

In $CeRh_{1-x}Co_xIn_5$ (Zapf et al. 2001) and in $CeRh_{1-x}Ir_xIn_5$ (Pagliuso et al. 2001), the coexistence of superconductivity and antiferromagnetism is claimed over a large doping range, respectively, $0.4<x<0.6$ for Co substitution and $0.3<x<0.6$ for Ir substitution. However, investigations of the amplitude of the specific heat anomalies and their broadening at T_N and T_C, as well as on their low-temperature behavior must be revisited before claiming that the derived phase diagrams are evidence of AF and S coexistence. Microscopically, one must worry about the effects of internal strain, mismatch between the a- and c-lattice parameters and their ratio c/a on doping (see discussion on URu_2Si_2 in Section 6). Even for the pure compound $CeIrIn_5$, the puzzle still remains why there is such a large difference between the value of $T_C(\rho) = 1.3\,K$ and $T_C(C) = 0.38\,K$ measured respectively by resistivity and specific heat (Thompson et al. 2001). At least it is obvious that the domain of AF and S coexistence deserves systematic study, careful analysis, and excellent hydrostatic conditions. The disorder by washing out the first-order transition seems to restore a homogeneous situation or at least the simultaneous observation of AF and S anomalies over a large doping region. Let us note that as for $CeRhIn_5$, discrepancies appear between NMR (collapse of M_o at P_C) (Mito et al. 2001) and magnetic neutron diffraction (discontinuity of M_o at P_C) (Llobet et al. 2003).

4.3.2. *$CeCoIn_5$: A new field induced superconducting phase*

A large variety of superconducting studies have been performed on $CeCoIn_5$. Although the d-singlet state is well established, the previous claim of a $d_{x^2-y^2}$ order parameter comes mainly from the anisotropy of the thermal conductivity in a rotating magnetic field (Izawa et al. 2001), which has been questioned in a field-angular study of the specific heat. The superconducting gap may be most probably of d_{xy} type (Aoki et al. 2004). This d_{xy} order parameter may be driven by valence fluctuations since AF fluctuations mainly in the (Π, Π) directions of the basal plane may induce a $d_{x^2-y^2}$ order parameter (Miyake 2004).

Special attention must be given to strong coupling effects since superconductivity appears at a value T_C higher than the Fermi liquid temperature T_I or T_{KL}, which can be attributed to FL low-energy excitations. As discussed in Section 3, the large specific heat jump $\Delta C/(C(T_C+\varepsilon))$ may not reflect a large strong coupling ($\lambda \gg 1$) but a delay in the achievement of the normal-phase Fermi liquid value which will be reached only at very low temperature (see Kos et al. 2003). Recent experiments made by Knebel et al. (2004) have completed the calorimetric studies of Sparn et al. (2001) as a function of pressure up to 1.5 GPa. The pressure variation of $T_C(P)$ and of the specific jump $\Delta C/C(T_C)$ is now established up to 3 GPa (fig. 35). $T_C(P)$ reaches its maximum for $P = 1.5$ GPa while the specific heat jump at T_C continuously decreases under pressure. Neglecting strong coupling effects (see Kos et al. 2003), the jump normalized to the value of m^* at $T = 0$ K must be universal with $\Delta C/m^* T_C$ constant. With this hypothesis the effective mass decreases gradually under pressure (a factor 4 between 0 and 3 GPa).

Two new features appear in magnetic field which have been predicted three decades ago (Saint James et al. 1969) for the specific case where the upper critical magnetic field is governed by Pauli limitation at 0 K:

(1) a crossover from second order to first order in H_{C2} (T) at T_0 (Izawa et al. 2001 , Bianchi et al. 2002 , Tatayama et al. 2002),
(2) a new high field phase reminiscent of the FFLO phase.

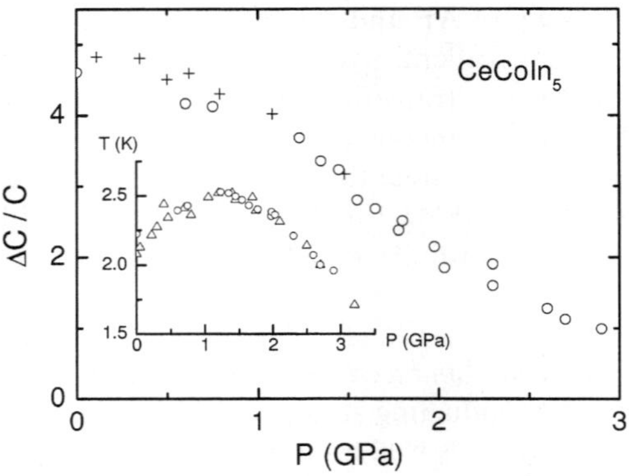

Fig. 35. $CeCoIn_5$. Jump of the superconducting specific anomaly normalized to the value just above T_C: (○) from Knebel et al. (2004); (+) from Sparn (2001). Inset: variation of T_C (P) measured by ac calorimetry (○) and ac susceptibility (△).

We will focus our discussion mainly on this last discovery. Figure 36 represents the domain of the new high magnetic field superconducting phase, which appears below a temperature $T_C^* < T_0$ in specific heat measurements (Bianchi et al. 2003a), other thermodynamic experiments (Radovan et al. 2003, Watanabe et al. 2004) and recently in thermal conductivity experiments (Capan et al. 2004). The FFLO state of $CeCoIn_5$ has been discussed theoretically by Won et al. (2003) . In FFLO, the Maki parameter depends only on the pressure variation of H_{C2} (orb)$\sim(m^* T_C)^2$. The initial pressure dependence of T_{FFLO}/T_C is weak (see Section 3). Only above 1.5 GPa, is the T_{FFLO} predicted to collapse (using the previous relative pressure variation of m^* and T_C).

As discussed in Section 2 for $CeRu_2Si_2$, the supplementary interesting features will be if the magnetic interaction is modified for $H < H_{C2}$ (0). In $CeCoIn_5$, a field-induced quantum critical point at H_M is suspected for $H = 5$ and 8.5 T, respectively, for $H//c$ and $H//a$ axis (Paglione et al. 2003, Radovan et al. 2003, Bianchi et al. 2003b). Extrapolating from $CeRu_2Si_2$, one can guess that (1) a strong dilatation at H_M will be a supplementary favorable factor to drive the phase transition at H_{C2} from second order below H_M to first order above H_M at T_0;(2) the new field induced superconducting phase may be boosted by the enhancement of the magnetization and also can appear when a strong coherence due to Cooper pairing exists in the polarized phase, i.e. below T_α (H). The role of the magnetic correlations in the field behavior of the superconducting phase of $CeCoIn_5$ may open other perspectives than for the classical FFLO frame work. Recently (Flouquet et al. 2005b), it was suggested that the main phenomena may be

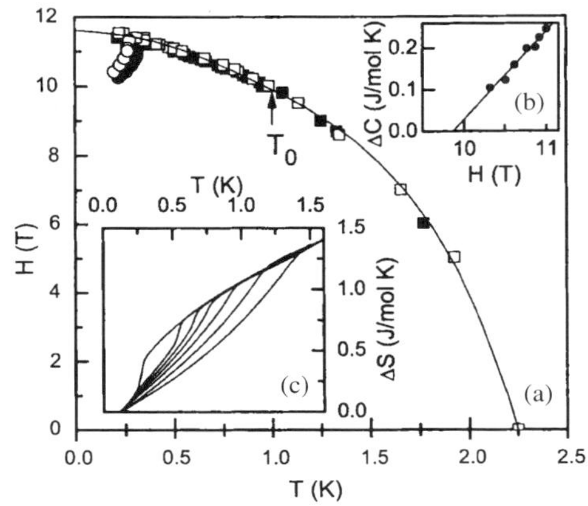

Fig. 36. (a) H–T phase diagram of $CeCoIn_5$ with both $H \parallel [110]$ (filled symbols) $H \parallel [100]$ (open symbols). (○) and (●) indicate the T_{FFLO} anomaly for $H \parallel [100]$ and $H \parallel [110]$, respectively. Inset (c): entropy gain from $T = 0.13$ K for fields of 11.4, 11, 10.8, 10.6, 10.22, 9.5 and 8.6 T (from left to right). Inset (b): specific heat jump at the T_{FFLO} (from Bianchi et al. 2003a).

the field switch of the AFM-SC boundary designed on fig. 33. When T_C (H) becomes lower than the linear P extrapolation of T_N, magnetism and superconductivity may coexist again.

Finally, a surprising result on superconductivity was found by extending the studies on 115 cerium compounds to 115 plutonium systems. The high T_C record in HFC was beaten recently with the discovery of superconductivity in $PuCoGa_5$: $T_C = 18.5$ K (Sarrao et al. 2002) again in Los Alamos. The heavy quasiparticle leads to $\gamma = 77\,\mathrm{mJ\,mol^{-1}\,K^{-2}}$ in their normal phase and a corresponding huge value of the initial slope of H_{C2} $(T) \sim -59\,\mathrm{kOe\,K^{-1}}$ leading to an orbital limit of 740 kOe at $T = 0$ K, roughly twice the estimated Pauli limit of 340 kOe. The superconducting parameters $\xi_0 = 21$ Å, $\lambda_L = 1240$ Å and $\kappa = 32$ are rather similar to those found in SCES superconductors. The role of spin fluctuations seems to be reflected in the low value of the Curie–Weiss temperature θ (2 K) of the susceptibility, which obeys roughly a Curie–Weiss law. The high T_C of $PuCoGa_5$ may be due to an increase in the effective Fermi temperature. A spectacular observation (Pagliuso et al. 2002, Kumar et al. 2004) is that the variation of T_C versus the ratio c/a of the lattice parameters is a beautiful unique straight line for the Ce and Pu 115 compounds. This large variation of T_C versus the separation of the basal plane provides convincing evidence of the importance of the magnetic and electronic anisotropies. It points out the key role of lattice deformations.

4.4. *Recent exotic superconductors: $CePt_3Si/PrOs_4Sb_{12}$*

A new unexpected result (Bauer et al. 2004) is the discovery of superconductivity in the non-centrosymmetric compound of $CePt_3Si$ since it was stressed that the lack of an inversion center may not be favorable for superconductivity. It was even proposed that a material without inversion center would be a bad candidate for spin-triplet pairing (Anderson 1984). Gorkov and Rashba (2001) clarify that the order parameter will be a mixture of spin-singlet and spin-triplet components. The supplementary interest of this new material is the coexistence of long-range magnetism at $T_N = 2.2\,K$ (Metoki et al. 2004) and of superconductivity at $T_C = 0.78\,K$, with clear heavy fermion properties ($\gamma \sim 400\,mJ\,mol^{-1}\,K^{-2}$). The occurrence of a hybrid situation with both singlet and triplet components is manifested in NMR properties which appear as a mixture of conventional and unconventional behaviors (Yogi et al. 2004): The nuclear relaxation rate T_1^{-1} shows a kind of Hebel–Slichter anomaly usually absent in unconventional superconductors and a temperature variation which cannot be described with either hypothesis. As in UPd_2Al_3, superconductivity appears to coexist with the AF state. The link with a AF QCP is not obvious. Even the origin of the pairing mechanism is not clear since, due to the lack of inversion center, ferromagnetic and antiferromagnetic fluctuations will inevitably be mixed. The discovery of superconductivity in $CePt_3Si$ appears during this revision of the review so I will not go further into the recent activity on this subject. An excellent report on the experimental status can be found in Bauer et al. (2005). Recent theoretical papers are Frigeri et al. (2004), Samokhin et al. (2004) and Mineev (2004a–2005).

Another interesting case has been provided by discovery of the superconductivity in the skutterudite heavy fermion compound $PrOs_4Sb_{12}$ with $T_C \sim 1.8\,K$ (Bauer et al. 2002). The new insight of the Pr case by comparison to the Ce or Yb ones (which are Kramer's ions with even magnetic degenerate crystal field levels) is that now the crystal field state can be a singlet. Thus, if an extra hybridization occurs, there are two competing mechanism for the formation of a singlet state: the crystal field and the Kondo effect. Furthermore, depending on the crystal field scheme, fancy couplings and ordered phases can occur with dipole, quadrupole, octopole order parameters. For general considerations, the reader can look at Maple et al. (2003), Aoki et al. (2005) or Sakakibara et al. (2005). Let us also point out that an extra source of the mass enhancement can also occur due to the dressing of the itinerant electron by the inelastic scattering from the singlet ground state Γ_1 to the excited triplet Γ_5 (Fulde and Jansen 1983). The feedback ingredient in this new toy (Frederick et al. 2004) is evident in the strong dependence of γ on C_{CF}. It is of course directly related to the weakness of $C_{CF} = 8\,K$ and the large value of $\gamma{\sim}350\,\mathrm{mJ\,mol^{-1}\,K^{-2}}$ (Bauer et al. 2002, Maple et al. 2003). Tiny variations of C_{CF} can drive the material to a multipolar instability.

After the discovery of superconductivity, the two surprises concerning superconductivity are:

(i) the observation of a double transition at T_C^A and T_C^B (Vollmer et al. 2003) and
(ii) the unusual weak sensitivity to doping as demonstrated by studies on $Pr_{1-x}La_xOs_4Sb_{12}$ and $PrOs_{4(1-x)}Ru_{4x}Sb_{12}$ (T_C of $LaOs_4Sb_{12}$ and $PrRu_4Sb_{12}$ are respectively equal to 0.74 K and 1.04 K) (see Rotundu et al. 2004, Frederick et al. 2004). It is quite different from the previous doping sensitivity of the unconventional superconductors of Section 4.

This suggests that we may need to search for arguments on the apparent double superconducting transition which may be not related to an exotic multicomponent order parameter as described in Section 6 for UPt_3. Two macroscopic observations lead to our proposal that the lowest temperature $T_C^B = 1.75\,\mathrm{K} < T_C^A = 1.85\,\mathrm{K}$ at $H = 0\,\mathrm{K}$ may be evidence for either a sharp crossover or a real condensation of the previous overdamped dispersive crystal field excitations (see Kuwahara 2004, 2005, and Raymond et al. 2005) referred to often as magnetic excitons:

(1) the magnetic fields, H_{C_2} (T_A) and H_{C_2} (T_B) are mainly parallel (Tayama et al. 2003 , Méasson et al. 2004),
(2) the splitting is roughly preserved independent of pressure (Measson 2005).

Strong support that T_C^B is associated with a feedback effect from superconductivity on the magnetic excitons was given by inelastic neutron scattering experiments on single crystals at T_C^A (Raymond et al. 2005, Kuwahara et al. 2005). The coherence of the Cooper pair drives a shift of C_{CF} to lower energy and also to the narrowing of its width Γ. The strong interplay of C_{CF} with Γ and γ may produce the emergence of a second anomaly at T_C^B. Tiny changes of C_{CF} at T_C^A can strongly affect the specific heat and may lead to a sharp crossover at T_C^B. The underlying question is if the overdamped magnetic excitons above T_C^A become real excitations at T_C^B associated with a very weak induced multipolar component. A tiny magnetic component was observed in μSR experiments below T_C and interpreted as evidence of unconventional superconductivity with broken time-reversal symmetry (Aoki et al. 2003). Experiments on thermal conductivity were analyzed in the multiple phases depending on temperature and external magnetic field (Izawa et al. 2003). We will argue that the reported four-fold anisotropy observed may be due to a weak change in C_{CF} as the field is rotated in the (0, 0, 1) plane.

The exotic situation of $PrOs_4Sb_{12}$ and the unusual properties of its superconductivity has led to many theoretical developments (see Miyake et al. 2003, Goryo 2003, Maki et al. 2003, Ichioka et al. 2003, Sergienko and Curnoe 2004) within the perspective of unconventional superconductivity. An interesting proposal is that superconductivity in $PrOs_4Sb_{12}$ may be mediated by the magnetic excitons as will be discussed later for UPd_2Al_3 (Matsumoto and Koga 2004).

In contrast to the debate on superconductivity, the physics of the field induced ordered phase in $PrOs_4Sb_{12}$ appears to be very well established by neutron scattering experiments (Khogi et al. 2003, Kuwahara et al. 2004) and to be well described by a theoretical model assuming an interaction of quadrupolar moments of Pr 4f electrons in the single–triplet crystal field levels (Shina et al. 2004).

5. Ferromagnetism and superconductivity

Essential points:

- For UGe_2: Dual aspect of ferromagnetism. Pressure switch from FM_2 to FM_1 (two successive ferromagnetic phases) at a first order transition pressure, P_X. Metamagnetism of FM_1 toward FM_2 and of the paramagnetic phase PM via two successive metamagnetic transitions to FM_1 and FM_2. Appearance of superconductivity centered on P_X with the consequence of the superconducting phase S2 being coupled to FM_2 and S1 coupled to FM_1.
- For URhGe, as in UGe_2, the superconductivity is not directly associated to a ferromagnetic quantum critical point.
- For $ZrZn_2$: extra proofs is required.
- For εFe: superconductivity appears unconventional induced by FM spin fluctuations.

5.1. *The Ferromagnetism of* UGe_2

UGe_2 is an orthorombic crystal, which was already studied for its ferromagnetic phase a decade ago (Onuki et al. 1992). The uranium atoms are arranged as zigzag chains of nearest neighbors that run along the crystallographic *a*-axis, which is the easy magnetization direction. The chains are stacked to form corrugated sheets as in α-uranium but with Ge atoms inserted along the *b*-axis. The difference between UGe_2 and α-uranium is that in UGe_2, the nearest-neighbor separation $d_{u-u} = 3.85\,Å$ is larger. This leads to greater localization of 5f electrons and much larger entropy at low temperature. At $P = 0$, UGe_2 is a ferromagnet below $T_{Curie} = 54\,K$. It has a moderate residual $\gamma = 35\,mJ\,mol^{-1}\,K^{-2}$ term as $T \rightarrow 0\,K$. The value of the ordered moment $M_o = 1.48\,\mu_B$/uranium is far lower than the full moment $3.5\,\mu_B$ of a free uranium ion and smaller than the Curie Weiss moment ($2.8\,\mu_B$) measured above T_{Curie} (Huxley et al. 2001). On cooling the resistivity first decreases slowly to $\rho\,(T_{Curie}) \sim 100\,\mu\Omega\,cm$ and then drops rapidly below T_{Curie}. This variation is reminiscent of properties found in the magnetically ordered Kondo lattice. Under pressure, T_{Curie} decreases; the critical pressure $P_C = 1.6\,GPa$ appears as a first-order transition. The (T_{Curie}, P) contour has been determined by different methods and notably, recently, by neutron diffraction. For ferromagnetism, magnetization measurements also give direct access to the order parameter (Pfleiderer and Huxley 2002, Huxley et al. 2003a).

The assertion that even at $P = 0$, UGe_2 is itinerant was based on the already reported large pressure dependence of T_{Curie} and M_o as well as on band calculations with their qualitative success to explain quantum oscillation frequencies found in dHvA experiments. Band structure analysis on UGe_2 may suggest some tendencies to quasi-two-dimensional majority carrier Fermi surface sheets below T_{Curie} with even sections which may be parallel, i.e. giving some 1D character. However the nesting predictions for the wavevector Q differ, either $Q \perp a$ for Yamagami (2003) or $Q // a$ for Shick and Picket (2001). The densities of states of spin-up and spin-down electrons have a large 5f contribution at the Fermi level.

Figure 37 shows the temperature dependence of the ordered moment $M(T)$ squared at different pressures as detected from the nuclear Bragg reflections by neutron diffraction. At a critical pressure $P_X = 1.20\,GPa$, a change occurs in the temperature dependence of $M^2(T)$ at T_X. The emergence of another temperature $T_X < T_{Curie}$ at pressure below P_X is also marked by an extra drop of the resistivity (fig. 38). Careful neutron diffraction measurements have failed to detect any additional reflections. Ferromagnetism persists through P_X. Magnetization experiments indicate that the extrapolation of the ordered moment M_o at $T = 0\,K$ leads to a discontinuity of M_o at P_X (Pfleiderer and Huxley 2002). Thus the transition from a high-moment ferromagnetic phase FM_2 to the other low-moment ferromagnetic phase FM_1 is first order at low temperature. As the signature of the $FM_1 \rightarrow FM_2$ transition disappears rapidly when P decreases from P_X, the critical endpoint at P_{cr} must be a few tenths of a GPa below P_X. However a crossover temperature T_X can be drawn down to $P = 0$ (where

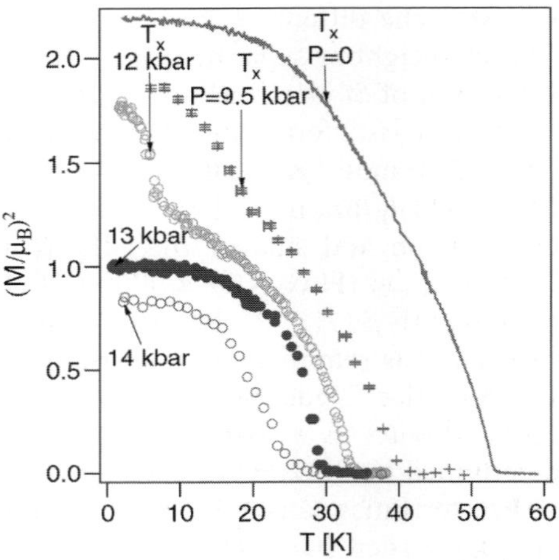

Fig. 37. The temperature dependence of the ordered moment squared at different pressure deduced from neutron scattering measurements (Huxley et al. 2003a).

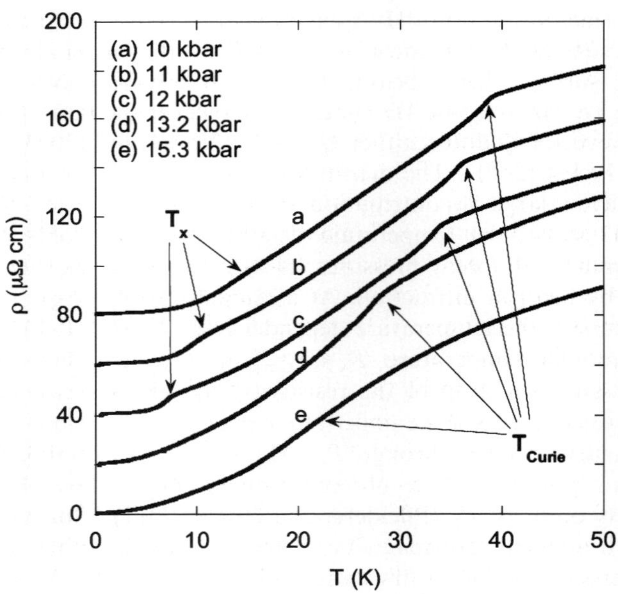

Fig. 38. Temperature variation of the resistivity of UGe_2 at different pressure with the emergence of the two anomalies at T_{Curie} and T_X below $P_X = 12.2$ kbar (Sheikin 2001b, Huxley et al. 2001).

$T_X = 28\,K$) in the temperature derivative of $\rho(T)$ as well as in thermal expansion experiments (Oomi et al. 1993).

Magnetic form factor experiments (Kernavanois et al. 2001) have been performed to check if there is a change in the orientation of the magnetic moment by comparing data on quite different wavevectors (0,4,0) and (0,0,1) for $P = 1.2\,GPa$. The excellent scaling through T_X supports the persistence of the same type of magnetic structure. The agreement between magnetic form factors and the square of the magnetization suggests a homogeneous scenario of the $FM_2 \rightarrow FM_1$ transition. However the weak difference between (0,0,1) and (1,1,1) reflections points out a slight modification in the origin of the magnetism, i.e. a difference in orbital component related to a change in electronic structure (tiny modification of the valence may lead to a drastic change in the electronic structure). This statement is supported by a recent band theory picture: over a range of volumes two nearly degenerate FM states exist which differ most strikingly in their orbital character on the uranium site (Shick et al. 2004).

Resistivity measurements show that the A coefficient of the AT^2 inelastic contribution jumps by a factor 4 above P_X. The related increase in the renormalized density of states is directly observed in specific heat experiments as a function of pressure as reported in fig. 39a (Tateiwa et al. 2003a, Fisher et al. 2005). At $T = 0$, γ ($P = 1.15\,GPa = P_X - \varepsilon$) is roughly half the value of γ at $P = 1.28\,GPa = P_X + \varepsilon$. At $P = 1.15\,GPa$, the transition at $T_X \sim 5\,K$ is characterized by a steep drop in C/T (Tateiwa et al. 2003a). Analysis of the low-temperature specific heat data (Fisher and Phillips 2005), in the form $\gamma T + \beta T^3$, shows a maximum of β at P_X (fig. 39b), i.e. where T_C also reaches its maximum. Quantum oscillation experiments show drastic changes between the FM_2 phase below P_X and the PM phase above P_C in agreement with band structure calculations (fig. 40) (Terashima et al. 2001, Settai et al. 2002). The dHvA signals between P_X and P_C are still controversial so we will not enter into this debate. Intrinsic inhomogeneity may wipe out the dHvA signal between P_X and P_C. However, we keep the message that superconductivity will be strongly influenced by FS reconstruction which occurs at P_X and P_C.

Similar to the transition from the FM_2 to the FM_1 ground state in zero field under pressure, the magnetic field can restore the FM_2 above P_X via a metamagnetic transition at H_X, and even can lead to the cascade $PM \rightarrow FM_1 \rightarrow FM_2$ through two successive metamagnetic transitions at H_C and H_X (fig. 41). As observed for $CeRu_2Si_2$, the transition at H_X corresponds to a critical value of the magnetization independent of pressure. This suggests that metamagnetism takes place when the Fermi energy crosses a sharp maximum in the density of states for one spin polarization. Analysis of the magnetization along the a- and c-axes show that, below P_X in the FM_2 state,

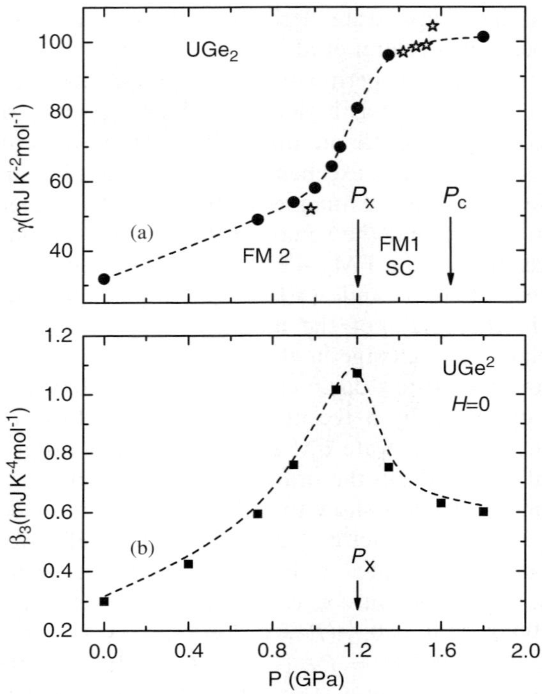

Fig. 39. (a) (●) Plot of γ versus P for UGe_2. Note the rapid rise in γ for P_X in the superconducting region of the phase diagram (Fisher et al. 2005). (b) β_3 versus P from fits to the low-temperature specific heat of UGe_2 in $H = 0$. It is unlikely that β_3 for the lattice would increase with pressure, or that it has any significant P dependence up to $P = 1.80$ GPa. Therefore, β_3 must be a composite of the weak T^3 lattice term plus an additional strong term. A low excitation mode can be represented by a T^3 term in the limited temperature range (15 K) of the fit. Furthermore, at 1.80 GPa, in the paramagnetic region of the phase diagram, a larger T^3 term than that at $P = 0$ indicates some additional contribution to the specific heat. (Fisher and Phillips 2005).

the Pauli susceptibility is quite isotropic while above P_X, $\chi_a \sim 4\chi_c$. This ratio increases slightly above P_C (Huxley et al. 2003a).

Spin dynamics have been studied recently by inelastic neutron scattering (Huxley et al. 2003b). In contrast to the case found for d-metal ferromagnets, the magnetic excitations at small q extend to higher energies in UGe_2. The relaxation rate Γ_q of the magnetization density does not vanish linearly as $q \to 0$ but has a strong independent q component. The product $\chi_q \Gamma_q$ is weakly q-dependent above T_{Curie} (fig. 42). For a clean metal not too close to T_C, in 3d intermetallic compounds, it depends linearly on q (Lonzarich and Taillefer 1985). Here $\chi_q \Gamma_q \sim 0.70$ meV as $q \to 0$. This large strength of $\chi_q \Gamma_q$

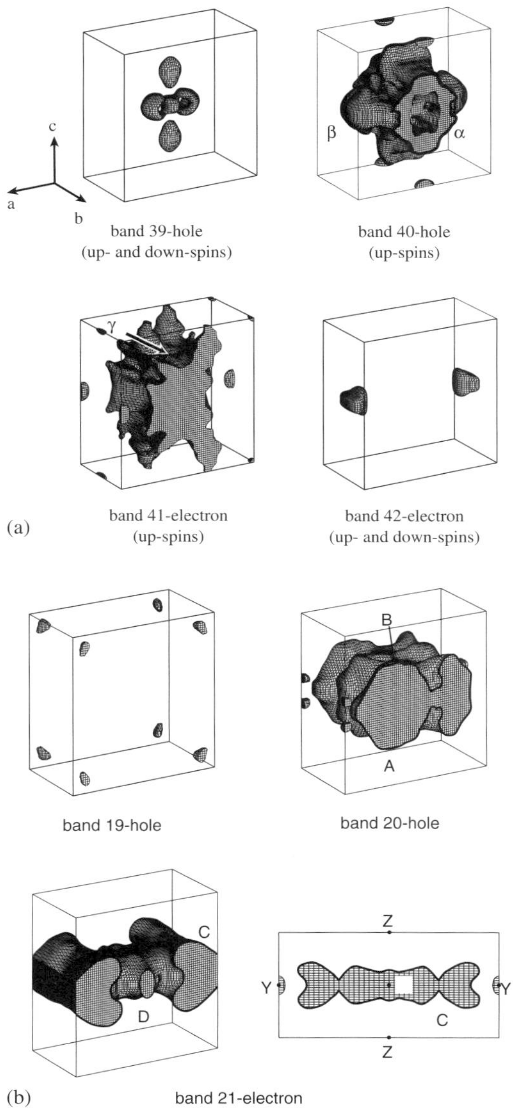

Fig. 40. (a) Fermi surfaces in the ferromagnetic state of UGe_2 (FM_2) (Settai et al. 2002), (b) Fermi surfaces in the paramagnetic state of UGe_2 (Settai et al. 2002).

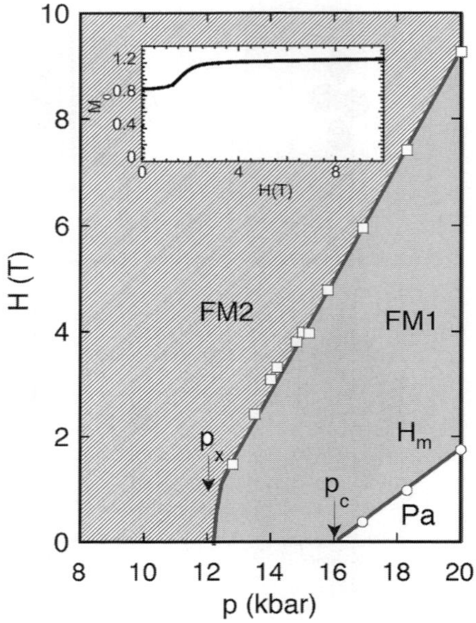

Fig. 41. (*H*, *P*) phase diagram of UGe_2. The insert shows the jump of M_o (in μ_B) at the transition $FM_1 \rightarrow FM_2$ at P_X for $T = 2.3$ K (Pfleiderer and Huxley 2002).

implies that long-wavelength fluctuations of magnetization relax rapidly. Thus the average magnetization density will no longer be a conserved quantity (similar phenomena are well known for liquid ^{3}He via the dipolar interaction). As for nearly ferromagnetic d compounds, $\chi_q \Gamma_q$ is T-independent above T_C. Below T_C, it decreases strongly. Temperature analysis of χ_q gives a finite magnetic coherence length ξ_m^o $(T \rightarrow 0) \sim 24$ Å larger than the typical values found in localized systems (~6 Å). Since the relaxation Γ_q is related with the energy scale for magnetically mediated superconductivity, the finite quasi-uniform damping of Γ_q may enhance T_C. It is worthwhile noting that positive muon spin relaxation measurements show magnetic correlations with tiny magnetic moments (Yaouanc et al. 2002).

5.2. *UGe_2 a Ferromagnetic superconductor*

The discovery of superconductivity (Saxena et al. 2000) just near 1.0 GPa and below P_C, i.e. inside the ferromagnetic domain (fig. 43) is remarkable as superconductivity occurs when the Curie temperature is still high. The maximum of T_C at $T_C^{max} \sim 700$ mK at $P \sim 12.5$ kbar appears far below the

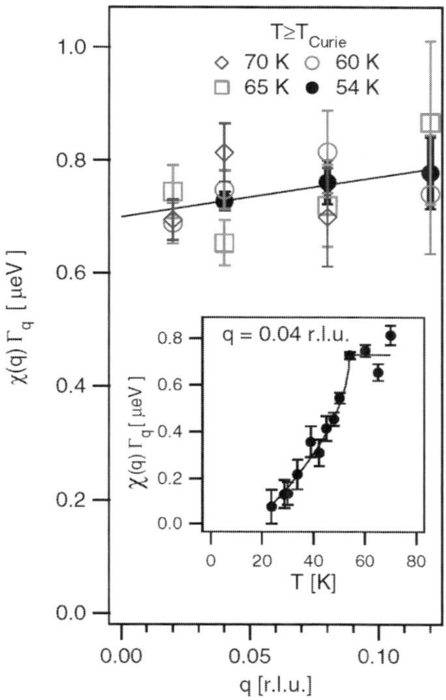

Fig. 42. The q dependence of the product $\chi(q)\Gamma_q$ at different temperatures above T_C. Inset: the temperatures dependences of $\chi(q)\Gamma q$ at $q = (0, 0, 0.04$ r.l.u) above and below T_C. The lines serve to guide the eye (Huxley et al. 2003b).

Curie temperature $T_{Curie}{\sim}35\,K$ and in a high moment $M_o{\sim}1.2\,\mu_B$ ferromagnetic state. As in a simple one electron picture, the exchange magnetic field (near 100 T assuming a simple formula $M_o \times H_{ex} = k_B T_{Curie}$ or $M_o = H_{ex}\chi_{pauli}$) exceeds the Pauli limit ($H_p{\sim}1$ T at $T_C^{max} = 0.7$ K), and so triplet superconductivity seems likely. The bulk nature of the superconductivity was suggested in flux flow resistivity experiments (Huxley et al. 2001) and established without ambiguity by the observation of a nearly 30% specific heat jump at T_C^{max} for $P = 1.22\,GPa{\sim}P_X$ (Tateiwa et al. 2001) and in the calorimetric experiment in Berkeley (Fisher and Phillips 2005). The estimate of the electronic mean-free path $\ell{\sim}1000\,Å$ from ρ_0, specific heat and H_{C_2} shows that the clean condition ($\ell > \xi_0{\sim}100\,Å$) is achieved in the Grenoble–Cambridge or Osaka experiments performed on high-quality single crystals ($\rho_0 < 1\,\mu\Omega\,cm$). T_C^{max} appears to coincide with the collapse of P_X so is probably strongly related to the FM_2–FM_1 transition at P_X.

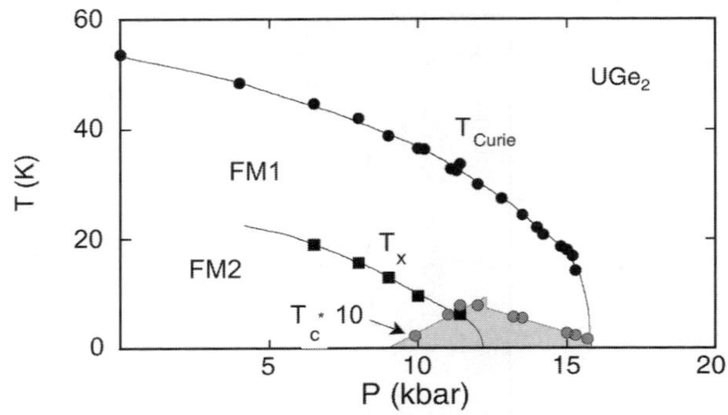

Fig. 43. The (T, P) phase diagram of UGe_2. The Curie temperature T_{Curie} (●), the supplementary characteristic temperature T_X (■), which leads to first-order transition at $T \to 0\,K$ and the superconducting temperature T_C (●) are shown.

After subtraction of a small temperature contribution due to nuclear scattering, the temperature dependence of the intensity of the (0, 0, 1) Bragg reflection measured by neutrons shows no reduction below $T_C = 0.7\,K$ due to superconductivity. The persistence of the same magnetic polarization is strong evidence of triplet pairing with the coexistence of ferromagnetism and superconductivity (Huxley et al. 2003a).

The transition from FM_1 to FM_2 at H_X is directly felt in the field variation of H_{C2} as shown at 13.5 kbar just above P_X in fig. 44 where the $H_{C2}(T)$ data at low fields below H_X in the FM_1 state appears to belong to a different phase than the points in the FM_2 high-field state. The FM_1 phase seems to have a lower maximum in the critical temperature than the FM_2 phase (Sheikin et al. 2001b) but also a lower upper critical field as $T \to 0\,K$. No direct accurate experiments on the pressure dependence of T_C near P_X have yet been performed. The puzzling result is that the specific heat jump $\Delta C/\gamma T_C$ at P_X is far below the BCS value, and secondly it drops rapidly on either side of P_X (fig. 45) (Tateiwa et al. 2003b, Tateiwa 2004). Of course, departure from T_C^{max} leads to a huge sensitivity to pressure gradients and impurities. A marked peak of $\Delta C/\gamma T_C$ near P_X deserves confirmation and explanation.

Evidence for microscope coexistence of ferromagnetism and unconventional superconductivity seem now well established by a ^{73}Ge NQR study under pressure (Kotegawa et al. 2003, Kotegawa et al. 2005). The NMR data are summarized in fig. 46. The nuclear spin lattice relaxation rate $1/T_1$ has probed the ferromagnetic transition exhibiting a peak at T_{Curie} as well as a decrease without a coherence peak below T_C. The coexistence of ferromagnetism and unconventional superconductivity, presumably with a linear gap node, is clear. By comparison to previous Tateiwa results, the P range of gapped superconductivity is larger: the residual density of states at $T = 0\,K$

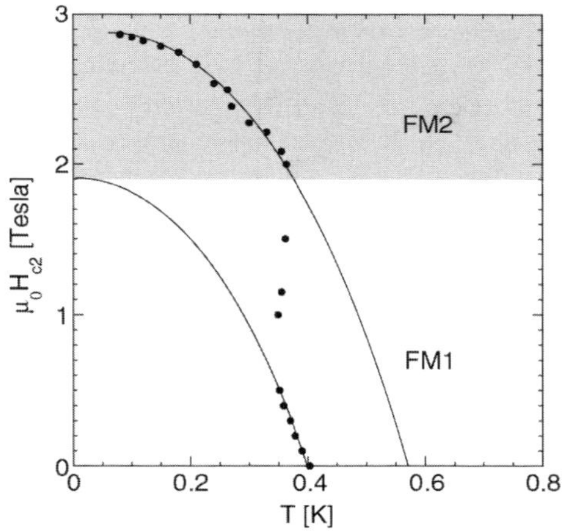

Fig. 44. The temperature field coordinates of the mid-points of the superconducting transition measured by resistivity are shown at a pressure of 13.5 kbar. The position of the onset of the bulk $FM_1 \rightarrow FM_2$ transition seen in the resistivity in the normal phase at 1 K is also indicated as the lower limit of the shaded region (Huxley et al. 2003a).

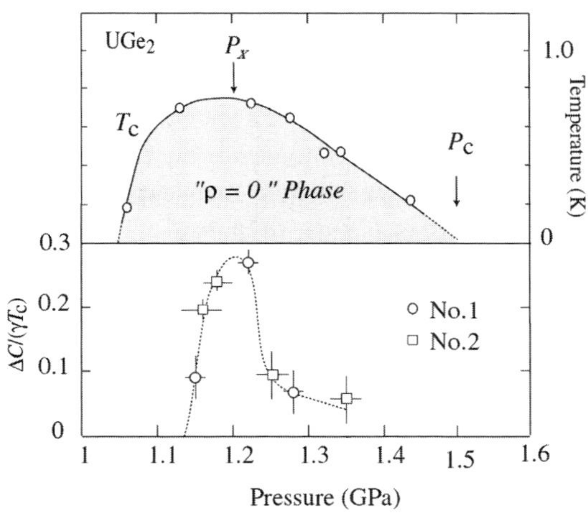

Fig. 45. Pressure dependence of T_C and the value $\Delta C/\gamma T_C$ (from Tateiwa et al. 2003b, Tateiwa 2004).

is respectively 65%, 37% and 30% of the normal phase density of states at $P = 1.15$, 1.2 and 1.3 GPa. By comparison with our conclusion from $H_{C2}(T)$ jump at $P\sim 1.3$ GPa, it is found that, just below P_X in the FM_2 phase, $T_C\sim 0.35$ K. Just above P_X, in the FM_1 phase, $T_C\sim 0.7$ K. This

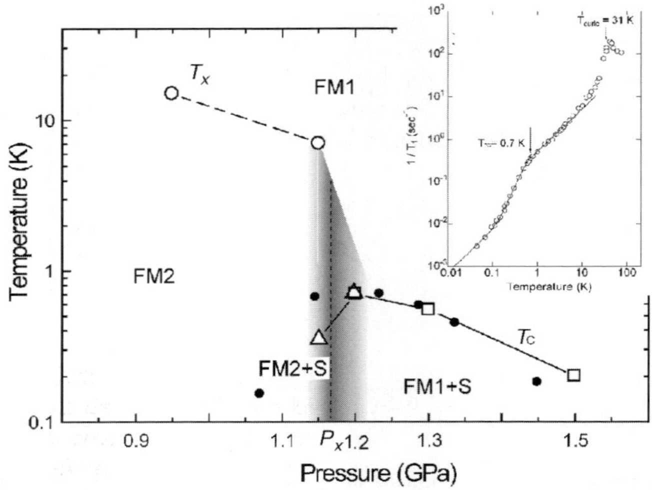

Fig. 46. (T, P) phase diagram of UGe_2 detected in NQR experiment (Kotegawa et al. 2003 and 2005): P_X is marked by the dashed line and the shaded region corresponds to phase separation. The solid wide refer to previous work, the NQR determination of T_X and T_C are respectively (○) and (Δ, •). Inset: temperature variation of $1/T_1$ with the clear signature of ferromagnetism and superconductivity. The solid line is a calculation assuming unconventional line gap nodes.

opposite conclusion shows that careful measurements are needed in the vicinity of P_X.

The reproducibility of superconductivity was first verified inside the Cambridge/Grenoble collaboration and then rapidly confirmed in Osaka (Tateiwa et al. 2001), Nagoya (Motoyama et al. 2001) and La Jolla (Bauer et al. 2001). The occurrence of superconductivity in UGe_2 is now well established. However, the homogeneity of the effect, as well as the conditions of its observation (clean or dirty limit) are still controversial. Experiments on UGe_2 polycrystals can lead to the conclusion that the clean limit requirement ($\ell > \xi_0$) may not be a necessary condition so implicitly opens the possibility of conventional pairing. To confirm the unconventional nature of superconductivity, a new generation of systematic measurements is clearly needed on different crystals.

It was claimed by Motoyama et al. (2003) that for $P < P_X$ a clear sharp anomaly can be observed in the ac magnetic susceptibility at T_{Curie} while an imperfect superconducting shielding effect occurs at T_C. As P increases away from P_X, the reverse is observed, i.e. the peak anomaly at T_{Curie} becomes broad while the diamagnetic susceptibility approaches perfect superconducting shielding. We have already stressed that the self-consistency between the square of the magnetization and the ferromagnetic intensity detected on Bragg reflection is an important proof of ferromagnetic homogeneity on both sides of P_X. Conclusions on the homogeneity of the superconductivity and ferromagnetism on each side of P_X and P_C may be premature as the respective broadening of the transitions are sensitive to the pressure dependence of the characteristic temperatures which are strong below P_X for T_C and above P_X for T_{Curie}. As described, the recent accurate NMR data support a homogeneous coexistence of FM and S.

5.3. *Ferromagnetism and superconductivity in URhGe and* $ZrZn_2$

URhGe appears to be a promising new material as the superconductivity at $T_C \sim 300\,mK$ is already achieved at $P = 0$. This favorable situation seems correlated with a smaller U–U distance, $d_{u-u} = 3.5\,Å$, than in UGe_2 corresponding to onset of ferromagnetism below $T_{Curie} = 10\,K$ with a residual $\gamma = 160\,mJ\,mol^{-1}\,K^{-2}$ and a sublattice magnetization of $0.4\,\mu_B/U$ atom (Aoki et al. 2001). The bulk nature of the superconductivity is demonstrated by the specific heat jump $\Delta C/\gamma T_C \approx 0.45$ at T_C . Of course, with this result for a polycrystal, one may speculate that the relative weak jump of $\Delta C/\gamma T_C$ and the high value of C/T at $T = 0\,K$ even in the superconducting state are indications for a zero gap for the minority spin while the majority spins may lead to a temperature dependence of C/T characteristic of a polar state (line of gap nodes).

However, further developments are required to succeed in the growth of excellent crystals and then to perform careful measurements as will be reported for UPt_3 or UPd_2Al_3. The growth of excellent crystals with convincing superconducting properties has been unsuccessful in different laboratories. Systematic measurements on polycrystals underline that the clean limit must be fulfilled for the occurrence of superconductivity. Very recently, superconducting single crystals have been produced in Grenoble (Huxley and Hardy 2004). The upper critical field on single crystals can be well fitted with triplet pairing (Hardy 2004, Hardy et al. 2005a).

Pressure experiments have been performed (Hardy et al. 2005b); contrary to UGe_2, T_{Curie} increases under pressure at least up to 10 GPa while a linear extrapolation of the pressure decrease of T_C suggests $P_S = 3\,GPa$. The URhGe (T, P) phase diagram is different from the UGe_2 one, $P_S \ll P_C > 10\,GPa$ (fig. 47). The driving mechanism for superconductivity may not be ferromagnetic spin fluctuations even if the superconductivity must adjust its pairing with the spin arrangement. T_C is field-invariant with low magnetic field ($H < 500\,Oe$) along the easy axis of magnetization (c-axis) which allows the creation of a single magnetic domain.

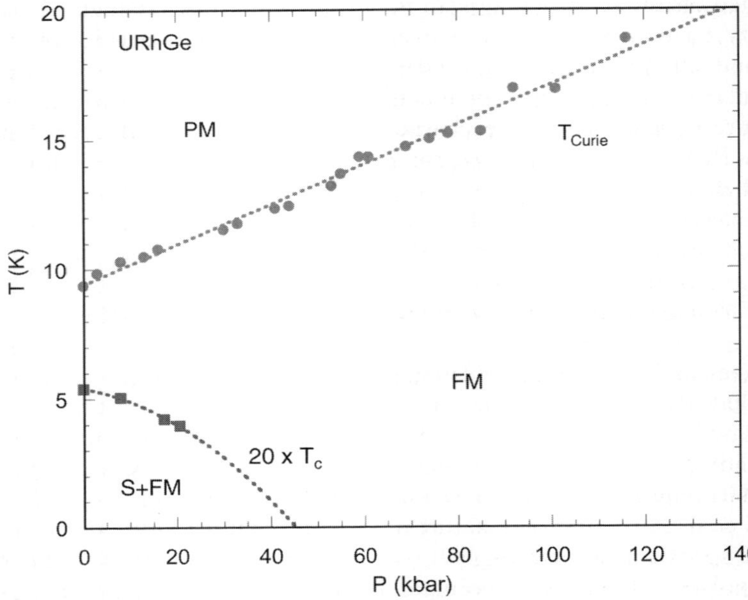

Fig. 47. Pressure dependence of T_{Curie}, T_C of URhGe (Hardy et al. 2005b).

It is of supplementary interest in URhGe that the field switches the easy axis of magnetization from *c*- to *b*-axes at $H_C\sim 12\,T$. Of course, this transition is associated with changes of the Fermi surface and related variations of the spin and charge dynamics. The big surprise is the H re-entrance of superconductivity (Levy et al. 2005).

Just before the report of superconductivity in URhGe, it was claimed that the well-known weak Heisenberg itinerant ferromagnet $ZrZn_2$ ($T_{Curie} = 30\,K$, $M_o = 0.17\mu_B$ at $P = 0$) may become a superconductor below $T_C = 0.2\,K$ (Pfleiderer et al. 2001). The unconvincing points were the non-vanishing value of the resistivity below T_C, the lack of a maximum in the imaginary part χ'' ($q = 0, \omega \rightarrow 0$) of the susceptibility at T_C and furthermore the absence of any specific heat jump at T_C. The main evidence of intrinsic phenomena was the apparent collapse of T_{Curie} and T_C at $P_C = 1.5\,GPa$. After this first report, a full superconductivity resistivity drop was observed below T_C as well as a smooth maximum in χ'' at T_C (see Pfleiderer et al. 2001, Hayden 2002, Pfleiderer 2003). However still no specific heat anomaly has been found at T_C and contradictory links are reported between the occurrence of superconductivity and sample purity. Recently, in Grenoble (Boursier 2005), for single crystals of $ZrZn_2$ with RRR $\sim$ 30, a complete resistive superconducting transition was detected with $T_C = 80\,\text{mK}$ and $H_{C2}(0) \sim 1$ kOe. In Sendai (Kimura 2004), no trace of superconductivity can be found even for RRR $\sim$ 60 and in Tokyo with RRR $\sim$ 140 (Takashima et al. 2004). Here new materials and careful tests are needed to understand the superconductivity of $ZrZn_2$ and even to assign its intrinsic character. A recent publication by Uhlarz et al. (2004) show that the collapse of ferromagnetism occurs at $P = 1.65\,\text{GPa}$ through a first-order transition.

Unsuccessful attempts to discover superconductivity have been made in different cerium ferromagnetic HFC ($CeRu_2Ge_2$, $CeRh_3B_2$, $CeAg_2Sb$) as well as in uranium ferromagnets UP and U_3P_4 and 3d systems such as Ni_3Al (Niklowitz 2003) or Y_2Ni_3. The problem of unconventional superconductivity and ferromagnetism is far less well documented than the previous one of unconventional superconductivity and antiferromagnetism.

5.4. *Ferromagnetic fluctuation and superconductivity in εFe?*

The other major breakthrough is the discovery of superconductivity in the high-pressure ($P > 13\,\text{GPa}$) crystallographic ε hexagonal compact phase of Fe, which is the material in the Earth's core: $T_C^{\max} = 3\,\text{K}$ at $P = 22\,\text{GPa}$ (Shimizu et al. 2001 (fig. 48). In this ε phase, the ground state is paramagnetic but with an enhanced density of states. Band calculations suggest a proximity to antiferromagnetism (Saxena and Littlewood 2001). Nevertheless, it has been predicted that in ε-Fe, ferromagnetic fluctuations will lead to higher T_C than the AF fluctuations (Jarlborg 2002). Of course, the martinsitic process of the α to ε conversion favors the persistence of a "ferromagnetic" memory over short distances.

Careful absolute resistivity measurements have been recently performed on different crystals of Fe. In good agreement with the previous remarks, the temperature variation of the resistivity just above T_C follows the $T^{5/3}$ law predicted for ferromagnetic fluctuations. The large slope $\partial H_{C2}/\partial T$ at T_C confirms that heavy particles are involved in the pairing ($m* \sim 10\text{m}_\text{o}$) (Jaccard et al. 2002, Holmes et al. 2004b, Jaccard and Holmes 2005). The necessity to respect the clean limit condition ($\ell > \xi_0$) points to unconventional pairing. By contrast to the previous examples of UGe_2, URhGe and $ZrZn_2$, here the superconductivity occurs in the PM state. So εFe, appears to have concomitant NFL properties and unconventional superconductivity near $P = 20\,\text{GPa}$. Furthermore the superconductivity may keep memory of the spin and charge dynamics of the low-pressure α cubic phase. The discovery of this new class of unconventional superconductors provides strong motivation for new theoretical developments.

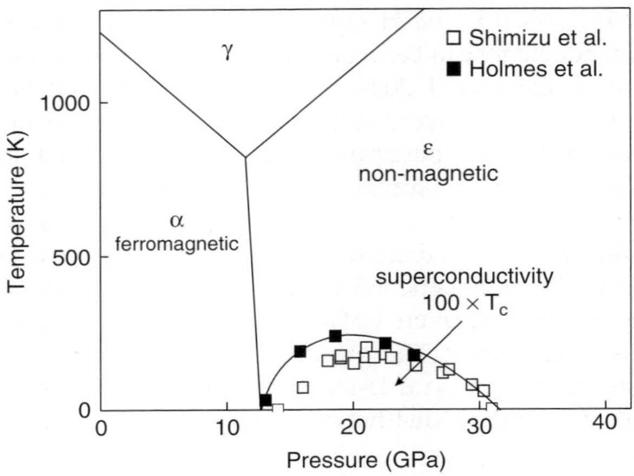

Fig. 48. The body-centered-cubic α phase is the well-known ferromagnet, If the high pressure and high temperature γ phase were stable at low temperatures they would form an antiferromagnet at 100 K at ambient pressure. The hexagonal-close-packed ε phase is non-magnetic and becomes superconducting at low temperatures. From (□), Shimizu et al. (2001); (■), Holmes et al. (2004b).

5.5. *Theory of ferromagnetic superconductors*

The different superconducting states in ferromagnetic phases for crystals with cubic (Samokhin and Walker 2002) and orthorhombic structure (Mineev 2002) have been classified on general symmetry arguments. For the $ZrZn_2$ cubic case (Walker and Samokhin 2002), it was predicted that the gap nodes change when the magnetization is rotated by the magnetic field. Tests can be easily made in ultrasound attenuation and thermal conductivity experiments. For an orthorhombic point group, only 1D representations are possible. This can lead to a magnetic superconducting phase with spontaneous magnetization when superconductivity occurs inside the ferromagnetic region ($T_{\mathrm{Curie}} > T_{\mathrm{C}}$) for the case of strong spin–orbit coupling (Fomin 2001).

In general, no symmetry nodes exist if the pairing amplitude with zero projection of the Cooper pair spin is not vanishing. If for some reason the pairing amplitude with zero projection of the Cooper pair spin (the $\hat{z}$ component of the order parameter) is absent then the superconducting gap Δ can have point nodes parallel to the magnetic axis at $k_x = k_y = 0$ (Fomin 2002). The possible absence of the z-component of the order parameter also practically eliminates any possibility of the occurrence of a FFLO type state (no paramagnetic limitation).

One question is the reason for the stabilization of superconductivity in the ferromagnetic domain. It was suggested by Walker and Samokin (2002) that a positive feedback occurs due to an exchange interaction between the magnetic moments of the Cooper pair and the magnetization density. For the previous orthorhombic case, the possible superconducting states are non-unitary and Cooper pairs have a spin momentum, which is proportional to the difference in the density of populations of pairs with spin-up and spin-down. The interaction of the spontaneous magnetic moment of the Cooper pair with the exchange field due to the ferromagnetism H_{ex} stimulates the superconducting state (Mineev 2002). This effect exists only due to the difference in the density of states on the Fermi surfaces for the spin-up and spin-down quasiparticles (Ambegaokar and Mermin 1973) and leads to an enhancement of the critical temperatures,

$$\frac{T_C(H_{ex}) - T_C}{T_C} \approx \frac{\mu_B H_{ex}}{\varepsilon_F}.$$

In parallel, the ferromagnetic magnetization creates a magnetic field H_{em} that acts on the orbital electron motion to suppress the superconducting state. The reduction of the critical temperature due to the orbital effect is:

$$\frac{T_C(H_{em}) - T_C}{T_C} \approx \frac{d^2 H_{em}}{\Phi_0}$$

where d is the characteristic length scale over which the order parameter changes near the upper critical field. For a pure superconductor, d simply coincides with the usual coherence length ξ_0. In the vicinity of the critical pressure P_S the coherence length $\xi_0 \sim v_F/T_C$ will eventually exceed the mean-free path ℓ. Thus close to P_S in this unconventional clean superconductor, d is given by ℓ. The comparison of the previous two expressions shows that triplet superconductivity can be stimulated by ferromagnetism.

An alternate idea for a weak Heisenberg ferromagnet such as $ZrZn_2$ is linked to the disappearance of the transverse magnetic fluctuation for coherent magnons below T_{Curie} with an enhancement of T_C on the ferromagnetic side because the coupling of magnons to the longitudinal magnetic susceptibility strongly enhances T_C in the FM state with respect to the paramagnetic state (Kirkpatrick et al. 2001). This possibility seems to be ruled out in the UGe_2 Ising ferromagnet as transverse modes like magnons will be missing.

Further, explanations have been proposed for UGe_2 on the remarkable coincidence that T_C is optimum just at P_X. The drop of the resistivity at T_X as well as the coincidence of the maximum in T_C when T_X collapses are reminiscent of the paramagnet α uranium where T_X is identified as the charge density wave temperature T_{CDW}. Furthermore, the common point between UGe_2 and α-uranium is their zigzag uranium chain (Huxley et al. 2001). These anologies plus the unusual temperature variation of the neutron intensity of ferromagnetic Bragg reflections (at T_X) and a bump in C/T near T_X give support for a model where a charge density wave CDW may occur below T_X (Watanabe and Miyake 2002a, b). This model is able to explain the field instability at H_X and the unusual shape of H_{C2} (T). However up to now, as indicated above, no lattice and magnetic superstructure has been detected. In the same spirit, a zero-temperature Stoner model (Sandeman et al. 2003) was proposed for a system which has a twin peak structure in the electronic density of states (DOS), i.e. the necessary ingredients for two metamagnetic instabilities at H_C and H_X. Triplet superconductivity is driven in the ferromagnetic phase by tuning the majority spin Fermi level through one of the two peaks. The maximum of T_C is found at P_X, i.e. at the magnetization jump.

Another possibility is to consider the possible occurrence of s-wave superconductivity by bypassing the argument on the strength of the exchange field seen by the conduction electrons. Such a possibility is considered mainly for UGe_2 since the ferromagnetism may come from the localized 5f part. It was shown that the coupling of two electrons via a localized spin can be attractive (Suhl 2001) and demonstrated that this s-wave attraction holds for the whole FS (Abrikosov 2002). The supplementary condition for the occurrence of superconductivity is a large density of states at Fermi level, i.e. the occurrence of heavy fermions. The applicability to UGe_2 is an open question as emphasized previously since its ferromagnetism has both localized and itinerant characters.

There are particular features associated with unconventional superconductivity in ferromagnets. It was recently emphasized that triplet ferromagnetic superconductors with up–up and down–down spin pairs with scattering at a finite spin–orbit coupling are two band superconductors. The consequence is that the dependence of T_C on the impurity content is non-universal but determined by two independent dimensionless parameters linked to the respective scattering time and to the pairing. The departure from the universal Abrikosov Gorkov law will be a qualitative demonstration of the two band character (Mineev and Champel 2004).

An interesting new feature is that the superconducting order parameter is coupled to the magnetization (Fomin 2001, Machida and Ohmi 2001). The spin direction of the dominant pairing amplitude is fixed by the magnetization direction. As this direction changes from one domain to another, the properties of this layered structure can differ from those of a domain structure with singlet pairing. One should expect here that domain walls play the role of weak links.

Another supplementary consideration (Buzdin and Mel'nikhov 2003), at least in resistivity experiments, is that of domain wall superconductivity. In each domain a finite average magnetic induction $4\pi M_o$ exists (near 2000 Oe for UGe_2). Assuming a thin domain wall ($\ll \xi$) and modelling the domain interface by a step-like function $\pm M_o$ on each side of the wall, the orbital effect is cancelled. On cooling, the superconductivity will first appear locally at a domain wall not inside a magnetic domain. Furthermore depending on the relative orientation of M with respect to H, different critical temperatures will occur between two opposite domains. This must lead to an unusual broadening at $H = 0$ near T_C which will disappear rapidly in magnetic field.

6. The four uranium heavy Fermion superconductors

Essential points:

- Diversity of the (P, T) superconducting phase with respect of universality, which may happen for a hypothetical quantum critical point.
- UPt_3: an example of the role of the Kondo lattice condensate to unconventional superconductivity. Careful experiments, which demonstrate the unusual low-temperature excitations of unconventional superconductors.
- UPd_2Al_3: evidence of d-wave pairing in inelastic neutron scattering and tunnelling experiments.
- URu_2Si_2: switch from SMAF or hidden order to large moment antiferromagnetism LMAF magnetism at P_X. Disappearance of superconductivity at P_{cr} the critical pressure where the first order transition line (T_X, P_X) meets or approaches the magnetic line (T_N, P).
- UBe_{13}: A superdense Kondo lattice due to the low carrier density?

6.1. *Generalities*

The Kondo picture cannot be applied straightforwardly to actinide compounds. In U systems, evidence of a $f^n \Leftrightarrow f^{n\pm1}$ transition is lost for example in photoemission spectra (Allen 2002). Even without renormalization, a rather broad fband exists. The valence is not near an integer number and furthermore the fluctuations will often occur between two magnetic configurations of U^{3+} and U^{4+}. For Sm and Yb the fluctuations happen between a trivalent configuration which is a Kramer's ion and a divalent non-magnetic configuration; for Tm, as for U, the valence fluctuation involves two magnetic configurations. The description of U compounds requires consideration and knowledge of their band structures while, for Ce, a HFC qualitative picture can be made from its local behavior (n_f, T_K, Δ_{CF}). For a view on the magnetism of U intermetallic systems the reader can consult the review article of Sechovsky and Havela (1998). We will focus mainly on the interplay between superconductivity and magnetism.

In previous work, either the requirement to apply a pressure or the difficulty to grow large single crystals has restricted experimental studies. Extensive data have been obtained on four different uranium heavy fermion superconductors UPt_3, UPd_2Al_3, URu_2Si_2 and UBe_{13}. The growth of large crystals has permitted combined studies, notably neutron scattering experiments, quantum oscillation (de Haas van Alphen), NMR and various macroscopic probes. Our main aim is to stress important features of these unconventional superconductors:

- the necessity to treat impurity scattering in the unitary limit,
- the consequence of lines of zeros or point nodes in the angular variation of the gap Δ_k (low-energy excitations, sensitivity to the Doppler shift even at low magnetic field, appearance of a normal fluid component at very low temperature directly linked with the change of sign of Δ_k)
- the relation between the spin pairing of the order parameter and $H_{C2}(T)$.

Let us summarize briefly the main phenomena at zero pressure. The occurrence of bulk superconductivity is proved with the observation of marked specific heat anomalies at T_C (fig. 49). Outside UBe_{13} , the three other compounds have specific heat jumps at $T_C(\Delta C/\gamma T_C \approx 1)$ often smaller than the usual BCS value of 1.43 observed for s-wave superconductors. The decrease of C below T_C is not exponential but follows mainly a T^2 power law for UPt_3, URu_2Si_2 and UPd_2Al_3 and T^3 for UBe_{13} (Brison et al. 1994a), that leads one to suspect a line of zeros for the first three cases and point nodes for UBe_{13}. UPt_3 is an intriguing superconductor having a double superconducting transition. This phenomenon appears highly reproducible and boosts the focus on UPt_3 with a large diversity of experiments (Joynt and Taillefer 2002). Such a multicomponent phase diagram cannot be explained in the framework of conventional superconductivity. We will see later that it needs also a new concept for SMAF at least for its dynamics. The main trend is to assume a bidimensional E_{2u} order parameter split by a symmetry-breaking field (SBF) (see fig. 50). The possibility of a multicomponent phase diagram was already stressed in the study of $U_{1-x}Th_xBe_{13}$ ($0.02<x<0.047$) (Ott et al. 1985) but the requirement of doping has

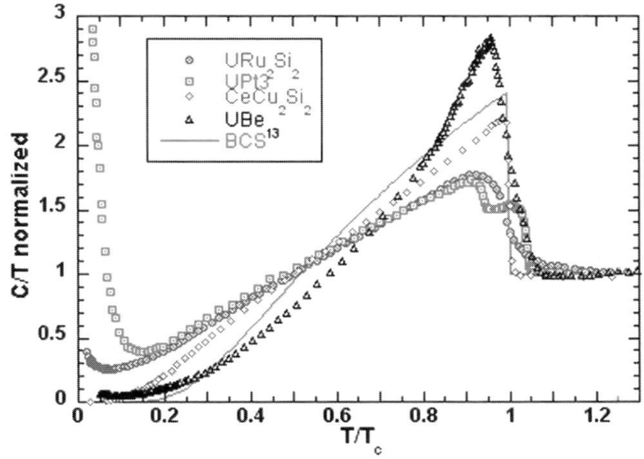

Fig. 49. Specific heat of various HFC normalized to the value of C/T in the normal phase just above T_C (Brison et al. 1994a and b).

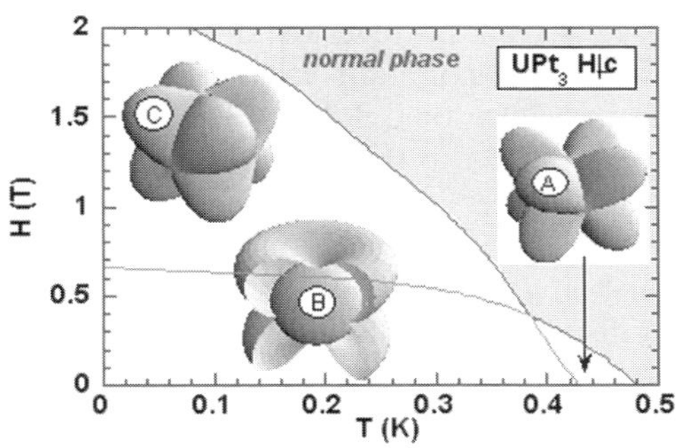

Fig. 50. Schematic phase diagram of the three superconducting phases of UPt_3, together with the hypothetical corresponding gap structures in the E_{2u} model for a spherical Fermi surface (apparent on the B phase gap scheme). The A and C phases differ by a rotation of the azimuthal line of nodes. Only a second-order point node on the c-axis remains in the B phase (Brison et al. 2000).

prevented careful studies in the low-temperature phase. The controversy still exists if the second low-temperature transition is linked or not with the onset of long-range magnetism.

In these four cases, the superconductivity is not obviously related to the proximity of a magnetic instability. In UPt_3, below $T_N \sim 6$ K neutron scattering experiments as well as X-ray scattering (Aeppli et al. 1988a, b, Isaacs et al. 1995) show the appearance of a tiny ordered moment ($M_o = 0.02\mu_B$/U atom) along the a^*-axis of the basal plane. No sign of static magnetic order has been found in macroscopic experiments or low-energy spectroscopy (NMR–μSR). In UPd_2Al_3, a clear AF ordering occurs below $T_N = 14\,K$ with a large sublattice magnetization $M_o \sim 1\mu_B$ (large moment antiferromagnetism LMAF) located in the basal plane and a propagation vector $Q_0 = (0, 0, 1/2)$ (Krimmel et al. 1992) along the c-axis of the hexagonal crystals. In URu_2Si_2 below $T_N \sim 17.2\,K$, a large specific heat anomaly characterizes the onset of long-range ordering; however only a small sublattice magnetization $M_o \sim 0.03\mu_B$ is detected along the c-axis with a propagation vector $Q_0 = (0, 0, 1)$ (Broholm et al. 1987, 1991). In this so-called SMAF phase, a hidden order (HO) must exist which has not yet been identified (Coleman 2002, Mydosh 2003,). In UBe_{13}, no long-range magnetism has been detected at least above T_C. The interesting feature is that when superconductivity appears, no simple Fermi liquid regime is yet established (as for $CeCoIn_5$). For example, the resistivity has reached its maximum only at $T_M \sim 2.5\,K$. At T_C, large inelastic scattering still occurs leading to $\rho(T_C) \sim 100\,\mu\Omega\,cm$.

Under pressure, the tiny ordered moment M_o in UPt_3 vanishes linearly with pressure at $P_X \sim 0.4$ Gpa; however the drop of M_o (Hayden et al. 1992) is not correlated with a concomitant variation of T_N as observed in the previous cerium case of $CeIn_3$ or $CePd_2Si_2$. The important point is that the collapse of M_o appears linked to the disappearance of the superconducting splitting with increase in pressure. The antiferromagnetism may be at the origin of the SBF (Behnia et al. 1990, Trappmann et al. 1991, de Visser et al. 2002). T_C may vanish at $P_S \sim 5\,GPa$ far above 0.4 GPa.

In UPd_2Al_3, T_N decreases initially strongly under pressure with a slope $\delta T_N/\delta p = 900$ mK GPa^{-1}. No magnetic signal can be detected in resistivity experiments above 7.5 GPa. A linear pressure extrapolation suggests $P_C \sim 15\,GPa$. T_C, by contrast, is first quite insensitive to pressure up to 7 GPa and then decreases at a rate of 90 mK GPa^{-1} (Link et al. 1993). Extrapolating to zero will give $P_S \sim 30\,GPa$ quite different from P_C. Up to now no data exist in the critical regime P_C or P_S. A high-pressure structural study points out a pressure-induced electronic transition at $P = 25\,GPa$ from hpc to orthorombic phase (Krimmel et al. 2000).

In URu_2Si_2, recent experiments show that above $P_X = 0.5\,GPa$ at low temperature, the HO phase with its weak antiferromagnetism and superimposed hidden order is replaced by a large-moment antiferromagnetic phase (LMAF) with the same Q wavevector but a sublattice magnetization (jump of M_o from $0.15\mu_B$ to $0.35\mu_B$) (Amitsuka et al. 1999). The superconductivity of URu_2Si_2 disappears at a pressure $P_S = 1.2\,GPa$ with a large pressure rate of 1 K/GPa (McElfresh et al. 1987, Schmidt 1993) while T_N

increases initially under pressure with a variation $\partial T_N/\partial p = 1.9\,\text{K GPa}^{-1}$. $P_S = 1.2\,\text{GPa}$ seems to be near the critical pressure where the first-order T_X line may end up or meet the two second-order HO and LMAF (T_N) lines.

In UBe_{13}, T_C decreases with a slope of $130\,\text{mK GPa}^{-1}$ suggesting $P_S = 8\,\text{GPa}$ (Chen et al. 1984, Thompson et al. 1987). Surprisingly, thermoelectric power experiments suggest that magnetic ordering may occur above 6.7 GPa (Mao et al. 1988).

It is interesting to compare the experimental and theoretical evaluation of the London penetration depth λ_L:

$$\lambda_L = \left(\frac{m^*}{\mu_0 n e^2}\right)^{1/2}$$

as it is a quantity sensitive to the ratio of the effective mass to the carrier number. In Table 7 the average measured values (λ_L^{exp})by muon experiments are compared with the results obtained from a free electron assumption assuming either three electron carrier per unit formula (λ_L^{3+}) or a number of carriers derived from band calculations (λ_L^{BC}) (see URu_2Si_2, UBe_{13} above):

The large values of $\lambda_L^{exp}$for URu_2Si_2 and UBe_{13} agree well with the low carrier case of these two materials. Their magnitude deserves further consideration as the dependence between the effective mass m^* and the number of carriers n_e may not be given by a free electron hypothesis (see Section 1). All these compounds are strong type II superconductors with a large parameter kappa ($\kappa \sim 50$ or 100).

In these complex materials, a large diversity of cases may occur. The relevance of spin dynamics to superconductivity is not obvious but at least inelastic neutron scattering measurements point out the occurrence of large magnetic responses. Subtle games between parameters may lead to unique situations well suited for a given property. This is the case for the dynamical magnetic susceptibility in UPd_2Al_3 where in the AF state, the normal phase is characterized by an important quasi-elastic contribution peaked around the AF wavevector Q_0 and by a spin wave or magnetic exciton dispersion. In

TABLE 7

For these four uranium HFS: molar volume in cm^3 /mole U, Sommerfeld coefficient γ in mJmole^{-1} K^{-2} , London penetration λ_L^{3+}, λ_L^{BC} and λ_L^{exp} in Å as defined in the text: λ_L^{exp} is the experimental value found in the reference.

	V_m	γ	λ_L^{3+}	λ_L^{BC}	λ_L^{exp}	Ref.
UPt_3	42	457	3600	3600	5200	Yaouanc et al. (1998)
UPd_2Al_3	62	140	2100	2100	4510	Knetsch et al. (1993)
URu_2Si_2	49	70	700	17,000	9000	Feyerherm et al. (1994)
UBe_{13}	81	1000	6000	90,000	12000	Dalmas et al. (2000)

contrast to the previous cerium HFC, there is a peaked dynamic component. This intense quasielastic signal is completely modified below T_C. The energy redistribution is a direct consequence of a gap opening characteristic of a d-wave even pairing (Bernhoeft 2000, Sato et al. 2001).

6.2. *UPt_3: multicomponent superconductivity and slow fluctuating magnetism*

The magnetism of U atoms is first dominated on cooling by a fluctuating local moment ($\sim 2\mu_B$) at a characteristic energy of 10 meV (Aeppli et al. 1988b); paramagnetic moments on nearest-neighbor sites in adjacent planes start to be antiferromagnetically coupled below $T = 20$ K at the wavevector $Q = (0, 0, 1)$ with a typical energy of 5 meV. At lower energies ($\sim$0.3 meV) another antiferromagnetic correlation appears at $Q_0 = (1/2, 0, 1)$ with a small moment $0.1\mu_B$; it corresponds to be AF coupling of neighboring sites on the a^*- or b-axes inside the basal plane (Aeppli et al. 1988a). A slow magnetic component of the itinerant quasiparticles was detected in time-of-flight neutron scattering experiments (Bernhoeft and Lonzarich 1995).

Below $T_N \sim 6$ K, a part of the AF fluctuations at Q_0 appears static in a neutron scattering experiment with a tiny ordered moment $M_o = 0.02\,\mu_B/$U atom as $T \to 0$ K. Comparable signals are absent in probes having a longer time scale (NMR–μSR) as well as in macroscopic properties (C, χ, ρ). The magnetic coherence length is finite (300 Å). A linear temperature dependence of the intensity of the magnetic Bragg peak in neutron scattering suggests an extrinsic origin for this SMAF or precursor effects linked to the slow down of the correlation while the real phase transition may appear at far lower temperatures. Evidences for this last possibility are the observation of a specific heat anomaly at $T_N \sim 18$ mK (Schuberth et al. 1992, Brison et al. 1994a) and the concomitant increase of the correlation length with a divergence near 20 mK (Metoki et al. 2000). The tiny moment antiferromagnetic correlation can be qualitatively interpreted with the crystal field singlet–doublet scheme as the thermal fluctuations can play a main role practically in the whole temperature region where correlations set in (Fomin and Flouquet 1996). This framework can also explain that under pressure M_o collapses linearly with pressure up to 0.5 GPa with no concomitant pressure dependence of T_N. The observation of uncorrelated pressure dependences of M_o and T_N can also be explained in an inhomogeneous scheme with the P disappearance of a fraction of a magnetic phase characterized by LMAF. However, since the same tiny ordered moment $M_o \sim 0.02\,\mu_B$ occurs in high-quality crystals, a self-consistent intrinsic mechanism should control SMAF. Attempts have been made to discover crystallographic structural modulations or trigonal distortions but deeper experiments (Dalmas de Réotier

2005) rule out the proposals (Midgley et al. 1993, Walko et al. 2001). The small-angle neutron scattering shows intense diffusion along the a^*- and c^*-axes with "rod" shapes suggesting planar defects (Huxley et al. 2000, van Dijk et al. 2002). Their implication in the magnetism of UPt_3 is open.

The claim of a double-intrinsic superconducting transition with our colleagues in Berkeley (Fisher et al. 1989) was the end of a long inquiry which:

- first started with the observation of a split specific heat transition (Sulpice et al. 1986) on a polycrystal but, unfortunately, was not associated with a unique diamagnetic shielding at T_C^A;
- second continued with the confirmation of the same features on another polycrystal (Ravex et al. 1987);
- third led to assign the phenomena to intrinsic properties as it survived in the specific heat measurement realized in Berkeley on a single crystal now available in Grenoble with the arrival of L. Taillefer from Cambridge (Fisher et al. 1989);
- fourth appeared to collapse in magnetic field at a tetracritical point (Hasselbach et al. 1989).

Accurate determinations of the (H, T) superconducting phase diagram with the three phases A, B, C were achieved rapidly by ultrasound measurements (Adenwalla et al. 1990, Bruls et al. 1990) by magnetocaloric effect (Bobenberger et al. 1993) and by magnetostriction (Van Dijk et al. 1993, 1994). Despite the reproductibility of the phenomena in different laboratories, for a decade I was not completely confident of its intrinsic nature, keeping in mind that for an experimentalist it is best to find one's own error. However, the intrinsic nature of the phenomena seems established.

Systematic studies were performed on $U(Pt_{1-x}Pd_x)_3$ alloys (de Visser 2002). The striking result is that the superconductivity collapses at the concentration $x{\sim}0.6\%$ above which the SMAF phase is replaced by a LMAF phase. In this last state, the link between T_N and M_o is recovered starting from zero at $x_c = 0.6\%$ reaching $T_N = 6\,K$ and $M_o = 0.6\,\mu_B$ for $x = 5\%$. The coincidence between the collapses of both T_C and T_N at x_c at least proves that T_C is not correlated with an AF QCP as in Section 4. It has been suggested that doping may be associated with a shift in the spectral weight from ferromagnetism to antiferromagnetism and thus with the disappearance of the odd-parity superconducting state. However a strong decrease of T_C will occur on doping in this unconventional superconductor. Substitutions on the U sites, whatever is the doping, (non-magnetic (La, Th) or magnetic (Gd)), destroy superconductivity above x_c. The absence of differences between paramagnetic substituents or non-magnetic impurities have suggested an odd-parity pairing (Dalichaouch et al. 1995).

We have emphasized that in unconventional superconductors, any type of impurity has pair-breaking effects. The reason is that in the scattering processes from impurities, the wavevector of the incident electrons may be changed to a position on the FS where the order parameter has a different sign. It was soon recognized that in order to explain the thermal transport in these exotic superconductors, a maximum of scattering must be realized corresponding to the unitary limit with a phase shift $\delta = \pi/2$ (Pethick and Pines 1986, Schmidt-Rink et al. 1986). Now it is taken as a general hypothesis for all non-conventional superconductors. For this pair breaking, magnetic and non-magnetic impurities have similar effects. An important consequence of the unitary limit is the creation of virtual bound states, which are revealed by a sharp resonance at the Fermi level. In the gapless regime of the unconventional superconductors the normal fluid component leads to residual linear T terms in specific heat and thermal conductivity (Hirschfeld et al. 1988,1986, Schmidt-Rink et al. 1986).

Several experiments (see Sauls 1994) favor the choice of the E_{2u} bi-dimensional order parameter reported in fig. 50. The first possibility of a E_{2u} state appears for the description of the upper critical field $H_{C2}(T)$ measured along the a- and c-axes. The absence of a Pauli limit for $H//a$ can be explained if the d-vector of an odd-parity order parameter is parallel to c. Indeed, no Pauli limit will occur when H is perpendicular to c as in this equal spin pairing scheme ($\uparrow\uparrow$, $\downarrow\downarrow$), the Pauli susceptibility is invariant through T_C (Shivaram et al. 1986, Piquemal et al. 1987). At least, in the strong spin-orbit (SO) limit a Pauli limit is predicted for $H//c$ as observed experimentally on $H_{C2}(T)$. The E_{2u} state of the B phase is called a hybrid state as lines of zeros exist in the basal plane and point nodes along the c-axis with an energy gap Δ_k vanishing as $\sin^2\theta$, θ being the azimutal angle between the k and c-vectors. The other bi-dimensional E_{1g} even state in the B phase has a line of zeros again in the basal plane but Δ_k vanishes as $\sin\theta$ as $\vec{k}$ approaches $\vec{c}$.

Thermal conductivity (κ) experiments are a powerful technique to test the hybrid state as a T^3 dependence of κ is predicted for both E_{1g} and E_{2u} states at least above the characteristic temperature T^* of the normal fluid component induced by the impurities. In the low-temperature regime $20\,\text{mK} < T < 70\,\text{mK}$ the thermal conductivity in both directions (κ_a and κ_c) shows the hybrid state T^3 variation (Suderow et al. 1997, 1998). Below 20 mK, down to the lowest measured temperature of 16 mK, where the linear T variation emerges. This component stems from the band of impurities present in the case of symmetry enforced gap nodes. A vanishing gap without a change of sign of the order parameter would be, in contrast, smeared out by impurities. It was stressed that, in the gapless regime for a E_{2u} state, κ/T may reach the same universal value along c as that along a

(Lee 1993, Graf et al. 1996) while for E_{1g} no unique extrapolation of κ/T will occur. With the present unique results on the thermal conductivity in the gapless regime of UPt_3, the data do not appear consistent with the E_{2u} choice (universal constant of κ/T along a). But, the validity of the present theoretical hypothesis (phase shift, isotropy of the relaxation state) has to be re-examined.

The low-energy excitations lead to a strong sensitivity to the magnetic field. The quasiparticle energies are Doppler shifted by the superfluid flow induced around each vortex. Unlike conventional superconductors, quasiparticles can be now excited even in small magnetic field. Volovik first pointed out that for lines of nodes a $\sqrt{H}$ contribution will dominate the usual linear H dependence of C/T produced by the creation of the normal core component (Volovik 1993). Furthermore, by considering combined H and thermal variations, a scaling relation with a single variable $x = T/T_C\sqrt{H_{c2}/H}$ (Kopnin and Volovik 1996, Simon and Lee 1997) must be respected in the specific heat and the thermal transport. The scaling is observed for both, a and c, directions in UPt_3 (Suderow et al. 1998). Thus field and thermal response agree with a hybrid gap at least in the B phase.

Another interesting test is to follow the thermal anisotropy κ_c/κ_a or κ_b normalized to their normal phase values. A first attempt (Flouquet et al. 1991) on quite different crystals failed to detect any anisotropy. However, new generations of experiments with better conditions (high quality crystal ($T_C \sim 530\,\mathrm{mK}$, $\rho_0 \sim 0.5\,\mu\Omega\,\mathrm{cm}$) and on adjacent pieces from the same batch) show a large temperature variation in the ratio κ_c/κ_a (Lussier et al. 1994, Huxley et al. 1995) The difficulty is that the comparison of the experiment with theoretical models depends on the parametrization of the order parameter close to the nodes. The initial weak temperature dependence of κ_c/κ_b close to T_C and the still large value of κ_c/κ_b at low temperature seems to favor the E_{2u} choice. Ultrasonic attenuation experiments have also been used to probe the quasiparticle spectrum. Combined measurements of transverse and longitudinal ultrasonic attenuation confirm the hybrid gap in the phase B and gap structure with additional nodal planes in the A phase as predicted for both E_{1g} and E_{2u} scenarios (Ellman et al. 1996).

Macroscopic evidence of the difference between the B and A order parameters was given in the observation of the flux lattice in small-angle neutron scattering experiments (fig. 51) (Huxley et al. 2000). In the high-temperature A phase, the flux lattice arrangement is governed by the strong gap anisotropy in the basal plane. It gives rise to different domains for the orientation of the vortex lattice. At lower temperature in the B phase, the gap with its line of zeros is isotropic in the basal plane. In the B state, the weak anisotropy of the Fermi surface will fix the vortex orientation. A theoretical analysis has been proposed by Champel and Mineev (2001). The

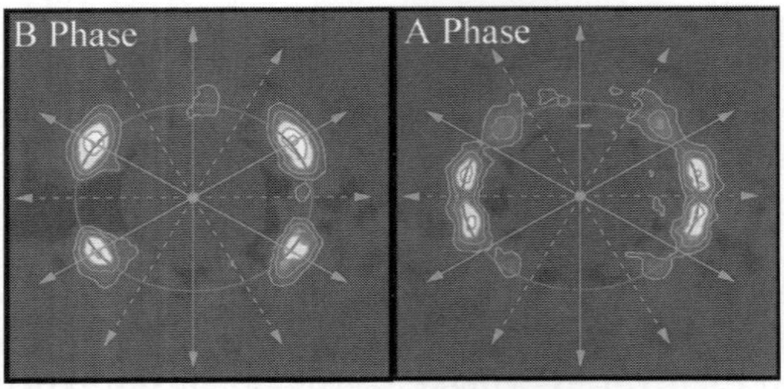

Fig. 51. Flux line lattice in the B and A phases of UPt_3 observed by small-angle neutron scattering experiments (Huxley et al. 2000).

non-local correction in the Landau free energy may not play a big role in the orientation of the vortex lattice in contrast to classical low-dimensional superconductors. Independent of the models used for the order parameter, the results in the A phase can be explained only if the order parameter belongs to a 2d irreducible representation (Rodière 2001).

It was hoped that confirmation of a E_{2u} odd-parity order parameter could be easily found in a NMR experiment. In the strong spin–orbit limit, assuming the Knight shift is dominated by the Pauli susceptibility, it is predicted that there would be no change through T_C for $H//a$ but that a substantial decrease for $H//c$ may occur. As tiny variations of the Knight shift (Tou et al. 1996, 1998) were found in each direction, different proposals have been given to understand the experiments. One of them suggests a drastic change in the hypothesis for spin–orbit coupling: switching from strong to weak limit will allow one to explain a field orientation of the d-vector. The proposed orbital order parameter is quite similar to that of the E_{2u} state (Ohmi and Machida 1996). Another remark starts with the consideration that the Van Vleck susceptibility may be large and thus also the contribution to the Knight shift. Furthermore it is shown that a new type of Fermi liquid state can be realized in the specific situation of $5f^2$ configuration for a singlet crystal field ground state with no enhancement of the Pauli susceptibility but enhancement of the density of states (Ikeda and Miyake 1997). Tiny temperature decreases of the Knight shift through T_C are in excellent agreement with this picture. Assuming an opposite hypothesis of Kramers ions with a doublet ground state, it was recently pointed out, that a self-consistent analysis of the Korringa relaxation, the γ Sommerfeld coefficient of C/T, and the Knight shift gives, in UPt_3, a large enough Pauli contribution to the Knight shift to be probed experimentally. The NMR results for different HFC leads to the classification of UPt_3, UNi_2Al_3 with odd-parity pairing and $CeCu_2Si_2$, $CeCoIn_5$, UPd_2Al_3 with even parity pairing (Tou et al. 2005). Figure 52 compares the Knight shift

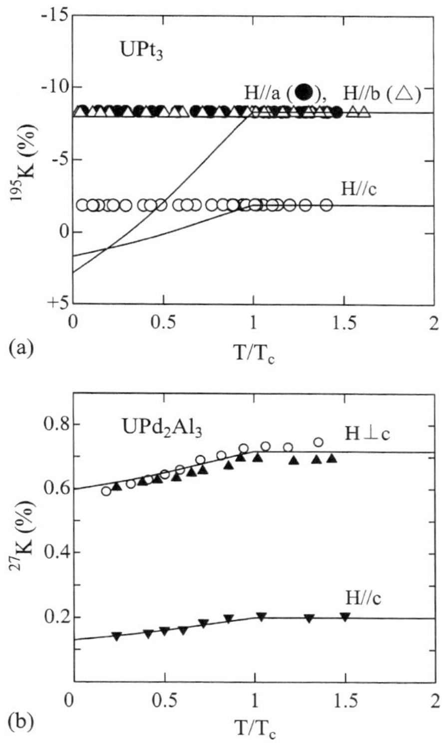

Fig. 52. *T* dependence of the Knight shift of UPt_3 and UPd_2Al_3 measured along the *a*- and *c*-axes. The solid lines in both figures are calculations using the Sommerfeld coefficient γ for a d-wave singlet model (Tou et al. 2005).

variation of UPt_3 and UPd_2Al_3 along the main orientation axes. The solid lines in both figures are calculations using the Sommerfeld coefficient and a d-wave singlet model.

In the complex UPt_3 material, a complete understanding of the different phases has not been achieved. The main puzzles are the link between SMAF which looks mainly like a slow fluctuating cluster above T_C and the underlying symmetry-breaking field (SBF) with the apparent necessity to revisit the strong spin-orbit hypothesis. If the magnetic fluctuations are slow ($\approx 10^{-6}$ s) enough by comparison to the superconducting lifetime of a Cooper pair $\tau \sim \xi_0/v_F \sim 2 \times 10^{-13}$ s they can be an efficient SBF (van Dijk et al. 2002). However, no calculation exists for their effect on δT_C. NMR and neutron scattering time scales are 10^{-6} s to 10^{-12} s. In the Kondo cloud image, it is quite probable that the anisotropy linked to the slow dynamics is completely different from that suspected from instantaneous pictures. Traveling along a loop path ℓ_{KL} the quasiparticle feels different atomic properties of U and Pt atoms and thus during this orbital motion the initial strong spin-orbit condition can be smeared out. An interesting observation is that two distinct and isotropic Knight shifts have been found by muon experiments for the field in the basal plane with drastic and opposite temperature dependence around T_N. Unfortunately, the threshold field for this two-component magnetic response is not yet determined (Yaouanc et al. 2000).

6.3. *UPd_2Al_3, localized and itinerant f-electrons: a magnetic exciton pairing*

Extensive discussion can be found in the recent review by Thalmeier et al. (2004) with emphasis on the dual model for U-based systems (Yotsubashi et al. 2001, Sato et al. 2001, Zwicknagl and Fulde 2003) and on nodal superconductivity mediated by magnetic excitons which originates from crystal field transitions (see $PrOs_4Sb_{12}$ in Section 4). The main difference with the spin fluctuation mechanism is that they require a localized electron component.

In UPd_2Al_3, a well-defined AF ordering occurs at $T_N = 14\,K^n$ with $M_o = 0.85\,\mu_B$ (Krimmel et al. 1992) and a moderate Sommerfeld coefficient $\gamma = 145\,mJ\,mole^{-1}\,K^{-2}$. At $T_C = 2\,K$, the large specific heat jump $\Delta C/\gamma T_c{\sim}1.2$ proves that the heavy fermion quasiparticles themselves condense in Cooper pairs (Geibel et al. 1991). As high quality single crystals have been obtained, excellent macroscopic and microscopic experiments have been performed. De Haas van Alphen, experiments succeed to detect eight kinds of dHvA branches, which have been obtained in band calculations based on different techniques (Inada et al. 1994, Knöpfle et al. 1996). In a so-called dual model, good agreement has been found by keeping two of the three electrons localized on the *U* sites and another delocalized. The same approach seems also capable of explaining the UPt_3 Fermi surface (Zwicknagl and Fulde 2003).

Concerning superconductivity, AF ordering may play a minor role as emphasized for conventional superconductivity. Observations of power laws notably in the nuclear relaxation time $T_1^{-1}{\sim}T^3$ suggests here also unconventional superconductivity with nodes (Kyogaku et al. 1993). The even nature of the pairing is well established from the drop of the Knight shift (see fig. 52) below T_C as well as from the Pauli limitation of $H_{C2}\,(T)$ for both directions.

As already pointed out, the unique feature of UPd_2Al_3 among HFC is its sharp signal in inelastic neutron experiments. Just above T_C, two contributions appear in the dynamical susceptibility $\chi''(q, \omega)$ (Metoki et al. 1998, Bernhoeft et al. 1998). The first one can be regarded as damped spin waves or magnetic excitons due to the action of the intersite exchange on the CEF excitations. It softens and becomes overdamped as T approaches T_N. These modes can extend up to 10 meV. But, for a wavevector along the c-axis, they became sharp and well defined (see Hiess et al. 2004). Of course, heavy fermions exist with a quasielastic component but it is strongly peaked around the ordered wavevector. Crossing through T_C leads to a drastic change with a resonance like structure of the low-energy response (simultaneous inelasticity and enhancement of the signal) (fig. 53). The feedback between magnetic excitations and the low-energy component illustrates the coupling between the modes. It has been argued that the magnetic excitations may produce the attractive interaction of the itinerant electrons. Quantitative analysis of the spectrum by two different approaches (Bernhoeft 2000, Sato et al. 2001) lead to the conclusion that an even-parity unconventional gap must occur along the c-axis:

$$\Delta_{SC}(k) = \Delta_0 \cos k_z c$$

where Δ_0 is the amplitude of the gap function and c the lattice constant (A_{1g} representation). A line of nodes will appear at the intersection of $k = \pm 0.5 Q_0$ with the Fermi surface. Furthermore, the gap function as well as the inelastic spectrum can explain self-consistently tunneling experiment results obtained along the c-axis in a N–I–S (Normal-Insulating-Superconducting) junction performed on a high-quality UPd_2Al_3 film (fig. 54) (Jourdan et al. 1999, Huth and Jourdan 1999, Huth et al. 2000). An anomaly in the tunneling spectrum occurs at the energy of the mediated boson (magnetic excitations),which is comparable to the gap. The conductivity modulation at about 1.2 meV is roughly the energy (1.5 meV) of the magnetic excitation at

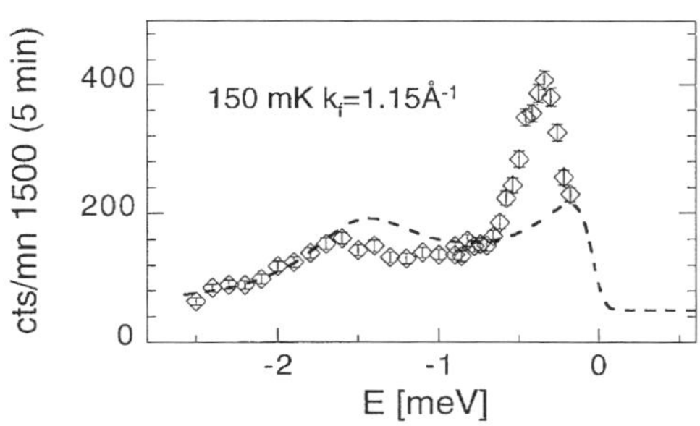

Fig. 53. Inelastic neutron data (◇) from UPd_2Al_3 at the AF wavevector at 150 mK. The dashed line is the extrapolated response for the normal antiferromagnetic state at this temperature (Bernhoeft 2000)

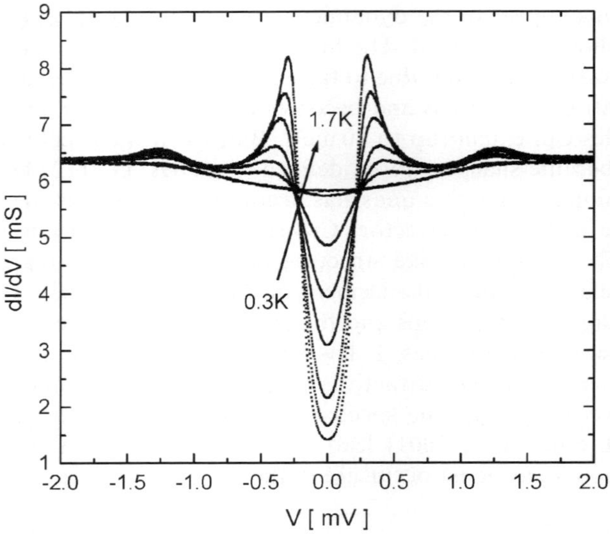

Fig. 54. Differential conductivity of a UPd_2Al_3, AlO_x, Pb junction as a function of the temperature (0.3/0.5/0.7/0.9/1.1/1.3/1.5, 1.7) (Jourdan et al. 1999).

the magnetic Bragg point (0, 0, 1/2),which is found in the normal phase. In a weak BCS limit, such a coincidence is not expected. In strong coupling, it reflects strong retardation effects which may be caused by a slow velocity of the magnetic excitons by comparison to the Fermi velocity of the heavy fermion quasiparticles (Sato et al. 2001). In agreement with a A_{1g} representation for the order parameter, there is no node of the gap function along the crystallographic *c*-axis normal here to the surface of the UPd_2Al_3 film.

UPd_2Al_3 appears as a model case where both the order parameter and the mechanism of superconductivity can be established. It is worthwhile to mention that a resonance spectrum below T_C was first observed for $YBa_2Cu_3O_7$ and is now considered as part of the evidence of $d_{x^2-y^2}$ singlet pairing (Rossat Mignod et al. 1992). By comparison to UPd_2Al_3, the complete inelastic spectrum is far more complex. So the discussion on high T_C mechanism is still controversial.

6.4. *URu_2Si_2: from hidden order to large moment*

New interest in the superconductivity of URu_2Si_2 follows from the fact that its collapse at $P_S \sim 1.2$ GPa may be related to recent features discovered on the pressure stability of its magnetic order. The nature of the phase transition at $T_N = 17.5$ K was enigmatic (see Mydosh et al. 2003) as the huge λ anomaly in the specific heat suggests the onset of a magnetic transition associated with a large sublattice magnetization ($M_o \sim 1\,\mu_B$/U atom) (Schlabitz et al. 1986). However the tiny AF sublattice magnetization ($M_o \sim 0.03\,\mu_B$/U atom) (Broholm et al. 1987, 1991) at the wavevector $Q_0 = (0, 0, 1)$ is too small to explain the amplitude of the specific heat anomaly (see Section 2) in a static picture. A hidden order (HO) must be coupled to this weak AF component. Below T_N, clear features of (HO) in inelastic neutron scattering experiments are two minima at Q_0 and Q_1 characterized respectively by the low-energy gaps $\omega_{Q_0} = 1.6$ meV and $\omega_{Q_1} = 4.5$ meV. Above $T_N = 17.5$ K, the excitation at Q_0 becomes quasielastic ($\Gamma = 1$ meV) at T_N and vanishes above 25 K while at Q_1 it stays inelastic but with a very strong damping ($\Gamma = 3$ meV) (see also Bourdarot 2003a). Taking into account this spin dynamic through T_N, there is no longer mystery in the specific heat anomaly but the nature of the order parameter is still open.

One theoretical proposal (Chandra et al. 2002) is that the HO corresponds to incommensurate orbital antiferromagnetism, which may be due to circulating current between the uranium ions. Recently, a specific search for hidden orbital currents by neutron scattering (Wiebe et al. 2004) was unsuccessful. The microscopic signatures of the HO are the previous gap ω_0 which explains basically the size of the specific heat anomaly at T_N and an observation by NMR on Si of a field-independent contribution to the linewidth (Bernal et al. 2001). As we will see there are controversies on the intrinsic character of the tiny ordered moment $M_o \sim 0.02\,\mu_B$. Restricting the problem to local order parameters of $5f^2$ shells, Kiss and Fazekas (2004) propose that the best candidate for HO may be a staggered order of octupoles. As we will see later, the duality between itinerant and localized states may lead to another issue.

The new highlight was the observation that the HO phase becomes unstable under pressure. At $P = P_X \sim 0.5$ GPa, the long-range order switches from "mysterious" HO to a "classical" LMAF (Amitsuka et al. 1999). Furthermore, using strain gauges to detect a volume discontinuity, Motoyama et al. (2003) found that at low temperature the transition is first order. Later it was confirmed in a neutron scattering experiment (Bourdarot et al. 2003a) that the line (T_X, P_X) of the phase transition approaches or ends up at (T_N, P) for $P = P_{cr} \sim 1.2$ GPa (fig. 55). The change from HO to LMAF is accompanied by important modifications of the inelastic neutron spectrum: disappearance of the excitation ω_{Q_0} at Q_0 , and the pressure increase of ω_{Q_1} at Q_1 which reaches 9.2 meV at 11 GPa. The issue on the dipole magnetic contribution at low pressure below P_X is open.

The first order nature of the transition can be invoked to keep a few percent of the LMAF phase even below P_X. The main support for this

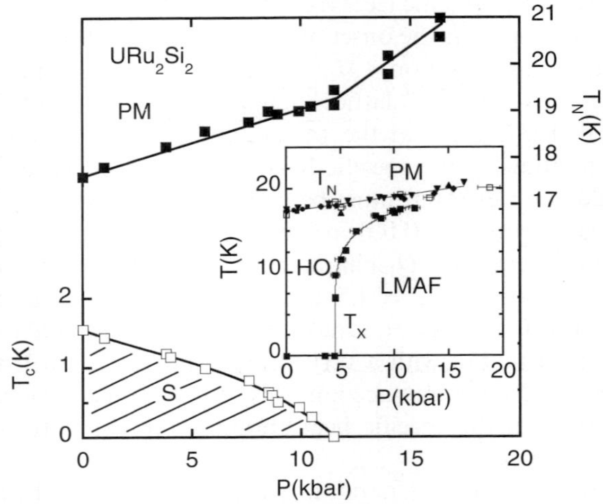

Fig. 55. (T, P) phase diagram of URu_2Si_2 with pressure variation of T_N and T_C (Schmidt 1993) measured by resistivity. P_X is the first-order transition between HO and LMAF phase at $T = 0\,K$ (Motoyama et al. 2003). Inset: domain of existence (T_X) of HO state and LMAF detected by neutron scattering below 1.2 GPa (Bourdarot et al. 2004)

possibility is that the NMR frequency of the Si site characteristic of a large ordered moment is constant whatever the pressure range but the fraction (f) of LMAF (i.e. the intensity of the NMR signal) increases continuously with pressure before reaching 1 at P_{cr} (fig. 56) (Matsuda et al. 2001, 2003). If the fraction f is a function of P and H, the system contains enough variables to explain NMR and neutron scattering.

In a dual model, Okuno and Miyake (1998), following the Ising Kondo lattice model of Sikkema et al. (1996), have proposed that a spin density wave (SDW) occurs in the itinerant system due to partial nesting of the Fermi surface with feedback to induce an extra, weakly ordered moment M_o on the localized singlet levels added to the tiny moment m created by the SDW. The SDW is also associated with a charge gap (Δ_G) in the electronic spectrum ($\Delta_G \sim k_B T_N$) (see recent developments by Fomin 2004, Mineev and Zhitomirsky 2004). The HO can be defined by m and M or the charge gap Δ_G and the spin gap ω_0. Unlike the other models, the SDW scenario is based on the coexistence of two orderings with the same AF dipole symmetry. It has been shown (Mineev and Zhitomirsky 2004) that the field dependences of the staggered magnetization and of the spin gap are in quantitative agreement with the experiments. In a magnetic field, Δ_G decreases while ω_{Q_0} increases.

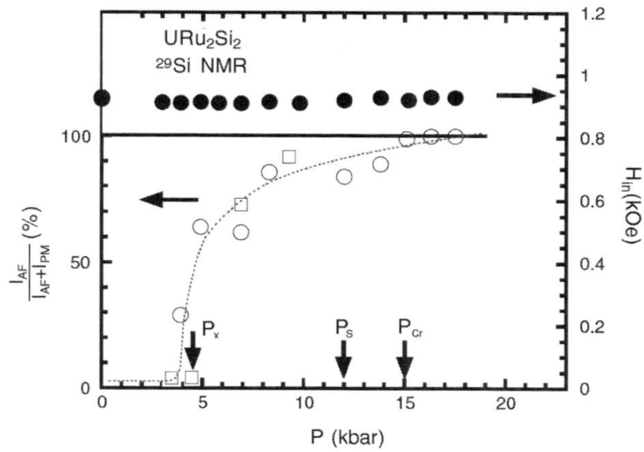

Fig. 56. The pressure dependence of internal field H_{in} and the normalized intensity (○) of the H_{in}-split line $I_{AF}/(I_{AF}+I_{PM})$ with phase separation between AF and PM signals (Matsuda et al. 2003). Full circles: the invariance of the NMR frequency at the Si sites; Open squares: the AF fraction estimated from muon experiments (Amato et al. 2004). The full line is a guide to the eye showing the 100% LMAF. 100% is achieved at $P_{cr}\sim 1.5\,\mathrm{GPa}$ not too far from P_S. A steep increase in the fraction f of LMAF state occurs at P_X.

Neutron scattering experiments in a magnetic field on a single crystal (Bourdarot et al. 2003b) suggest that the weak magnetic moment of the HO phase is intrinsic. The reproductibility of the weak magnetic moment at $P=0$ on single crystals suggests an intrinsic origin. An explanation for weak magnetism may be linked with the microscopic nature of the magnetism itself. The spin and orbital contribution can change with magnetic field and pressure. B and calculations point out that the tiny ordered moment at the uranium site may come from a cancellation between spin and orbital 5f moments respectively $-0.75\,\mu_B$ and $+0.86\,\mu_B$ while the d component has a weak ($-0.01\,\mu_B$) equal weight of spin and orbital parts (Yamagami and Hamada 2000). It is worthwhile to remark that this d component has the magnitude of M_o at $P=0$. In HFC, there is the underlying possibility that a singular behavior may be due to d–f correlations. The possibility of SDW was suggested first by Maple et al. (1986a) and Schoenes et al. (1987) and confirmed recently by thermal transport (Bel et al. 2004b, Behnia et al. 2005).

A phenomenological approach, in a full homogeneous frame for the stability of two phases is to assume that the hidden order parameter $\psi\sim m$ is coupled to the ordered moment M_o, with M_o and ψ of the same symmetry. The two states above and below $T_X(P)$ are phases with reversed size of ψ and M_0, i.e. large ψ coupled with small M_o or small ψ coupled with large

M_0. The predicted first order line will end up at P_{cr}, T_{cr} (Bourdarot et al. 2003a, Mineev and Zhitomirsky 2004). In this scheme, the inhomogeneous effects detected in NMR or muon experiments may be due to supplementary specific experimental conditions (for example powdering for NMR, supercooling or superheating at the first order transition T_X). A careful verification must be performed now to demonstrate if T_{cr} ends up on the T_N line.

Let us stress that URu_2Si_2 has rich multiple magnetic field phases which may be the combination of a quantum critical field (basically like that found in $CeRu_2Si_2$ at H_M) and magnetic phase diagrams (Kim et al. 2003, Harrison et al. 2003). It was proposed that a reentrant HO phase is created in the vicinity of a quantum critical endpoint for $H = 35\,T$. We will not go further into this complex interplay. The pressure evolution of these rich (H, T) phases will give a key to the understanding of the HO phase.

As already emphasized, T_C drops rapidly under pressure as it vanishes near 12 kbar (Mc Elfresh et al. 1987, Schmibt 1993). Thus T_C collapses at the pressure $P_S \sim P_{cr}$ where the magnetic order switches from HO to LMAF (fig. 55). The apparent survival of superconductivity observed by resistivity up to $P \sim P_{cr}$ may be a consequence of the slow disappearance of the HO fraction and the smooth increase of the LMAF fraction. So one may suspect that superconductivity is associated only with the HO phase. The bulk nature of superconductivity may disappear at P_X. New experiments on superconductivity will certainly clarify the nature of the normal phase and notably will confirm or not if there is an intrinsic phase separation between HO and LMAF order. By comparison to UPt_3 and UPd_2Al_3, the superconductivity of URu_2Si_2 has been far less studied due to the difficulty to obtain high-quality crystals. For example, the superconducting transition measured by specific heat does not coincide with the resistive transition and a strong broadening occurs in magnetic fields for the determination of H_{C2} (T) (Brison et al. 1995). The striking result is the quasi-absence of Pauli limitation along the a-axis which suggests an odd-parity pairing with d-vector along the c-axis. Careful studies of the superconducting transition by specific heat rule out the occurrence of a double transition (Hasselbach et al. 1991, Ramirez et al. 1991). At least, the SMAF, with the propagation vector ($Q_0 = (0, 0, 1)$) which preserves the lattice symmetry, cannot play the role of a SBF.

Quantum oscillations (dHvA, de Haas Shubnikov) (Bergmann et al. 1997, Ohkuni et al. 1997, 1999) were successfully observed in the normal phase. In agreement with a gap opening at T_N, only a few orbits have been detected. One may expect that the Fermi surface will react to the change of magnetic ground state at P_X. But only slight increases in the dHvA frequency occur up to 2 GPa (Nakashima et al. 2003). For the spherical α branch, the effective mass changes drastically going from $m^* \sim 17 m_o$ at $P = 0\,\text{kbar}$ to

$m^* = 8m_o$ at $P = 18$ kbar in agreement with a drop in the A coefficient ($A \sim m^{*2}$) roughly from $0.12\,\mu\Omega\,\mathrm{cm\,K^{-2}}$ at $P = 0$ to $0.03\,\mu\Omega\,\mathrm{cm\,K^{-2}}$ at $P =$ 20 kbar (Schmidt 1993). This dependence may be related to the collapse of T_C.

The interesting new ingredient is that URu_2Si_2, in its AF phase, can be classified as a compensated a semi-metal with a total carrier content near 0.07 per formula unit (Yamagami and Hamada 2000). Thus the effective mass per carrier may be comparable to that of UPt_3 (Maple et al. 1986b, Schoenes et al. 1987, Bel et al. 2004b). Another example of a semi-metal is UBe_{13}. The drastic difference between the two cases is that, in URu_2Si_2, the Fermi liquid regime seems well established above T_C since the onset of long-range ordering occur at high temperature ($T_N/T_C \sim 10$), while in UBe_{13} the effective Fermi temperature is comparable to T_C and no long-range ordering is detected above T_C. However, in high magnetic field the feedback of the magnetostriction on the carrier number may be an important parameter.

URu_2Si_2 is an illustrative example of the variation of T_N or T_C by the application of uniaxial strain σ along the a- and c-axes or hydrostatic pressure P. According to the Ehrenfest relation, the σ variation of the critical temperature is linked to the corresponding jump of the longitudinal thermal expansion ($\Delta\alpha_i$) and of the specific heat jump (ΔC) by the relation

$$\frac{\partial T_C}{\partial \sigma_i} = \frac{V_m \Delta\alpha}{\Delta C / T_c}$$

For URu_2Si_2 the corresponding variations are (van Dijk et al. 1995, de Visser et al. 1986, Guillaume 1999)

$$\frac{\partial T_N}{\partial \sigma_a} = +900\ \mathrm{mK\ GPa^{-1}} \qquad \frac{\partial T_N}{\partial \sigma_c} = -410\ \mathrm{mK\ GPa^{-1}}$$

$$\frac{\partial T_C}{\partial \sigma_a} = -620\ \mathrm{mK\ GPa^{-1}} \qquad \frac{\partial T_C}{\partial \sigma_c} = +430\ \mathrm{mK\ GPa^{-1}}$$

A uniaxial strain along a increases T_N and decreases T_C. Along c, σ_c will decrease T_N and increase T_C. The high strain sensitivity of HFC puts experimental constraints on the hydrostatic pressure conditions and emphasizes the possibility of filamentary or induced phenomena near dislocations or stacking faults where a pressure difference of order kbar can occur over a few atomic distances. The relevance of excellent hydrostatic conditions was already reported for $CePd_2Si_2$ where it was known from thermal expansion experiments that σ_a and σ_c have opposite effect on T_N (van Dijk et al. 1995). The possibility of a mismatch inside $CeIrIn_5$ was also emphasized; recent experiments also show that strong opposite uniaxial effects occur on T_C (Oeschler et al. 2003a).

6.5. *The UBe_{13} enigma: a low-density carrier?*

As pointed out, at $P = 0$, superconductivity appears in UBe_{13} at T_C from a paramagnetic metallic phase far above the temperature of the Fermi liquid regime (Ott et al. 1983). Similar situations may occur in $CeCoIn_5$, $PuCoGa_5$ or $CeCu_2Si_2$ at $P = 0$. The link between superconductivity of UBe_{13} with its heavy quasiparticles (suggested early on for $CeCu_2Si_2$) may not be a strange exotic case but another example of a new general class of unconventional superconductors.

The interest in UBe_{13} was reinforced when it was shown that under doping with Th ($U_{1-x}Th_xBe_{13}$ alloys) (Ott et al. 1985), two successive phase transitions occur at T_{C_1} and T_{C_2} in the concentration range $0.020 = x_1 < x < 0.042 = x_2$ as shown in fig. 57. The transition at higher temperature (T_{C_1} for $x_1 < x < x_2$) corresponds to a superconducting transition. It was questioned if the second transition truly coincides with a phase transition to another superconducting order parameter since a magnetic component ($M_o = 0.01\ \mu_B$) has been discovered by μSR below T_{C_2} (Heffner et al. 1987). The second phase transition may indicate long range magnetic ordering or the entrance to a new superconducting phase. Most of the proposals assume a change in the superconducting order parameter. For example, an initial even d-wave A phase ($x < x_1$) strongly suppressed by doping would be supplemented by a s-wave at T_{C_1} and then a s+id state at T_{C_2} (Kumar and Wölfle 1987). The cascade may also be from a single odd component to a multicomponent odd order parameter on cooling (Sigrist and Rice 1989). Recently, a theory called "ferrisuperconductivity"

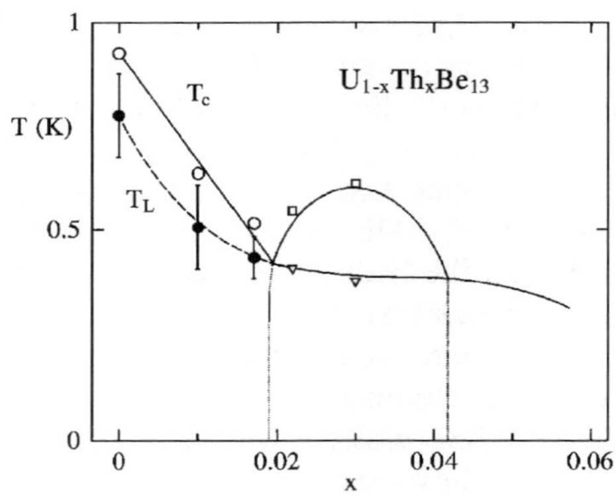

Fig. 57. Schematic T–x phase diagram of $U_{1-x}Th_xBe_{13}$. Full lines: transitions after the recent work of Schreiner et al. (1999); broken line: T_L (x) corresponds to a crossover reported in Kromer et al. (2000).

(Martisovitz et al. 2000) has been proposed with coherent pair motion and incoherent quasiparticles. In this approach a yet unobserved charge density wave (CDW) is predicted below T_{C_2}.

Experimentally, the $U_{1-x}Th_xBe_{13}$ phase diagram has been re-examined by macroscopic measurements (Oeschler et al. 2003b). At low temperature, the study of the residual γT term in the specific heat (which is sensitive to resonant defect scattering) shows structures at x_1 and x_2 Thus it gives evidence of changes in unconventional pairing at these boundaries (Schreiner et al. 1999). Simultaneous specific heat and thermal expansion measurements on pure UBe_{13} point out a crossover temperature T_L below T_C, which is a precursor of T_{C_2}. As its field response at H_L differs from H_{C2}, it has been suggested that T_L is linked to magnetic correlations.

Among the four archetype superconductors, UBe_{13} was the only one where AF correlations were not detected early on. Experiments on polycrystals show two quasi-elastic q independent responses, one broad and one narrow, with $\Gamma/2$ half linewidth of 13 meV and 1.5 meV, respectively (Goldman et al. 1986, Lander et al. 1992). New measurements on UBe_{13} ($T_C = 0.9$ K) and $U_{1-x}Th_xBe_{13}$ ($x = 3.5\%$, $T_{C_1} = 0.55$ K, $T_{C_2} = 0.4$ K) single crystals allow detection of longitudinal AF fluctuations with the wavevector (1/2, 1/2, 0); for both cases, the energy window extends near 1–2 meV while the correlation length is quite short. The coupling is restricted to the next-nearest-neighbour uranium ions (Coad et al. 2000, Hiess et al. 2002). Above 20 K, this AF magnetic fluctuation disappears. The search for CDW for $x = 0.35\%$ below T_{C_2} was unsuccessful despite the great neutron diffraction sensitivity to any movement of the Be cages. The experimental limit of 0.003 Å for the displacement is smaller than that (0.10 Å) proposed in the theory of Martisovitz et al. (2000) . Furthermore, no elastic AF magnetic diffraction line was detected below T_{C_2}, i.e. any ordered moment will be lower than 0.025 μ_B.

Another striking effect in UBe_{13} at $P = 0$ is the unusual temperature dependence of H_{C2} (T) with a huge initial slope $\partial H_{C2}/\partial T \approx -55$ T/K, with a strong negative curvature and a convex shape at intermediate temperatures (fig. 58) (Thomas et al. 1995). The temperature and pressure dependence of H_{C2} has been described in a simple strong coupling model for even-parity superconductivity (Glemot et al. 1999, Brison et al. 2000). The conflict with the paramagnetic limit at $T = 0$ seems resolved by its enhancement due both to direct strong coupling effects (increase of the ratio of the gap energy by $k_B T_C$) and to the formation of a spatially modulated superconducting FFLO state favored by the dominance of the paramagnetic limit. The values of $\lambda{\sim}15$ at $P = 0$ (Table 8) are difficult to justify and far above the usual value ($\lambda{\sim}1.4$) found in other HFC where similar fits have been performed ($CeCu_2Si_2$, $CeIn_3$ and $CePd_2Si_2$). At least the derived relative

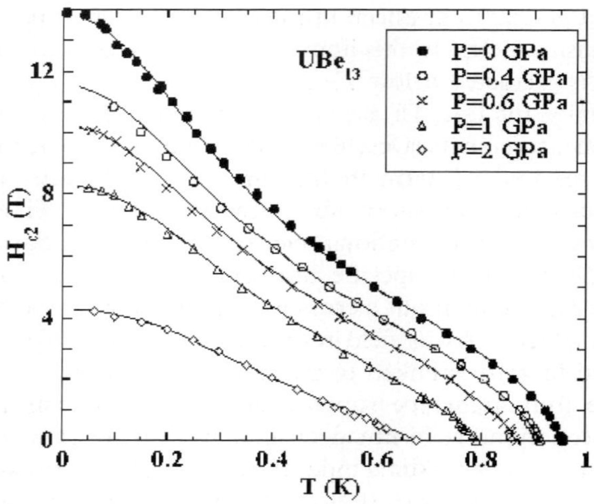

Fig. 58. UBe_{13}: the upper critical field at different pressures from Glémot et al. (1999).

TABLE 8
Pressure dependence of the parameters used in the fit of H_{C2} (T) for UBe_{13}. (Glémot et al. 1999).

P (GPa)	λ	$(\partial H_{C2}/\partial T)T_C$ (T/K)	T_{FFLO}/T_C
0	15	−55	0.45
0.4	13	−42	0.42
0.6	12	−32	0.37
1.0	11	−21	0.26
2.0	6.5	−8.5	0.10

pressure dependence of $\lambda \sim m^*/m$ agrees well with the pressure dependence of the effective mass obtained from the specific heat or the slope $\partial H_{C2}/\partial T$. The possibility of a FFLO state comes only from the unusual temperature dependence of H_{C2} (T). Contrary to $CeCoIn_5$ there is no confirmation of FFLO states by other techniques.

It has been proposed that an alternate route is a model with a field-induced mixture of two odd-parity irreducible representations (a mixture of A_{1u} and E_u) (Fomin and Brison 2000). The agreement is less satisfactory than in the even pairing case with strong coupling but inclusion of other effects can correct the discrepancies: mixing with other odd representations, mixture between odd and even representations and introduction of strong coupling. In this model, the second phase (E_u) does not appear in zero field at zero pressure; the magnetic field introduces mixtures of the representations. Here one may expect a pressure change in the pressure variation of T_C depending on whether a A_{1u} or E_u phase is achieved. Extrapolation suggests that the E_u representation will appear first in zero field above 30 kbar. As the UBe_{13} situation appears unique, a singular point may occur for this system.

One paradox is that γ and $\chi = \partial M/\partial H$ both have a weak magnetic field dependence. So one expects that the A coefficient of the T^2 resistivity law should also have a weak field variation (see below $CeRu_2Si_2$ and UPt_3). As shown in fig. 59, this is not the case (Brison et al. 1989). Both A and ρ_0 decrease strongly with field, however in a first approximation ρ_0/A are weakly field dependent. The main field effect is a change in the carrier number. It looks as though the carrier is released by magnetic field.

It was already suggested that, in UBe_{13}, the carrier number may be low (Takegahara et al. 1986, Norman et al. 1987, Brison et al. 1989). Recent band calculations (Takegahara and Harima et al. 2000) assuming the 5f electron to be itinerant indicate that the FS is remarkably simple (fig. 60) with only $n_h = 2.87\%$ holes and $n_e = 2.87\%$ electrons per atomic volume in this compensated metal. So UBe_{13} fulfills the condition of a superdense Kondo lattice with a deficiency of charge carriers per atomic site. In this superdense Kondo lattice, simple arguments with a rigid normalized density of states will never work. As $\rho_0 \sim (n_e + n_h)^{-1}$ decreases with magnetic field for a given mean-free path, the magnetic field seems to act as a carrier pump.

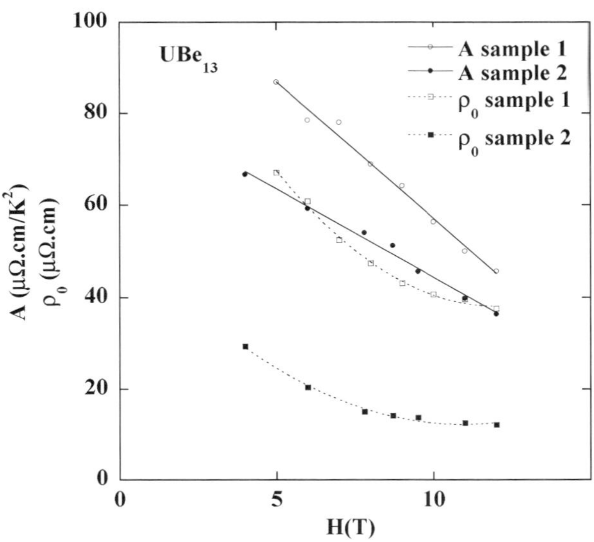

Fig. 59. Magnetic field variation of ρ_0 (H) and $A_H T^2$ of UBe_{13} at $P = 0$ (Brison et al. 1989).

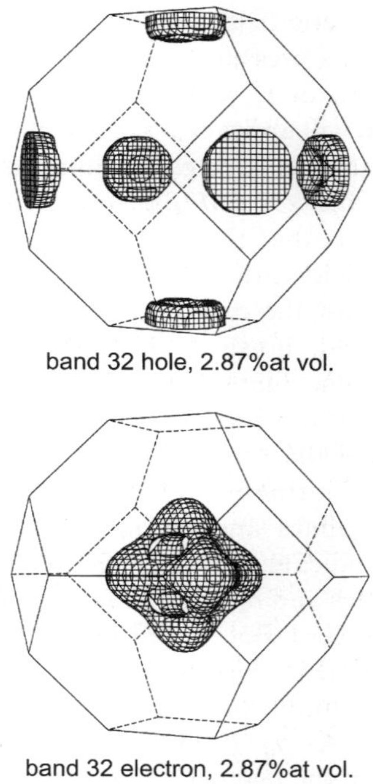

Fig. 60. Fermi surface of UBe_{13} (Harima 2004, private communication).

This pumping may stop near the field called H_L in the Dresden experiment (Oeschler et al. 2003b).

The magnetostriction induces a crystal change from cubic to tetragonal (van Dijk et al. 1994). The deformation parallel to the field is opposite to that in the plane perpendicular to the field. Again weak effects in ordinary metals will be magnified by the huge Grüneisen parameter. Tiny effects with a change of symmetry are reinforced by the exceptional condition of a semi-metal. In this picture, at each field H, there corresponds a carrier concentration $n(H)$ and thus an extrapolated value T_C ($n(H)$, $H = 0$). The field variation of carriers occurs mainly up to $H_L \sim 6\,T$ above which UBe_{13} becomes a normal metal. In doping with $ThBe_{13}$ the striking point in this "composite" system is that the double transition occurs basically near 3%, i.e. roughly for the concentration where the number of charges doubles by comparison to the pure compound UBe_{13}. $ThBe_{13}$ is a good metal with a large number of carriers $n_e \sim 1$ per atomic volume (Harrison et al. 2000). Under pressure, as the wavefunctions overlap more, the numbers of carriers will increase. The field and pressure effects are nicely evident in the 3d plot of $\rho_0(P, H)$ (by Aronson et al. 1990). For example, for $P > 20$ kbar, the monotonic variation of $\rho_0(H)$ suggests that the field pump of the carrier is no longer efficient, a pressure induced carrier concentration is a better tool.

By contrast to UBe_{13}, in $CeRu_2Si_2$ and UPt_3 γ, χ and $\sqrt{A}$ are field-insensitive below their pseudo-metamagnetic field H_M, respectively equal to 7.8 and 21 T, while $\rho_0(H)$ has a strong positive linear residual magnetoresistivity:

$$\rho_0(H) = \rho_0 + \mathrm{bH}$$

at least extrapolated from $T > T_C(0) \sim 0.55$ K for UPt_3 (Taillefer et al. 1988). Below T_C (0), in UPt_3, a different regime occurs as the quasiparticle enters in a collisionless regime ($\omega_c \tau \sim 1$, where ω_c is the cyclotron frequency, and τ the relaxation time). The appealing idea is that the field behavior in the collision regime ($T > T_C(0)$) may be a specific reaction of the lattice with the creation of a longitudinal orbital voltage antagonist to the normal current flow. If we assume as for a normal metal that the magnetoresistivity is a function of the product of the cyclotron frequency ($\omega_c \sim H$) by the collision time τ: $\Delta\rho_0(H)/\rho_0 \approx \omega_c \tau$, then $\Delta\rho$ becomes independent of ρ_0: $\Delta\rho_0(H) \sim R_L H$. The question of a collective field response (here $R_L H$) of the heavy particle is open. A classical interpretation of the H linear residual magnetoresistivity was given by Ohkawa (1990); the field reveals the local disorder of the ligand which induces a distribution in the Kondo temperature. At $H = 0$ all Kondo centers will be equivalent (unitary limit). Our message is that a deep look at simple transport data may lead to unexpected insights.

7. Conclusion and perspectives

The main trends are:

Section 1: Important progress has been made in the self-consistent treatment of the correlations including feedback from the crystal field. Basic questions emerge notably consequences for localized-itinerant duality of 4f or 5f electrons. At $T = 0$ K, our attempt to classify HFC by three pressures P_{KL}, P_C and P_V (respectively pressure switch from localized to itinerant 4f description, from AF to PM, and to valence or large orbital fluctuation) is a simplification extrapolating the phase diagram of Ce metal to a firstorder transition down to 0 K. It is also possible to consider that another critical point (P_{KL}, T_{KL}) will appear at low temperature.

Section 2: Careful experiments have been realized on normal phase properties ($CeRu_2Si_2$, $CeCu_6$, $CeNi_2Ge_2$, $YbRh_2Si_2$). An interesting point is where the itinerant picture of the 4f Yb electron will be recovered. It may happen that very low temperature is needed for this to be achieved as the

Fermi liquid regime below T_I may depend not only on $P - P_C$ but also on a renormalization parameter such as $\Delta/k_B T_K$. The key question is the location of P_{KL} with respect to P_C and the coincidence or not of P_C and P_V. The increase in competitive studies by quantum oscillations will boost our understanding soon. The effects of the magnetic field are rich as it acts on the intersite correlations, not only on the Kondo effect but also on the mixing or decoupling of the crystal field level.

Section 3: The interplay of spin dynamics and density fluctuations are strong in HFS due to the huge electronic Grüneisen parameter. The weight of each contribution in Cooper pairing depends on the relative position of P_{+S}, P_{-S} for the onset and disappearance of superconductivity with respect to P_C and P_V. Other sources of pairing such as magnetic excitons can occur.

Section 4: There is already an excellent basis for HFS where AF fluctuations play a key role: $CeIn_3$, $CePd_2Si_2$, $CeRh_2Si_2$. The canonical case of $CeCu_2Si_2$ seems now to be fully understood with the direct observation of the magnetic structure at $P = 0$ and also with a clear appearance under pressure of two different superconducting domains. The vitality of research on HFC is continuously reactivated by the discovery of new materials: a few years ago the 115 Ce HFC, recently $CePt_3Si_2$ and $PrOs_4Sb_{12}$. It is important to elucidate the roles of dimensionality, of crystal symmetry, and of combined effects of nesting and multipolar ordering.

Section 5: The growth of excellent crystals of UGe_2 and recently of URhGe has led to the discovery of ferromagnetic superconductors. Triplet superconductivity is suspected in this case as it breaks the conventional wisdom from s-wave superconductors that ferromagnetic and superconductivity are antagonistic. In this recent subject, new windows appear for a careful study of superconductivity through the first order pressure P_X from FM_2 to FM_1 in UGe_2, a fine analysis of reentrant superconductivity in URhGe including the magnetostriction effect, the resolution of the mysterious superconductivity in $ZrZn_2$ and the hope for new examples.

Section 6: Despite a large activity on UPt_3 there are still mysteries concerning the symmetry breaking field of the multidimensional order parameter, i.e. basically what is the origin of the weak antiferromagnetic component. For URu_2Si_2, the link between the low pressure HO phase and superconductivity will be certainly confirmed soon. For UPd_2Al_3 after the nice results obtained by tunneling and neutron scattering experiments, a new generation of experiments will be designed to define the interplay between magnetic nodes (spin wave or exciton) and the Cooper pair. Finally for UBe_{13} progress in high-pressure techniques will allow one to follow the change in strong coupling conditions for superconductivity, the restoration of Fermi liquid properties, and the possible emergence of long-range antiferromagnetism.

Here we have focused mainly on paramagnetic normal-phase properties and their link with superconductivity. A missing important domain concerns a careful analysis of the magnetic or multipolar order parameter as well as their excitations. A highly documented case is CeB_6 with its two successive quadrupolar and dipolar ordering (see Shiina et al. 1977). The URu_2Si_2 discussion on a HO parameter gives at least an idea of the large variety of different possibilities. The new skutterudite family also opens a wide domain.

It is important to make precise the characteristic lengths involved in HFC physics. Up to now, almost no direct derivation has been given on the magnetic correlation length close to P_C. Indirect observations point out that a weakly first order transition often may be achieved close to P_C. The generality of HFC in condensed matter is that the forces are atomic (Å) the correlations are often nanometric (Kondo length, superconducting coherence length, magnetic correlation). An electronic mean free path of a tenth of micron can be achieved, a nice achievement of the clean limit condition, i.e. physics governed by the correlations is realized. Furthermore, the experimentalists have made major progress in the handling of micrometric crystals. In this process, low cost investment has revealed fascinating and unexpected horizons.

Developments often coincide with the discovery of new materials, mastery in their microhandling, the increase in the sensitivity of measurement, as well as in the feasibility of the measurements (appearance of cantilevers for quantum oscillation, pressure tuning *in situ*) and also the combination of different probes (see the realization of excellent pressure or magnetic field experiments by macroscopic techniques as well as microscopic ones: quantum oscillations, NMR, neutron scattering and synchrotron radiation). Direct imaging by tunneling methods will be a tremendous advance. Young physicists will certainly be able to discover unexpected facets of heavy fermion matter.

Acknowledgements

I express my gratitude to J. Friedel, R. Tournier and F. Holtzberg. K. Behnia and K. Miyake who have suggested many improvements in the manuscript. J.P. Brison, P. Haen and G. Knebel helped me by their comments but also by using different results some not yet unpublished. I thank F. Bourdarot, D. Braithwaite, M. Continentino, B. Fak, D. Jaccard, J. Thompson, A. Huxley, Ph. Niklowitz, F. Lapierre, P. Lejay, G. Lapertot, S. Raymond, L.P. Regnault, H. Suderow, J.L. Tholence, J.P. Sanchez for their suggestions. I have greatly enjoyed the hospitality of Berkeley (R.A.

Fischer and N. Phillips), Cambridge (G. Lonzarich and S.S. Saxena), Osaka (K. Amaya, K. Asayama, K. Miyake, Y. Miyako and Y. Kitaoka), Tokkai (S. Kambe, Y. Haga and D. Aoki), Toyama (J. Sakurai), Geneva (J. Sierro and D. Jaccard), Moscow (A. Buzdin, I. Fomin, Y. Gaidukov, V. Mineev and G. Volovik) and IBM Yorktown Heights (F. Holtzberg and S. von Molnar). The new Ph.D. students R. Bel, R. Boursier, J. Derr, F. Hardy, A. Holmes, S. Kawasaki, W. Knafo and M.A. Measson, gave a strong stimulation to break away from the usual consensus. I also thank the young researchers and Ph.D. students A. Demuer, P. Rodière and I. Sheikin. My participation to the recent workshops organized in the Lorentz Center of Leiden by G. Stewart, A. de Visser and Q. Si, and in Santa Fe by D. Pines, J. Thompson and J. Sarrao (ICAM) was very stimulating as well as meetings from Center of Excellence in Osaka University and Tokyo Metropolitan University. Inside Europe, heavy fermion matter has profited from the FERLIN network of the European Science Foundation (ESF). M. Perrier has transformed my old hand fashioned manuscript to the present document. J.P. Brison and G. Knebel have played a major role for the improvement of the figures.

A new version of the manuscript was corrected during my stay in Tokyo Metropolitan University in November 2004. I thank my host, Professor H. Sato and his collaborators, (Prof. Y. Aoki, M. Khogi, O. Sakai and Dr. K. Kuwahara) for stimulating discussions. Prof. H. Harima was an excellent teacher in the new subject of skutterudites.

References

Abragam, A. and B. Bleaney, 1970, Electron Paramagnetic Resonance of Transition Ions (Clarendon Press, Oxford).
Abrahams, E., et al., 1995, Phys. Rev. B **52**, 1271.
Abrikosov, A.A., 1988, Fundamentals of the Theory of Metals (North-Holland, Amsterdam).
Abrikosov, A.A., 2002, J. Phys.: Condens. Mat. **13**, 943.
Abrikosov, A.A. and A.I. Buzdin, 1988, Pis'mev Zh. Eksp & Teor Fiz **47**, 204.
Abrikosov, A.A. and L.P. Gorkov, 1961, Sov. Phys. JETP **12**, 1243.
Adenwalla, S., et al., 1990, Phys. Rev. Lett. **65**, 2298.
Aeppli, G. and Z. Fisk, 1992, Comments Condens. Matter Phys. **16**, 155.
Aeppli, G., et al., 1986, Phys. Rev. Lett. **57**, 122.
Aeppli, G., et al., 1988a, Phys. Rev. Lett. **60**, 615.
Aeppli, G., et al., 1988b, J. Magn. Magn. Mater. 76–77, 385.
Aliev, F.G., et al., 2001, Some Modern Aspects of the Physics of Strongly Correlated Electron Systems (Institute Universitario de Cienca de Materials Nicolas Cabrera).
Allen, J.W., 2002, Solid State Commun. **123**, 469.
Allen, J.W. and R.M. Martin, 1982, Phys. Rev. Lett. **49**, 1106.
Amato, A., 2004, J. Phys.: Condens. Mat. **16**, S4403.
Amato, A., et al., 1987, J. Low Temp. Phys. **68**, 371.

Amato, A., et al., 1989, J. Low Temp. Phys. **77**, 195.
Amato, A., et al., 1993, Physica B 186–188, 276.
Amato, A., et al., 1997, Rev. Mod. Phys. **69**, 1119.
Ambegaokar, V. and N.D. Mermin, 1973, Phys. Rev. Lett. **30**, 81.
Ambrumenil, N. and P. Fulde, 1985, J. Magn. Magn. Mater. 47–48, 1.
Amitsuka, H., et al., 1999, Phys. Rev. Lett. **83**, 5114.
Anderson, P.W., 1967, in: Problème à N Corps – Many Body Physics, ed C. Hewitt, and R. Balian (Gordon and Breach, New York).
Anderson, P.W. and W.F. Brinkman, 1973, Phys. Rev. Lett. **30**, 170.
Anderson, P.W. and W.F. Brinkman, 1975, in: The Helium Liquids, ed G.M. Armitage, and I.E. Farguhar (Academic Press, London) pp. 315–416.
Anderson, P.W., et al., 1984, Phys. Rev. B **30**, 4000.
Andres, K., et al., 1975, Phys. Rev. Lett. **43**, 1779.
Aoki, D., et al., 2001, Nature **413**, 613.
Aoki, H., et al., 1993, J. Phys. Soc. Jpn. **62**, 3157.
Aoki, H., et al., 1995, Physica B 206–207, 26.
Aoki, H., et al., 2004, J. Phys.: Condens. Mat. **36**, L13.
Aoki, Y., et al., 1998, J. Magn. Magn. Mater. 177–181, 271.
Aoki, Y., et al., 2003, Phys. Rev. Lett. **91**, 067003.
Aoki, Y., et al., 2005, J. Phys. Soc. Jpn. **74**, 209.
Araki, S., et al., 2002, J. Phys.: Condens. Mat. **14**, 377.
Aronson, M.C., et al., 1990, Physica B **163**, 515.
Asayama, K., 2002, Nuclear Magnetic Resonance in Electron System (Shokabo, Tokyo).
Aschcroft, N.W. and N.D. Mermin, 1976, Solid State Physics (Harcourt College Publishers, Phildelphia).
Baber, W.G., 1937, Proc. Roy. Soc. A **158**, 3838.
Barbara, J., et al., 1979, J. Appl. Phys. **50**, 2300.
Barla, A., et al., 2004, Phys. Rev. Lett. **92**, 066401.
Barla, A., et al., 2005. Phys. Rev. Lett. **94,** 166401.
Bauer, E., et al., 2004, Phys. Rev. Lett. **92**, 027003.
Bauer, E., et al., 2005, Physica B, 359–361, 360.
Bauer, E.D., et al., 2001, J. Phys.: Condens. Mat. **13**, L759.
Bauer, E.D., et al., 2002, Phys. Rev. B **65**, 100506R.
Beal Monod, M.T. and J.M. Lawrence, 1980, Phys. Rev. B **21**, 5400.
Behnia, K., et al., 1990, J. Appl. Phys. **67**, 5200.
Behnia, K., et al., 2004, J. Phys.: Condens. Mat. **16**, 5187.
Behnia, K., et al., 2005. J. Phys.: Rev. Lett. 94, 156405.
Bel, R., et al., 2004a, Phys. Rev. Lett. **92**, 217002.
Bel, R., et al., 2004b, Phys. Rev. B **70**, 220501R.
Belitz, T.R., et al., 1999, Phys. Rev. Lett. **82**, 4707.
Bellarbi, B., et al., 1984, Phys. Rev. B **30**, 1182.
Benoit, A., et al., 1978, J. Phys. Lett. **39**, 94.
Benoit, A., et al., 1979, J. Phys. C5 **40**, 328.
Benoit, A., et al., 1980, Solid State Commun. **34**, 293.
Benoit, A., et al., 1981, in: Valence Fluctuations in Solids, ed L.M. Falicov, W. Hanke, and M.P. Maple (North-Holland Publishing Company, Amsterdam) 283.
Benoit, A., et al., 1985, J. Phys. Lett. **46**, L923.
Bergmann, C., et al., 1997, Physica B 230–232, 348.
Bergmann, G., 1991, Phys. Rev. Lett. **67**, 2545.
Bernal, O.O., et al., 2001, Phys. Rev. Lett. **87**, 196402.

Bernhard, C., et al., 1999, Phys. Rev. B **59**, 14099.
Bernhoeft, N., 2000, Eur. Phys. J. B **13**, 685.
Bernhoeft, N., 2001, J. Phys.: Condens. Mat. **13**, R771.
Bernhoeft, N., et al., 1998, Phys. Rev. Lett. **81**, 4244.
Bernhoeft, N.R. and G.G. Lonzarich, 1995, J. Phys.: Condens. Mat. **7**, 7325.
Besnus, M.J., et al., 1985, Solid State Commun. **55**, 779.
Bianchi, A., et al., 2002, Phys. Rev. Lett. **89**, 137002.
Bianchi, A., et al., 2003a, Phys. Rev. Lett. **91**, 187004.
Bianchi, A., et al., 2003b, Phys. Rev. Lett. **91**, 257001.
Bickers, N.E., et al., 1985, Phys. Rev. Lett. **54**, 230.
Blandin, A., 1973, in: H Suhl, ed V. Magnetism (Academic Press, New York) 57.
Bobenberger, B., et al., 1993, Physica B 186–188, 248.
Bourdarot, F., et al., 2003a, Phys. Rev. Lett. **90**, 067203.
Bourdarot, F., et al., 2003b, Cond-Mat, 0312206.
Boursier, R., et al., 2005, Thèse, Université de Grenoble.
Braithwaite, D., et al., 2000, J. Phys.: Condens. Mat. **12**, 1339.
Brandt, N.B. and V.V. Moschalkov, 1984, Adv. Phys. **33**, 373.
Bredl, C.D., et al., 1978, Z. Phys. B **29**, 327.
Brison, J.P., 2004, Private communication.
Brison, J.P., et al., 1989, J. Phys. **50**, 2795.
Brison, J.P., et al., 1994a, Physica B 199–200, 70.
Brison, J.P., et al., 1994b, J. Low Temp. Phys. **95**, 145.
Brison, J.P., et al., 1995, Physica C **256**, 128.
Brison, J.P., et al., 1997, Physica 230–232, 406.
Brison, J.P., et al., 2000, Physica B **280**, 165.
Brodale, G.E., et al., 1985, Phys. Rev. Lett. **56**, 390.
Broholm, C., et al., 1987, Phys. Rev. Lett. **58**, 1467.
Broholm, C., et al., 1991, Phys. Rev. B **43**, 12809.
Bruls, G., et al., 1990, Phys. Rev. Lett. **65**, 2294.
Burdin, S., et al., 2000, Phys. Rev. Lett. **85**, 1048.
Buzdin, A.I. and A.S. Mel'nikov, 2003, Phys. Rev. **67**, 020503R.
Canfield, P.C., et al., 1998, Phys. Today **51**, 40.
Capan, C., et al., 2004, Phys. Rev. **70**, 134513.
Castro Neto, A.H. and B.A. Jones, 2000, Phys. Rev. B **62**, 14975.
Champel, T. and V. Mineev, 2001, Phys. Rev. Lett. **86**, 4903.
Chandra, P., et al., 2002, Nature **417**, 831.
Chen, J.W., et al., 1984. Proceedings of the 17th International Conference on low temperature physics, ed U. Eckern, A. Schmidt, W. Weber and H. Wuehl Elsevier Science publishers BV, 325.
Cichorek, T., et al., 2003, Acta Phys. Pol. B **34**, 371.
Choi, S.M., et al., 2001, Phys. Rev. Lett. **87**, 107001.
Chubukov, A.V., et al., 2002, ArXiv: Cond-Mat, 0201140.
Chubukov, A.V., et al., 2003, ArXiv: Cond-Mat, 0311420V1.
Cibin, R., 1990, Thesis, University of Geneve.
Coad, S., et al., 2000, Physica B 276–278, 764.
Coleman, P., 1999, Physica B 259–261, 353.
Coleman, P., 2002, in: Local Moment Physics in Heavy Fermion Electron Systems, ed F. Mancini (American Institute of Physics, Washington) pp. 79–160.
Coleman, P. and C. Pépin, 2003, Phys. Rev. B **68**, 220405R.
Coleman, P., et al., 2001, J. Phys.: Condens. Mat. **13**, R723.

Continentino, M.A., 2001, Quantum Scaling in Many Body Systems. World Scientific, Lecture Notes in Physics (World Scientific, New York) p. 67.
Continentino, M.A., 2004a, Brazilian J. Phys., Cond-mat 041231 and 2005 35, 197.
Continentino, M.A. and A.S. Ferreira, 2004, Phys. Rev. B **69**, 233104.
Continentino, M.A., et al., 2005, Physica B, 359–361, 744.
Coqblin, B. and J.R. Schrieffer, 1969, Phys. Rev. **185**, 847.
Coqblin, B., et al., 2003, Phys. Rev. B **67**, 064417.
Cornut, B. and B. Coqblin, 1972, Phys. Rev. B **5**, 4541.
Cox, D. and M. Maple, 1995, Phys. Today **48**, 232.
Curro, N.J., et al., 2004, Cond-Mat, 0402179V2.
Custers, C.V., et al., 2003, Nature (London) **424**, 524.
Dalichaouch, V., et al., 1995, Phys. Rev. Lett. **75**, 3938.
Dalmas de Réotier, P., 2005, Private Communication.
De Visser, A., 1986, Ph.D., University of Amsterdam, Amsterdam.
De Visser, A., 2002, Physica B **319**, 233.
Deflets, C., et al., 2003, Cond-Mat, 0306742.
Degiorgi, L., 1999, Rev. Mod. Phys. **71**, 687.
Demuer, A., 2000, Ph.D., université Joseph Fourier Grenoble.
Demuer, A., et al., 2002a, J. Phys.: Condens. Mat. **14**, L529.
Demuer, A., et al., 2002b, J. Phys.: Condens. Mat. **13**, 9335.
Denlinger, J.D., et al., 2000, Journal of Electron Spectroscopy and related Phenomena, Vol. 117–118, p. 347.
Dionicio, G., et al., 2005, Physica B. 359–361, 50.
Doiraud, N., et al., 2003, Nature **425**, 595.
Doniach, S., 1977, Physica B **91**, 231.
Ebihara, T., et al., 2004, Phys. Rev. Lett. **93**, 246401.
Edwards, D.O., Proceedings of 1970 Ultralow Temperature Symposium, Naval Research Laboratory, Washington, DC NRL Report, Vol. 7133, p. 27.
Elliot, R.J., 1972, Magnetic Properties of Rare Earth Metal (Plenum Press, New York).
Ellman, B., et al., 1996, Phys. Rev. B **54**, 5043.
Emery, W.J., 1983, J. Phys. (Paris) Colloq. 44 C3, 977.
Endo, M., 2004, Phys. Rev. Lett. **93**, 247003.
Eremets, M., et al., 1996, High Pressure Experimental Methods (Oxford University Press, Oxford).
Evans, S.M.M., 1992, J. Magn. Magn. Mater. **108**, 135.
Fak, B., et al., 1998, J. Low Temp. Phys. **110**, 417.
Fak, B., et al., 2004, J. Magn. Magn. Mater. 272–271 S1, E13.
Fak, B., et al., 2000, J. Phys.: Condens. Mat. **12**, 5423.
Fay, D. and J. Appel, 1980, Phys. Rev. B **22**, 3173.
Fazekas, P., 1999, Lectures Notes on Electron Correlation and Magnetism (World Scientific, Singapore).
Felner, I., et al., 1997, Phys. Rev. B **55**, 3374.
Feyerherm, A., et al., 1994, Phys. Rev. Lett. **73**, 1849.
Fischer, Ø., 1978, Appl. Phys. **16**, 1.
Fischer, Ø., 1990, in: Magnetic Superconductors in Ferromagnetic Materials, ed K.H.A. Buschow, and E.P. Wohlforth (Science Publishers BV, Amsterdam) pp. 1–5.
Fischer, Ø. and M. Peter, 1973, in: H. Suhl, ed V. Magnetism (Academic Press, New York) 327.
Fisher, R.A., et al., 1989, Phys. Rev. Lett. **62**, 1411.
Fisher, R.A., et al., 1991, J. Low Temp. Phys **84**, 49.
Fisher, R.A., et al., 2002, Phys. Rev. **65**, 224509.

Fisher, R.E. and N. Phillips, 2005, Private Communication.
Fisk, Z., et al., 1998, Science **239**, 33.
Flouquet, J., 1978. Prog. Low Temp. Phys. VII, eds Brewer (North-Holland, Amsterdam).
Flouquet, J. and Buzdin, A. 2002. Phys. World **41**, January.
Flouquet, J., et al., 1982, J. Appl. Phys. **53**, 2126.
Flouquet, J., et al., 1988, J. Magn. Magn. Mater. 76–77, 285.
Flouquet, J., et al., 1991, Physica C 185–189, 372.
Flouquet, J., et al., 2002, Physica B **319**, 251.
Flouquet, J., et al., 2004, J. Magn. Magn. Mater. 272–271, 27.
Fomin, I.A., 2001, JETP Lett. **74**, 116.
Fomin, I.A., 2002, Sov. Phys. JETP **95**, 940.
Fomin, I.A., 2004, Private communication.
Fomin, I.A. and J. Flouquet, 1996, Solid State Commun. **99**, 795.
Fomin, I.A. and J.P. Brison, 2000, J. Low Temp. **119**, 627.
Forgan, E.M., et al., 1990, J. Phys.: Condens. Mat. **2**, 10211.
Frederick, N.A., et al., 2004, Phys. Rev. **69**, 024523.
Frigeri, P.A., et al., 2004, Phys. Rev. Lett. **92**, 097001.
Frossati, G., et al., 1976, Phys. Rev. Lett. **36**, 203.
Fukazawa, H. and K. Yamada, 2003, J. Phys. Soc. Jpn. **72**, 2449.
Fukuhara, T., et al., 1996, J. Phys. Soc. Jpn. **65**, 1559.
Fulde, P., 1994, J. Low Temp. Phys. **95**, 45.
Fulde, P. and R.A. Ferrel, 1964, Phys Rev. A **135**, 550.
Fulde, P. and J. Jensen, 1983, Phys. Rev. B, **27**, 4085.
Fulde, P., et al., 1988, F. Seitz, D. Turnbull, and H. Ehreureich, eds, Solid State Physics Vol. 41 (Academic Press, New York).
Fuseya, Y., et al., 2003, J. Phys. Soc. Jpn. **72**, 2914.
Gammel, P.L., et al., 2000, Phys. Rev. Lett. **84**, 2497.
Gegenwart, P., et al., 1999, Phys. Rev. Lett. **82**, 1293.
Gegenwart, P., et al., 2002, Phys. Rev. Lett. **89**, 056402.
Geibel, G., et al., 1991, Z. Phys. B **84**, 1.
Georges, A., et al., 1996, Rev. Mod. Phys. **68**, 13.
Georges, A., et al., 2004, Lectures on the Physics of Highly Correlated Electron Systems VIII, Vol 75, American Institute, Physics Conference Proceedings.
Ginzburg, V., 1957, Sov. Phys. JETP **4**, 153.
Glemot, L., et al., 1999, Phys. Rev. Lett. **82**, 169.
Gloos, K., et al., 1993, Phys. Rev. Lett. **70**, 501.
Glyde, H.R., et al., 2000, Phys. Rev. B **61**, 1421.
Goldman, A., et al., 1986, Phys. Rev. B **33**, 1626.
Gorkov, L., 1987, Sov. Sci. Rev. Sect A **9**, 1.
Gorkov, L.P. and E.I. Rashba, 2001, Phys. Rev. Lett. **87**, 037004.
Goryo, J., 2003, Phys. Rev. B **67**, 184511.
Graf, M.J., et al., 1996, J. Low Temp. Phys. **102**, 367.
Grempel, D.R. and Q. Si, 2003, Phys. Rev. Lett. **91**, 156404.
Grewe, M. and F. Steglich, 1991, K.A. Gschneidner, and L. Eyring, eds, Handbook on the Physics and the Chemistry of Rare Earth Vol. 14 343.
Grosche, F.M., et al., 2000, J. Phys.: Condens. Mat. **12**, L533.
Grosche, F.M., et al., 2001, J. Phys.: Condens. Mat. **13**, 2845.
Grüneisen, E., 1912, Ann. Phys. Leipzig **39**, 257.
Guillaume, A., 1999. Thesis, Université de Grenoble.
Haen, P., et al., 1996, J. Phys. Soc. Jpn. **65**(Suppl. B), 27.

Haen, P., et al., 1987, J. Low Temp. Phys. **67**, 391.
Haga, Y., 2005, Private Communucation.
Hanzawa, K., et al., 1987, J. Phys. Soc. Jpn. **56**, 678.
Hardy, F., 2004, Thesis, Université de Grenoble.
Hardy, F., et al., 2005a, to be published Phys. Rev. Letters.
Hardy, F., et al., 2005b, Physica B 359–361, 1111.
Harima, H., 2004, Private communication.
Harrison, N., et al., 2000, Phys. Rev. B **61**, 1779.
Harrison, N., et al., 2003, Phys. Rev. Lett. **90**, 096402.
Hasegawa, H. and T. Moriya, 1974, J. Phys. Soc. Jpn. **36**, 1542.
Hasselbach, K., et al., 1989, Phys. Rev. Lett. **63**, 93.
Hasselbach, K., et al., 1991, Phys. Lett. A **156**, 313.
Hayden, S., 2002, Private communication.
Hayden, S., et al., 1992, Phys. Rev. B **46**, 8675.
Heffner, R.H. and M.R. Norman, 1996, Comments Condens. Mat. Phys. **17**, 361.
Heffner, R.H., et al., 1987, Europhys. Lett. **3**, 751.
Hegger, H., et al., 2000, Phys Rev. Lett. **84**, 4986.
Held, K., et al., 2001, Phys. Rev. Lett. **87**, 276404.
Helfrich, R., 1995, Ph.D. Thesis, University of Darmstadt.
Hertz, J.A., 1976, Phys. Rev. B **14**, 1165.
Hess, D.W., et al., 1993, Encyclopedia of Applied Physics, Vol. 7 (VCH, Weinheim) p. 435.
Hewson, A.C., 1992, The Kondo Problem to Heavy Fermions (Cambridge University Press, Cambridge).
Hiess, A., et al., 2002, Phys. Rev. B **66**, 064531.
Hiess, A., et al., 2004, Cond-Mat, 0411041.
Hirsch, J.E., 1985, Phys. Rev. Lett. **54**, 1317.
Hirschfeld, P.J., et al., 1986, Solid State Commun. **59**, 111.
Hirschfeld, P.J., et al., 1988, Phys. Rev. B **37**, 83.
Holmes, A.T., et al., 2004a, Phys. Rev. B **69**, 024508.
Holmes, A.T., et al., 2004b, J. Phys.: Condens. Mat. **16**, S1121.
Holtmeier, S., 1994, Ph.D. Thesis, Université Joseph Fouier, Grenoble.
Hotta, T. and K. Ueda, 2003, Phys. Rev. B **67**, 104518.
Houghton, A., et al., 1987, Phys. Rev. B **35**, 5123.
Huth, M. and M. Jourdan, 1999, Adv. Solid State Phys. **39**, 351.
Huth, M., et al., 2000, Physica B 281–282, 882.
Huxley, A. and F. Hardy, 2004, Private communication.
Huxley, A., et al., 1995, Phys. Lett. A **209**, 365.
Huxley, A., et al., 2000, Nature **406**, 160.
Huxley, A., et al., 2001, Phys. Rev. B **63**, 144519.
Huxley, A., et al., 2003a, J. Phys.: Condens. Mat. **5**, S1945.
Huxley, A., et al., 2003b, Phys. Rev. Lett. **91**, 207201.
Ichioka, M., et al., 2003, J. Phys. Soc. Jpn. **72**, 1322.
Ikeda, H., 1997, Ph.D. Thesis, University of Osaka.
Ikeda, H. and K. Miyake, 1997, J. Phys. Soc. Jpn. **66**, 3714.
Ikezawa, I., et al., 1997, Physica B 237–238, 210.
Inada, Y., et al., 1994, Physica B 199–200, 119.
Isaacs, E.D., et al., 1995, Phys. Rev. Lett. **75**, 1178.
Ishida, K., et al., 1999, Phys. Rev. Lett. **82**, 5353.
Ishida, K., et al., 2002, Phys. Rev. Lett. **89** 167202–1.
Ishida, K., et al., 2003, Phys. Rev. B **68**, 184401.

Izawa, K., et al., 2001, Phys. Rev. Lett. **87** 057002–1.
Izawa, K., et al., 2003, Phys. Rev. Lett. **90**, 117001.
Jaccard, D. and J. Flouquet, 1985, J. Magn. Magn. Mater. 47–48, 45.
Jaccard, D., et al., 1985, J. Magn. Magn. Mater. 47–48, 23.
Jaccard, D., et al., 1987, J. Appl. Phys. **26**, 517.
Jaccard, D., et al., 1988, J. Magn. Magn. Mater. 76–77, 255.
Jaccard, D., et al., 1992, Phys. Lett. A **163**, 475.
Jaccard, D., et al., 2002, Phys. Lett. A **299**, 282.
Jaccard, D. and H. Holmes, 2005, Physica B 359–361, 333.
Jacoud, J.L., 1989, Thèse, Univsersité Joseph Fourier, Grenoble.
Jarlborg, T., et al., 2002, Phys. Lett. A, **300**, 518.
Jayaraman, A., 1965, Phys. Rev. A **137**, 179.
Jones, B.A., et al., 1988, Phys. Rev. Lett. **61**, 125.
Jourdan, M., et al., 1999, Nature **398**, 47.
Joynt, R. and L. Taillefer, 2002, Rev. Mod. Phys. **74**, 235.
Julian, S.R., et al., 1994, Physica B **199**(200), 36.
Jullien, R., et al., 1979, J. Appl. Phys. **50**, 7555.
Kadowaki, K. and S.B. Woods, 1986, Solid State Commun. **58**, 507.
Kadowaki, H., et al., 2003, Phys. Rev. B **68**, 140402.
Kadowaki, H., et al., 2004, Phys. Rev. Lett. **092**, 097204.
Kambe, S. and J. Flouquet, 1997, Solid State Commun. **103**, 551.
Kambe, S., et al., 1995, Solid State Commun. **95**, 449.
Kambe, S., et al., 1996, J. Low Temp. Phys. **102**, 477.
Kambe, S., et al., 1997, J. Low Temp. Phys. **108**, 383.
Kambe, S., et al., 1999, J. Phys. Low Temp. **117**, 101.
Kawarasaki, et al., 2000, Phys. Rev. B **61**, 4167.
Kawasaki, S., et al., 2001, Phys. Rev. B **65**, 020504R.
Kawasaki, S., et al., 2003b, Phys. Rev. Lett. **91** 137001–1.
Kawasaki, S., et al., 2004, J. Phys. Soc. Jpn. **73**, 1647.
Kawasaki, Y., et al., 1998, Phys. Rev. B **58**, 8634.
Kawasaki, Y., et al., 2002, Phys. Rev. B **66**, 224502.
Kawasaki, Y., et al., 2003a, J. Phys. Soc. Jpn. **72**, 2308.
Kernavanois, N., et al., 2001, Phys. Rev. B **64**, 174500.
Kernavanois, N., et al., 2003, Acta Phys. Pol. **34**, 721.
Kim, K.H., et al., 2003, Phys. Rev. Lett. **91**, 256401.
Kimura, N., 2004, Private communication.
King, C.A. and G.G. Lonzarich, 1991, Physica B **171**, 161.
Kirkpatrick, T.R. and D. Belitz, 2003, Phys. Rev. B **67**, 024515.
Kirkpatrick, T.R., et al., 2001, Phys. Rev. Lett. **87**, 127003.
Kiss, A. and P. Fazekas, 2004, Cond-Mat, 0411029.
Kitaoka, Y., et al., 2001, J. Phys.: Condens. Mat. **13**, L79.
Knafo, W., 2003, J. Phys.: Condens. Mat. **15**, 3741.
Knafo, W., 2004, Phys. Rev. B **70**, 174401.
Knebel, G., et al., 1999, Phys. Rev. B **59**, 12390.
Knebel, G., et al., 2002, Phys. Rev. B **65**, 024425.
Knebel, G., et al., 2004, J. Phys. Cond. Matter **16**, 8905.
Knebel, G., et al., 2005, Physica B 359–361, 20.
Knetsch, A., et al., 1993, Physica B 186–188, 300.
Knöpfle, K., et al., 1996, J. Phys.: Condens. Mat. **8**, 901.
Kobayashi, T.C., 2001, Private communication.

Koehler, W.C., et al., 1966, Phys. Rev. **151**, 405.
Kohori, Y., et al., 2000a, Physica B 281–282, 12.
Kohori, Y., et al., 2000b, Eur. Phys. B **18**, 601.
Kohori, Y., et al., 2001, Phys. Rev. B **64**, 134526.
Kohori, Y., et al., 2002, Physica B 312–313, 126.
Kondo, J., 2005, J. Phys. Soc. Jpn. **74**, 1–238.
Kopnin, N.B. and G.E. Volovik, 1996, JETP Lett. **64**, 690.
Kos, S., 2003, Cond-Mat. 0302089.
Kotegawa, H., et al., 2003, J. Phys.: Condens. Mat. **15**, S2043.
Kouroudis, I., et al., 1987, Phys. Rev. Lett. **58**, 820.
Krimmel, A., et al., 1992, Z. Phys. B **86**, 161.
Krimmel, A., et al., 2000, J. Phys.: Condens. Mat. **12**, 8801.
Kromer, F., et al., 2000, Phys. Rev. B **62**, 12477.
Küchler, S., et al., 2003, Phys. Rev. Lett. **91**, 066405.
Kulic, M., et al., 1997, Phys. Rev. B **56**, R11415.
Kumar, P. and P. Wölfle, 1987, Phys. Rev. Lett. **59**, 1954.
Kumar, R.S., et al., 2004, Cond-Mat. 0405043.
Kuramoto, Y. and Y. Kitaoka, 2000, Dynamics of Heavy Electrons (Clarendon Press, Oxford).
Kuramoto, Y. and K. Miyake, 1990, J. Phys. Soc. Jpn. **59**, 2831.
Kuwahara, K., et al., 2004, J. Phys. Soc. Jpn. **73**, 1438.
Kuwahara, K., et al., 2005, Cond-Mat 0504655.
Kyogaku, M., et al., 1993, J. Phys. Soc. Jpn. **62**, 4016.
Lacerda, A., et al., 1989, Phys. Rev.B **40**, 11429.
Lander, G., et al., 1992, Phys. Rev. B **46**, 5387.
Lapertot, G., et al., 1993, Physica B 186–188, 454.
Larkin, A.I. and Y.N. Ovchinnikov, 1965, Sov. Phys. JETP, **20**, 762.
Laughlin, R.D. and D. Pines, 2000, PNAS **97**, 27.
Lavagna, M., et al., 1982, Phys. Lett. 90A, 210.
Lawrence, J. and S. Shapiro, 1980, Phys. Rev. B **22**, 4379.
Layzer, A. and D. Fay, 1971, Int. J. Magn. **1**, 135.
Lee, P.A., 1993, Phys. Rev. Lett. **71**, 1887.
Leggett, A.J., 1975, Rev. Mod. Phys. **47**, 331.
Leggett, A.J., 1987, J. Magn. Magn. Mater. 63–64, 406.
Levanyuk, A.P., 2001, in: Some Modern Aspects of the Physics of Strongly Correlated Systems, ed F.G. Aliev, J.C. Gomez-sal, H. Suderow, and R. Villar (Instituto Universitario de Ciencia de Materials, Nicolas Cabrera UAM) 15.
Levy, F., et al., 2005, to be published.
Lin, C.L., et al., 1985, Phys. Rev. Lett. **54**, 2541.
Link, P., et al., 1993, J. Phys.: Condens. Mat. **7**, 373.
Lloblet, A., et al., 2003, Phys. Rev. B **69**, 024403.
Lonzarich, G.G., 1988, J. Magn. Magn. Mater. 76,77, 1.
Lonzarich, G.G. and L. Taillefer, 1985, J. Phys. C **18**, 4339.
Lussier, B., et al., 1994, Phys. Rev. Lett. **73**, 3294.
Machida, K. and T. Ohmi, 2001, Phys. Rev. Lett. **86**, 850.
Maki, K., et al., 2003, Europhys. Lett. **64**, 496.
Makoshi, K. and T. Moriya, 1975, J. Phys. Soc. Jpn. **38**, 10.
Malterre, D., et al., 1996, Adv. Phys. **45**, 299.
Mao, S.Y., et al., 1988, J. Magn. Magn. Mater, 76–77, 241.
Maple, B., 1973, in: H Suhl, ed V. Magnetism (Academic Press, New York) 289.
Maple, M.B., et al., 1986a, Phys. Rev. Lett. **56**, 185.

Maple, M.B., et al., 1986b, Acta Phys. Pol. B **56**, 185.
Marcenat, C., et al., 1988, J. Magn. Magn. Mater. 76–77, 115.
Martisovitz, V., et al., 2000, Phys. Rev. Lett. **84**, 5872.
Mathur, N.D., et al., 1998, Nature **394**, 39.
Matsuda, K., et al., 2001, Phys. Rev. Lett. **87**, 087203.
Matsuda, K., et al., 2003, J. Phys.: Condens. Mat. **15**, 2363.
Matsumoto, M. and M. Koga, 2004, J. Phys. Soc. Jpn. **73**, 1135.
Mc Elfresh, M.W., et al., 1987, Phys. Rev. B **35**, 43.
Measson, M.A., 2005, Private comumnication.
Metoki, N., et al., 1998, Phys. Rev. Lett. **80**, 5417.
Metoki, N., et al., 2000, Physica B **280**, 362.
Metoki, N., et al., 2004, J. Phys.: Condens. Mat. **16**, L207.
Midgley, P.A., et al., 1993, Phys. Rev. Lett. **70**, 678.
Millis, A., 1993, Phys. Rev. B **48**, 7183.
Mineev, V., 2004a, Int. J. Mod. Phys. B **18**, 2963.
Mineev, V., 2005. Phys. Rev. B, 71, 012509.
Mineev, V. and T. Champel, 2004, Phys. Rev. B **69**, 054508.
Mineev, V. and K.V. Samokhin, 1999, Introduction to Unconventional Superconductivity (Gordon and Breach SV Publishers, New York).
Mineev, V.P., 2002, Phys. Rev. B **66**, 134504.
Mineev, V.P. and M. Zhitomirsky, 2004, Cond-Mat, 0412055.
Miranda, E., et al., 1997, Phys. Rev. Lett. **78**, 290.
Mito, T., et al., 2001, Phys. Rev. B **63**, 220507R.
Miyake, K., 2004, Private communication.
Miyake, K. and H. Kohno, 2005, J. Phys. Soc. Jpn. **75**, 254.
Miyake, K. and Y. Kuramoto, 1990, J. Magn. Magn. Mater. 90–91, 438.
Miyake, K. and H. Maebashi, 2002, J. Phys. Soc. Jpn. **71**, 1007.
Miyake, K. and O. Narikiyo, 2002, J. Phys. Soc. Jpn. **71**, 867.
Miyake, K., et al., 1989, Solid State Commun. **71**, 1149.
Miyake, K., et al., 1986, Phys. Rev. **34**, 6554.
Miyake, K., et al., 2003, J. Phys.: Condens. Mat. **15**, L275.
Monthoux, P. and G. Lonzarich, 2001, Phys. Rev. B **63**, 054529.
Monthoux, P. and G. Lonzarich, 2003, Phys. Rev. B **66**, 224.
Mori, H., et al., 1999, Physica B 259–261, 58.
Morin, P., et al., 1988, J. Low Temp. Phys. **70**, 377.
Moriya, T., 1985, Spin Fluctuation in Itinerant Electron Magnetism (Springer, Berlin).
Moriya, T., 2003a, Acta Phys. Pol. B **34**, 287.
Moriya, T. and T. Takimoto, 1995, J. Phys. Soc. Jpn. **64**, 960.
Moriya, T. and K. Ueda, 2000, Adv. Phys. **49**, 555.
Moriya, T. and U. Ueda, 2003b, Rep. Prog. Phys. **66**, 1299.
Motoyama, G., et al., 2001, Phys. Rev. B **65**, 020510R.
Motoyama, G., et al., 2003, Phys. Rev. Lett. **90**, 166402.
Mott, N.F., 1935, Pro. Phys. Soc. **47**, 571.
Movshovich, R., et al., 1996, Physica B 223–224, 126.
Movshovich, V., et al., 2001, Phys. Rev. Lett. **86**, 5152.
Müller, K.H., 2001, Rep. Prog. Phys. **64**, 943.
Müller-Hartmann, E. and J. Zittartz, 1971, Phys. Rev. Lett. **26**, 428.
Mydosh, J.A., et al., 2003, Acta Phys. Pol. B **34**, 659.
Naidyuk Yu, G. and I.K. Yanson, 1998, J. Phys.: Condens. Mat. **10**, 8905.
Nakajma, S., 1973, Prog. Theor. Phys. **50**, 1101.

Nakashima, N., et al., 2003, J. Phys.: Condens. Mat. **15**, S2011.
Newns, D.C. and N. Read, 1987, Advs. Phys. **36**, 799.
Ng, N.K. and C.M. Varma, 1997, Phys. Rev. Lett. **78**, 330.
Niklowitz, Ph., 2003, Ph.D. Thesis, Cambridge.
Norman, M.R., 1993, Phys. Rev. Lett. **71**, 3391.
Norman, M.R., et al., 1987, Phys. Rev. B **36**, 4058.
Norman, M.R., et al., 2003, Phys. Rev. Lett. **90**, 16601.
Nozières, P., 1974, J. Low Temp. Phys. **17**, 31.
Nozières, P., 1985, Ann. Phys. Fr, **10**, 19.
Nozières, P., 1986. Cours, Collège de France, Magnetisme et localisation dans les liquides de Fermi.
Nozières, P., 1998, Eur. Phys. B, **6**, 447.
Oeschler, N., et al., 2003a, Phys. Rev. Lett. **91**, 076402.
Oeschler, N., et al., 2003b, Acta Phys. Pol. **34**, 255.
Ohashi, M., et al., 2003, Phys. Rev. **68**, 144.428.
Ohkawa, F., 1990, Phys. Rev. Lett. **64**, 2300.
Ohkuni, H., et al., 1997, J. Phys. Soc. Jpn. **66**, 945.
Ohmi, T. and K. Machida, 1996, J. Phys. Soc. Jpn. **65**, 4018.
Ohkuni, H., et al., 1999, Philos. Mag. **79**, 1045.
Okuno, Y. and K. Miyake, 1998, J. Phys. Soc. Jpn. **67**, 2469.
Ong, N.P. and Y. Wang, 2004, Physica C 408–410, 11.
Onishi, Y. and K. Miyake, 2000, J. Phys. Soc. Jpn. **69**, 3955.
Onuki, Y. and T. Komatsubara, 1987, J. Magn. Magn. Mater. 63,64, 281.
Onuki, Y., et al., 1992, J. Phys. Soc. Jpn. **61**, 293.
Onuki, Y., et al., 2003, Acta Phys. Pol. B **34**, 667.
Oomi, G., et al., 1993, Physica B 186–188, 758.
Ormeno, R.J., et al., 2002, Phys. Rev. Lett. **88**, 047005.
Osheroff, D., et al., 1980, Phys. Rev. Lett. **44**, 792.
Ott, H.R., 1987, in: Prog Low Temp Phys XI, ed D.F. Brewer (Elsevier Science Publishers, Amsterdam) pp. 215–289.
Ott, H.R., et al., 1985, Phys. Rev. B **11**, 1651.
Ott, H.R., et al., 1983, Phys. Rev. Lett. **74**, 4734.
Paglione, S., et al., 2003, Phys. Rev. Lett. **91**, 246405.
Pagliuso, P.G., et al., 2001, Phys. Rev. B **64**, 100503R.
Pagluiso, P.G., et al., 2002, Physica B **312**, 129.
Pankov, S., et al., 2004, Phys. Rev. B **69**, 054426.
Paschen, S., et al., 2004, Nature **432**, 881.
Paulsen, X., et al., 1990, J. Low Temp. Phys. **81**, 317.
Payer, K., et al., 1993, Physica B 186–188, 503.
Pépin, C., 2004, Cond-Mat, 0407155.
Pethick, C.J. and D. Pines, 1986, Phys. Rev. Lett. **57**, 118.
Petrovic, C., et al., 2001, J. Phys.: Condens. Mat. **13**, 337.
Peyrard, J., 1980, Ph.D. Université Joseph Fourier, Grenoble.
Pfleiderer, C., 2003, Acta Phys. Pol. B **34**, 679.
Pfleiderer, C. and A. Huxley, 2002, Phys. Rev. Lett. **89**, 147005.
Pfleiderer, C., et al., 2001, Nature **412**, 58.
Pfleiderer, C., et al., 2004, Nature **427**, 227.
Pietri, R. and B. Andraka, 2000, Phys. Rev. B **62**, 8619.
Piquemal, F., et al., 1987, J. Magn. Magn. Mater. 63–64, 469.
Plessel, J., et al., 2003, Phys. Rev. B **67**, 180403R.

Radovan, H.A., et al., 2003, Nature **425**, 51.
Ramirez, A.P., et al., 1991, Phys. Rev. B **44**, 5392.
Ravex, A., et al., 1987, J. Magn. Magn. Mater. 63–64, 400.
Raymond, S. and D. Jaccard, 2000, Phys Rev. B **61**, 8679.
Raymond, S., et al., 1997, J. Low. Temp. Phys. **109**, 205.
Raymond, S., et al., 1999a, Physica B 259–261, 48.
Raymond, S., et al., 1999b, J. Phys.: Condens. Mat. **11**, 5547.
Raymond, S., et al., 2001, J. Phys.: Condens. Mat. **13**, 8303.
Raymond, S., et al., 2005, Physica B 359–361, 898.
Razafidandimby, H., et al., 1984, Z. Phys., B **54**, 111.
Regnault, L.P., et al., 1988, Phys. Rev. B **38**, 4481.
Rehmann, S., et al., 1997, Phys. Rev. Lett. **78**, 1122.
Reinders, P.H.P., et al., 1987, J. Magn. Magn. Mat. 63–64, 297.
Ribault, M., et al., 1979, J. Phys. Lett. **40**, 413.
Rodière, P., 2001, Thèse, Université de Grenoble.
Roehler, J., et al., 1988, J. Magn. Magn. Mater. 76–77, 340.
Rosch, A., 1999, Phys. Rev. Lett. **82**, 4280.
Rosch, A., 2000, Phys. Rev. B **62**, 4945.
Rosch, A., et al., 1997, Phys. Rev. Lett. **79**, 159.
Rossat-Mignod, M.J., 1986, in: Neutron Scattering in Condensed Matter Research, Ch. 20, eds K.L. Skold and D.L. Price (Academic Press, New York).
Rossat-Mignod, M.J., et al., 1988, J. Magn. Magn. Mater. 76–77, 376.
Rossat-Mignod, M.J., et al., 1992, Physica B 180–181, 383.
Rotundu, C.R., et al., 2004, Cond-Mat, 042599.
Roussev, R. and A.J. Millis, 2001, Phys. Rev. B **63**, 140504R.
Sachdev, S., 1999, Quantum Phase Transitions (Cambridge University Press, Cambrigde).
Saint James, J.D., et al., 1969, Type II Superconductivity (Pergamon Press, Oxford).
Sakai, O. et al., 2005, Cond-Mat 0503675.
Sakakibara, T., et al., 2005, Physica B 359–361, 836.
Sakurai, J., 1994, J. Low Temp. **95**, 165.
Sakurai, J., 2001, Encyclopedia Science and Technology ISPN 0.0804311526 8018 (Elsevier Science Ltd, Amsterdam).
Salce, B., et al., 2000, Rev. Sci Instru. **71**, 2461.
Samokhin, K.V. and M.B. Walker, 2002, Phys. Rev. B **66**, 024512.
Samokhin, K.V., et al., 2004, Phys. Rev. **69**, 094514.
Sandeman, K., et al., 2003, Phys. Rev. Lett. **90**, 167005.
Sarrao, J.L., et al., 2002, Nature **420**, 297.
Sato, M., et al., 2004, J. Phys. Soc. Jpn. **73**, 3418.
Sato, N.K., et al., 2001, Nature **410**, 340.
Satoh, H. and F.J. Ohkawa, 2001, Phys. Rev. B **63**, 184401.
Sauls, J.A., 1994, Ad. Phys. **43**, 113.
Saxena, S.S. and P.B. Littlewood, 2001, Nature **412**, 290.
Saxena, S., et al., 2000, Nature **406**, 587.
Scharnberg, K. and R.A. Klemm, 1988, Phys. Rev. B **22**, 11290.
Schilling, J.S., 1979, Adv. Phys. **28**, 657.
Schlabitz, W., et al., 1986, Z. Phys. B **62**, 171.
Schmidt, L., 1993, Ph.D. Thesis, UJF Grenoble.
Schmidt-Rink, S., et al., 1986, Phys. Rev. Lett. **57**, 2575.
Schoenes, et al., 1987, Phys. Rev. B **35**, 5375.
Schreiner, T., et al., 1999, Europhys. Lett. **48**, 568.

Schröder, A., et al., 1998, Phys. Rev. Lett. **80**, 5623.
Schröder, O., et al., 2000, Nature (London) **407**, 351.
Schuberth, E.A., et al., 1992, Phys. Rev. Lett. **63**, 117.
Sechovsky, V. and L. Havela, 1998, Buschow, eds, Handbook of Magnetic Materials Vol. 11 (North-Holland, Amsterdam) 309.
Senthil, T., et al., 2004a, Phys. Rev. B **69**, 035111.
Senthil, T., et al., 2004b, Cond-Mat, 0409033.
Sergienko, I.A. and S.H. Curnoe, 2004, Phys. Rev. B 70, 144522.
Settai, R., et al., 2002, J. Phys.: Condens. Mat. **14**, L29.
Settai, R., et al., 2003, J. Magn. Magn. Mater. **272276**, 223.
Settai, R., et al., 2005, Physica B 359–361, 317.
Sheikin, I., et al., 2001a, J. Low Temp. Phys. **122**, 591.
Sheikin, I., et al., 2001b, Phys. Rev. B **64**, 220504R.
Sheikin, I., et al., 2003, Phys. Rev. B **67**, 094420.
Shick, A.B.S. and W.E. Picket, 2001, Phys. Rev. Lett. **86**, 300.
Shick, A.B.S., et al., 2004, Phys. Rev. B **70**, 134506.
Shiina, R., et al., 2004, J. Phys. Soc. Jpn. **73**, 3453.
Shimizu, K., et al., 2001, Nature (London) **406**, 316.
Shishido, H., et al., 2002, J. Phys. Soc. Jpn. **71**, 162.
Shishido, H., et al., 2005, J. Phys. Soc. Jpn. **74**, 1103.
Shivaram, B.S., et al., 1986, Phys. Rev. Lett. **57**, 1259.
Si, Q., 2003a, Cond-Mat, 0302110V1.
Si, Q., et al., 2001, Nature **413**, 804.
Si, Q., et al., 2003b, Phys. Rev. B **68**, 115103.
Sichelschmidt, J., et al., 2003, Phys. Rev. Lett. **91**, 156401.
Sigrist, M. and T.M. Rice, 1989, Phys. Rev. B **39**, 2200.
Sikkema, A.E., et al., 1996, Phys. Rev. B **54**, 9322.
Simon, H. and P.A. Lee, 1997, Phys. Rev. Lett. **78**, 1548.
Smith, T.F. and E.S. Fisher, 1973, J. Low Temp. Phys. **12**, 631.
Sparn, G., et al., 2001. in: H.D. Hochkeimer,et al., eds, Frontier of High Pressure Research II (Kluwer Academic Publishers, The Netherlands) Amsterdam. 413.
Springford, M., 1997, Electron (Cambridge University Press, New York) Cambridge.
Steglich, F., 1977, Treusch, ed., Festkörperproblem Advance in Solid State Physics Vol. XVII (Vieneg, Braunschweig) 319.
Steglich, F., et al., 1979, Phys. Rev. Lett. **43**, 1892.
Steglich, F., et al., 1996, in: Proceedings of Physical Phenomena at High Magnetic Fields II, ed Z. Fisk, L. Gorkov, D. Metzer, and R. Schieffer (World Scientific, Singapore) 125.
Stewart, G., 2001, Rev. Mod. Phys. **73**, 797.
Stewart, G., et al., 1984, Phys. Rev. Lett. **52**, 679.
Stockert, O., et al., 1998, Phys. Rev. Lett. **80**, 5627.
Stockert, O., et al., 2004, Phys. Rev. Lett. **92**, 136401.
Suderow, H., et al., 1997, J. Low Temp. Phys. **108**, 11.
Suderow, H., et al., 1998, Phys. Rev. Lett. **80**, 165.
Suhl, H., 2001, Phys. Rev. Lett. **87**, 167007.
Sulpice, A., et al., 1986, J. Low Temp. Phys. **62**, 39.
Sun, P. and G. Kotliar, 2003, Phys. Rev. Lett. **91**, 037209.
Tachiki, M. and S. Maekawa, 1984, Phys. Rev. B **29**, 2497.
Taillefer, L. and G. Lonzarich, 1988, Phys. Rev. Lett. **60**, 1570.
Taillefer, L., et al., 1988, Physica C 153–155, 451.
Takabatake, T., et al., 1998, J. Magn. Magn. Mater. 177–181, 277.

Takashima, S., et al., 2004, Private communication.
Takegahara, K. and H. Harima, 2000, Physica B 281–282, 764.
Takegahara, K., et al., 1986, J. Phys. F Metal Phys. **16**, 1961.
Takimoto, T. and T. Moriya, 1996, Solid State Commun. **99**, 457.
Takimoto, T. and T. Moriya, 2002, Phys. Rev. B **66**, 134516.
Takimoto, T., et al., 2003, Cond-Mat, 0212467V1.
Takke, R., et al., 1981, Z. Phys. B **44**, 33.
Tatayama, T., et al., 2002, Phys. Rev. Lett. B **65**, 180504.
Tateiwa, N., 2004, Phys. Rev. B **69**, 180513.
Tateiwa, N., et al., 2001, J. Phys.: Condens. Mat. **13**, L17.
Tateiwa, N., et al., 2003a, Acta Phys. Pol. B **34**, 515.
Tateiwa, N., et al., 2003b, Physica C 388–389, 527.
Tayama, T., et al., 2003, J. Phys. Soc. Jpn. **72**, 1516.
Terashima, T., et al., 2001, Phys. Rev. Lett. **87**, 166401.
Thalmeier, P. and B. Lüthi, 1991, K.A. Gschneider, Jr., and L. Eyring, eds, Handbook on the Physics and Chemistry of Rare Earths Vol. 14 (Elsevier Science Publishers BV) New York. 225.
Thalmeier, P. and G. Zwicknagl, 2004, Handbook on the Physics and Chemistry of Rare Earth (Elsevier) Vol. 34, chap 219, 135 and Cond-Mat 0312540.
Thalmeier, P. et al., 2004, in: Frontiers in Superconducting Material, ed A. Narlikaa (Springer-Verlag, Berlin). Cond-Mat 0409363.
Thessieu, C., et al., 1995, Solid State Commun. **95**, 707.
Thessieu, C., et al., 1998, J. Magn. Magn. Mater. 177–181, 609.
Thessieu, C., et al., 1998, Physica B 259–261, 847.
Thomas, F., et al., 1996, J. Phys.: Condens. Mat. **8**, 51.
Thomas, F., et al., 1995, J. Low Temp. Phys. **102**, 117.
Thompson, J.D., et al., 1987, Phys. Rev. B **35**, 48.
Thompson, J.D. and J.M. Lawrence, 1994, in: Handbook on the Physics and Chemistry of Rare Earths and Actinides, ed K.A. Gschneider, et al., (Elsevier, Amsterdam) 19.
Thompson, J.D., et al., 2001, J. Magn. Magn. Mater. 226–230, 5.
Tokiwa, W., et al., 2004, J. Magn. Magn. Mater. 272–276, 87.
Tou, H., et al., 1996, Phys. Rev. Lett. **77**, 1374.
Tou, H., et al., 1998, Phys. Rev. Lett. **80**, 3129.
Tou, H., et al., 2005, J. Phys. Soc. Jpn. **74**, 1245.
Trappmann, T., et al., 1991, Phys. Rev. B **43**, 13174.
Trovarelli, O., et al., 2000, Phys. Rev. Lett. **85**, 626.
Tsuji, N., et al., 2003, J. Phys.: Condens. Mat. **15**, 1993.
Tsunetsugu, H., et al., 1988, Rev. Mod. Phys. **69**, 809.
Ueda, K. and Y. Onuki, 1998, Physics of Heavy Electron Systems (Shokadu, Tokyo).
Uhlarz, M., et al., 2004, Phys. Rev. Lett. **93**, 256404.
Van Dijk, N., et al., 2000, Phys. Rev. B **61**, 8922.
Van Dijk, N., et al., 2002, Physica B **319**, 220.
Van Dijk, N.H., et al., 1993, Physica B 186/188, 267.
Van Dijk, N.H. et al., 1994, Ph.D. Thesis, University of Amsterdam, Amsterdam.
Van Dijk, N.H., et al., 1995, Phys. Rev. B **51**, 12665.
Vargoz, E., et al., 1998, Solid State Commun. **106**, 631.
Varma, C.M., et al., 2002, Phys. Rep. **361**, 267.
Vettier, C., et al., 1986, Phys. Rev. Lett. **56**, 1980.
Vojta, M., 2003, Rep. Prog. Phys. **66**, 2069.

Vollhardt, D., Wölfle, P., 1990. The Superfluid Phases of Helium, Vol. 3, (Taylor and Francis, London).
Vollmer, Y., et al., 2003, Phys. Rev. Lett. **90**, 05707.
Volovik, G.E., 1993, JETP Lett. **58**, 469.
von Löhneysen, H., et al., 1994, Phys. Rev. Lett. **72**, 3262.
von Löhneysen, H., et al., 1996, J. Phys.: Condens. Mat. **8**, 9689.
von Löhneysen, H., et al., 2000, J. Phys. Soc. Jpn. **69**(Suppl. A), 63.
Wachter, P., 1993, K.A. Gscheidner, et al., eds, Handbook of Physics and Chemistry of Rare Earths Vol. 19 (North-Holland, Amsterdam) Amsterdam 177.
Waler, J.T. and T. Cromer, 1965, J. Chem. Phys. **42**, 4112.
Walker, M.B. and K.V. Samokhin, 2002, Phys. Rev. Lett. **88**, 207001.
Walker, D.A., et al., 2001, Phys. Rev. B **63**, 054522.
Wang, Y. and B.R. Cooper, 1968, Phys. Rev. **172**, 539.
Wang, Y.L. and B.R. Cooper, 1969, Phys. Rev. **185**, 696.
Wang, Y., et al., 2002, Phys. Rev. Lett. **88**, 257003.
Watanabe, S. and K. Miyake, 2002a, Physica B 312–313, 115.
Watanabe, S. and K. Miyake, 2002b, J. Phys. Chem. Solids **63**, 1465.
Watanabe, T., et al., 2004, Phys. Rev., B **70**, 020506R.
Welp, U., 1988, Ph.D. Thesis, University of Konstanz, Konstanz.
Welp, U., et al., 1987, J. Magn. Magn. Mater. 63–64, 28.
Wertharmer, N.R., et al., 1966, Phys. Rev. **147**, 295.
Wiebe, C.R., et al., 2004, Phys. Rev. B **69**, 132418.
Wilhelm, H. and D. Jaccard, 2004, Phys. Rev. B **69**, 214408.
Wilhelm, H., et al., 2001, J. Phys.: Condens. Mat. **13**, L329.
Wilson, A.H., 1938, Proc. Roy. Soc. A **67**, 580.
Wilson, K.G., 1975, Rev. Mod. Phys. **47**, 773.
Won, H., et al., 2003, Cond-Mat., 0306548.
Yamada, K., 1975, Prog. Theor. Phys. **53**, 970.
Yamagami, H., 2003, J. Phys.: Condens. Mat. **15**, S2271.
Yamagami, H. and N. Hamada, 2000, Physica B **384**, 1295.
Yaouanc, A., et al., 1998, J. Phys.: Condens. Mat. **10**, 9791.
Yaouanc, A., et al., 2000, Phys. Rev. Lett. **84**, 2702.
Yaouanc, A., et al., 2002, Phys. Rev. Lett. **89**, 147001.
Yeh, A., et al., 2002, Nature (London) **419**, 459.
Yogi, M., et al., Phys. Rev. Lett. 93, 027003 (2004).
Yotsubashi, S., et al., 2001, J. Phys. Soc. Jpn. **70**, 186.
Yu, W., et al., 2004, Phys. Rev. Lett. **92**, 086403.
Yuan, H.Q., et al., 2003, Science **302**, 2104.
Zaanen, J., 2001, Phil. Mag. **81**, 1485.
Zapf, V.S., et al., 2001, Phys. Rev. B **65**, 014506.
Zhu, J.X., et al., 2003a, Phys. Rev. Lett. **91**, 156404.
Zhu, L., et al., 2003b, Phys. Rev. Lett. **91**, 066404.
Zlatic, V., et al., 2003, Phys. Rev. **68**, 104432.
Zülicke, U. and H.J. Millis, 1995, Phys. Rev. B **51**, 8996.
Zwicknagl, G., 1992, Adv. Phys. **41**, 203.
Zwicknagl, G. and P. Fulde, 2003, J. Phys.: Condens. Mat. **15**, S1911.
Zwicknagl, G. and V. Pulst, 1993, Physica B **186**, 895.

CHAPTER 3

THERMODYNAMICS AND TRANSPORT IN SPIN-POLARIZED LIQUID ^{3}HE: SOME RECENT EXPERIMENTS

BY

O. BUU, L. PUECH and P. E. WOLF

Centre de Recherches sur les Très Basses Températures, CNRS, BP 166, F-38042 Grenoble Cédex 9, France

Progress in Low Temperature Physics, Volume XV

ISSN: 0079-6417 DOI: 10.1016/S0079-6417(05)15003-3

Contents

1. Introduction

Strongly correlated fermionic systems are common in nature, ranging from condensed matter physics (heavy fermions systems, high critical temperature superconductors, so on) to nuclear physics (nuclear matter) and astrophysics (neutrons stars), or, very recently, atomic physics (ultracold degenerate Fermi gases). Among these systems, liquid ^{3}He is probably the simplest stable example one can dream of: it is a pure system, with no underlying lattice, and the interactions between the spin $\frac{1}{2}$ atoms are simple and accurately known, consisting mainly of hard core repulsion at short distance and weak long-range attraction, as described by the Aziz potential (Aziz et al. 1979). Since its first liquefaction in the late 1940s, ^{3}He has been experimentally studied in great detail, the thermodynamic and transport properties having been measured over the whole range of temperature and pressure (<3 K, and 0–30 bar) in the normal phase as well as in the superfluid phases. As such, liquid ^{3}He is ideal for testing our understanding of strongly correlated fermions.

At low temperature ($T<200$ mK), Landau theory offers a satisfying phenomenological description, but, due to the many-body nature of the problem, Landau parameters cannot be rigorously derived from a first-principles calculation. This led to the development of a variety of models, trying to account for the value and the density dependence of Landau parameters. In particular, two opposing views picture ^{3}He as being close to a ferromagnetic instability ('nearly ferromagnetic' model) or to a Mott localization transition ('nearly localized' model). Both provide a more or less satisfying description of the properties of the unpolarized liquid (see Bonfait et al. 1990 for a review). Another issue is the description of ^{3}He beyond the Landau Fermi liquid regime. Close to the critical point, the experimental transport properties resemble those of a quantum gas of hard spheres (Wilks 1967). The transition regime, from 200 mK to 2 K, has been little studied theoretically, most often at a qualitative level only (see Lhuillier 1991 for a review).

In the low temperature, degenerate, regime, the properties of liquid ^{3}He are expected to be quite sensitive to the spin polarization as a consequence of the Pauli principle. The theory of thermodynamics and transport phenomena in spin-polarized liquid ^{3}He within the framework of the Landau theory have been reviewed by Meyerovich (1987, 1990). Beyond the Landau theory, Castaing and Nozières suggested in 1979 that the effects of polarization should allow one to test the 'nearly ferromagnetic' versus the 'nearly localized' pictures (Castaing and Nozières 1979). More generally, the extra control parameter provided by the spin polarization could open a new realm of properties – including the possibility of new phases (Bashkin 1984a,

Lhuillier and Laloë 1979, Stringari et al. 1987, Pickett 1989), where different theories could be compared.

These possibilities provided the initial motivation for producing and studying spin-polarized liquid ^{3}He. As the equilibrium polarization is quite small in currently available static magnetic fields, one has to use out-of-equilibrium methods. Castaing and Nozières thus proposed in 1979 to obtain transiently polarized liquid by melting a strongly polarized ^{3}He solid (Castaing and Nozières 1979). This rapid melting technique was soon demonstrated by a number of groups (Schumacher et al. 1979, Chapellier et al. 1979, Bonfait et al. 1984, Dutta and Archie 1985). The first experiments conducted using this method have been reviewed in the 1990 review paper of Bonfait et al. (1990). At that time, there was a contradiction between the polarization dependence of the melting pressure, which suggested the occurrence of a metamagnetic transition, in agreement with the 'nearly localized' picture, and that of the sound velocity, which showed no anomaly. The difficulties of interpretation associated with the two-phase coexistence in the first type of experiment led the Grenoble group to undertake a measurement of the magnetization curve in a fully liquid sample, where the polarization can be determined unambiguously. This measurement rested on an elaboration of the rapid melting technique, where the polarized ^{3}He is obtained inside a silver sinter heat exchanger. This was an experimental breakthrough, as this modification allowed one to obtain strongly polarized ^{3}He at controllable temperature and pressure, opening the way to measurements, which would have been impossible otherwise. It is the purpose of the present paper to describe these experiments, which study the polarization dependence of thermodynamic (effective field, specific heat) and transport (viscosity, thermal conductivity) properties. The results provide a means to test the different theoretical descriptions of liquid ^{3}He. The thermodynamic measurements rule out any metamagnetic transition and show that the 'nearly ferromagnetic' and 'nearly localized' pictures are probably too extreme. The transport measurements give surprizing results. On the one hand, the polarization dependence of the viscosity resembles that predicted for a dilute gas of fermions (lack of density dependence at low temperature, persistence of the effect beyond the degenerate regime), which we did not expect for a dense liquid. On the other hand, the polarization dependence of the thermal conductivity is found to be much weaker than predicted for a dilute gas of fermions.

We have chosen to give a self-contained presentation of our experimental results and their theoretical context. It is our belief that such a synthesis might be useful beyond the field of liquid ^{3}He physics. For example, the Hubbard model discussed here is relevant to recent work on ^{3}He films (Casey et al. 2003), and applies as well to the case of cold bosons (Greiner et al. 2002) or fermions (Modugno et al. 2003) in optical lattices. Also, the transport mechanisms in liquid ^{3}He might have some relevance in the new field of strongly interacting cold fermions gases, in these special circumstances where the s- and p-wave scattering cross-sections would be simultaneously large (Regal et al. 2004). However, it should be stressed that our presentation of the theoretical background mainly focuses on the two 'extreme' points of view represented by the 'nearly ferromagnetic' model and 'nearly localized' model. Other relevant approaches (e.g. density-functional calculations Barranco et al. 1996, Gatica et al. 1998, induced interaction model (Sanchez-Castro and Bedell 1989a,b)) are only mentioned in the context of our experimental results.

Although we will not discuss them, we would finally like to mention a number of other recent developments in the field of spin-polarized pure and dilute ^{3}He. For pure liquid ^{3}He, these are the study of the vapor pressure (Villard et al. 2000) and the analysis of the melting of polarized solid ^{3}He (Marchenko et al. 1999). For spin-polarized ^{3}He–^{4}He mixtures, we quote the recent viscosity measurements by Owers-Bradley (1997), Owers-Bradley et al. (1998), Woerkens (1998), Akimoto et al. (2002), and the measurement of the osmotic pressure of saturated ^{3}He–^{4}He mixture (Rodrigues and Vermeulen 1997). Finally, a good deal of work deals with the transverse spin dynamics in polarized Fermi systems: the non-linear phenomena induced by the dipolar or demagnetizing fields have been studied by a number of authors (Nacher and Stoltz 1995, Fomin and Vermeulen 1997, Owers-Bradley et al. 2000, Villard and Nacher 2000, Nacher et al. 2002a, b, Krotkov et al. 2002). The possible damping of spin waves at zero temperature is also a subject of present theoretical (Meyerovich 1985, Mullin and Jeon 1992, Meyerovich and Musaelian 1993, Fomin 1997, Mineev 2004) and experimental (Wei et al. 1993, Ager et al. 1995, Roni and Vermeulen 2000, Buu et al. 2002c, Akimoto et al. 2003) controversy.

The outline of this chapter is as follows: we first present the initial theoretical issues by describing the Landau approach, the 'nearly ferromagnetic' and 'nearly localized' pictures, both for the unpolarized and spin-polarized liquid ^{3}He (Sections 2 and 3). Next, we describe in Section 4 the various techniques used to polarize liquid ^{3}He, putting the emphasis on the method of rapid melting inside a silver sinter. The success of this method relies on the combination of a small Kapitza resistance of the silver sinter, and a relatively long magnetic surface relaxation time. As silver sinters are widely used in the low-temperature community, we chose to describe in detail the measurements of these quantities in Appendix A.

Section 5 presents the first experiment that we performed; that is the determination of the susceptibility up to an effective field of 200 T, and an analysis of its results in relation to Section 3. The next three sections concern the measurements of the viscosity (Section 6), the thermal conductivity (Section 7), and the specific heat (Section 8) of the polarized liquid. All these measurements use the same experimental set up, based on a vibrating wire viscometer. Operating this viscometer required that we open a small gap inside the sinter. The influence of this gap on the polarization and temperature gradients inside the cell is of central importance for a correct interpretation of our experimental results. However, the corresponding discussion is rather technical. This is why we have specifically devoted two appendices (B and C) to these problems. Finally, Appendix D gives the theoretical background necessary for the discussion of our transport measurements.

2. Normal liquid ^{3}He

2.1. *Landau theory*

Experimentally, at low enough temperature ($T < 200$ mK), the properties of liquid ^{3}He mimic those of a degenerate system of free fermions. Quantitatively, however, the properties are strongly renormalized compared to their Fermi gas values. The Landau theory of Fermi liquids forms the basis of the current understanding of ^{3}He in the degenerate regime. In this section, we review the main features of this theory and its relevance to liquid ^{3}He. For the details, we refer the reader to the classic textbook by Pines and Nozières (1966), or the more recent book by Baym and Pethick (1991).

2.1.1. *Background*

At the heart of Landau theory lies the concept of *quasiparticle*: by definition, an interacting Fermi system is said to be 'normal' if the spectrum of its excited states is similar to that of the non-interacting Fermi gas. In particular, this definition excludes the presence of a gap in the spectrum. The low-energy excitations of a normal Fermi liquid, reminiscent of the single particles excitations in the ideal Fermi gas, are called quasiparticles.

At $T = 0$ K, all the quasiparticles sit inside a Fermi sphere in the momentum space, whose radius k_F is the same as that of the non-interacting system. At finite temperature, quasiparticles are excited out of the Fermi sphere over a width in energy of order $k_B T$. Collisions between quasiparticles give rise to their finite lifetime τ, given by

$$\frac{1}{\tau} \propto \frac{m^{*3}}{\hbar^6} \bar{W}(k_B T)^2, \tag{2.1}$$

where m^* is the quasiparticle effective mass and $\bar{W}$ the averaged collision probability. This implies that the quasiparticles are well-defined excitations (i.e. their lifetime is longer than $\hbar/(k_B T)$) only at low temperature, when the excited quasiparticles lie in the vicinity of the Fermi surface. Hence, the range of applicability of Landau theory is restricted to the degenerate regime.

The existence of a Fermi sphere and the phase space restriction imposed by the Pauli principle upon the quasiparticles yield the usual temperature dependence for the thermodynamic and transport quantities:

- Specific heat: $C_v \propto T$
- Magnetic susceptibility: $\chi =$ constant
- Viscosity: $\eta \propto 1/T^2$

These temperature dependences are actually observed in liquid ^{3}He below 100 mK (Wheatley 1975). We turn now to the quantitative calculation of these quantities in Landau theory.

2.1.2. *Thermodynamic properties*

Landau theory accounts for the renormalization of the thermodynamic quantities by an effective interaction between the quasiparticles and the Fermi-liquid function (Landau 1958). For an unpolarized system, this interaction depends on the spins of the two quasiparticles, and on the angle θ between their wavevectors of amplitude k_F. The coefficients of its expansion over Legendre polynomials are the 'Landau interactions' $f_l^{\sigma\sigma'}$, with $l = 0, 1, \ldots, \sigma$ and σ' being spin indices. These parameters, made dimensionless by multiplying by the density of states, and grouped into spin-symmetric, and spin-antisymmetric terms are the Landau parameters F_l^a and F_l^s. The renormalization factors of the thermodynamic quantities can be expressed in terms of these Landau parameters. For example, the specific heat and the magnetic susceptibility are given by

$$\frac{C_v}{C_{v0}} = 1 + \frac{F_1^s}{3} = \frac{m^*}{m}, \tag{2.2}$$

$$\frac{\chi}{\chi_0} = \frac{m^*}{m}\frac{1}{1 + F_0^a}, \tag{2.3}$$

where the index 0 on the left-hand side refers to the values calculated for the ideal gas of same density. The measured enhancement factors for the specific heat and the magnetic susceptibility, and the corresponding Landau parameters are listed in Table D.1 for liquid ^{3}He. The corrections to the ideal gas model are large, reflecting the strength of the interactions between the ^{3}He atoms; the effective mass enhancement factor, for instance, reaching ≈ 6 at high pressure. They vary strongly with the density through the

TABLE D.1
Enhancement factor of the thermodynamic quantities due to the interactions.

P (bar)	$v(P, T = 0.1\ \text{K})$	χ/χ_0	$C_v/C_{v0} = m^*/m$	F_0^s	F_0^a	F_1^s	F_1^a
S.v.p.	36.84	9.14	2.76	9.15	−0.698	5.27	−0.55
3	33.87	11.34	3.13	15.83	−0.724	6.40	−0.73
6	32.07	12.93	3.44	22.22	−0.734	7.32	−0.79
9	30.76	14.36	3.72	28.61	−0.737	8.15	−0.86
12	29.71	15.67	3.98	34.97	−0.741	8.95	−0.90
15	28.86	17.02	4.24	41.33	−0.746	9.71	−0.95
18	28.13	18.25	4.49	48.03	−0.754	10.47	−0.99
21	27.56	19.30	4.71	54.37	−0.756	11.14	−0.99
24	27.06	20.29	4.93	61.02	−0.757	11.80	−1.00
27	26.58	21.27	5.17	68.22	−0.757	12.50	−0.99
30	26.14	22.31	5.40	75.60	−0.758	13.20	−0.98
33	25.71	23.44	5.65	83.44	−0.759	13.96	−1.01
34.36	25.54	23.70	5.76	87.09	−0.757	14.28	−0.99

Note: The specific heat enhancement factors and the Landau coefficients F_0^s, F_1^s, and F_1^a have been tabulated by Greywall (1983). The values of F_0^a between 0 and 27 bar come from more recent measurement by Hensley et al. (1993). The values at higher pressure are taken from Greywall (1983).

dependence of the repulsive interaction with the distance between the atoms. This fact raises different interpretations in the microscopic models.

2.1.3. *Transport properties*

Transport phenomena in Landau theory are described in terms of fluxes of quasiparticles. The complex problem of calculating transport properties in a dense system is replaced by the simpler case of weakly interacting fermions. For example, the shear viscosity η is given by

$$\eta = \frac{1}{15} N(0) p_F^2 v_F^2 \tau_\eta, \tag{2.4}$$

where $N(0)$ is the density of states at the Fermi level, $v_F = p_F/m^*$ is the Fermi velocity, and τ_η is the 'viscosity' collision time (see Appendix D for details). This expression can be rewritten as $\eta = \frac{1}{5} n p_F v_F \tau_\eta$, n being the particle density, similar to the usual expression for the kinetic theory of gases.

At low temperature, only two-body collisions have to be taken into account, and the collision time τ_η can be expressed in terms of angular averages of the scattering amplitude for two colliding quasiparticles (see Appendix D for details). In contrast to the Fermi liquid function, this scattering amplitude depends not only on the angle θ between the two incident quasiparticles, but also on the scattering angle ϕ in the center of mass reference frame. Qualitatively, the energy conservation and the phase space restrictions due to the Pauli principle lead to the well-known $1/T^2$ dependence of the collision time. Quantitatively, the calculation of τ_η requires a model for the scattering amplitude.

The so-called s–p approximation consists of expressing this amplitude in terms of the Landau parameters. Quite generally, there is an exact relationship between the forward scattering amplitude at $\phi = 0$ and the Fermi-liquid function. The s–p approximation extends this relationship to arbitrary angles ϕ by making the simplest assumptions respecting the symmetry requirements imposed by the Pauli principle, which are that singlet and triplet scattering takes place in the s-wave and p-wave channels only, respectively. This fully determines the angular dependence of the scattering amplitude. Dy and Pethick (1969) have calculated the low-temperature limit of ηT^2 within this s–p approximation, retaining only the Landau coefficients $l = 0$ and 1 for the Fermi-liquid function. We have reproduced these calculations (see Appendix D) using the more recent values of the Landau coefficients quoted in Table D.1. Table D.2 compares the predicted values to those experimentally measured by Parpia et al., and Carless et al., reported by Carless et al. (1983). As noted by the later authors, the experimental pressure dependence is reproduced, although the experimental values are nearly twice the predicted ones.

A better agreement between experiment and theory is obtained by Ainsworth and Bedell (1987). These authors have used the so-called induced interaction model (see Quader and Bedell 1985 and references therein) to compute the dependence of the collision probability on the momentum transfer, retaining the Landau coefficients up to $l = 4$ in the calculation (the coefficients for $l = 2$–4 being computed within the model). They attribute the improvement to the difference between the dependences of the collision probability on the scattering angle ϕ for head-on collisions ($\theta \approx \pi$) given by their model and the s–p approximation (see Appendix D).

2.1.4. *Shortcomings of Landau theory*

Landau theory of Fermi liquids correctly describes the temperature dependence of the thermodynamic and transport properties of degenerate liquid ^{3}He. However, neither the Landau parameters, nor the scattering amplitude, can be calculated within the theory. In particular, the experimental dependence of the Landau parameters on density, and more importantly, the mechanisms underlying that dependence, are beyond the Landau approach. We shall now describe two simplified microscopic models, which aim to explain the density dependence of the Landau parameters by the proximity of ^{3}He to a phase transition: the 'nearly ferromagnetic' model and the 'nearly localized' model.

TABLE D.2

The low-temperature limit of ηT^2, interpolated from Fig. 6 of Carless et al. (1983) is compared to the prediction of the s–p approximation, as computed using the Landau coefficients given in Table D.1.

P (bar)	$\eta T^2(\mu P\,\mathrm{K}^2)$	s–p approximation	Induced interaction model
S.v.p.	0.235	0.153	0.28
3	0.21	0.125	0.24
9	0.16	0.094	0.20
15	0.14	0.072	0.16
21	0.12	0.074	0.13
27	0.11	0.066	0.115
34	–	0.059	0.095

Note: The values computed at s.v.p. and 27 bar differ from those of Dy and Pethick (1969) (also quoted in Baym and Pethick 1991) only due to a different choice of values for the Landau parameters. The results calculated with the induced interaction model (Ainsworth and Bedell, 1987) were obtained with the values of Landau parameters tabulated in Wheatley (1975).

2.2. *The 'nearly ferromagnetic' model*

2.2.1. *Stoner model*

The simplest way to include hard core repulsion between the atoms is to add a contact interaction term $U\delta(\vec{r}-\vec{r})$ to the kinetic energy of the particles.[1] The resulting hamiltonian can be written as

$$H = H_0 + U\int n_\uparrow(\vec{r})n_\downarrow(\vec{r})\,\mathrm{d}\vec{r}, \tag{2.5}$$

where H_0 is the kinetic energy of the 'bare' fermions and $n_\sigma(\vec{r})$ the density operator for particles of spin σ (the interaction is restricted to particles of unlike spin because of the Pauli principle). In spite of its simplicity, this hamiltonian has only been solved exactly in one dimension (Lieb and Wu 1968). We focus here on the Stoner model (see White 1970 for reference) in which the interaction term is treated within the mean-field (Hartree–Fock) approximation. The ground state energy (per unit volume) as function of the spin-polarization $m = (\langle n_\uparrow\rangle - \langle n_\downarrow\rangle)/n$, n being the particle density, is given by

$$E = \frac{n^2}{4}(am^2 + bm^4 + U(1-m^2)), \tag{2.6}$$

where $a = 1/N(0)$; $N(0)$ is the density of states of the bare particles at the Fermi level (for one spin species), and b a constant depending on the band

[1] In this expression, U is a phenomenological parameter representing the interaction strength between two ^{3}He atoms at their average separation in the liquid.

structure ($b>0$ for a parabolic band). The first two terms correspond to the small m expansion of the kinetic energy of the free fermions. The last term is the mean-field expression of the interaction term, that is $\langle n_\uparrow n_\downarrow \rangle \approx \langle n_\uparrow \rangle \langle n_\downarrow \rangle = (1-m^2)n^2/4$. As the low-field susceptibility χ is the inverse of the second derivative of the energy with respect to magnetization, we can see that the negative $-Um^2$ term enhances the susceptibility with respect to its $U=0$ value, χ_0, by the Stoner factor S given by

$$S = \frac{\chi}{\chi_0} = \frac{1}{1-N(0)U}. \tag{2.7}$$

For strong enough interactions ($N(0)U=1$), the susceptibility diverges, and the system undergoes a phase transition to a ferromagnetic state. In liquid ^{3}He at high pressure, the enhancement factor reaches $S=24$ yielding an interaction parameter $N(0)U=0.96$. From this point of view, liquid ^{3}He can be said to be 'nearly ferromagnetic'.[2]

The Stoner model, as a mean-field theory, neglects fluctuations. Yet, in the vicinity of the transition, large spin fluctuations are expected to dominate the behavior of the system. They are indeed observed clearly in degenerate liquid ^{3}He in neutron scattering experiments (Fåk et al. 1994, Glyde et al. 2000). This is the main point of the paramagnon theory, which we describe next.

2.2.2. *Paramagnons*

In the ferromagnetic phase, the low-energy excitations are spin waves or 'magnons.' In the paramagnetic phase, these modes are broadened and take the name 'paramagnons' (see Levin and Valls 1983 for a review). In the critical region, the thermal excitation of paramagnons at finite temperature has two important consequences: first, the resulting contribution to the specific heat adds to that of the bare particles, yielding a renormalization of the effective mass, which, for ^{3}He, is comparable to that experimentally measured (Béal-Monod et al. 1968, Bonfait et al. 1990). Second, for increasing temperature, it makes the magnetization decrease faster than for the free fermion gas. The resulting temperature dependence of the magnetic susceptibility obeys the scaling law:

$$\chi = S\chi_0\left(1-\alpha_1\left(\frac{T}{T_{\mathrm{SF}}}\right)^2\right) \tag{2.8}$$

[2] In this context, we mention the initial suggestion by Bashkin of a high temperature ($\simeq 0.3$ K) spontaneous ferromagnetism, due to the decrease in free energy induced by the thermal excitation of spin waves, which could offset the increase in the kinetic energy (Bashkin 1984). This ferromagnetism could however only occur on a microscopic scale (Bashkin 1984).

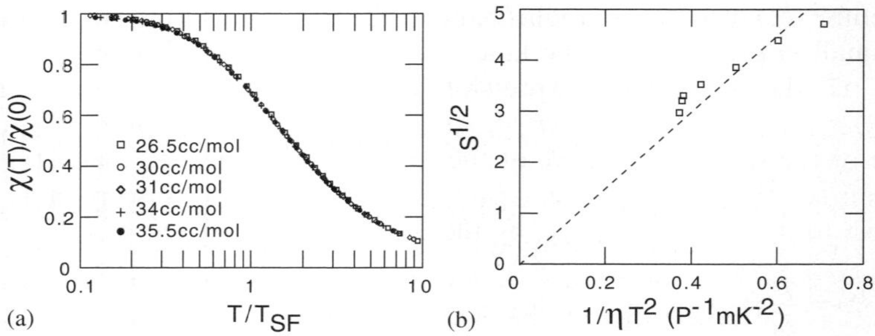

Fig. 1. (a) Temperature dependence of the magnetic susceptibility χ normalized by its value at $T = 0$, $S\chi_0$ (data from Ramm et al. 1970); the scaling law with $T_{FS} = T_F/S$ is shown for five different densities (corresponding to S varying from 10 to 24). (b) Pressure dependence of the viscosity (data from Carless et al. 1983); the dashed line is a guide for the eye.

with $T_{SF} = T_F/S$, and $\alpha_1 \approx \pi^2/6$ for a parabolic band. This temperature dependence resembles[3] that of the free fermion gas, but the temperature scale is reduced by the large factor S with respect to the latter case. As shown in fig. 1, the scaling law is well verified experimentally for ^{3}He (Ramm et al. 1970), the temperature scale being correctly given by $T_{SF} = T_F/S$ and the value of the coefficient α_1 being 1.24, close to the value predicted by the theory.

Paramagnons also contribute to transport properties. Taking the extreme point of view that the transport properties uniquely arise from the scattering of fermions by paramagnons, Rice has related the viscosity to the dynamical structure factor of paramagnons (Rice 1967). As shown by Béal-Monod, such an approach implies general relations between the viscosity and the magnetic susceptibility (Béal-Monod 1988):

$$\eta T^2 \propto \chi^{-2} \quad \text{for } 0.5 < N(0)U < 0.75, \tag{2.9}$$

$$\eta T^2 \propto \chi^{-1/2} \quad \text{for } 0.75 < N(0)U < 1. \tag{2.10}$$

These relations are valid, not only for zero field, but for finite field as well, provided one takes $\chi(H) = M/H$. As in Landau theory, the shear viscosity coefficient is found to vary as $1/T^2$ at low temperature. Moreover, for unpolarized ^{3}He, relations (2.9) and (2.10) predict a strong pressure dependence mainly through the Stoner enhancement factor S:

[3]For free fermions, $\alpha_1 \approx \pi^2/12$, and $S = 1$.

$$\eta T^2 \propto 1/S^2 \quad \text{for } 0.5 < N(0)U < 0.75, \tag{2.11}$$

$$\eta T^2 \propto 1/\sqrt{S} \quad \text{for } 0.75 < N(0)U < 1. \tag{2.12}$$

The latter case, which is supposed to hold for liquid ^{3}He, is consistent with the experiment (see fig. 1).

Overall, the paramagnon model provides a reasonable description of liquid ^{3}He. In particular, the prediction of a scaling law for the magnetic susceptibility is its major success. On the other hand, the paramagnon theory fails to describe the 'mechanical' properties of liquid ^{3}He. The strong pressure dependence of the compressibility, for example, is not predicted by the model. On theoretical grounds, the paramagnon model has often been considered as unsuitable to describe strongly interacting Fermi systems (Lhuillier 1991). In particular, Sanchez-Castro and Bedell (1989b) have argued that relations (2.9) and (2.10) between the viscosity and the magnetic susceptibility hold only for weakly interacting systems.

2.3. *The 'nearly localized' model*

In Landau theory, the magnetic susceptibility enhancement stems from the interaction parameter $(1 + F_0^a)^{-1}$ as well as from the renormalization of the effective mass m^*/m. Anderson and Brinkman pointed out that the increase in susceptibility with increasing pressure was not due to the coefficient F_0^a, which saturates at a constant value -0.75 (see Table D.1), but to the effective mass enhancement (Anderson and Brinkman 1975). They interpreted the strong increase of the effective mass with the pressure as a sign of an incipient localization in the system.

2.3.1. *Lattice model of 'nearly localized' ^{3}He*

The 'nearly localized' model is a simplified model where the particle motion is restricted on a cubic lattice. The dynamics are given by the lattice version[4] of the contact–interaction hamiltonian, which we introduced in the previous section (Hubbard hamiltonian):

$$H = H_0 + U \sum_{\mathrm{i}} n_{\mathrm{i}\uparrow} n_{\mathrm{i}\downarrow} = t \sum_{\langle \mathrm{i,j} \rangle, \sigma} c^+_{\mathrm{i},\sigma} c_{\mathrm{j},\sigma} + U \sum_{\mathrm{i}} n_{\mathrm{i}\uparrow} n_{\mathrm{i}\downarrow}, \tag{2.13}$$

where $n_{\mathrm{i}\sigma}$ is the number of particles of spin σ at site i, and t is the hopping amplitude. This hamiltonian also describes the dynamics of electrons on a lattice, and has been much studied in the context of the Mott metal insulator

[4]Note that U is now an energy, instead of an energy times an unit volume in the continuous hamiltonian.

transition (MIT). The mean-field solution of the problem gives the same result as the Stoner model described in Section 2.2.1. Beyond mean field, localization (i.e. MIT in the context of electrons) shows up for a half-filled band (one fermion per site) and large enough U/t, when the on-site repulsion is so large that it forbids double occupancy of a site, hence hopping. In the context of ^{3}He, the insulator phase would correspond to the solid, and the metallic phase to the liquid. An explicit calculation of the transition is based on Gutzwiller's variational ansatz, which is reviewed by Vollhardt (1984). Gutzwiller's basic idea is to reduce the interaction term of the hamiltonian by decreasing the number of doubly occupied states, $d = \langle n_{i\uparrow} n_{i\downarrow} \rangle$ with respect to its Stoner mean field value $d = \langle n_{i\uparrow} \rangle \langle n_{i\downarrow} \rangle$. This reduction occurs at the expense of a larger kinetic energy, because decreasing the number of empty sites makes hopping less favorable. The ground state energy (per site) is obtained by minimizing the following expression (Vollhardt 1984) with respect to the number of doubly occupied sites d:

$$E_g = q_\uparrow \varepsilon_\uparrow + q_\downarrow \varepsilon_\downarrow + Ud, \tag{2.14}$$

where $\varepsilon_\uparrow$ ($\varepsilon_\downarrow$) is the average kinetic energy of the up (down) spins in the non-interacting case. In this ansatz, the interaction term Ud is treated exactly, while the modification of the kinetic energy is approximated by the factors $q_\uparrow$ and $q_\downarrow$. These factors are given by

$$q_\uparrow = \frac{(n_\uparrow - d)(1 - n + d) + 2\sqrt{d(1 - n + d)(n_\uparrow - d)(n_\downarrow - d)} + d(n_\downarrow - d)}{(1 - n_\uparrow)n_\uparrow}, \tag{2.15}$$

where $n_\uparrow = \langle n_{i\uparrow} \rangle$, $n_\downarrow = \langle n_{i\downarrow} \rangle$ and $n = n_\uparrow + n_\downarrow$ ($q_\downarrow$ is obtained by permutation of $\uparrow$ and $\downarrow$). This expression can be viewed as an average of the hopping of (say) an up spin to a neighboring site over the four possible down-spin configurations of the considered pair of sites, neglecting the correlations and considering the down spins as fixed (Nozières 1986). Each configuration is weighted by the geometric mean of the probabilities of the initial and final states, normalized by the probability $(1 - n_\uparrow)n_\uparrow$ of the pair of sites being occupied by one up-spin only, independently of the occupancy of down spins. For example, the last term of eq. (2.15) corresponds to the case where both sites are occupied by down spins, the probabilities of both the final and initial states being equal to $d(n_\downarrow - d)$, as one site is doubly occupied and the other is singly occupied by a down spin. As discussed by Vollhardt, the factor q can be identified as the inverse of the effective mass enhancement factors m^*/m, i.e.

$$\frac{m^*_\sigma}{m} = \frac{1}{q_\sigma}. \tag{2.16}$$

In the absence of interactions, $d = n_\uparrow n_\downarrow$ and $q = 1$ (Vollhardt 1984). The repulsive interactions give $q < 1$, so that the kinetic energy is indeed increased.[5]

2.3.2. *Predictions of the 'nearly localized' model*

Brinkman and Rice (1970) have shown that the Gutzwiller approximation indeed leads to a phase transition, in the special case of a half-filled band (one particle per site) for the unpolarized system ($n_\uparrow = n_\downarrow = 1/2$). From eq. (2.15), we get $q_\uparrow = q_\downarrow = 8d(1 - 2d)$. Minimization of eq. (2.14) then leads to

$$d = \frac{1}{4}\left[1 - \frac{U}{U_\mathrm{c}}\right] \tag{2.17}$$

with $U_\mathrm{c} = 8|\varepsilon_0|$, ε_0 being the average particle kinetic energy per site. This implies that, above the finite value U_c of the interaction parameter, $d = 0$, so that the system is localized. Localization gives rise to a divergence of the effective mass:

$$\frac{m^*}{m} = \frac{1}{1 - (U/U_\mathrm{c})^2}. \tag{2.18}$$

As shown by Brinkman and Rice, close to the transition, the magnetic susceptibility diverges as m^*/m, with F_0^a approaching a constant limit, similar to what is observed for liquid ^{3}He as the pressure is increased. Vollhardt applied this model to liquid ^{3}He and showed that its results compare favorably to the experimental values: the ratio U/U_c determined from the specific heat enhancement is found to be 0.8 at low pressure and 0.9 near the melting pressure, consistent with liquid ^{3}He being close to the localization transition. The pressure dependence of the susceptibility and compressibility enhancement factors predicted from this 'nearly localized' model are in qualitative agreement with the experimental values. Strikingly, the limiting value of $F_0^\mathrm{a} \approx -3/4$ is accounted for by the model within a few percent.

The 'nearly localized' model has some limitations: it is a $T = 0$ K calculation. There has been an attempt to extend the model to finite temperature and polarization (Seiler et al. 1986), but, according to Nozières (1986), this extension relies on disputable assumptions (for example, the polarization dependence of the double occupancy of sites is neglected, and the entropy is arbitrarily assumed to be reduced by a constant factor with respect to its expression for free fermions). The second limitation

[5] The *average* kinetic energy being referred to the Fermi level of non-interacting spins, is *negative*!

comes from the Gutzwiller ansatz, which is essentially a 'static' approach and therefore unsuitable for the calculation of the transport coefficients (Vollhardt, private communication). This is why, unfortunately for the experimentalists, the 'nearly solid' model makes no specific prediction about the viscosity behavior. Finally, one may object to the use of a lattice model to describe a liquid system. Vollhardt (1984) argued that Gutzwiller's approach was particularly suitable for liquid ^{3}He since it is independent of the topology of the lattice, the only quantity related to the existence of a lattice being the parameter ε_0.

It is important to note that the choice of a half-filled band is crucial in order to obtain localization at a finite interaction. If vacancies are present, an infinite U is required to obtain localization. An alternative description of the liquid could then be that it is a nearly half-filled band, with a small concentration of vacancies, and a large interaction ($U/U_c \gg 1$). Such a model has been proposed by Vollhardt et al. (1987). In their model, the lattice density is taken as that of the solid ^{3}He on the melting curve, and the filling fraction is chosen so as to match the experimental liquid density (strictly speaking, this is only imposed at saturated vapour pressure and 34 bar, the pressure being the true control parameter). With respect to the half-filling situation, the concentration of vacancies varies between 34 and 5% for pressures between s.v.p. and 34 bar. The (constant) on-site interaction is fixed by the requirement to give the proper effective mass near solidification, which is obtained, as in the original model, within the Gutzwiller approach. For an elliptical density of states, this interaction is found to be $U/U_c = 1.3$, so that, for the half-filled band, the system would be fully localized. Note that the situation is reversed with respect to the original model where the approach to transition with increasing pressure is driven by a variable interaction, the filling being fixed. Due to the smaller number of free parameters, the agreement of the model with the experimental susceptibilities and effective masses is not as good as in the original Vollhardt model, where the interaction strength is determined independently for each pressure. In particular, the predicted effective mass is too small at zero pressure, and grows too fast when the solidification is approached.

Finally, we note that, beyond Gutzwiller's variational calculation, the existence of the Mott transition within the Hubbard model can be exactly demonstrated in the limit of infinite spatial dimensionality (Metzner and Vollhardt 1989) by the so-called dynamical mean field method, which consists in mapping the problem onto a single impurity Anderson problem (see Georges et al. 1996 for review). We will come back to this point in Section 5.

2.4. *Liquid ^{3}He at finite temperature*

As the temperature is increased, the properties of liquid ^{3}He deviate from the simple dependences discussed in Section 2.1.1. The initial deviations, which occur faster than would be expected for a free fermion gas, can be understood within the Landau theory by taking into account the coupling of the quasiparticles to long-wavelength spin fluctuations (Baym and Pethick 1991). At larger temperatures, the quasiparticle lifetime becomes shorter, and eventually, quasiparticles are no longer well-defined excitations in the system, resulting in a more drastic change of behavior. Experimentally, strong deviations from Landau theory appear at $T > 100$–150 mK, when the specific heat goes through a plateau. As this temperature is much smaller than the bare Fermi temperature ($T_F = 3$–5 K, depending on pressure), it is not obvious whether the underlying Fermi statistics may manifest itself or not beyond the range of validity of the Landau theory. The magnetic susceptibility, for example, crosses over from the Pauli behavior to the Curie behavior around the 'magnetic Fermi temperature' $T_{SF} = 160$–200 mK (Ramm et al. 1970). From this point of view, one may argue as Lhuillier (1991) that, in contrast to the case of a gas (Meyerovich 1978, Lhuillier and Laloë 1982), all Fermi statistics effects could disappear above this 'effective' degeneracy temperature.

This view is supported by two papers by Andreev. Castaing and Nozières (1979) have pointed out that this degeneracy temperature T_{SF} is much smaller than the temperature scale Θ_D of the vibrational motion in liquid ^{3}He (the equivalent of the Debye temperature $\Theta_D \approx 14$ K at 0 bar). Based on this remark, Andreev (1978) argued that there is a high-temperature regime $T_{SF} \ll T \ll \Theta_D$ where the system can be viewed as a glass, whose excitations are small oscillations of the atoms about their local equilibrium positions. In this regime, the quantum zero-point motion is still dominant (otherwise the system would freeze!), but the Fermi statistics plays no role; for this reason it is called 'semi-quantum' regime. Andreev has calculated that the entropy should be linear in temperature, and that the viscosity should be inversely proportional to the temperature (Andreev and Kosevich 1979).

Only the first prediction is supported by experiment. The linear temperature dependence of the entropy is indeed well verified experimentally above 1.5 K (Greywall 1983). Moreover, the extrapolation of this linear behavior to $T \to 0$ gives a limiting value of $R \ln 2$, which corresponds to the entropy of the non-degenerate spin degrees of freedom. However, the viscosity at low pressure is roughly constant between 1.5 and 3 K (Black et al. 1971). In fact, the first experimentalists have noted the similarity between ^{3}He at high temperature and a quantum gas, rather than a glass (Taylor and Dash

1957). For example, the Prandtl ratio $\kappa/\eta C_v$ (κ being the thermal conductivity) is close to 2.5, that is the value expected for an ideal gas (Wilks 1967). This shows that the understanding of the high-temperature behavior of liquid ^{3}He is far from complete.

3. Spin-polarized liquid ^{3}He

As discussed in the previous section, different models have been introduced to describe the temperature and pressure dependence of the thermodynamic quantities of liquid ^{3}He with various success. In 1979, Castaing and Nozières suggested that experiments carried out in spin-polarized ^{3}He could decide between these models, since they predict very contrasting behaviors for ^{3}He at high magnetic field (Castaing and Nozières 1979). In this section, we elaborate on this matter, by describing the predictions for ^{3}He at high-magnetic field, based on three different models: the model of Sanchez-Castro and Bedell, which implements an extension of the Landau theory to polarized systems, and the aforementioned 'nearly ferromagnetic' and 'nearly localized' models.

3.1. *Landau theory for spin-polarized systems*

The Landau theory described in Section 2.1 is valid only at low magnetic field ($\mu^3 B \ll k_B T_F$, where μ_3 is the magnetic moment of ^{3}He). Building on general principles set by Abrikosov and Dzyaloshinskii (Abrikosov and Dzyaloshinskii 1958, Bashkin and Meyerovich 1978, Meyerovich 1987), Sanchez-Castro and co-workers have considered an explicit extension of Landau theory to spin-polarized ^{3}He (Bedell and Sanchez-Castro 1986, Sanchez-Castro et al. 1989). This extension involves the introduction of polarization-dependent Landau interactions $\tilde{f}_l^{\sigma\sigma'}(m)$ (with, in particular, $\tilde{f}_l^{\uparrow\uparrow} \neq \tilde{f}_l^{\downarrow\downarrow}$). In the following, we express these interactions in terms of modified Landau parameters, defined as $\tilde{F}_l^{\sigma\sigma'} = N(0)\tilde{f}_l^{\sigma\sigma'}$.

As such, the extended theory has no predictive power, since it relies on new adjustable parameters. In order to obtain a predictive theory, Bedell and Sanchez-Castro have introduced a model for the polarization dependence of the Landau coefficients (Bedell and Sanchez-Castro 1986): the coefficients $l \geqslant 2$ are assumed to be zero. The polarization dependence of the remaining coefficients are given by polynomial expressions, of second degree for $l = 0$, and of fourth degree for $l = 1$. The nine adjustable parameters required by this model can then be determined from the experimental values of the thermodynamic quantities of non-polarized liquid ^{3}He, with the help of sum rules and Maxwell relations. These parameters can then be used to calculate the properties of ^{3}He as function of the polarization (Sanchez-Castro et al. 1989).

3.1.1. *Thermodynamic properties*

The expressions for the magnetic susceptibility and the heat capacity per unit volume for a polarized Fermi liquid are (Bedell and Sanchez-Castro 1986)

$$\chi^{-1} = \frac{1}{4\mu_3^2}(N_\uparrow(0)^{-1} + N_\downarrow(0)^{-1} + N(0)^{-1}(\tilde{F}_0^{\uparrow\uparrow} + \tilde{F}_0^{\downarrow\downarrow} - 2\tilde{F}_0^{\downarrow\uparrow})), \tag{3.1}$$

$$C_v = \frac{\pi^2 k_B^2}{3}(N_\uparrow(0) + N_\downarrow(0))T, \tag{3.2}$$

where $N_\sigma(0)$ is the density of states at Fermi level of spin σ. Note that eq. (3.1) is the simple generalization of eq. (2.3) to the case of a polarization-dependent interaction between unlike spins.

The model of Bedell and Sanchez-Castro predicts a steady decrease of the density of states at the Fermi level with the polarization. This is mirrored by a monotonic decrease of the specific heat as a function of the polarization (see fig. 2), the decrease reaching 90% of the initial heat capacity at full polarization. Note that the initial $\propto m^2$ decrease at low polarization is simply an input to the determination of the parameters of the model, the slope being computed through a Maxwell relation from the experimental temperature dependence of the magnetic susceptibility.

The decrease in density of states with the polarization alone would decrease the magnetic susceptibility. However, the combination of

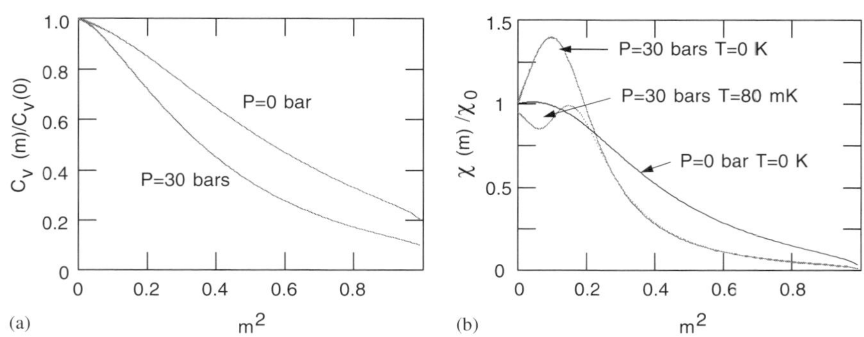

Fig. 2. Predictions of the Bedell and Sanchez-Castro model, computed as in Sanchez-Castro et al. (1989). (a) Polarization dependence of the specific heat at constant magnetization; (b) Polarization dependence of the magnetic susceptibility, showing a 'nearly metamagnetic' peak.

Landau interactions, which enters the susceptibility also depends on polarization. Using Bedell and Sanchez-Castro's notations, one has $\tilde{F}_0^{\uparrow\uparrow} + \tilde{F}_0^{\downarrow\downarrow} - 2\tilde{F}_0^{\downarrow\uparrow} = 2F_0^{a} + (b_1 F_0^{\uparrow\uparrow} - c_1 F_0^{\downarrow\uparrow})m^2 + \cdots$. With the values of b_1 and c_1 taken from Sanchez-Castro et al. (1989), this interaction becomes more negative with polarization, resulting in a peak of the susceptibility at $m \approx 0.3$. Sanchez-Castro et al. interpreted this feature as a tendency of ^{3}He to come closer to a magnetic instability when a field is applied. For this reason, they called their model 'nearly metamagnetic' model. They observed, however, that finite temperature corrections tend to wash out the susceptibility peak. As we shall see in Section 3.3, the 'nearly localized' model also exhibits a metamagnetic behavior. In the latter case, however, this behavior gives rise to a first-order transition, and the specific heat increases as the transition is approached, reflecting the tendency of atoms to localize. It is important to point out that there is no such physical picture in the case of the 'pseudo transition' of Sanchez-Castro et al.

3.1.2. *Transport properties*

In dilute, degenerate, Fermi gases, the transport coefficients should strongly increase with the spin polarization (Bashkin and Meyerovich 1977, 1978). The argument is the following: as the interaction between atoms is short-ranged, the collisions are expected to be mainly s-wave at low density enough for the Fermi wavelength to be larger than the ranger of interaction. Due to the Pauli principle, collisions then take place only between atoms of unlike spins. Hence, increasing the spin-polarization decreases the number of collisions and leads to a strong enhancement of the transport coefficients, which has been calculated by Bashkin and Meyerovich (1977, 1978) and Mullin and Miyake (1983), in the case where both spin species are degenerate. For very large polarizations, and finite temperature, one might reach the case where the minority spins become non-degenerate, which affects the polarization dependence of the transport coefficients. This has been considered by Meyerovich (1978) and Mullin and Miyake (1986).

The same argument also applies to non-degenerate Fermi gases, as long as the temperature is low enough for the collisions to be mainly s-wave (which corresponds to the de Broglie wavelength being much larger than the range of interaction) (Meyerovich 1978, Lhuillier and Laloë 1982). At larger temperatures, where higher values of the orbital angular momentum l become important, the scattering amplitude between atoms remains sensitive to their fermionic character, i.e. a collision between parallel spins can only occur for odd values of l. As shown by Lhuillier, this leads to a complex temperature dependence of the effect of polarization on transport, which depends on the exact shape of the interaction potential between atoms. For gaseous ^{3}He, a 'bump,' as a function of temperature, of the polarization-induced thermal conductivity enhancement is thus predicted (Lhuillier 1983) and observed (Leduc et al. 1987) around 2 K. For spin-polarized ^{3}He–^{4}He mixtures, the transport coefficients at arbitrary temperatures have been computed by Hampson et al. (1988) (viscosity, thermal conductivity) and Jeon and Mullin (1987) (spin diffusion). The first paper accounts for the polarization dependence of the viscosity observed in dilute mixtures ^{3}He in ^{4}He (Greywall and Paalanen 1981, Owers-Bradley et al. 1988, Candela et al. 1991, Owers-Bradley 1997, Akimoto et al. 2002). In particular, a large growth of the viscosity with spin polarization is observed at low temperature, in agreement with the general argument developed above.

This general argument does not apply directly to dense systems like liquid ^{3}He, where binary collisions between atoms involve a large number of spherical harmonics and many-body collisions also occur. However, one should expect an effect of polarization at low temperature enough for transport to be described in terms of collisions of quasiparticles (Section 2.1.3). It is however not obvious that polarization should enhance the transport coefficients, because the situation differs from a dilute, degenerate, Fermi gas in two respects. First, the Fermi momentum in liquid ^{3}He is such that collisions with angular momentum higher than $l = 0$ are not unlikely. Second, as we deal with quasiparticles (and not bare particles), the scattering amplitudes might as well depend on the polarization. However, in the specific case of fully polarized liquid ^{3}He, Bashkin and Meyerovich have argued that, due to the disappearance of the s-wave channel, the Fermi-liquid interaction could be decreased, leading to an increase of the transport coefficients (Bashkin and Meyerovich 1981).

In its full generality, the polarization dependence of the transport coefficients in the Fermi-liquid regime arises through the lifetimes τ_σ, the Fermi momenta p_σ, the Fermi velocities v_σ, and the densities of states $N_\sigma(0)$ of the two spin species $\sigma = \uparrow$ and $\downarrow$. Anderson et al. (1987) have provided exact analytical expressions for the transport coefficients as functions of the probability of collision. For example, the viscosity has the form

$$\eta = \frac{1}{15} \sum_{\substack{\sigma=\uparrow,\downarrow \\ \sigma'=\uparrow,\downarrow \\ t=1,2}} \sqrt{N_\sigma(0)\tau_\sigma N_{\sigma'}(0)\tau_{\sigma'} p_\sigma v_\sigma p_{\sigma'} v_{\sigma'}}\, S_{\sigma t} R^+(\lambda_t) \tilde{S}_{\sigma' t}, \tag{3.3}$$

where $S_{\sigma t}$ and $\tilde{S}_{\sigma' t}$ are 2×2 matrices, which diagonalize the coupled transport equations for spin $\uparrow$ and $\downarrow$ quasiparticles; the index t is a summation over the two associated eigenvalues, and the function R^+ is defined in Appendix D. Anderson et al. (1987) give the expression of the matrices $S_{\sigma t}$ and $\tilde{S}_{\sigma' t}$ and the coefficients λ_t as functions of angular averages of the collision probability.

As in the non-polarized case, a model has to be introduced to calculate the collision probability between quasiparticles. Hess and Quader (1987) have used the s–p approximation to express this collision probability as function of the Landau parameters, whose polarization dependence was taken from the model of Sanchez-Castro et al. (1989). However, one parameter (the spin-flip scattering amplitude) is not determined by this model, and its polarization dependence had to be set arbitrarily. Historically, this model was introduced to explain the experimental results of Kopietz et al., which showed a *decrease* of the viscosity with the polarization, in contrast to the case of a gas (Kopietz et al. 1986). When Vermeulen (Vermeulen et al. 1988) and Kranenburg (Kranenburg et al. 1989) subsequently proved that the results of Kopietz et al. were flawed, a controversy started about the origin of the decrease predicted by Kopietz et al. (see the discussion in Béal-Monod 1988): the question was to know whether this behavior came from the particular choice of Kopietz et al. for the spin-flip term or from the model of Sanchez-Castro et al. for the polarization-dependent Landau parameters; at that time, the latter case was felt to have a larger implication, since it would have questioned the validity of the 'nearly metamagnetic' model. We refer to Appendix D for a more detailed discussion of this problem in view of our own, recent, results.

3.2. *The 'nearly ferromagnetic' model at high field*

For free fermions, the zero-temperature susceptibility decreases with polarization (b is positive in eq. (2.6)). The effect is more marked in the Stoner model, since, as m increases, the relative effect of the interaction term becomes smaller, thus reducing the difference between the interacting and the non-interacting systems. The decrease of susceptibility implies that spin fluctuations are reduced at high polarization. In this sense, polarization may be said to make the system less critical, i.e. farther from the ferromagnetic transition. The reduction of the number of paramagnons also decreases the specific heat and increases the viscosity. These effects can be quantified using the expression derived by Béal-Monod (1983) for the magnetization at low temperature ($T \ll T_{SF} = T_F/S$) and moderate magnetic field ($\mu_3 B \ll k_B T_F/S^{3/2}$):

$$\frac{m}{B} = S\chi_0\left(1 - \alpha_1\left(\frac{T}{T_{SF}}\right)^2 - \beta_0\left(\frac{B_S}{T_{SF}}\right)^2 + \alpha_2\left(\frac{T}{T_{SF}}\right)^2\left(\frac{B_S}{T_{SF}}\right)^2\right), \tag{3.4}$$

noting $B_S = \mu_3 B S^{1/2}$. For a parabolic band, the expansion coefficients are $\alpha_1 = \pi^2/6$, $\beta_0 = 1/6$, and $\alpha_2 \approx 1.5$. At zero temperature, the susceptibility

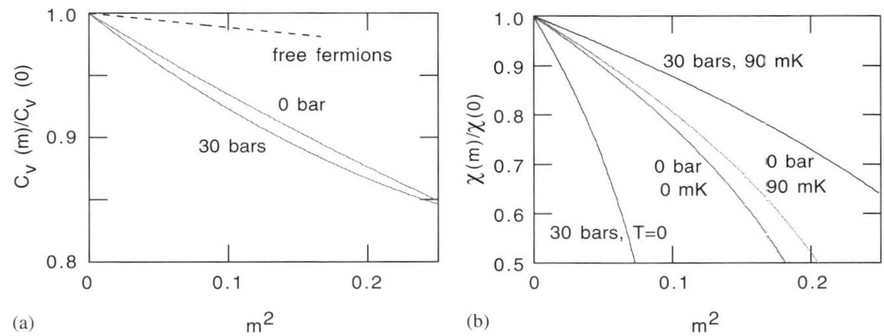

Fig. 3. Polarization dependence in the paramagnon model of (a) the specific heat at constant magnetization at 0 and 30 bar. The dashed line is the lowest-order term for free fermions; (b) the magnetic susceptibility $\chi = (\partial M/\partial B)_T$ at $T = 0$ (Stoner model) and $T = 90\,\mathrm{mK}$. The values used for the Stoner enhancement factor and the effective mass are 8.6 and 2.8 (21.5 and 5.5) at 0 (30) bar. The range of validity of the calculation is $m^2 < 1/S$.

decreases with applied field, the field-scale being $T_F/S^{3/2}$. The latter result may be directly obtained from eq. (2.6), noting that b does not depend on S and is positive. At zero field, the temperature dependence is due to the thermal excitation of paramagnons, as described in Section 2.2.2. At finite temperature and finite field, the cross term $\propto B^2T^2$ tends to decrease the reduction of susceptibility by the applied field (see fig. 3b), the effect being more marked at high pressure, where the low-field susceptibility is the most enhanced. This behavior is physically explained by the fact that the thermal reduction of the susceptibility caused by the paramagnons decreases as the spin fluctuations are inhibited by the magnetic field. This partly offsets the reduction of the susceptibility by the field, and makes the relation $m(B)$ more linear as the temperature increases.

The decrease of specific heat may then be obtained using Maxwell's relation

$$\left(\frac{\partial S}{\partial M}\right)_T = -\left(\frac{\partial B}{\partial T}\right)_M \rightarrow \left(\frac{\partial \gamma}{\partial m}\right)_T = -R\frac{\mu_3}{k_B}\left(\frac{\partial^2 B}{\partial T^2}\right)_M, \tag{3.5}$$

where $\gamma = C_m/T$ is the linear coefficient of the molar specific heat at constant magnetization,[6] and R the gas constant. The relation between the magnetic field and the polarization is obtained by inverting the expression

[6]Which is the relevant case for our experiments, see Section 8.

(3.4) to third-order in m and second-order in T:

$$\frac{3\mu_3 B}{2k_B T_{SF}} = \left(1 + \alpha_1 \left(\frac{T}{T_{SF}}\right)^2\right) m + \left(-\beta_0 + (\alpha_2 - 4\beta_0\alpha_1)\left(\frac{T}{T_{SF}}\right)^2\right) m^3. \tag{3.6}$$

Inserting expression (3.6) into relation (3.5) yields, after integration,

$$\gamma(m) - \gamma(0) = -\frac{2}{3}\frac{R}{T_{SF}}\left(\alpha_1 m^2 - \frac{2}{9} S(\alpha_2 - 4\beta_0\alpha_1) m^4\right) \tag{3.7}$$

(a similar expression can be found in Béal-Monod and Daniel (1983) for the specific heat at constant magnetic field). The result, normalized by the experimental values of Greywall, that is $\gamma(0) = \pi^2 R/T_F\, m^*/m$ (Greywall 1983), is shown in fig. 3a. As $T_{SF} = T_F/S$, and α_1 is twice the free fermions value $\pi^2/12$, the initial decrease of $\gamma(m)/\gamma(0)$ is larger than in the non-interacting case by a factor $2(1 + F_0^a)^{-1} \approx 8$ at large pressure. In the case of ^{3}He (parabolic band), $\alpha_2 > 4\beta_0\alpha_1$, and the initial decrease is weakened by the m^4 correction term.

The viscosity is expected to be reduced at high field since the 'quench' of the spin fluctuations implies a decrease of the number of collisions experienced by the fermions. A detailed calculation by Béal-Monod (1988) supports this argument: the general relations (2.9) and (2.10) show that the viscosity and the susceptibility vary in opposite directions with the magnetic field. In the specific case of liquid ^{3}He, the polarization-induced decrease of the susceptibility predicted by the Stoner model implies an increase in viscosity. From eq. (3.4), the correction factor is found to be proportional to m^2 for small polarization:

$$\frac{\eta(m) - \eta(0)}{\eta(0)} = \frac{2}{9}\beta_0 S m^2 \quad \text{for } T \to 0, \tag{3.8}$$

where the pre-factor, of order of unity at 30 bar, is strongly pressure-dependent through the Stoner factor S. These features can be tested experimentally.

3.3. *The 'nearly localized' model at high field*

In contrast to the paramagnon model, spin polarization drives the system closer toward the transition in the 'nearly localized' model. However, we will see that the localization transition is masked by a first-order transition to a ferromagnetic state ('metamagnetic' transition) when the magnetic field is increased. Before the transition, both the magnetic susceptibility and the heat capacity are found to increase with the polarization. While the increase of the susceptibility is a general feature for 'nearly localized' systems, the increase of heat capacity is specific to the half-filled Vollhardt model.

3.3.1. *Metamagnetic transition*

Gutzwiller's solution of the Hubbard hamiltonian can be extended to arbitrary polarization straightforwardly. This has been done by Vollhardt (1984). Here we reproduce his main results and extend his analysis of the metamagnetic transition: at arbitrary polarization m (and half-filling) we have $n_\uparrow = (1+m)/2$ and $n_\downarrow = (1-m)/2$, so that the factor q as a function of the polarization is

$$q_\uparrow = q_\downarrow = q = 4d\frac{1-2d+\sqrt{(1-2d)^2-m^2}}{(1-m^2)}. \tag{3.9}$$

The energy per site becomes

$$E(m) = q\varepsilon_0(m) + Ud. \tag{3.10}$$

Minimizing eq. (3.10) with respect to d gives the number of doubly occupied sites as a function of the polarization and the ground state variational energy. Figure 4 shows $d(m)$ and $q(m)$ for several values of U/U_c, in the case of the elliptical density of states considered by Vollhardt. As in the non-interacting case, the spin polarization decreases the number of doubly occupied lattice sites. As a result, a fixed polarization m reduces the critical interaction where the localization transition would take place (i.e. where d vanishes) to $\approx 0.6U_c$ for $m \to 1$.

Thus, one would expect that, by increasing the polarization, a continuous transition from a delocalized to a localized state would occur for $0.6 < U/U_c < 1$. In fact, this continuous transition is always masked by a first-order transition to a ferromagnetic (localized) state. As shown by Vollhardt (1984), a metamagnetic transition occurs in an even larger interaction range, corresponding to $U/U_c > 0.44$ (in liquid ^{3}He, this condition is fulfilled at all pressures). The phenomenon, illustrated in figs. 5a and b for $U/U_c = 0.9$ (corresponding to ^{3}He at 27 bar), follows the pattern common to first-order transitions: for a given magnetic field B, the equilibrium polarization is obtained by minimizing the 'magnetic enthalpy' $E(m) - \mu_3 Bm$. As shown in fig. 5a, this enthalpy has two extrema at low fields (a minimum and a maximum), in addition to a minimum at $m = 1$. The polarizations of the two extrema are plotted in fig. 4b. At low field, the equilibrium polarization corresponds to the first minimum of the enthalpy (lower branch in fig. 5b). At a high enough magnetic field ($B = 35$ T in the example shown in fig. 5), the enthalpies at the first minimum ($m < 1$) and in the fully polarized

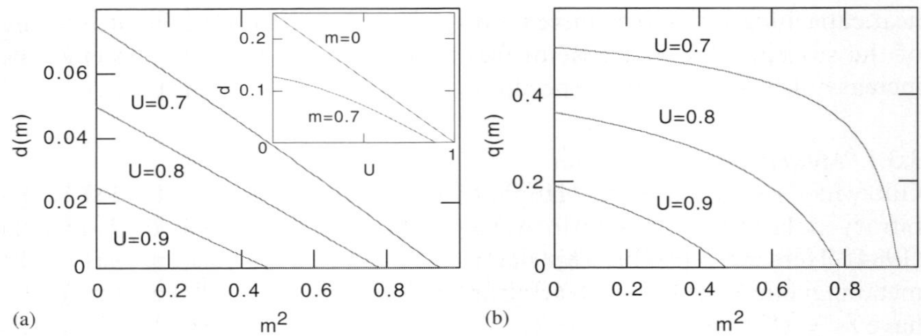

Fig. 4. Polarization dependences of the double occupancy d and discontinuity q in the 'nearly localized' model, for three values of the ratio U/U_c. $U/U_c = 0.9$ and 0.8 correspond to 27 and 0 bar, respectively. The inset shows how a 70% polarization reduces the critical interaction for localization ($d = 0$).

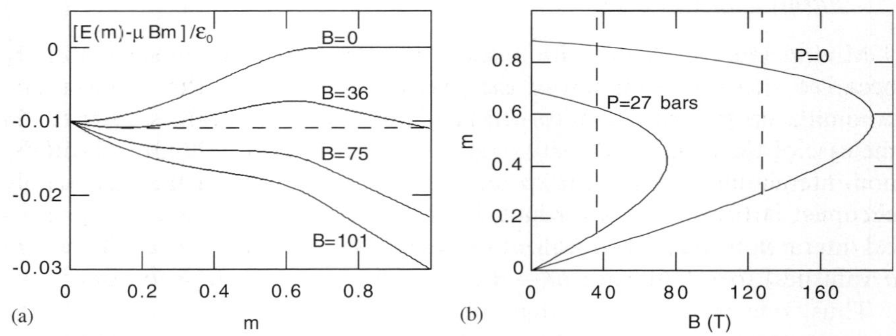

Fig. 5. (a) Gutzwiller magnetic enthalpy as a function of polarization for different fields ($U/U_c = 0.9$): $B = 0$, 36.3 T (where the enthalpies of the paramagnetic and fully polarized phases are equal), $B = 75.2$ T (spinodal point), and $B = 101$ T (beyond the metamagnetic transition). We used $\varepsilon_0(0) = 4/3\pi E_F$, appropriate for an elliptic density of states, with the Fermi energy given by $E_F/\mu_3 = 7930$ T ($P = 27$ bar, 26.5 cm^3/mole). (b) Field dependence of the polarization, given by the extrema of the enthalpy in fig. 5a; the lower (upper) branch correspond to the local minimum (resp. maximum). They merge at the spinodal point, above which the only minimum is $m = 1$. The vertical line is the Maxwell plateau where the minima at $m < 1$ and $m = 1$ have the same enthalpy.

state ($m = 1$) are equal, so that the system separates into two phases. Beyond this threshold, the paramagnetic phase becomes unstable relative to the fully polarized state. Eventually, the two extrema $m < 1$ of the enthalpy merge at the spinodal point, and the only stable phase is the fully polarized state. Again, the spinodal point occurs at a polarization smaller than the polarization for which d vanishes (corresponding to the $B = 0$ point in fig. 5b), that is, the metamagnetic transition takes place before the localization transition.

3.3.2. *Behavior at low m*

This metamagnetic transition causes the magnetic susceptibility and the specific heat to diverge. In the following, we concentrate on the behavior of these quantities for smaller polarization m.

The non-linear susceptibility has been computed by Vollhardt, who has found that the susceptibility increases with magnetization, the non-linearity being stronger as the interaction U increases. This behavior is illustrated in fig. 6a for an elliptical density of states. As discussed by Nozières, this behavior does not depend on the band structure as soon as U is large enough. Indeed, the energy at polarization m, $E(d(m), m)$, can be computed from

$$E(d_0, m) = E(d, m) + \frac{\partial^2 E}{\partial d^2}\frac{(d - d_0)^2}{2}, \tag{3.11}$$

where d_0 is the double occupancy at zero polarization, and where we have expanded $E(d, m)$ around its minimum $d(m)$. Close to the localization threshold, $d_0 \approx 0$ and $E(d_0, m)$, which depends on the band structure through $\varepsilon_0(m)$, is proportional to d_0 (eqs. (3.9) and (3.10)). On the other hand, the leading-order contribution to the second term on the right-hand side of eq. (3.10) varies as m^4, since $d - d_0$ is proportional to m^2 (see fig. 4a), and does not critically depend on d_0. Hence, the m^4 term in $E(d, m)$ will be negative and independent of d_0, while the m^2 term will be proportional to

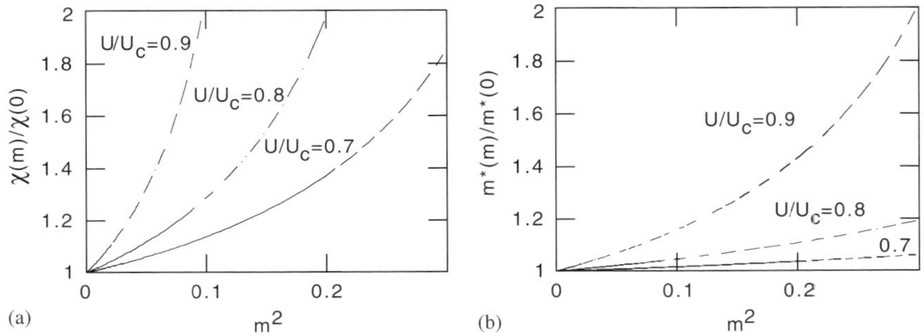

Fig. 6. Initial polarization dependence, for three values of U/U_c, of (a) the differential susceptibility χ, normalized by the low-field susceptibility; (b) the specific heat, normalized by the low-field specific heat. The curves are dashed beyond the metamagnetic transition. The dependence becomes more pronounced with the interaction. For the specific heat, this mainly originates in the larger effective mass at zero polarization.

d_0. This implies that the susceptibility (the inverse of the second derivative of the energy) increases with polarization, the increase being stronger as the localization is approached ($d_0 \to 0$), as illustrated in fig. 6a.

Figure 6b shows the simultaneous enhancement of effective mass as a function of polarization. Its increase is much larger than the corresponding decrease of the density of states at the Fermi energy,[7] which implies that the specific heat also increases with polarization. The initial linear increase of m^* with m^2 is connected to $q \propto d$ (eq. (3.9)) close to localization, and d decreasing as m^2. As $q \approx 8d + \mathrm{O}(m^2)$ for $d_0 \to 0$, the slope of $q(m^2)$ does not critically depend on U (see fig. 4). But, as $q(0)$ vanishes for $U \to U_c$, the *relative* effect of polarization on specific heat $(\partial(m^*(m)/m^*(0))/\partial m^2)$ strongly increases with the interaction U. Note that, unlike the increase of susceptibility, which would occur even if d would *increase* with m^2 (as only $(d - d_0)^2$ enters eq. (3.11)), the increase of specific heat directly reflects the decrease of double occupancy as the polarization is increased. However, this relationship between the specific heat and double occupancy only holds within Vollhardt's model. As we shall see in Section 8, a simultaneous decrease with polarization of double occupancy and of specific heat is possible within the model of Vollhardt, Wölfle and Anderson (VWA).

3.4. *Experimental tests*

The previous sections have underlined the importance of carrying out experiments at finite spin polarization for elucidating the behavior of liquid ^{3}He. At low temperature, they provide a good test for the models. In particular, measurements of the heat capacity and the magnetic susceptibility could decide between the 'nearly localized' model, the 'nearly ferromagnetic' model, or the 'nearly metamagnetic' model. On the other hand, at high temperature, such experiments can directly probe the effect of Fermi statistics in the system.

However, to be conclusive, the experiments have to be performed at very high polarization: for example, in the paramagnon model, the expected change in heat capacity at 27 bar is only 0.2% between 0 and 10 T, and the susceptibility variation is −0.3% over this range of magnetic field. In the 'nearly localized' model, the onset of the metamagnetic transition occurs at $B \approx 35$ T, and the spinodal point is at $B \approx 75$ T. Such a large (static)

[7] The density of states is proportional to the average of the Fermi wave vectors $p_\uparrow$ and $p_\downarrow$, which varies as $1/2((1+m)^{1/3} + (1-m)^{1/3}) \approx (1 - m^2/9 + \cdots)$, and decreases by only 3% between $m = 0$ and 0.4.

magnetic field is not presently accessible. The next section describes a way around this problem with the use of non-equilibrium methods.

4. Production of highly polarized degenerate liquid ^{3}He

Back in 1980, when interest in polarized liquid ^{3}He began to surge, obtaining a strongly magnetized sample was an experimental challenge in itself. It is therefore no surprise that the experimental results have been determined by progress in polarization production techniques. In this section, we give a short review of the existing polarization methods. We then focus on the technique of rapid melting of a polarized solid, emphasizing the experimental constraints associated with its use. We discuss the latest development of this method, which consists in melting the solid inside a silver sinter. Owing to this technique, we were able to obtain a strongly polarized liquid a wide range of pressures and temperatures, which was the key to the experiments presented in this section.

4.1. *Review of polarization techniques*

Due to the small ^{3}He magnetic moment ($\mu_3/k_B = 0.778$ mK/T) and the comparatively large 'magnetic Fermi temperature' ($T_{SF} \approx 160$–200 mK), the Pauli susceptibility of liquid ^{3}He is small. For example, a static field of 10 T produces a polarization $m = \frac{3}{2}\frac{\mu_3}{k_B T_{SF}}$ of only $m = 0.04$. Therefore, very large polarizations can be achieved only in situations where the magnetization is not in equilibrium with the applied field.

Non-equilibrium methods rely on the fact that the time constant of the magnetization relaxation is large. In bulk liquid ^{3}He, magnetization relaxation occurs via the weak dipole–dipole interaction. T_1 varies as $1/T^2$ at low temperature, and reaches 4000 s at 0.05 K (Rodrigues and Vermeulen 1995, Van Steenbergen et al. 1998). Practically, the relaxation of magnetization occurs mainly on the walls of the experimental cell. Fortunately, this extrinsic relaxation process, which depends on the substrate as well as the interface between the bulk ^{3}He and the walls, is slow at high field (Van Steenbergen et al. 1998, Bravin et al. 1998). The magnetic relaxation time being much longer than all other microscopic relaxation times, polarization is a slow variable, allowing meaningful measurements of (equilibrium) thermodynamic and transport properties as a function of polarization, rather than of magnetic field.

One of the polarization methods proposed originally is based on optical pumping (Lhuillier and Laloë 1979). Optical pumping provides up to $m = 0.85$ polarized ^{3}He gas, the condensation of which yields transiently spin-polarized liquid. However, the time required to liquefy the sample is comparable to the intrinsic T_1 in the liquid phase, and a significant fraction of the polarization is lost during this stage. With a single shot method, the Paris group obtained $m = 0.44$ at 0.4 K (Tastevin et al. 1988) and $m = 0.30$ at 1 K (Tastevin 1992) in pure ^{3}He. In a later development, the magnetization relaxation was partly overcome by circulating spin-polarized ^{3}He atoms in a U-shaped cell from a gas buffer (Candela et al. 1994). This technique achieved a steady-state enhanced polarization of $m = 0.56$ at 0.2 K in ^{3}He–^{4}He mixtures. So far, the base temperature has proved to be the main limitation of the technique, and, unless the sample can be cooled down more efficiently, the degenerate regime in liquid ^{3}He seems out of reach of optical pumping-based methods.

Another out-of-equilibrium technique is 'dynamical polarization,' where the polarization is transferred from a RF-pumped paramagnetic substrate to the liquid (Saito et al. 1985, Schuhl et al. 1987), but the polarizations achieved are very small. A recent and efficient method is 'spin distillation' (Nacher et al. 1991, Rodrigues and Vermeulen 1997): dilution of ^{3}He in ^{4}He is a well-known cooling technique. In the presence of a magnetic field, if the dilution process takes place on a timescale shorter than T_1, the total polarization of the sample is conserved. At $P > 2.6$ bar, the magnetic susceptibility of the concentrated phase is larger than that of the saturated dilute phase. Consequently, a fast dilution process increases the polarization of the concentrated phase and decreases the temperature. This method has been implemented successfully in a closed-cycle scheme by Rodrigues and Vermeulen using a Leiden-type refrigerator (Rodrigues and Vermeulen 1997). They have produced a steady-state enhanced polarization of the ^{3}He-rich phase, seven times larger than the equilibrium value. Up to now, only moderate polarizations have been obtained ($m = 0.15$ at 6 T), but the major advantage of the technique is that it makes it possible to carry out experiments at low temperature ($T \approx 15$ mK), allowing investigation of the possible existence of a zero-temperature damping of spin waves (Roni and Vermeulen 2000).

The only method currently available to produce highly polarized degenerate liquid ^{3}He is the technique of rapid melting of a magnetized solid. Before presenting it in detail in the next section, we emphasize that a common question to these various polarization techniques is whether the results obtained, at a given polarization, depend only on the effective field (which is the equilibrium field corresponding to this polarization), or whether the applied field B_a also plays a role. For thermodynamic measurements, the applied field only comes in through the Zeeman energy. Thus, the polarization dependence of the free energy will depend on B_a, but in a known way (see Section 5.2), while the specific heat at constant polarization will not

(Section 8). For non-equilibrium properties, the situation is more subtle. The difference between the Fermi levels for up and down spins, that is the difference between the effective and applied fields, provides an extra phase space for non-spin conserving collisions (Castaing and Nozières 1979). This implies for example that, for non-zero polarization, the longitudinal spin relaxation rate remains finite at zero temperature and zero applied field, so that there is a fundamental difference between the out-of-equilibrium situation and the equilibrium one in this case (Nelson and Mullin 1994). However, we do not expect this extra phase space to be important for the transport properties (viscosity, thermal conductivity) described in this paper, which mainly involve spin-conserving collisions. This means that only the polarization (or the effective field), and not the applied field, will determine these quantities.

4.2. *Rapid melting of a polarized solid ^{3}He*

The 'rapid melting' technique, originally proposed by Castaing and Nozières (1979), provides transiently a high polarization in liquid ^{3}He at low temperature. It relies on the large magnetic susceptibility of solid ^{3}He with respect to that of the liquid phase.

Solid ^{3}He behaves as a Curie paramagnet down to approximately 20 mK. Below this temperature, due to quantum exchange between the spins, deviations occur, which eventually lead to magnetic ordering (at $T \approx 3$ mK for $B = 7$ T (Godfrin et al. 1980)). For 5 mK $< T <$ 20 mK, the magnetization of the solid can be inferred from the depression of the melting curve. Measurements by Kranenburg et al. (1987) up to 9.3 T have shown good agreement with the magnetization computed in a mean-field approximation by Stipdonk et al. (1985). For our experimental field ($B = 11$ T), the solid polarization computed from this approximation varies from $m = 0.82$ to 0.58 between 5 and 10 mK, corresponding to an effective Néel temperature of 2.7 mK. This Néel temperature also correctly accounts for the polarization inferred from the depression of the melting curve measured by Yawata (2001) from 5 up to 100 mK and up to 25 T. If a magnetized crystal is melted in a time shorter than T_1, the magnetization is conserved so that a strong magnetization can be initially achieved in the liquid. However, the temperature of polarized liquid produced by rapid melting turns out to be high. This is due to two effects:

- The melting process increases the temperature of the sample.
- After melting, the spin polarization decays irreversibly toward its equilibrium value, which releases heat.

Concerning the first effect, even reversible adiabatic melting of ^{3}He increases the temperature because (at a given temperature) the entropy of the solid phase is larger than the entropy of the liquid phase. This is the reverse Pomeranchuk effect. For instance, starting with a solid at $T = 5\,\text{mK}$ and $m = 0.80$ yields a final temperature of $T = 100\,\text{mK}$. Moreover, in a rapid melting experiment, the melting is generally carried out irreversibly, so the final temperature is higher: overall, the heat released is ≈ 0.9 J/mol of ^{3}He liquid at 27 bar, which would increase the temperature up to ≈ 400 mK under adiabatic conditions (Buu et al. 1998b).

Concerning the second effect, the power released during the magnetization relaxation is, at low temperature and per unit volume, $\dot{Q}_{\text{spins}} = B\dot{M}$, B being the magnetic field one should apply at equilibrium to obtain the total magnetization M in liquid ^{3}He (see Section 5.5.2 below for details). Assuming the polarization to be linear in field, this expression can be rewritten (see Section 6.3.3) as

$$\dot{Q}_{\text{spins}} = \mu_0 \frac{M_{\text{sat}}^2}{\chi} m\dot{m}, \tag{4.1}$$

χ being the magnetic susceptibility of liquid ^{3}He, M_{sat} the saturation magnetization per unit volume and μ_0 the vacuum permeability. For an initial polarization of $m = 0.7$, the total energy released amounts to 0.4 J/mol at 27 bar. In an adiabatic cell, this would increase the temperature further from 400 to 550 mK.

In order to obtain polarized liquid ^{3}He at lower temperature, one must cool the warm sample within a time shorter than T_1. This is the topic of the next section.

4.3. *Cooling polarized liquid ^{3}He*

Cooling polarized liquid ^{3}He first requires an efficient low-temperature thermal reservoir, due to the very large specific heat of the liquid. Furthermore, liquid ^{3}He has to be efficiently coupled to the reservoir, first, to ensure a thermalization time shorter than the magnetic relaxation time and second, to limit the temperature gradients generated by the power released by the magnetic relaxation. In the following, we analyze these problems in an idealized situation, before describing two solutions actually used.

4.3.1. *Thermal coupling of ^{3}He*

Postponing the description of the thermal reservoir to the next section, we first illustrate the importance of a good thermal coupling by considering the simple situation of a warm sample of bulk liquid ^{3}He enclosed between two parallel walls heat sunk to the cold reservoir. If the Kapitza resistance is negligible, the characteristic time required to cool down the sample is given by $\tau = C_3/\kappa_3 L^2/(4\pi^2)$, with C_3 the heat capacity of ^{3}He per unit of volume, κ_3 the thermal conductivity, and L the spacing between the walls. Taking $L = 4$ mm and values from the literature for the properties of liquid ^{3}He at 27 bar and 0.1 K, we find $\tau \approx 8$ s. For bulk liquid, this is much smaller than T_1. From this standpoint, the thermal conductivity of the liquid is not a severe limitation.

The real problem is the thermal homogeneity of the sample, a basic requirement in most experiments. Large thermal gradients occur during the cooling, but their lifetime is also τ, i.e. they only affect the initial (i.e. the largest) polarizations. This is to be contrasted to the thermal gradients caused by the irreversible magnetization decay, the lifetime of which is $\approx T_1/2$, i.e. covers the whole experiment. As $\tau \ll T_1/2$, the temperature difference between the center of the cell and the walls can be computed from eq. (4.1) and the thermal conductivity κ_3. It can be expressed as a function of the same τ as above as $\Delta T = \mu_0(\pi M_{\mathrm{sat}})^2/(2\chi C_3)(\tau/T_1)m^2$. With a typical value $T_1 \approx 300$ s in a bulk rapid melting experiment (Kranenburg et al. 1989), the temperature difference at the beginning of the magnetic relaxation with $m = 0.7$ initial polarization is $\Delta T \approx 50$ mK. Thus, the thermal gradients caused by the polarization decay are large enough to bias measurements. This remark is particularly true for viscosity measurements, since the viscosity of ^{3}He decreases by a factor 2 between 100 and 150 mK.

Here, we assumed the thermal coupling to be controlled by the weak thermal conductivity of bulk liquid ^{3}He only. At low temperature, the thermal conductivity improves, but the Kapitza resistance cannot be neglected any longer. Although the temperature gradients are then smaller, it becomes impossible to cool down the ^{3}He sample on a timescale shorter than T_1 using the simple geometry above. The next paragraph describes two solutions to overcome these problems.

4.3.2. *Strategies*

One method to produce polarized liquid ^{3}He at low temperature is to start with a partially solidified sample (Kranenburg et al. 1989). The initial liquid fraction is then used as the cold reservoir and is directly coupled to the polarized liquid after the melting. This suppresses the problem of Kapitza resistance. The drawback of this method is that the initial polarization gets spread out into the cold source. So, there is a trade off between the polarization available and the temperature.

The other approach, which we used for the experiments presented in this section, consists in confining the ^{3}He sample in a silver sinter, itself thermally connected to a large heat tank (Puech et al. 1989). The sinter pore size is small enough ($\approx$700 Å) to allow fast heat diffusion to the silver grains, so the temperature remains uniform within the pores. Moreover, this material has a large specific area ($\approx 6\ \text{m}^2/\text{cc}$), so that the Kapitza resistance is negligible above 80 mK. Owing to the thermal conductivity of silver, the ^{3}He 'effective' thermal diffusivity is enhanced by a factor of about 20 when confined in a sinter with respect to the bulk. The large specific area increases the magnetization relaxation rate, but, fortunately, the characteristic time is long enough for the experiments to be carried out ($T_1 = 30$–100 s; see Appendix A). If we consider an experimental cell of ^{3}He confined in silver sinter and go through the analysis of the last paragraph again, we find that the cooling time drops to $\tau = 0.4$ s ($\ll T_1$) and that the temperature difference between the center and the walls at the beginning of the experiment is only $\Delta T = 7$ mK (for $T_1 = 80$ s), a much more reasonable number.

4.3.3. *Polarization homogeneity*

Finally, one may be worried about the homogeneity of the magnetization during rapid melting experiments. In the method of 'dilution,' we expect large polarization gradients to set in because of the spin diffusion from the polarized liquid ^{3}He to the cold non-polarized liquid. In a silver sinter, the spin diffusion inside the pores is very fast ($\tau_{\text{spin}} \approx 4\ \mu\text{s}$) compared to T_1, so that the magnetization is uniform within the pores. Thus, assuming that the initial magnetization after the melting is the same in each pore, we expect no macroscopic magnetization gradients. The interface between the silver sinter and the liquid is formed by a few dense ^{3}He solid layers adsorbed on the walls, in which the magnetization relaxation is supposed to take place (Hammel and Richardson 1984). It has to be emphasized that this fraction of the sample is only of the order of a few percent of the total number of ^{3}He atoms, so the macroscopic magnetization measured in a rapid melting experiment is very close to the magnetization of the liquid.

5. Magnetic susceptibility

5.1. *Pre-1990 situation*

In their landmark paper, Castaing and Nozières (1979) suggested that the measurement of the magnetic susceptibility at high polarization would settle the debate between the 'nearly ferromagnetic' model and the 'nearly localized' model. Following their suggestion, various groups undertook experiments using the rapid melting method. The aim of the first experiment was to measure the melting pressure of ^{3}He at high polarization and to extract the susceptibility by using the Clausius–Clapeyron relation (for a review of these experiments, see Bonfait et al. (1990) and references therein).

The interpretation of these experiments was complicated, since the magnetization signal measured by NMR had contributions from both the liquid and the solid phases. In one of these experiments, the authors concluded to the possible existence of a peak in the magnetic susceptibility (Bonfait et al. 1984). This conclusion was in qualitative agreement with the predictions of the 'nearly localized' model and the 'nearly metamagnetic' model, although the peak was detected at a lower polarization ($m \approx 0.2$ at 100 mK) than predicted by the models. Subsequent sound–velocity measurements that were carried out to check this result failed to detect the anomaly (Bonfait et al. 1987). However, the authors argued the sound–velocity experiments could not exclude the presence of metamagnetism in liquid ^{3}He, since, first, they were carried out at high temperature (200 mK), second, they measured the sound velocity at constant magnetization, and not at constant magnetic field, the two velocities being expected to behave differently at a metamagnetic transition (Puech 1987).

The aim of the experiment we describe next was to settle the question by providing a direct measure of the magnetization curve (the magnetization as a function of the magnetic field) in a 100% liquid sample (Wiegers et al. 1991, Bravin et al. 1993, 1994). This experiment, which requires a high polarization, a low temperature, and a short thermal relaxation time, crucially relies on the technique of rapid melting in a silver sinter we described in Section 4.

5.2. *Method*

In a rapid melting experiment, the magnetization is out-of-equilibrium during its relaxation, and therefore bears no relation with the applied magnetic field. The method to measure the magnetization curve in this situation relies on the fact that the magnetization decay time T_1 is long compared to all other microscopic equilibrium times. Consequently, we can still apply equilibrium thermodynamics to the system, the magnetization being, at each time of the relaxation, a constrained quantity (Castaing and Nozières 1979). As we will show, this allows one to determine the magnetization curve from the simultaneous measurements of the temperature and the magnetization during an adiabatic relaxation.

At equilibrium, the applied magnetic field B is related to the magnetization M through

$$B = \left(\frac{\partial U}{\partial M}\right)_S, \tag{5.1}$$

U and S being respectively the internal energy and the entropy of the system. During a rapid melting experiment, the magnetization is out-of-equilibrium, but, on a timescale short compared to T_1, the internal energy is still a function of the instantaneous values of M and S. Now, eq. (5.1) does not give the applied field anymore, but the *effective* magnetic field instead, that is the field one would need to apply to obtain the magnetization M in an equilibrium situation (Castaing and Nozières 1979).

In the presence of an applied magnetic field B_a, the thermodynamic potential, $U - MB_a$, is conserved during an adiabatic relaxation. This condition can be expressed as

$$T\,\mathrm{d}S + (B - B_a)\,\mathrm{d}M = 0. \tag{5.2}$$

This equation shows that the ^{3}He magnetic energy is converted into heat during the magnetization decay. A fraction of the heat released by the relaxation increases the temperature of the system, while the remaining part is absorbed by the magnetization dependence of the entropy:

$$T\,\mathrm{d}S = C_m\,\mathrm{d}T + T\left(\frac{\partial S}{\partial M}\right)_T \mathrm{d}M, \tag{5.3}$$

C_m being the heat capacity at constant magnetization. Using Maxwell's relation $(\partial S/\partial M)_T = -(\partial B/\partial T)_M$, we rewrite eq. (5.2) as

$$C_m\,\frac{\mathrm{d}T}{\mathrm{d}t} = -(B_T - B_a)\,\frac{\mathrm{d}M}{\mathrm{d}t} \tag{5.4}$$

with $B_T = B - T(\partial B/\partial T)_M$.

Equation (5.4) is the cornerstone of the method. It shows that the relation between the magnetization and the effective field can be obtained from the simultaneous measurement of T and M during an adiabatic polarization relaxation, provided that the relaxation is slow enough for these quantities to be uniform across the cell. The effective field measured by this method contains a correction term $(B_T - B)$. At low temperature, when $S \propto T$, this correction is of order T^2, hence small.

5.3. *Experimental cell*

Our experimental setup, shown in fig. 7, was designed to fulfill the conditions required for the measurement of the magnetization curve, that is, good thermal insulation during the experiment and a short internal thermal time constant. It consists of the experimental compartment containing the ^{3}He sample, a separate chamber containing a large amount of unpolarized ^{3}He acting as a heat tank (HT), and a heat switch, which connects these elements to the mixing chamber of a dilution refrigerator (see fig. 7).

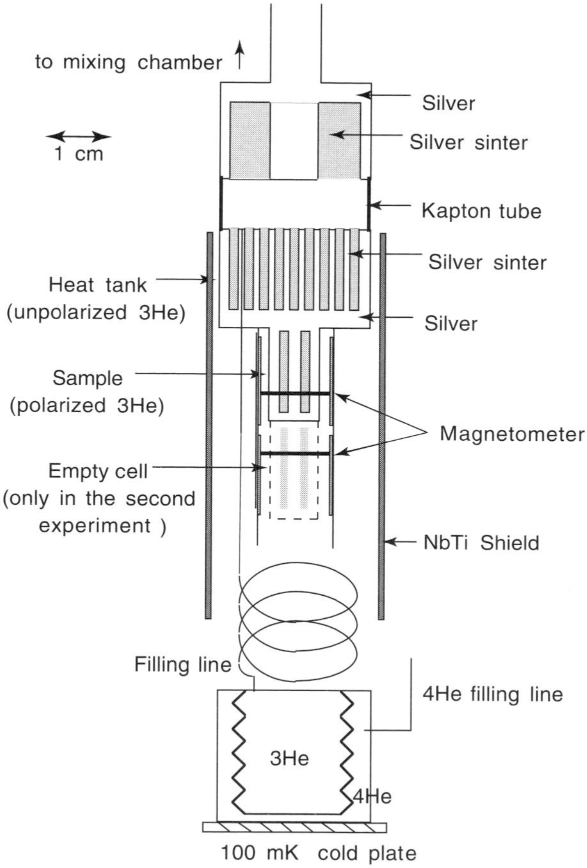

Fig. 7. Experimental setup

In order to maximize the thermal coupling of the HT and the sample, both the experimental compartment and the HT chamber are machined in a single piece of high-purity (4N) silver, which is then annealed to increase its thermal conductivity. The lower part (the cell itself) contains the experimental ^{3}He in four sintered[8] compartments, 2.5 mm in diameter and

[8]The silver powder used is of type *Ulvac* 700 Å from Inabata (Japan). The powder is packed at 45% of its initial volume, and baked at 200°C for 1 h under 1 bar of argon with a small amount of hydrogen.

20 mm in length (amounting to 8.1×10^{-3} mol of ^{3}He at 26 bar). These compartments are connected by pairs, each pair being filled by a separate filling line. The upper part (the HT) contains 77×10^{-3} mol of unpolarized ^{3}He at saturated vapor pressure in 50 similar sintered compartments, which provides a large heat capacity (about eight times larger than that of the cell at 80 mK). The reason for using small diameter compartments is that the thermal conductivity of the sinter is typically two orders of magnitude smaller than that of bulk silver. Owing to this design, the thermal equilibration time of the whole cell is only ≈ 1 s.

Because a conventional superconducting heat switch would not work in the large magnetic field required for the experiment, we use instead ^{3}He as the switching element (Wiegers et al. 1990). The upper end of the heat switch is a silver sinter anchored to the mixing chamber of the dilution refrigerator. It is connected to the HT through a 0.2 mm-thick kapton tube, both sinters being open to the volume enclosed. In the closed state, the 2.5 mm gap between the sinters is filled with ^{3}He. The level of ^{3}He inside the kapton tube can be varied by a bellow actuated by a separate ^{4}He circuit. The opening of the heat switch leaves the two sinters filled with ^{3}He (the lower one because the inlet tube is level with its upper end, the upper one owing to capillarity). In the open state, the heat leak between the cell at 90 mK and the mixing chamber at 5 mK is then only 10 nW (mainly through the kapton), that is 10^{-4}–10^{-2} of the power released during the magnetization relaxation.

The temperature is measured by thinned Speer carbon thermometers glued below the HT and the cell. A heater is also glued to the bottom of the heat tank for calibration purposes. The polarization is measured by a high-field SQUID magnetometer (see Bravin et al. 1992 for details). The pick-up loops are wound on a mandrel to form a gradiometer. The mandrel is attached to the top part of the heat switch, avoiding any thermal or mechanical contact with the cell. A superconducting NbTi shield is used to stabilize the applied magnetic field in the gradiometer region. Unlike NMR, which could not be used due to the silver walls, the SQUID detection is not selective, so great care has been taken to minimize any background magnetic contribution. In particular, the inlet capillary of the cell was made out of silver rather than CuNi.

The first cell build on this principle was used for the experiments already published. Recently, we repeated these experiments with a second cell, which corrected some of the problems encountered with the first one. In particular, the second set up was fitted with a dummy cell at its lower end (sitting in the middle of the second pick-up loop of the gradiometer) to cancel the magnetic signal coming from the silver walls. This provided a check of the validity of our original conclusions.

5.4. *Rapid melting experiments*

The ^{3}He solid is prepared by the 'blocked capillary' technique. The loading density (at ≈ 1.2 K) is chosen such that the sample pressure lies just above the melting curve, once at low temperature (≈ 34.4 bar). The solid is cooled for 24–48 h down to 5–6 mK under a 11 T magnetic field yielding a polarization $m \approx 0.8$. After opening the heat switch, which takes about 5 min, the melting of the magnetized solid is triggered by decompressing the sample to a dead volume at room temperature. The minimum in the melting curve limits the final pressure to values smaller than 29 bar. Most of the data we present have been taken at 26 bar.

The signal recorded during a rapid melting experiment using the first cell is shown in fig. 8. The melting process occurs separately in the two pairs of compartments, due to different flow impedances of the filling lines. In total, it lasts for 5 s and increases the cell temperature to 75 mK (fig. 8a). At the same time, the magnetization drops sharply, due to the decrease of ^{3}He density caused by the melting. The melting is followed by a quasi-exponential decay of the polarization with a time constant $T_1 \approx 70$ s (fig. 8b). Associated with this decay, the heat released by the magnetic relaxation causes the temperature to increase from 75 to 95 mK, corresponding to a total energy of ≈ 3 mJ. Because of the finite thermal conductivity of the silver, this also generates a small temperature difference between the cell and the HT on the same timescale. For longer times (beyond 100 s), a long-lived contribution to this temperature difference is also observed. We think that this contribution is due to overcooling of the thermometers by nuclear demagnetization of the silver paint used for the connections following the rapid temperature increase after the melting for the following reasons: first,

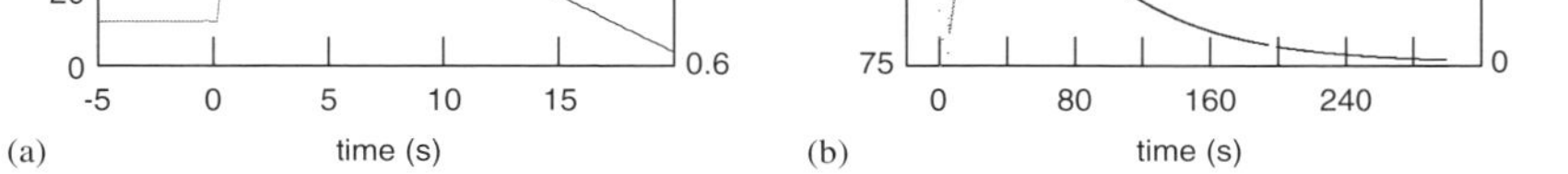

Fig. 8. Thermal and magnetic signals recorded; (a) during the first 20 s after the rapid melting, and (b) over a longer timescale. Both the HT and the cell temperatures are shown. The ^{3}He polarization is obtained from the raw signal by subtracting the cell background contribution as described in the text. The thermal effect of the magnetic relaxation is obvious.

this effect is not specific to the polarized liquid, but occurs as well following a fast heating of the cell filled with liquid at magnetic equilibrium up to 100 mK, starting from below 10 mK. Second, it is not observed when working at low field. Finally, it is not observed either with the second cell (fig. 10), where we used gold paint instead of silver paint to connect the thermometers.

In principle, the data shown in fig. 8 are all that is needed to obtain the magnetization curve of liquid ^{3}He. In fact, both the magnetization signal and the temperature signal have to be corrected for parasitic effects.

5.5. *Analysis*

5.5.1. *Magnetization signal*

As a SQUID magnetometer only measures changes of magnetization, a baseline has to be subtracted from its raw signal. In our experiment, this baseline is linear in time, due to the non-perfect (1%) shielding of the slow drift (5 G/h) of the applied 11 T magnetic field. This baseline, which, in terms of polarization, corresponds to $\delta m \approx 0.2\%$/h, is determined using times between 1000 and 2000 s after the melting.

Curve (A) of fig. 9a shows the signal measured during a rapid melting experiment. The signal, which has been corrected for the baseline, is expressed directly in terms of ^{3}He polarization. A fraction of this signal is not due to liquid ^{3}He, but to the relaxation of other paramagnetic moments in the walls of the cell, following the increase of the silver electronic temperature from <10 to ≈ 100 mK. This is demonstrated by mimicking the temperature step arising from the melting of the solid by a heat pulse applied when the cell filled with liquid ^{3}He at magnetic equilibrium and 26 bar, and starting from the same initial temperature. As shown by fig. 9a, the corresponding background signal represents about 10% of the rapid melting experiment signal. Half of it is due to the nuclear spins of silver (the magnetic moment of these is 18 times less than that of the ^{3}He) and relaxes with the time constant given by the Korringa law (≈ 100 s at 100 mK). The remaining contribution is partly due to the solid ^{3}He signal discussed below, and partly of unknown origin.

Curve (C) of fig. 9a shows the signal measured after rapid melting corrected for the parasitic contribution coming from the walls of the cell. From $m \approx 0.40$ to 0.02, the relaxation is exponential. The slower decrease at larger polarizations is discussed in Appendix A.2. With our second cell, which included an additional silver extension to cancel the background wall signal, the parasitic contribution is only 2% of the total signal (see fig. 9b). The corrected magnetization signal is nearly identical to that obtained with the

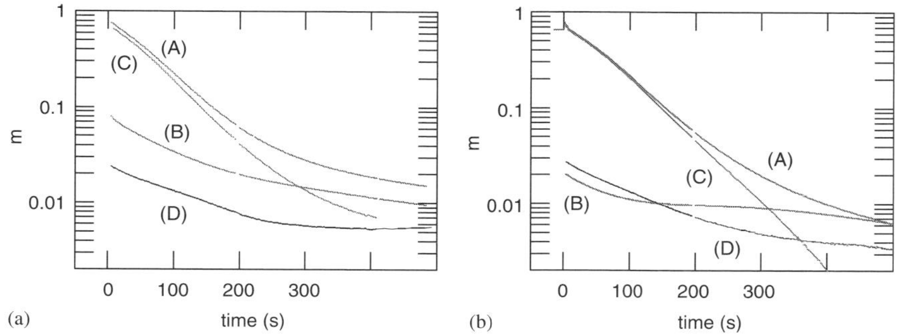

Fig. 9. Magnetic signals for the first (a) and the second (b) cells, recorded during: (A) a rapid melting experiment, (B) a heat pulse experiment at the same 11 T field. The difference of the two signals (C) is the magnetic contribution of the polarized liquid, (D) is the contribution arising from the ^{3}He solid layer measured during a heat pulse experiment (see text).

first cell, which proves the validity of the correcting procedure in the first case.

In fact, the corrected signal is not exclusively due to the magnetization of the polarized liquid ^{3}He. A small fraction of the signal originates in the solid ^{3}He adsorbed on the walls of the sinter. To single out this contribution, we have repeated the heat pulse experiment with an empty cell. The signal difference between the two heat pulse experiments (cell filled with liquid at equilibrium magnetization or empty) is the curve (D) of figs. 9a and b. This signal has a time dependence similar to that measured with the cell filled with solid ^{3}He (Buu et al. 1998b), but an amplitude about 30 times smaller, i.e. about 3%. This supports our assumption that it is mainly[9] due to the adsorbed solid ^{3}He. The magnitude of the effect corresponds to about three layers of solid ^{3}He (obeying the Curie law); in agreement with Fukushima et al. (1992) if we take the same specific area of the sinter ($\approx 20\ m^2/cm^3$) as these authors do (rather than our measured $\approx 10\ m^2/cm^3$). For this contribution, the correction procedure described above will be partly incorrect, as the relaxation of the solid layer does depend on the liquid polarization (see Appendix A.2). However, the resulting error in m will be less than 3% anyway.

5.5.2. *Power released*

The silver demagnetization not only contributes to the magnetic signal, but also absorbs heat during a rapid melting experiment. The whole silver HT

[9]The contribution of the bulk liquid, computed from Ramm et al. (1970), amounts to only 0.5%.

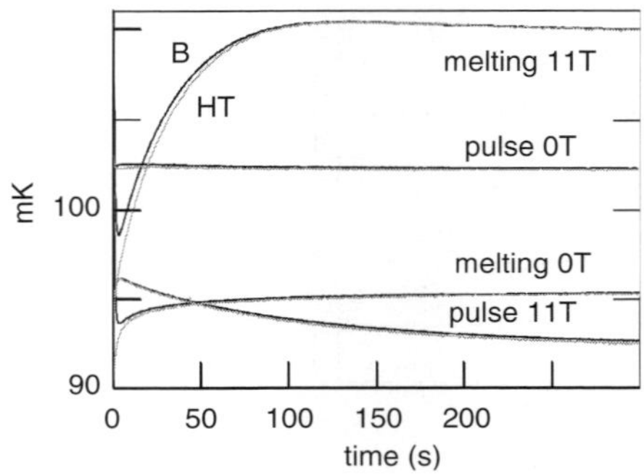

Fig. 10. Temperature evolution of the second cell, for the rapid melting experiments and the pulse experiments in 11 and 0 T. The gray lines correspond to the temperature of the HT, the black ones to that of the cell(B). The time origin is taken at the maximal temperature of the cell thermometer. The small temperature difference visible during the relaxation of magnetization of ^{3}He is due to the finite conductivity of the silver.

contributes to this parasitic (negative) power, in contrast to the case of the magnetic signal which, due to the gradiometer scheme, is only sensitive to the silver cell. This effect is corrected for by subtracting from the measured heating power in the rapid melting experiment the cooling power measured during the pulse experiment of fig. 10. As shown in fig. 11, this contribution amounts initially to about 10% of the heating power due to the ^{3}He relaxation. For the first cell, this subtraction has further advantage that it eliminates the effect of the long-lived difference between the two thermometers discussed in Section 5.4.

In fact, a small fraction of the cooling power recorded in the pulse experiment comes from a heat leak to the ^{3}He liquid in the filling line, which is poorly coupled to the cell and stays cold after the application of the heat pulse. This contribution is separately measured by repeating the pulse experiment with no applied field (fig. 10), so that the silver spins do not contribute to the cooling power. The corresponding power is subtracted (in algebraic value) from that of the high-field pulse experiment, or, equivalently, added to the power measured in a rapid melting experiment.

A third parasitic contribution to the thermal signal has the opposite effect of the cold filling line: the liquid produced by the melting, which is connected to the filling line stays warm (300–400 mK). The power leaking to the cell from this warm liquid, which has to be subtracted from the raw data, is determined in a rapid melting experiment carried out with no applied field (fig. 10). In absolute value, it is comparable to that of the silver contribution (fig. 11).

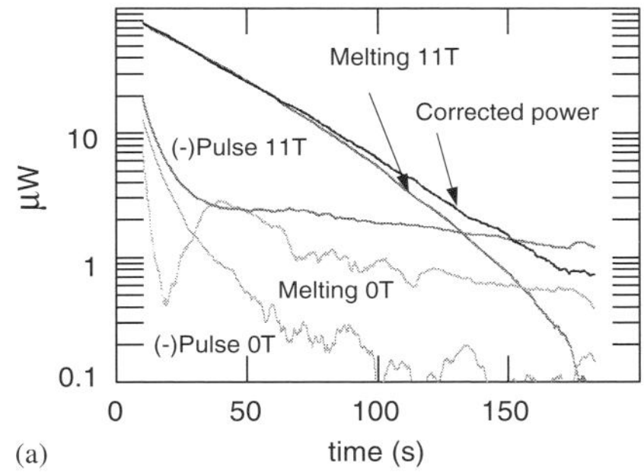

(a)

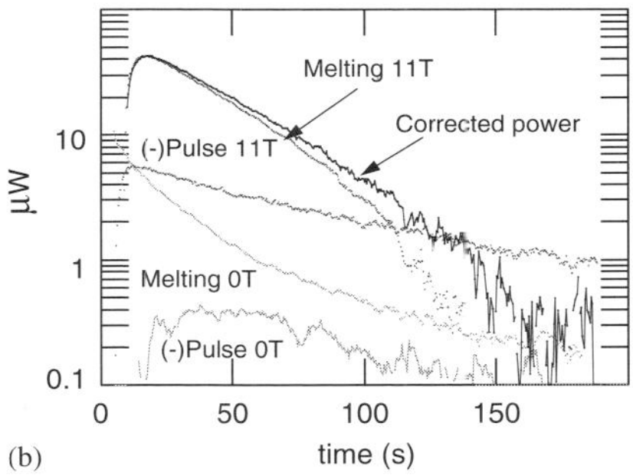

(b)

Fig. 11. Comparison for the first (a) and the second (b) cells of the power released during the magnetic relaxation and the various corrections: power leaked into the cell from the 'hot' ^{3}He in the filling line measured during a rapid melting experiment at 0 T; power absorbed (<0) by the silver spin relaxation measured during a heat pulse experiment carried out at 11 T; power leaking from the cell to the cold ^{3}He in the filling line (<0) measured during a heat pulse experiment at 0 T. The time origin is taken at the maximal temperature of the cell thermometer for each experiment. In figures (a) and (b), the raw powers are measured with the HT and the cell thermometers, respectively. The corrected power is computed using both thermometers.

To summarize, three experiments are necessary to correct the thermal signal, the results of which are compared for the two cells in fig. 11. The accuracy of the silver thermal correction has been checked in two test experiments. In the first one, the thermal effect of the liquid magnetic relaxation was simulated by a heater (Wiegers et al. 1991). In the second one, we applied a temperature step from 5 to 80 mK with a cell filled with solid ^{3}He and monitored the following cooling by adiabatic demagnetization (Bravin et al. 1993). The correction for the power leaked by the warm ^{3}He in the capillary, which cannot be directly tested, is also more uncertain, since the heat released by the melting depends on the field. In the second cell, we reduced the amplitude of this correction by minimizing the amount of 'hot' ^{3}He close to the cell, with the use of a thin (0.15 mm inner diameter) filling line over a longer distance than in the first cell. Accordingly, the filling line corrections are smaller in this case. The fully corrected thermal signal, computed from either the HT or the cell thermometer, is also displayed in fig. 11 for the two cells. In both cases, it is roughly exponential with a time constant of $36\,\mathrm{s} \approx T_1/2$.

5.5.3. *Magnetization curve of liquid ^{3}He*

The magnetization curve is obtained from the corrected signals using relation (5.4), the time derivatives being computed numerically. The polarization scale, which is calibrated by measuring the magnetization[10] of solid ^{3}He as a function of temperature between 20 and 150 mK, is determined to better than 10%.

To measure the power released during the relaxation, the thermometers are calibrated directly against the enthalpy of the cell in order to avoid a possible inaccuracy of our temperature scale. The procedure consists in reading the value of the carbon resistors during a series of calibrated heat pulses applied to the cell filled with liquid ^{3}He at 26 bar and at equilibrium magnetization. However, two sources of error in the power measurement have to be considered: a first source of error is the temperature gradient across the double cell induced by the heat released by the magnetic relaxation. As shown by fig. 10, the silver body of the cell is slightly warmer than that of the HT, due to the finite silver thermal conductivity. Because the HT total specific heat is much larger than that of the cell, this will result in a small error provided we take T in eq. (5.4) as the HT temperature. Furthermore, ^{3}He inside the pores of the cell or of the HT is respectively warmer or colder than the silver, the temperature difference being typically of several millikelvin for the present geometry (Section 4.3.1). Thanks to the symmetrical design of the HT and of the cell, the (measured) temperature of the silver is the average temperature in the system cell–HT.[11] However, the specific heat per unit volume of the sample and the HT differ, due to their different pressures. Adapting the results of Appendix C to the present case of small diameter (2.5 mm) sintered holes, this leads to a 1% difference at most between the computed power and the actual one. A second, and more serious, source of error is the polarization dependence of the specific heat of the sample. As shown by our following measurements (see Section 8), the specific heat of liquid ^{3}He at 27 bar decreases by 30–40% at $m = 0.7$, so that the power released during the ^{3}He magnetization decay, and hence the field, is overestimated by $1/8 \times 40\% = 5\%$ at most at this polarization.

5.6. *Discussion*

Figure 12a shows the magnetization curve at 26 bar, as measured in five different sets of experiments using the first cell. The main feature is the *downward* curvature showing up at large effective field. At $m = 0.6$, the effective field B_T is reproducibly 30% higher than its extrapolation from the

[10] We assumed the Curie law to hold above 20 mK. This is not strictly correct due to the exchange interaction. Using an effective Néel temperature of 2.7 mK (consistent with the discussion of Section 4.2), would result in a ≈ 10% decrease of our polarization scale.

[11] This assumes that the Kapitza resistance does not strongly depend on pressure.

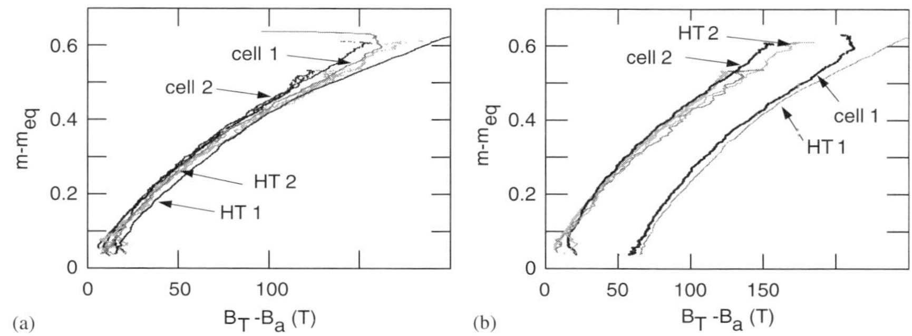

Fig. 12. The magnetization curve of liquid ^{3}He as function of $B_T - B_a$, $B_a = 11$ T being the applied field, and $B_T = B - T(\partial B/\partial T)_M$ the effective field with an entropic correction: (a) as determined in the first cell, at 26 and 10 bar. The dashed lines indicate the expected low-field behavior. (b) Comparison of the two cells at 26 bar. Measurements using the cell thermometer (thick black lines) slightly underestimate the field at a given polarization with respect to those based on the HT thermometer (thin gray lines). The curves for the first cell are shifted by +50 T for clarity. For the second cell, the HT thermometer was more noisy, which is reflected by the excess noise on the effective field. In all cases, the downward curvature is clear.

linear behavior. This difference is much larger than the error discussed above. Furthermore, this feature does not critically depend on the various corrections we have discussed in Section 5.5.2. Changing these corrections by $\pm 20\%$ preserves the downward curvature (Wiegers et al. 1991).

At low polarization, our results agree with the low-field susceptibility measured by Ramm et al. (1970). The finite temperature entropic correction can be computed from their data by taking: $B_T/B(T=0) = \chi(T=0)/\chi(T) - \chi(T=0)T\,\partial(1/\chi)/\partial T$. As the susceptibility decreases as $\propto T^2$, this factor is approximately equal to $\chi(T)/\chi(T=0)$, that is about 0.91 at 90 mK and 26 bar. This leads to an *increase* of the initial slope of the magnetization curve by a factor 1.09 with respect to its 0 K value, yielding the dashed line in fig. 12. The corrections of the thermal and magnetic signals are important to obtain this good agreement, which provides a sensitive test of their accuracy.

The magnetization curve at 26 bar, measured in three experiments with the second cell (with two different initial polarizations, corresponding to different starting temperatures), is compared in fig. 12b to the curve measured with the first cell. In each case, the magnetization curve was computed using either the HT or the cell thermometer. For the second cell, as for the first one, we find that using the cell thermometer underestimates the field at a given polarization. This is a direct consequence of the underestimation of the power released (fig. 11) due to the vertical temperature gradient (fig. 10) discussed above. With either thermometer, the magnetization curve obtained with the second cell closely follows the corresponding one obtained with the first cell, within a 5–10 T shift to lower fields. Based on the sensitivity of the magnetization curve to the choice of the thermometer, we speculate that this shift could be due to a small vertical temperature gradient along the HT in the second cell. However, the downward curvature is conspicuous in all cases. The fact that we observe the same curvature in two different cells, with different corrections (see figs. 11a and b) is a proof of its intrinsic character.

This downward curvature implies a monotonic decrease of the susceptibility with the magnetic field. This is in disagreement with the previous experiment of Bonfait et al. at 100 mK, where the observed 'plateau' of the melting pressure at $m \approx 0.2$ had been analyzed in terms of a peak of the susceptibility (Bonfait et al. 1984). In order to show that our experiment is sensitive enough to rule out such a behavior, we compare in fig. 13a the data of (Bonfait et al. 1984) to the pressure lowering predicted from our data. The pressure lowering is computed from the Clausius–Clapeyron relation

$$\left(\frac{\partial P}{\partial B}\right)_T = \frac{(M_L - M_S)}{(V_L - V_S)}, \tag{5.5}$$

where M_L (M_S) and V_L (V_S) are the molar volume and magnetization of the liquid (solid). Integrating over B, and using the Curie law for the solid,

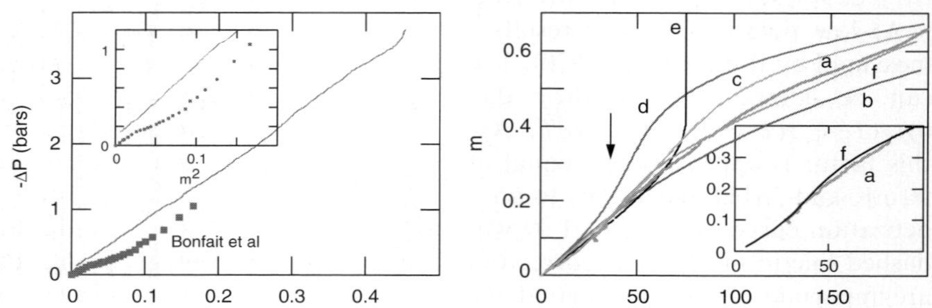

Fig. 13. (a) Polarization dependence of the melting pressure. The points are values of $P(0) - P(m)$ measured by Bonfait ($T = 100$ mK). The curve (vertically shifted for clarity) is predicted from our measurements of the magnetization curve ($T = 90$ mK), using the Clausius–Clapeyron relation. The precision of our measurements allows us to rule out the plateau observed by Bonfait. (b) Absolute polarization m in liquid ^{3}He as function of $B_T = B - T(\partial B/\partial T)_M$, the effective field with an entropic correction. (a) experimental curve (black); (b) Stoner model with $S = 21$; (c) 'nearly metamagnetic' model at $T = 0$; (d) 'nearly metamagnetic' model at $T = 80$ mK; (e) 'nearly localized' model with $U/U_c = 0.9$; the arrow marks the limit of stability of the paramagnetic phase with respect to the fully polarized phase in the 'nearly localized' model; (f) density functional at 0 K. The inset is a blown-up view of (a) and (f).

allows one to express the pressure $P(m)$ as a function of $B(m)$ and $\int_0^m B(m)\,dm$. The effective field $B(m) = B_T(m) - T(\partial S/\partial m)$ is computed from the measured B_T by assuming the entropy S to vary as $\propto m^2$, consistent with our measurements of the specific heat (Section 8). This gives $B(m) - B_T(m) \propto m$, the proportionality coefficient being determined from the known low-field polarization. The predicted pressure depression is linear in m^2 up to $m \approx 0.5$, and reaches 3 bar at $m \approx 0.6$. The slope is similar to that measured by Bonfait et al. at low polarization. This is expected, as the latter authors point out that their low-polarization behavior is consistent with that predicted from the low-field susceptibility of the liquid. In contrast, the plateau observed by Bonfait et al. around $m \approx 0.2$ is clearly inconsistent with our prediction. As recent investigations of the rapid melting process (Marchenkov et al. 1999) support the hypothesis Bonfait et al. made to extract the liquid polarization from the total magnetization, it suggests that the problem in these results originates in the pressure measurement itself.

In order to compare our finite temperature data to the prediction of the Stoner model at zero temperature, we normalized the field scale of the latter to reproduce the measured initial slope. As shown by fig. 13b, the result is in qualitative agreement with the measured magnetization curve. Unfortunately, it is not possible to compare our data to the (temperature dependent) prediction of the paramagnon model[12] as this prediction is restricted to $m<0.6$, and our error bars in this range are too large (Béal-Monod 1991). Moreover, the temperature is beyond the range of validity of the paramagnon model. Static field measurements at 30–50 mK up to 22 T have been undertaken to test the paramagnon model, but failed to reach a definitive conclusion (Buu et al. 1998a).

In contrast, our result is in clear disagreement with the models predicting a metamagnetic behavior for liquid ^{3}He. In particular, the 'nearly localized' model predicts a first-order transition to a fully polarized state at $m \approx 0.2$ (Vollhardt 1984). In such a case, the ^{3}He within each pore of the sinter would split into two phases at a given effective field B_T, and the transition would give rise to a Maxwell plateau. We see no sign of such a plateau. Metastability problems cannot be invoked either since our data extend far above the spinodal point ($m \approx 0.4$). It has been suggested that the absence of metamagnetism could be due to the confinement of our ^{3}He sample in the sinter. We believe that it is not the case, since the pores are large (hundreds

[12]Note that, for such a comparison, the fact that we measure B_T rather than B would have a crucial role. While, in the paramagnon model, $m(B)$ becomes less curved as the temperature increases (3.2), $m(B_T)$ becomes more curved, for a similar reason to that discussed for the nearly metamagnetic model below.

of Å) compared to any microscopic scale in the liquid. Second, as we shall see in Section 6, the viscosity of bulk ^{3}He measured over the same range of polarization does not present any anomaly.

Our result may be interpreted as evidence against the nearly localized model. At the time of these experiments, it was not clear whether the metamagnetic transition arose from the use of Gutzwiller's approximation or if it was an intrinsic feature of the model itself. More recently, Georges et al. investigated this question, using the method of dynamical mean-field (Section 2.3.2) instead of the less sophisticated Gutzwiller approximation, and found that the metamagnetic transition was still present in the 'nearly localized' model (Laloux et al. 1994). Georges and Laloux (1997) further suggested that the original model should be modified to account for the correct magnetic correlations in the system. This is achieved by including a nearest-neighbors ferromagnetic exchange term in the hamiltonian. Comparing the predictions of their modified model to the zero-field properties of ^{3}He, they concluded that liquid ^{3}He is close to *both* a localization transition (induced by the on-site repulsion) and a ferromagnetic instability (induced by the ferromagnetic exchange). For this reason, their model was named the 'Mott–Stoner' model. Even in such a model, a metamagnetic transition still occurs at $m = 0.4$ for the half-filled band. The only way to reconcile the model with the experimental behavior of the magnetic susceptibility was to assume a vacancy concentration of order 10%. The studies of Georges et al. thus suggest that a lattice model with a half-filled band is too 'rigid' to describe a liquid system such as ^{3}He. A qualitatively similar conclusion was reached by Laukamp and Vollhardt (1994) within the Vollhardt, Wölfle and Anderson model.

The other model predicting a metamagnetic behavior is the 'nearly metamagnetic' model of Bedell and Sanchez-Castro (Sanchez-Castro et al. 1989). In this phenomenological model, there is no divergence of the magnetic susceptibility, but a 'pseudo transition,' which should show up as a peak in the susceptibility at $m \approx 0.3$ at $T = 0$. This peak would give rise to an inflexion point in the curve $m(B)$ (curve (c) in fig. 13), which we do not observe. If we had measured the effective field B without the entropic correction, the absence of an anomaly in the curve $m(B)$ would have been expected in the 'nearly metamagnetic model,' since the peak of susceptibility is washed out at finite temperature (see fig. 2b). However, we measure $m(B_T)$, for which the effect should become more pronounced. This behavior simply follows from the fact that, up to second-order in T, the finite temperature correction to the effective field has its sign reversed when going from $B(m, T)$ to $B_T(m, T) = B - T\partial B/\partial T$. The consequence is an enhancement of the anomaly in the curve $m(B_T)$ at finite temperature, as shown in fig. 13. This anomaly is not consistent with our data, contradicting the main prediction of the 'nearly metamagnetic' model. It could seem surprising that such a phenomenological model could fail. In fact, as shown by eq. (3.1), the predicted peak originates from the fact that the interaction $(\tilde{F}_0^{\uparrow\uparrow} + \tilde{F}_0^{\downarrow\downarrow} - 2\tilde{F}_0^{\downarrow\uparrow}) = 2(F_0^{\mathrm{a}} + (b_1 F_0^{\uparrow\uparrow} - c_1 F_0^{\downarrow\uparrow})m^2 + \cdots)$ becomes more negative with polarization. As the unpolarized liquid interactions $F_0^{\uparrow\uparrow}$ and $F_0^{\downarrow\uparrow}$ are much larger than their difference F_0^{a} (i.e. $|F_0^{\mathrm{a}}| \ll F_0^s$), this behavior is crucially sensitive to the exact values of the small parameters $b_1 = -0.01$ and $c_1 = 0.04$ taken by Bedell. We believe that the accuracy in the determination of these parameters is too crude for the predicted behavior to be reliable. Reciprocally, our result could be used to improve the determination of these parameters. It would remain to be seen whether, with these new parameters, the model would still provide a good description of the other experimental data.

Our results have been compared by Gatica et al. (1998) to their prediction based on the density functional theory. Using the Barranco–Hernandez–Navarro density functional (Barranco et al. 1996), these authors predict a zero temperature relation $m(B)$, which is in agreement with our data above $m = 0.3$, but presents a local twofold increase of susceptibility near $m = 0.2$. The corresponding anomaly of $m(B)$ is weak, as shown in the inset of fig. 13(b). As pointed out by these authors, this prediction does not fall far outside the error bars we presented in Wiegers et al. (1991). However, it should be stressed that these error bars, which reflect the effect of a 20% change in the magnitude of the thermal corrections, are *not* independent. If the global thermal correction is, say, overestimated, all of our curve would be shifted to lower fields, but not distorted in such a way that it could be fitted by the prediction of Gatica et al. (1998). We therefore believe that, if the inflection predicted at 40 T were real, we would have observed it. A definite answer, however, would require higher resolution measurements, difficult with our method. A possibility could be to use the torque magnetometer described in Buu et al. (1998a), combined with the 45 T field now available in the US National High Magnetic Field Laboratory.

Finally, we conclude with a remark about the pressure dependence of the magnetization curve. In this experiment, we concentrated on the data at 26 bar for practical and theoretical reasons. From the theoretical standpoint, the effect of interactions is expected to be larger at larger pressure. Experimentally, 26 bar is close to the maximal pressure of 29 bar, which can be obtained using the blocked capillary method. An attempt to reach larger pressures by using a Pomeranchuk cell failed (Buu et al. 1998b). On the other hand, performing the same experiments at lower pressures turns out to be difficult: due to the heat release by the decompression of the liquid, the final temperature is higher (slightly above 100 mK at 10 bar), and the correction due to the warm capillary becomes larger. This makes the determination of the magnetization curve less reliable than at 26 bar. At 10 bar, the general trend was found to be similar, the magnetization curve being also curved downwards at this pressure (fig. 12a). As for pressures larger than 26 bar, we cannot exclude a metamagnetic behavior of ^{3}He. However, one has to bear in mind that this range of unexplored pressure is narrow: the melting pressure at $m = 0.5$ computed from our magnetization curve measurement is 32 bar.

6. Viscosity

6.1. *Motivation*

Originally, the motivation behind viscosity measurements in spin-polarized liquid ^{3}He was to measure the effect of the polarization on the quasiparticle mean free path. It was generally anticipated that the viscosity would increase because the polarization is expected to decrease the number of collisions. However, as we have seen in Section 3.1.2, it is not obvious that this argument should apply to liquid ^{3}He.

The difficulty of these measurements lies in the severe requirements in terms of thermal homogeneity: on one hand, the viscosity varies strongly with the temperature (as $1/T^2$); on the other, the effect of the spin polarization is moderate (this is why the viscosity of liquid ^{3}He is often used for high-field thermometry). So, in order to determine reliably the effect of polarization on viscosity, the temperature close to the viscometer must be measured with good accuracy. To the knowledge of the authors, five groups – including ours – have carried out experiments.

The first measurement, by Kopietz et al. (1986), gave a surprising result, since they observed a *decrease* of the viscosity with the polarization. In contrast, a second experiment, carried out by Vermeulen et al. (1988), showed an increase of viscosity, which cast some doubt on the first result. The two experiments, however, were difficult to compare, because Kopietz et al. used the rapid melting technique, whereas Vermeulen et al. used NMR pulses to vary the polarization produced by a static field. Finally, a series of careful experiments by Kranenburg et al. (1988, 1989) proved that the result of Kopietz et al. was an artifact due to the thermal gradients caused by the melting of the solid, and confirmed the result of Vermeulen et al.

In the experiments of Vermeulen et al. and Kranenburg et al., and also in the recent work of van Woerkens (1998), the largest polarization achieved was $m = 0.2$, due to the fact that, initially, the cell is only partly filled with solid (see Section 4.3). In all cases, the viscosity increase as a function of the polarization was well fitted by the empirical expression

$$\frac{\eta(m)}{\eta(0)} = 1 + \alpha m^2. \tag{6.1}$$

Vermeulen et al. measured a pre-factor $\alpha = 2 \pm 1$ at 30 bar and 45 mK. The rapid melting experiments of van Woerkens, carried out at 30 bar but at somewhat lower temperatures (30–60 mK) gave $\alpha \approx 6$–8, that is substantially higher than the value measured by Vermeulen et al. Finally, Kranenburg et al. measured a pre-factor $\alpha = 3.5 \pm 1.5$ independent of the temperature within their error bars between 80 and 340 mK, with pressures ranging from 24 to 27 bar.

Despite some quantitative discrepancies, the previous experimental results consistently show that the viscosity is increased by the polarization. The goal of the experiments we describe in this section was to use the advantages of our technique of rapid melting inside a sinter (namely, large polarizations and the ability to control the polarized ^{3}He temperature) to extend these results in three directions.

- At low temperature, one would like to measure the pressure dependence of the viscosity enhancement, since it provides information about the quasiparticle interactions, and also gives a direct way to check the prediction of the paramagnon model (cf. Section 3.2).
- At high temperature, viscosity measurements in spin-polarized liquid can be used as a probe of the degeneracy effects. From this point of view, the experiments of Kranenburg et al. at $T > 150$ mK are important, since they suggest that degeneracy effects persist beyond the degenerate regime. In our work, we have explored this high-temperature region more systematically to measure the disappearance of the degeneracy effects.
- Finally, the viscosity enhancement has been measured previously with polarization up to $m = 0.2$. In the experiments we describe next, the highest polarization achieved was $m = 0.7$.

6.2. *Experimental setup*

The setup we used for the viscosity measurements (Buu et al. 1999) is similar in design to that devised for the susceptibility measurements (see Section 5.3), but with two important changes. First, the HT is now thermally coupled to the mixing chamber, so as to allow regulation of its temperature. Second, because the cell has to accommodate a vibrating wire viscometer, a slit is made in the sinter of the experimental compartment. The

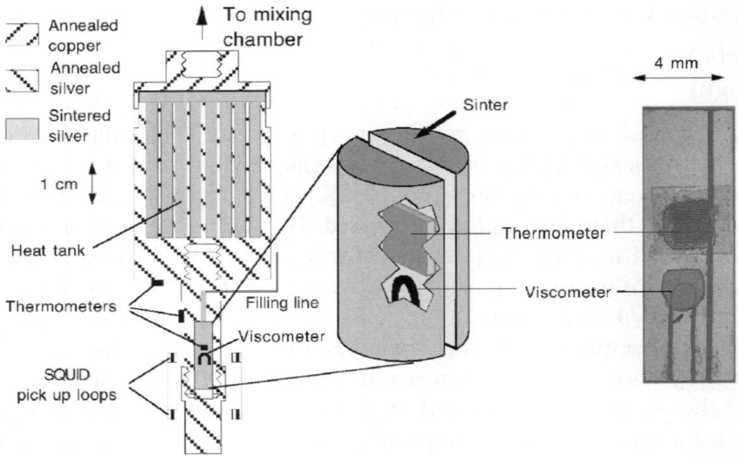

Fig. 14. Experimental setup (see text for details); the sinter containing the experimental ^{3}He is expanded, and the photograph shows the vibrating wire viscometer and the carbon thermometer mounted on the sapphire frame.

consequences of this pocket of bulk ^{3}He on the thermal and magnetic homogeneities will be discussed later on. Here, we focus on the cell design, and on the operation of the vibrating wire viscometer in our particular experiment.

6.2.1. *Experimental cell*

The setup is shown in fig. 14: The ^{3}He sample is contained in a 15 mm long, 4 mm diameter, sintered cylindrical cell. The sinter is cleaved by a 0.2 mm gap that leaves the necessary room for a vibrating wire viscometer and a thermometer (this is made by baking the silver powder with a 0.2 mm stainless-steel blade placed in the middle of the cell, and pulling out the blade after the sintering process). The choice of a 0.2 mm width was dictated by practical feasibility, and also the requirement to minimize the effect of the sinter walls on the measurement of viscosity (see Section 6.2.2). This experimental cell is connected to a large HT, similar to that used for the susceptibility experiment, and consisting of 0.1 mol of unpolarized liquid ^{3}He at saturated vapor pressure, stored in 46 sinter-filled holes, 2.6 mm in diameter and 31 mm in length.[13] The walls of the cell and the HT are made of 4N annealed silver to achieve fast thermal coupling between the two elements. As in the susceptibility cell, the sample is inserted into the pick-up

[13]The sinters in the cell and the HT have been made with the same silver powder as for the susceptibility cell and have been prepared following the same procedure.

coil of a SQUID magnetometer (Bravin et al. 1992). In order to minimize spurious magnetic signals, the nuclear magnetic signal of silver is compensated for by a dummy silver cell, attached below the actual cell inside the gradiometer compensation coil, and the cupronickel filling line enters the cell well above the gradiometer coils. Experiments with the empty cell showed the background magnetism to be less than 1% of the ^{3}He contribution. As in the susceptibility experiment, the polarization scale is calibrated in solid ^{3}He by using the Curie law (see note 10). Finally, the whole setup is directly connected to the mixing chamber of a dilution refrigerator through a rod of annealed copper.

The viscometer is a 1.5 mm-diameter loop-shaped manganin vibrating wire. The 'legs' of the wire are glued into grooves ultrasound-machined on a 0.17 mm thick, 4 × 14 mm sapphire frame, so as to achieve a correct centering inside the slit. Due to the large diameter of the sinter compartment (4 mm), the radial thermal gradient generated by the magnetic relaxation is significant, making it necessary to measure the viscometer temperature with a thermometer also located inside the slit. To this aim, we mounted a 0.09 mm thick, 1.7 × 1.7 mm carbon thermometer sliced from a 100 Ω speer resistor on the sapphire frame, 2 mm above the viscometer. Only the thermometer edges are glued to the frame so that it is in direct contact with the liquid. In the final stage of the assembling, the frame is inserted in the slit and the cell is sealed with a silver plug glued at the end of the experimental compartment. Additional carbon resistor thermometers and heaters are glued on the bottom of the HT, the walls of the cell, and the compensation cell. These are used to regulate the temperature of the HT during rapid melting experiments, and to characterize the thermal behavior of the cell (Appendix C).

6.2.2. *Measurement of the viscosity*

6.2.2.1. *The viscometer.* Our viscometer is similar to those widely used in ^{3}He physics. It consists of a loop of metallic wire immersed in the liquid, driven at resonance by the Lorentz force generated by an ac current in the presence of the vertical magnetic field (11 T) used to polarize the solid ^{3}He. The motion of the wire is detected through the induced voltage at its terminals, and the viscosity is measured from the damping of the oscillations.

A small diameter wire (32 μm, measured using a microscope) is chosen to keep the finite size corrections to the damping (see Section 6.2.2) to a minimum; the varnish has been removed from its active part (with the help of a chemical 'stripper'). This allows to increase the mass per unit of volume of the wire so that the viscometer damping is not too high at low temperature. The resulting density, determined by weighting a given length of stripped wire, is $\rho_{\text{wire}} = 7960\ \text{kg/m}^3$. We use a manganin, rather than a type II superconducting, wire, so as to avoid the complex extra damping due to the possible motion of the vortices in the latter case. The eddy current damping is quite small, thanks to the small diameter of the wire and its high resistivity, resulting in a quality factor, $Q \approx 6700$ in 11 T at the resonance frequency (10,210 Hz in vacuum). The resistive background ($\approx 45\,\Omega$, mainly due to the long 'legs') is accurately measured when the wire has been immobilized in solid ^{3}He, and subtracted from the measured voltage in the liquid to yield the motion-induced voltage. In most experiments, we use an excitation current of 4 μA_{rms}. The induced voltage at resonance then ranges from 40 to 400 μV_{rms}, corresponding to Q values[14] ranging from 50 to 500, for temperatures between 20 mK and 1 K (and a pressure of 27 bar). In these measurements, the corresponding dissipated power remains small enough to avoid local overheating of the ^{3}He sample. In our setup, a detected voltage of 1 mV_{rms} corresponds typically to an excursion of 10% of the wire radius, so that the force exerted by the fluid on the wire remains linear in the displacement amplitude (a necessary condition for the validity of the theoretical expression (6.2) below). Note that, with an excitation current 10 times larger, this would not be the case and the resonance quality factor would then depend on the drive level (Buu et al. 2000a).

6.2.2.2. *Finite size effects.* Let us now discuss how the damping of the wire is related to the viscosity (for a thorough discussion of the hydrodynamics of the vibrating wire viscometer, see Carless et al. (1983). The radius of the loop being much larger than both the wire diameter and the viscous penetration depth introduced below, it is legitimate to approximate the wire as an infinite straight cylinder. When a cylinder of radius a moves back and forth in liquid ^{3}He, it experiences from the surrounding fluid a force proportional to the mass of liquid it displaces. For an amplitude y_0 of the wire oscillations, this force is, per unit length of the wire

$$F = \omega^2 y_0 \pi a^2 \rho (k + ik'),$$

where the dimensionless Stokes' coefficients k and k' give respectively the reactive and the dissipative components of the hydrodynamic force. In an infinite medium, k and k' depend only on the ratio of the radius a of the wire to the viscous penetration depth δ, where $\delta = \sqrt{2\eta/\rho\omega}$, with $\omega/2\pi$ the frequency of the oscillations, ρ and η the mass per unit volume and the viscosity of ^{3}He. The exact expression is (Stokes 1901)

$$(k - 1) + ik' = -\left(\frac{4}{qa}\right)\left(\frac{H_1^{(1)}(qa)}{H_0^{(1)}(qa)}\right) \tag{6.2}$$

[14]The use of a heavier PtW wire (Voncken et al. 1998) would have resulted in lower viscous damping less convenient to measure.

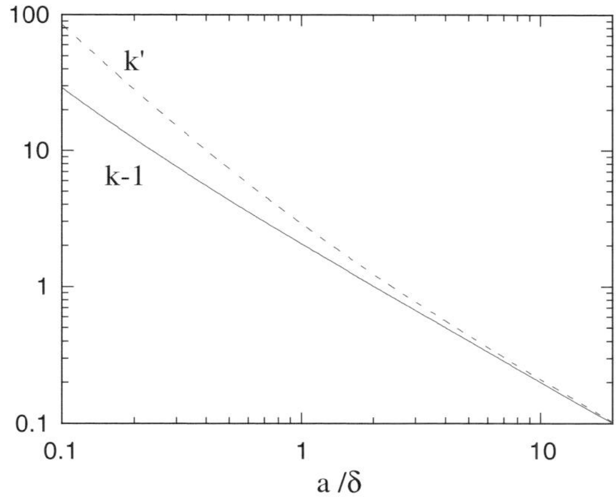

Fig. 15. The dependence of the Stokes coefficients k and k' on the ratio a/δ of the wire radius to the viscous penetration depth.

with $q = (1 + i)/\delta$, and $H_n^{(1)}$ being the nth-order Hankel function of the first kind. Figure 15 shows the resulting dependence of k and k' on the parameter a/δ. In the low viscosity regime ($k' < 1$, corresponding typically to $T > 30\,\mathrm{mK}$ in our situation), $k - 1 \approx k' \approx 2\delta/a$. The dependence of the parameters k and k' on the viscosity is more complex if the fluid is bounded, since the walls will modify the flow pattern. In this case, the drag force will also depend on the typical length scale of the medium (the width of the slit in our case). We can estimate the correction to the force brought about by the presence of a plane wall at a distance b from the wire in the following way: the velocity field generated by the wire motion can be written as the sum of two terms, a potential component, and a diffusive component, which extends only over a distance δ from the wire. In an infinite medium, the potential component is the 2D dipolar field $v_{\mathrm{potential}} \simeq v_{\mathrm{wire}}(a/r)^2$. Now, in the presence of a wall, the velocity at the wall is zero. If $\delta \ll a < b$ (low viscosity), only the potential component will extend to the wall, and its component perpendicular to the wall will be zero. The modified potential component (with respect to the infinite medium) can be calculated by the method of images, by adding a velocity field generated by a fictitious wire at a distance $2b$ from the actual wire and oscillating in opposite direction (see fig. 16a). In the vicinity of the actual wire, the image flow varies slowly in space, the average velocity being $\approx -v_{\mathrm{wire}}(a/2b)^2$, so that the total velocity of the wire with respect to the fluid is: $v \simeq v_{\mathrm{wire}}(1 + (a/2b)^2)$. The hydrodynamic force acting on the wire is thus increased by a factor $(1 + (a/2b)^2)$

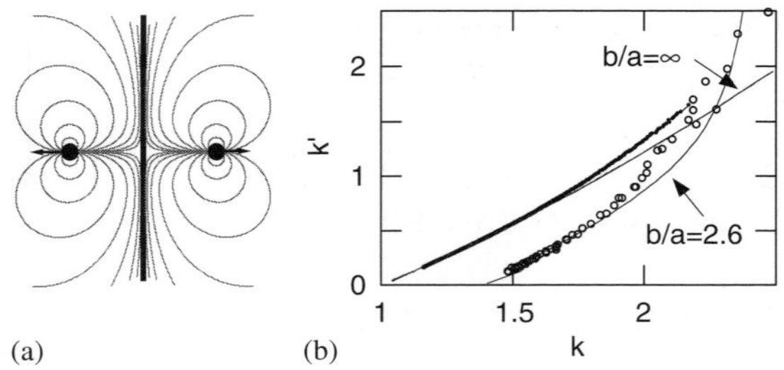

(a) (b)

Fig. 16. (a) Origin of the finite size effect. In a non-viscous fluid, the potential flow generated by a wire oscillating perpendicular to a wall at a distance b is found by the method of images. (b) A plot of k versus k' can reveal a finite size effect. Thin continuous lines: theory calculated from Carless et al. (1983) for a vibrating wire of radius a along the axis of a cylindrical container of radius b, for $b/a = \infty$ (infinite medium) and $b/a = 2.6$. Circles: vibrating wire off the center of the slit in a preliminary experiment ($a = 33$ μm); thick line: vibrating wire correctly centered in the final experiment. Both experiments were performed at 27 bar.

with respect to the case of an infinite medium. In our case of two parallel walls, the correcting factor of k and k' will be[15] $(1 + \frac{1}{2}(a/b)^2)$.

For a very viscous fluid ($\delta > b$), finite size effects are more difficult to compute, because the boundary layer is now also disturbed by the walls. Qualitatively, one expects the reactive component k to saturate at a finite value because the mass of the fluid dragged along by the wire is limited by the finite extent of the fluid medium (in an infinite medium, this component would diverge as $\delta/a \to \infty$). This behavior (and the correction term $(a/b)^2$ for $\delta/a \ll 1$) can be calculated exactly for the case of a wire coaxial to cylindrical walls (Carless et al. 1983).

Experimentally, we have access to the reactive and dissipative components of the force acting on the wire. A plot of k versus k' (obtained from eq. (6.6) below) allows one to detect finite size effects (fig. 15): in the low viscosity regime ($\delta/a \to 0$), $k \to (1 + \frac{1}{2}(a/b)^2)$. In the opposite limit $\delta/a \to \infty$, the saturation of the coefficient k is a clear manifestation of finite size effects. This behavior was observed on a 66 μm diameter viscometer used in a preliminary experiment (circles in fig. 16b). In contrast, in the final cell, thanks to the thinner wire and a better centering, the viscometer showed no sign of finite size effects up to $k' \approx 1$, corresponding to a temperature of 30 mK in unpolarized ^{3}He (thick line). As the k' values measured in

[15]Close to the wall, the potential flow has a non-zero parallel component, which gives rise to a local diffusive component. The associated extra dissipation is negligible with respect to that close to the wire (it is of order $v_{\mathrm{wire}}^2(a/b)^4$ per unit length, i.e. the integrated dissipation is a/b smaller than that close to the wire).

polarized ^{3}He do not exceed 1 (see fig. 18), we used expression (6.2) to extract the viscosity from the measured k'.

6.2.2.3. *Fast viscosity measurement.* From eq. (6.2), we compute the motion-induced voltage $V(\omega)$ at drive frequency ω:

$$V(\omega) \propto \left(\frac{IB^2 L_{\text{wire}}^2}{2\pi a^2 \rho_{\text{wire}}}\right) \frac{\mathrm{i}\omega}{\omega_0^2 - \omega^2(1 + (\rho/\rho_{\text{wire}})k) - \mathrm{i}\omega^2(\rho/\rho_{\text{wire}})k'}, \tag{6.3}$$

B being the magnetic field, I the amplitude of the excitation current, L_{wire}, ρ_{wire}, and ω_0 being respectively the length, the mass per unit volume and the resonance frequency of the wire in vacuum (which depends on the geometry and the mechanical properties of the wire (Carless et al. 1983). In the low damping regime ($Q \gg 1$), this reduces to a Lorentzian line:

$$V(\omega) \propto \frac{1/\Delta\omega}{1 + \mathrm{i}\left(\frac{\omega - \omega_{\text{res}}}{\Delta\omega/2}\right)} \quad \text{for } \omega - \omega_{\text{res}}/\omega_0 \ll 1, \tag{6.4}$$

where ω_{res} and $\Delta\omega$ are the frequency at maximum and the half-width at half-maximum of the resonance peak. They are related to k and k' through:

$$\omega_{\text{res}} = \omega_0\left(1 - \frac{\rho}{2\rho_{\text{wire}}}k\right), \tag{6.5}$$

$$\frac{\Delta\omega}{\omega_{\text{res}}} = \frac{\rho}{\rho_{\text{wire}}}k' \tag{6.6}$$

in the practical limit $(\rho/\rho_{\text{wire}})k \ll 1$. Note that as the viscosity increases, the resonance widens and shifts to lower frequencies.

Thus, measuring the resonance width gives k' (after due subtraction of the residual width measured in vacuum). From eq. (6.2), we obtain δ/a and therefore η. However, in a rapid melting experiment, recording the full resonance curve would take a time not negligible with respect to T_1. The usual solution is to track the resonance, which shifts as the viscosity changes, the in-phase signal at resonance being directly proportional to the quality factor $Q = \Delta\omega/\omega_{\text{res}}$ (eq. (6.4)). In our experiments, we chose to operate at a fixed frequency ω_1 and measure the in- and out-of-phase components of the voltage, which avoids the finite response time induced by the phase-locked loop in the first solution. Writing the motion-induced voltage $V = 1/(\tilde{V}_x + \mathrm{i}\tilde{V}_y)$, eq. (6.4) gives $\tilde{V}_x = A\Delta\omega$ and $\tilde{V}_y = 2A(\omega_1 - \omega_{\text{res}})$. So, the in-phase and quadrature components of the voltage yield directly k and k', the proportionality constant A being calibrated beforehand (which is done using a full resonance curve). In principle, the excitation frequency can be set arbitrarily provided that it lies within the resonance width. In practice, it is suitable to choose the frequency close to the 'magic' frequency, equal to the resonance frequency in the limit of vanishing viscosity (Carless et al. 1983). At this frequency, the response voltage remains approximately at 45° with respect to the drive current (in the low viscosity limit ($k' < 1$), $k - 1 \approx k'$), so that both in-phase and out-of-phase components carry some signal over the full viscosity range explored.

6.3. *Results*

In this section, we present measurements of the viscosity of liquid ^{3}He as a function of spin polarization. The experiments have been carried out between 40 mK and 1 K, with pressures ranging from 2 to 27 bar. We first show that spin polarization enhances the viscosity of the liquid, and then discuss the polarization dependence of the enhancement, as a function of temperature and pressure. The same experiments also allowed us to study how the relaxation of magnetization depends on pressure and temperature. This is separately discussed in Appendix A.2.

6.3.1. *Experimental procedure*

The production of polarized liquid ^{3}He for viscosity measurements follows the method used for the experiments on the magnetic susceptibility (see Section 5). The 'blocked capillary' technique is used to obtain a ^{3}He solid sample. The sample is cooled down to 7 mK in a 11 T magnetic field. The magnetized solid is melted by decompressing the sample to a dead volume at room temperature. The final pressure of the sample can be tuned by adjusting the initial pressure of the dead volume before the melting. The signals from the viscometer, the thermometers, and the SQUID are simultaneously recorded during the magnetic relaxation following the melting (at a sampling rate of one point/0.15 s during typically 1200 s, i.e. long enough to determine the baseline of the SQUID signal, see Section 5.5). A calibration run to measure the relation between the viscosity at equilibrium polarization as a function of the temperature is systematically carried out immediately after the magnetic relaxation.

Figure 17a shows the signals recorded during the first few seconds of a rapid melting experiment. The filling line is decompressed at time $t = 0$, and the solid in the cell melts about 2 s later, as indicated by the apparition of a motion-induced in-phase voltage across the vibrating wire (the initial voltage is due to the resistive background). Starting from 7 mK, the cell temperature reaches $\approx$ 220 mK during the melting, while the HT temperature stays below 80 mK. The cell is then cooled down by the HT. Due to the thermal resistance between the slit and the walls of the cell, the inner

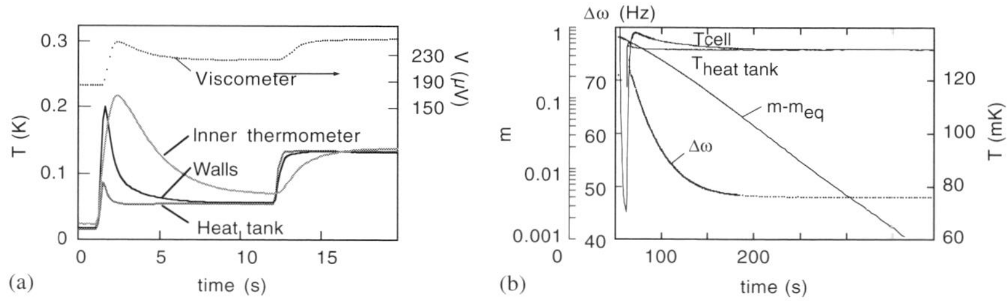

Fig. 17. A rapid melting experiment at a final pressure of 10 bar (a) Signals on a short timescale: melting inside the cell occurs at $t = 2$ s (note that the temperature values before melting are overestimated, due to the saturation of the carbon thermometers). After melting, the cell is cooled down by the HT. The thermal regulation is turned on at time $t = 12$ s. During these first few seconds, the evolution of the viscometer in-phase signal is driven by the changes of the inner temperature. (b) Signals on a longer timescale: the magnetization relaxes with a time constant $T_1 \approx 60$ s. The heat released by the polarization decay induces a temperature difference between the inner thermometer and the HT. The viscometer signal (here converted into the resonance width) shows a large decrease of the viscosity, which is mainly due to the relaxation of polarization.

temperature lags behind the wall temperature during this stage. When the cell temperature is low enough ($\approx$11 s after the melting), the thermal regulation is turned on in order to stabilize the HT temperature at a chosen value (130 mK in the example as shown in fig. 17a). During this rapid evolution, the viscometer signal is similar to that of the inner thermometer, reflecting the fast temperature changes of the ^{3}He sample, the polarization being essentially constant on this short timescale.

The effect of polarization is observable on the longer timescale of fig. 17b. After the short transient ($\approx$3 s, fig. 17a) following the regulation of the HT, the inner temperature reaches a quasi-stationary regime, where it slowly decreases to the HT temperature. The decaying temperature difference is due to the heat released by the irreversible magnetization relaxation of the ^{3}He sample, which occurs on the slow timescale $T_1/2$. Simultaneously, we observe a large decrease of the viscosity (the resonance width plotted in fig. 17b decreases). If the viscometer variations were due only to the inner temperature evolution, one would observe a moderate increase of the viscosity. Thus, the observed decrease of the viscosity is unambiguous evidence of the effect of spin polarization.

This effect is also seen directly in fig. 18a, where we compare the evolution of the viscosity during a rapid melting experiment, to the calibration curve measured in unpolarized ^{3}He, in the plane $(\Delta\omega, T_{\text{inner}})$ (resonance width versus temperature). The data at $t > 15$ s are plotted for four experiments at 27 bar, with similar initial polarizations ($m \approx 0.75$) but different final

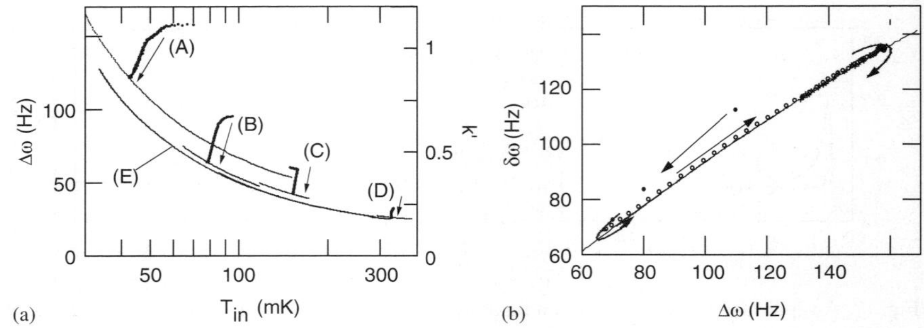

Fig. 18. (a) The relationship between the width $\Delta\omega$ of the viscometer resonance and the inner thermometer temperature, as recorded during the magnetic relaxation following rapid melting experiments (A–D) at four different regulation temperatures, is compared to that measured during following calibration runs at magnetic equilibrium. The arrows indicate the direction of increasing time. The increased width of the viscometer resonance during the magnetic relaxation indicates an enhancement of viscosity by the polarization. The calibration curves (A)–(C) lie above the intrinsic calibration curve (E), due to some extra damping (see text). (b) For the rapid melting experiment (A), the relationship between the width $\Delta\omega = \tilde{V}_x/A$ and the shift multiplied by 2 (referred to the operating frequency 10140 Hz), $\delta\omega = \tilde{V}_y/A$ is the same as during the calibration (continuous curve). The full circles correspond to the first second following the melting around the viscometer, the open circles to the fast cooling during the next 10 s, and the crosses to the relaxation of polarization.

regulation temperatures. Except for the first few data points, which are recorded while the temperature changes rapidly, the inner temperature decreases to the regulated HT temperature (as already seen in fig. 17b). Simultaneously, the resonance width decays onto the equilibrium curve. The resonance width is larger than its equilibrium value at the same temperature, showing that the spin polarization systematically reduces the viscosity. This effect has been observed at all the temperatures and all the pressures tested.

The behavior of the resonance shift during the rapid melting experiments is consistent with that of the resonance width. This is illustrated by plotting in fig. 18b the relation between the shift and the width of the resonance, for the case of the rapid melting experiment at 40 mK presented in fig. 18a. Except during the first second, when the viscometer amplitude builds up, the data agree with those obtained during the slow temperature sweep in the unpolarized liquid carried out after the rapid melting experiment. The agreement includes the data recorded during the increase of width due to the fast cooling as well as the data taken during the following decrease due to the relaxation of polarization. This shows that the behaviors of the shift and of the width in polarized ^{3}He are perfectly consistent with one another. However, the reader may wonder why the different calibration curves in fig. 18a do not coincide, but are vertically translated with respect to one another. We discuss this point in Section 6.3.2.

6.3.2. *Polarization-induced viscosity enhancement*

During the magnetic relaxation, the viscosity evolves due to the decrease of polarization, and, to a smaller extent, to the decrease of the slit temperature. To single out the effect of the spin polarization, we normalize the measured viscosity at given time and temperature (read by the inner thermometer) to the viscosity measured in non-polarized ^{3}He at the same temperature (see fig. 18a). As we just mentioned, fig. 18a shows that the calibration curves recorded after the melting experiments at a given pressure do not coincide. The difference between the curves is in fact due to a variable extra damping. In the following analysis, we have subtracted this extra damping, i.e. we have shifted vertically each set of data and the corresponding calibration curve in fig. 18a, so as to make the latter curve to coincide with the intrinsic' calibration curve. The description of the determination of this intrinsic calibration curve, the origin of the extra damping, and the adequacy of this procedure are deferred to Section 6.3.3. The polarized and unpolarized viscosities used below are obtained from the corrected width.

The resulting normalized viscosity is shown as a function of m^2 in fig. 19, for pressures of (a) 27, (b) 10, and (c) 2 bar.[16] The absolute polarization m is obtained by adding the calculated equilibrium polarization m_{eq} at the given temperature and pressure to the deviation from equilibrium measured by the SQUID magnetometer. In addition, fig. 19d gives a comparison of the data at 80 mK for the different pressures. In the same figure, we also plot the calculated viscosity enhancement, assuming the viscometer temperature to be equal to that of the walls. This allows one to visualize the influence of the radial thermal gradient on our measurements.

These figures make obvious our most salient results:

- The viscosity increases with the spin polarization proportional to m^2.
- The polarization-induced viscosity enhancement decreases with increasing temperature, the decrease being more marked at elevated pressure.
- At low temperature, the polarization-induced viscosity enhancement depends weakly upon pressure.

[16] In a few cases, the data have been recorded for a reduced range of polarization. This happened when the solid cooling time was not long enough to reach the maximum accessible polarization, or when we changed the temperature during the course of the magnetization relaxation.

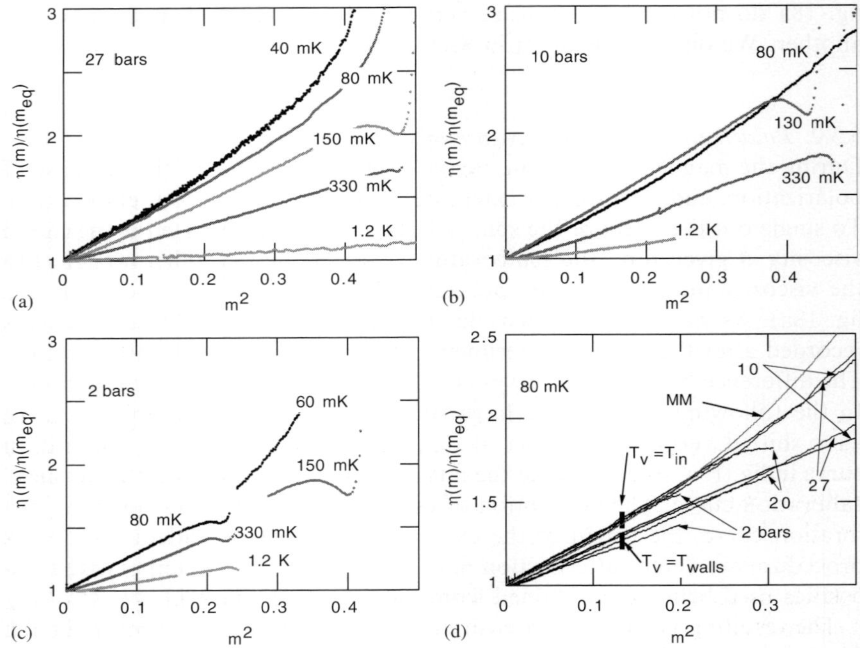

Fig. 19. Viscosity enhancement at (a) 27 bar for temperatures of 40, 80, 150, 330, and 1.15 K; (b) 10 bar for 80, 130, 330, and 1.15 K; (c) 2 bar and different temperatures; in order to limit the number of experiments, the temperature was changed from 60 to 80 mK during one relaxation, and from 150 to 330 mK for another, with temperature transients resulting in the observed wiggles after each change; (d) 80 mK for pressures of 2, 10, 20, and 27 bar, illustrating that the effect of polarization barely depends on pressure. This remains true even if we choose the wall temperature as an estimate of the viscometer temperature. The Mullin–Miyake (MM) prediction for s-wave scattering is shown for comparison.

Before quantifying these effects, it is necessary to analyze in detail the possible sources of error in our measurements. This is the topic of the following section. The reader only interested in the final discussion may skip this technical discussion and go directly to Section 6.3.4.

6.3.3. *Analysis of systematic errors*

Three factors may affect the validity of our measurements: an error in the viscosity measurement, a difference between the viscometer and inner thermometer temperatures, and a difference between the average cell polarization measured by the magnetometer and the local polarization close to the viscometer. We discuss these factors in the next three paragraphs.

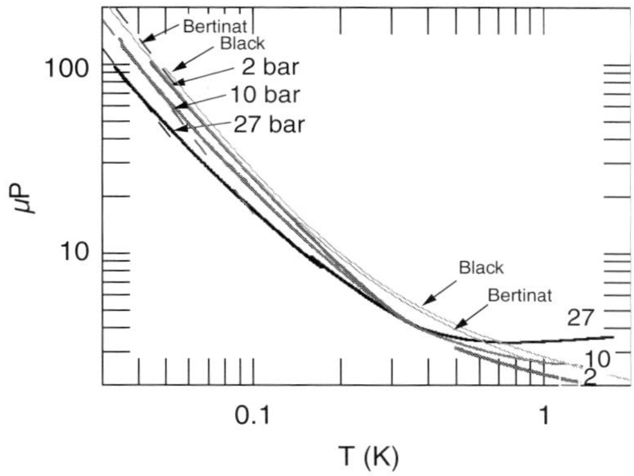

Fig. 20. Viscosity of non-polarized ^3He as a function of the temperature (in μP, i.e. in SI units). Measurements are performed in the 'normal' state of the viscometer. The thick lines are our data at 2, 10, and 27 bar. Below 150 mK, the 2 bar data are close to the s.v.p. measurements of Black et al. (1971) and Bertinat et al. (1963) (thin lines). Our measurements are also consistent with extrapolations of the low temperature $1/T^2$ behavior measured by Carless et al. (1983) and quoted in Table D.1 (dashed lines).

6.3.3.1. *Reliability of the viscosity measurement.* Errors in the viscosity measurement may stem from two reasons: error in the conversion of the resonance width into a viscosity, or error in the determination of the resonance width itself.

To correctly translate the measured resonance width into a viscosity, we have to make sure that the size of vibrating wire and the width of the slit are such that eq. (6.2) remains valid with good accuracy (i.e. negligible finite size effects, wire curvature radius larger than the viscous penetration depth). As discussed in Section 6.2.2, our geometrical parameters (slit width, loop radius) have been chosen to match this condition. Experimentally, this is supported by the measured relation between k and k', which agrees with that predicted from eq. (6.2). The measurement of an absolute viscosity requires an accurate determination of the wire radius a and wire density ρ_{wire}. Using the parameters given in Section 6.2.2, our viscosity data in unpolarized[17] ^3He at low pressure and below 150 mK are in agreement with those of Black et al. (1971), and about 10% larger than those of Bertinat et al. (1963) (see fig. 20). At higher pressure, our data are consistent with extrapolation of the low-temperature data of Parpia (Carless et al. 1983). The polarization-induced viscosity enhancement we are interested in is a relative effect. Therefore, it is much less sensitive to the actual values of the parameters a and

[17] Strictly speaking, the measured viscosity is at equilibrium polarization in our 11 T field. But this brings a negligible (< 1%) correction.

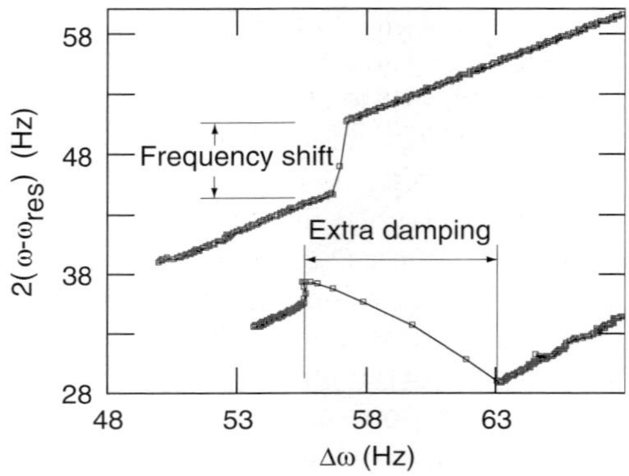

Fig. 21. Two examples of viscometer 'jumps' occurring during rapid melting experiments, affecting the resonance frequency shift and/or damping.

ρ_{wire} than an absolute measurement (we work in a regime where $k' \propto \delta/a$), so the error bars caused by the uncertainty in these parameters is negligible.

The determination of the resonance width can cause a more serious error. We mentioned at the end of Section 6.3.1 that the different calibration curves of the resonance width as a function of temperature obtained in unpolarized ^{3}He (fig. 18a) do not coincide. This means that the viscometer can be in different 'states,' corresponding to different resonance frequencies and/or widths at a given temperature. Between two given states, the difference of resonance width only weakly varies with temperature (fig. 18a). Random jumps from one 'state' to another were observed either during a melting experiment, or during a calibration run (see fig. 21), affecting the width and/or the shift. The amplitude of the jumps is usually of the order of a few hertz. We observed that overdriving the wire (with a 40 μA current) for a few seconds resulted in the lowest possible damping at a given temperature, once the excitation current was returned to its nominal value. We shall refer to the corresponding viscometer state as 'normal.' We speculate that silver dust coming from the sinter, and sticking temporarily to the wire, is at the origin of the observed extra damping, and that this dust is shaken off by overdriving the wire. Once we became aware of the problem, we systematically overdrove the wire during the first seconds after melting, giving data generally free of any jump, and a minimal width at the end of the relaxation. The other data, (usually presenting at most 2–3 jumps) were treated in the following way: starting from the end of relaxation, we first eliminate any jump during the magnetic relaxation (resp. the following calibration curve) by matching the width before and after the jump, proceeding backward in time (resp. forward). We thus obtain curves similar to those in fig. 18a. Second, we subtract from the data so obtained a constant width, such that the resulting width at the end of relaxation lies on the 'intrinsic' calibration curve, obtained in the 'normal' state of the viscometer. The data are then converted into viscosity, as discussed in Section 6.3.2. In unpolarized ^{3}He, this yields the curves of fig. 20.

This procedure assumes that the extra width is constant over the range of widths explored during the relaxation and that the changes in width only occur by sudden jumps. Experimentally, these assumptions seem to be well verified: in each case where we duplicated a corrected experiment by an experiment where the viscometer was in its 'normal' state, the measured polarization dependence of the viscosity was found to be the same (an extreme example being an experiment at 300 mK where the viscometer happened to have an extra damping of 20 Hz, comparable to the intrinsic width at this temperature!). However, a progressive change of the viscometer damping would remain undetected by our procedure. This effect (which, in our interpretation of the origin of the extra damping, would be unlikely) should not exceed a few hertz. This would be negligible for low-temperature data (a 1 Hz shift is equivalent to 1.5% error on the ratio $\eta(m)/\eta(m_{\mathrm{eq}})$ at 80 mK). The situation is more critical for high-temperature data, because of the narrower viscometer resonance (20 Hz at 1 K) and the weaker effect of the spin polarization (2 Hz at 27 bar). However, the fact that we indeed observe an approximate m^2 dependence of $\eta(m)/\eta(m_{\mathrm{eq}})$ at high temperature, makes it likely that no such progressive change occurs in that case.

6.3.3.2. *Magnetization gradients.* The magnetometer used for the rapid melting experiments measures the average polarization of the sample, while the vibrating wire probes the viscosity on the local scale of δ ($\delta < a$). Hence, magnetization gradients arising during the polarization decay can cause systematic errors. In a fully sintered cell, magnetization gradients will not occur if the polarization is homogeneous just after the melting. But, in the viscosity cell, the magnetization of the liquid located inside the slit relaxes only through diffusion to the sinter, causing the measured average polarization to be smaller than the relevant slit polarization. On the other hand, the relaxation on the surface of the vibrating wire may deplete the magnetization locally. The first mechanism is theoretically discussed in Appendix B, where we reach the conclusion that it should affect our measurements. The second one is discussed in the thesis of van Woerkens (Woerkens, PhD Thesis 1998), who concludes that the effect is negligible in his case. It is however difficult to translate directly this conclusion to our case, as it depends on the relaxing properties of the wire, which may be different.

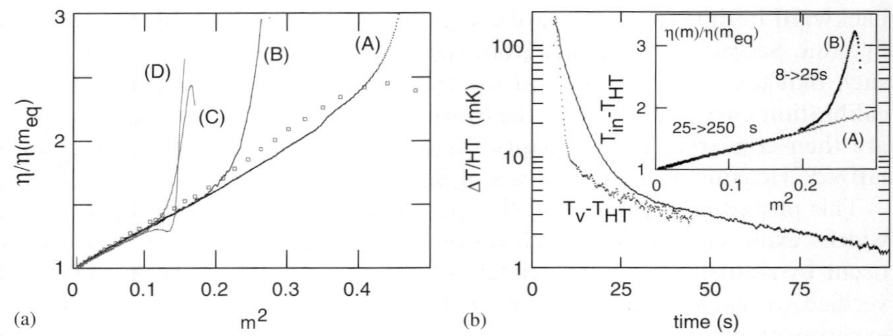

Fig. 22. (a) Four experiments at 27 bar and 80 mK with different initial polarization. The collapse of the traces on a single curve after the decay of the thermal transient proves that the magnetization gradient inside the sinter is negligible. The agreement with a preliminary experiment (squares) using a varnished wire suggests that the magnetization relaxation on the vibrating wire is negligible (for the preliminary experiment, the viscosity has been corrected for the finite size effect visible in Fig. 16). (b) Temperature difference $T_{\text{inner}} - T_{\text{ht}}$ between the inner thermometer and the HT as a function of time after the rapid melting of a 50% polarized solid. The inset is a detailed view of Fig. 22a. During the initial fast cooling by the HT (regulated at 80 mK), i.e. up to 25 s after decompression at time zero, the thermometer cools more slowly than the viscometer, causing the upward curvature of curve (B). The viscometer temperature T_{V} is corrected for the effect of polarization using curve (A) of the inset. Beyond $t = 25$ s, the transient gradients have relaxed, and the slow evolution of $T_{\text{inner}} - T_{\text{ht}}$ is due to the magnetic relaxation.

Experimentally, a direct way to probe the occurrence of the first mechanism is to compare a set of experiments carried out with different initial polarizations (i.e. different initial temperatures of the solid). If polarization gradients develop with time, different polarization profiles will correspond to the same average polarization, and different polarizations inside the slit, therefore different viscosities (see fig. B.2). Such a comparison is displayed in fig. 22, which shows the results of four different experiments at 27 bar and 80 mK. The four curves coincide after the thermal transients have relaxed (see Section 6.3.3), showing that the magnetization gradients are in practice negligible, which, although not understood yet, is fortunate. As for the relaxation on the surface of the vibrating wire, this effect is probably negligible. As shown in fig. 22, a preliminary experiment, where the vibrating wire had not been stripped from its polyethylene coating, yields identical results to those obtained with the present stripped wire, while the surface relaxation is most likely to be different in the two cases.

6.3.3.3. *Thermal gradients.* The last possible cause of error is a temperature difference between the viscometer and the inner thermometer. In principle, the cell design ensures that the temperature gradients are only along the radial direction: due to the high conductivity of the bulk silver, the walls of the cell are nearly isothermal. So, there should not be any significant vertical temperature gradient inside the slit provided that the properties of the silver sinter are uniform and that the temperature after the melting and the heat released during the magnetization relaxation are homogeneous along the cell axis. Consequently, the inner thermometer should be at the same temperature as the viscometer during the polarization decay, although it is located 2 mm above it.

However, inspection of fig. 22a shows that this is not always true. In each instance, there is a large upward curvature at the largest polarizations. This upward curvature is not related to the absolute polarization, but to the time elapsed since the melting. Analysis of the time evolution of the inner thermometer reveals that the upward curvature corresponds to times where the thermometer temperature decreases rapidly. This is illustrated in fig. 22b for the experiment (B) of fig. 22a. Here, the initial polarization was about 50%, i.e. small enough to know unambiguously the polarization dependence of the viscosity (from experiment (A) of fig. 22a). This allows us to correct for this dependence and compute the viscometer temperature. This representation makes explicit the fact that, initially, the inner thermometer cools more slowly than the viscometer. In fact, such an anomalously slow response of the thermometer is also observed in non-polarized ^{3}He, as discussed in Appendix C (see fig. C.1). We speculate that it is due to a pocket of ^{3}He trapped in the exfoliated structure of the carbon resistor. As a consequence, the thermometer does not properly read the viscometer temperature during typically 5–10 s after a rapid change of temperature, either due to the initial melting or to a regulation step. This explains the anomalies visible after such changes in fig. 19a–c, which have to be discarded for the analysis.

The obvious question is whether this anomaly will perturb the temperature reading beyond the transients, in the regime where the slit temperature differs from that of the HT temperature, due to the heat released by the magnetic relaxation (figs. 23a and b).

This question cannot be addressed by an experiment similar to that of fig. 19: indeed, to a first approximation, the same average polarization will correspond to the same power released, and the same reading of the inner thermometer, be it right or wrong. In other words, the agreement of the viscometer and thermometer for times $t > 25$ s in fig. 19b is meaningless (it is *a priori* assumed in order to calculate the polarization dependence of the viscosity!). Since we did not have any experimental means to address the problem, we may turn to a theoretical estimate of the temperature difference between the slit and the HT.

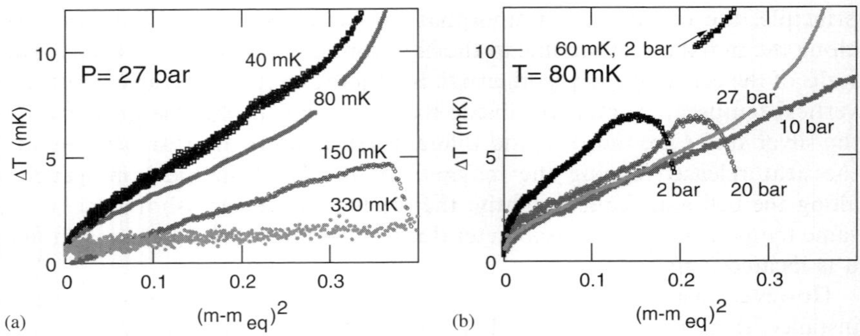

Fig. 23. Temperature difference $T_{\text{inner}} - T_{\text{ht}}$ between the inner thermometer and the HT: (a) for experiments of Fig. 19a, at 27 bar and different temperatures; (b) for experiments of Fig. 19d, at 80 mK and different pressures. Except during the temperature transients, the temperature difference is controlled by the magnetic relaxation and varies essentially as $(m - m_{\text{eq}})^2$ (deviations at small polarizations may be due to long drift times).

As the released heat evolves on a timescale ($\approx T_1/2$) much longer than that of the cell thermal response, the temperature difference between the slit and the HT can be obtained from a quasi-static calculation, by multiplying the slowly varying heating power by an appropriate thermal resistance, which depends on the spatial distribution of the heat released. This problem is dealt with in detail in Appendix C. We show that a lower bound of the temperature difference can be found by assuming the polarization to be uniform inside the whole cell. The heating power is then dissipated in the sinter only, and, assuming the polarization to be proportional to the effective field, this can be written as (per unit volume)[18]

$$\dot{Q}_{\text{sinter}} = \frac{\mu_0 M_{\text{sat}}^2}{\chi_T} \frac{\langle m \rangle^2}{T_1} \left(1 + \frac{V_{\text{slit}}}{V_{\text{sinter}}}\right) \tag{6.7}$$

with M_{sat} the saturation magnetization per unit volume, and, from eq. (5.4) $1/\chi_T = B_T/(\mu_0 M) = 1/\chi - T\partial(1/\chi)/\partial T$.

The ratio $V_{\text{slit}}/V_{\text{sinter}}$ of the volumes of ^{3}He contained in the slit and the sinter comes in because, in our approximation, the energy released by the relaxation inside the slit is evenly dissipated inside the sinter. The thermal resistance R_{eff},[19] defined by the relation $\Delta T = R_{\text{eff}} \dot{Q}_{\text{sinter}} V_{\text{sinter}}$, is the sum of the thermal resistance R_{sinter} between ^{3}He inside the sinter and the cell walls, and of the resistance R_{contact} of the metallic contact between

[18]Note that T_1, the measured relaxation time is denoted T_1^{eff} in Appendix C.

[19]As R_{eff} is measured for unpolarized ^{3}He, we implicitly assume it to be polarization independent. We show in Appendix C that this is a good approximation.

the cell walls and the HT (see Appendix C). These contributions can be expressed in terms of thermal relaxation times $\tau_{\text{sinter}} = R_{\text{sinter}} C_3 V_{\text{sinter}}$ and $\tau_{\text{contact}} = R_{\text{contact}} C_3 (V_{\text{sinter}} + V_{\text{slit}})$, with C_3, the (unpolarized) ^{3}He specific heat per unit volume of ^{3}He. Writing the molar susceptibility as $\chi_{\text{mol}} = \mathcal{N} \mu_3^2 / k T_F^{**}$, where T_F^{**} is a temperature-dependent magnetic Fermi temperature and introducing $T^{**} = T_F^{**} - T(\partial T_F^{**}/\partial T)$, the temperature difference between the slit and the HT is finally predicted to be

$$\Delta T = T_0 \frac{\tau_{\text{relax}}}{T_1} m^2 \tag{6.8}$$

with

$$\tau_{\text{relax}} = \tau_{\text{sinter}} \left(1 + \frac{V_{\text{slit}}}{V_{\text{sinter}}} \right) + \tau_{\text{contact}} \tag{6.9}$$

and $T_0 = T^{**}/(C_3^{\text{mol}}/R)$, C_3^{mol} being the ^{3}He heat capacity per mole.

This equation shows that the temperature difference across a cell filled with polarized ^{3}He scales as the product of the temperature- and pressure-dependent characteristic temperature T_0 by the ratio of the thermal relaxation time of the cell to the magnetic relaxation time. As such, it may be used in various experimental situations. In our case, the qualitative features of the observed temperature difference are well accounted for by eq. (6.8): after the relaxation of the transient gradients, the temperature difference is proportional to m^2 (see fig. 23b). Moreover, the temperature difference decreases at high temperature, because both the thermal resistance τ_{relax}/C_3 and the heat released by the spin system ($\propto T^{**}$) decrease ($T^{**} = 0$ in the Curie limit). On the other hand, the temperature difference increases at low pressure, since the amount of heat released is larger, the magnetic susceptibility of ^{3}He being smaller.

The values of the temperature difference between the HT and the viscometer computed from eq. (6.8) are listed in Table D.3. C_3 and T^{**} are obtained from Greywall's data on specific heat (Greywall 1983) and Ramm's magnetic susceptibility data (Ramm et al. 1970), respectively. The combination $\tau_{\text{relax}} = \tau_{\text{sinter}}(1 + V_{\text{slit}}/V_{\text{sinter}}) + \tau_{\text{contact}}$ is obtained in Appendix C from measurements of the thermal response time of the cell, together with a thermal model, the estimated precision ranging from ± 0.15 to ± 0.3 s as the temperature decreases. Since the magnetic relaxation is non-exponential, we take the average value of the decay rate ($1/m \, \mathrm{d}m/\mathrm{d}t$)) in the range of polarizations explored, the error bars giving the deviations from this averaged value. This error adds to that of τ_{relax} to give the error on the calculated temperature difference listed in the table.

Examination of Table D.3 shows that the calculated values (determined within 10%), are in reasonable agreement with the pressured ones, except at

TABLE D.3
Temperature difference between the viscometer and the HT during the polarization decay: experimental values: $(\Delta T/m^2)_{exp} = (T_{inner} - T_{ht})/m^2)_{exp}$.

T (mK)	P bar	τ_{relax} (S)	T_0 (mK)	T_1 (s)	$(\Delta T/m^2)_{theory}$(mK)	$(\Delta T/m^2)_{exp}$ (mK)
42	27	2.4±0.3	1200	80±20	36±14	23
78	27	1.9±0.25	780	75±15	21±7	18
152	27	1.4±0.25	520	75±5	9±2.5	12
327	27	0.55±0.20	360	65±5	3±1.5	2.5
78	10	1.4±0.25	1160	50±10	33±12.5	19
131	10	1.1±0.25	800	50±10	18±8	15
327	10	0.7±0.20	520	60±5	6±2	0
78	2	1.2±0.25	1580	35±5	56±19	36
152	2	0.9±0.25	960	55±5	18±6	17
326	2	0.6±0.20	660	55±5	8±3	5

Note: The theoretical values are computed from expression (6.8) with $T_0 = T^{**}/(C_3^{mol}/R)$. The relevant thermal resistance is expressed in terms of the thermal relaxation time τ_{relax}.

the lowest temperatures, where they are systematically larger, the difference remaining however consistent with the error bars. This implies that the inner thermometer does not overestimate the temperature difference measured during the magnetization relaxation, as we could have feared due to its anomalously long thermal response time. Therefore, this validates our determination of the polarization-induced viscosity enhancement. Conversely, one could hope to measure the power released by the relaxation, and hence, the magnetization curve of ^{3}He from knowledge of $\Delta T(t)$. An advantage with respect to the adiabatic method described in Section 5 would be that the thermal corrections due to the warm filling line, and the silver spins relaxation, are smaller (the large amount of silver in the HT does not contribute any longer since the temperature is regulated, and the heat injected from the capillary only comes in through the contact resistance between the cell walls and the HT, which is small). However, it turns out that this method gives a poor signal to noise ratio, due to the fact that the temperature difference is small, and that different thermometers have to be compared. The only point we can make is that the observation of a $\Delta T \propto m^2$ is not inconsistent with a downward curvature of $m(B)$, as the latter would be compensated for by the increase of $T_1(m)$ with polarization (Appendix A.2).

6.3.4. *Quantitative analysis of the effect of the polarization*

We are now in a position to quantify the polarization dependence of the viscosity: as shown in figs. 19a–c, the increase of viscosity is observed to be roughly proportional to m^2. We have examined the deviations from this simple behavior by plotting, for each rapid melting experiment, the quantity[20] $R_\eta = (\eta(m)/\eta(0) - 1)/m^2$ as a function of m^2. This quantity strongly varies at the highest polarizations, which results from the inaccurate reading of the thermometer during the initial temperature transients (discussed in Section 6.3.3 and fig. 23). Accordingly, this range of polarizations was discarded from the analysis. In the remaining range of polarizations (typically $m_{eq} < m < 0.6$, for an initial polarization of 70–75%), R_η is generally not constant, but we find no systematic correlation between the deviations from a constant and the pressure or temperature. It is possible that these deviations are dominated by the systematic errors discussed in Section 6.3.3. Note however that, if we would represent these deviations by a βm^4 term in an expansion

$$\frac{\eta(m)}{\eta(0)} \approx 1 + \alpha m^2 + \beta m^4 + \cdots, \tag{6.10}$$

$|\beta|$ would be less than 3 in any of our experiments. Moreover, in the particular case of the coldest experiment (27 bar, 40 mK) which we repeated twice, we find consistently a β of 2, in the polarization range $0.2 < m^2 < 0.3$, where the initial temperature gradients are negligibly small (see fig. 19a).

In the absence of any systematic deviation, we chose to estimate α by taking the average of the quantity R_η. For all experiments where this was possible, the average was taken over the range of polarizations $m_{eq} < m < 0.45$ (or less for small initial polarizations). The value[21] of α for each experiment is reported in Table D.4. The error bar corresponds to the maximal deviation of R_η from $\langle R_\eta \rangle$.

Additionally, one must take into account the error induced by a possible difference between the inner thermometer and viscometer temperatures during the magnetization relaxation. In order to give an idea of the sensitivity of our measurements to such an error, we included in Table D.4 the change of α, $\Delta\alpha_{therm}$, which would result from taking the temperature of the viscometer equal to the calculated value in Table D.3. $\Delta\alpha_{therm}$ is obtained by multiplying the difference between the calculated and measured slopes $(\Delta T/m^2)$ by $1/T\, \partial \ln \eta / \partial \ln T$. The change is larger at low temperature, reaching 15% at 40 mK. Above 300 mK, it is very small because the heat released during the polarization decay is negligible.

Figure 24 shows α as function of the temperature for the different pressures investigated. We confirm the trends previously observed in Section

[20] $\eta(m)/\eta(0)$ is obtained by multiplying the measured quantity $\eta(m)/\eta(m_{eq})$ by $\eta(m_{eq})/\eta(0) = 1 + \alpha m_{eq}^2$, using a first estimate of α.

[21] Note that the values of α depend on our polarization scale. Changing it by the estimated uncertainty of 10% (see note 10) would result in a 20% change of α, identical for all temperatures and pressures.

TABLE D.4
Summary of viscosity measurements.

T (mK)	P (bar)	α	$\partial \ln \eta / \partial \ln T$	$\Delta\alpha_{therm}$	Range
42	27.6	3.40 ± 0.10	1.9	+0.5	m<0.45
42*	27.2	3.40 ± 0.10	1.9	+0.5	m<0.45
78	27.0	3.20 ± 0.05	1.8	+0.1	m<0.45
78	27.0	2.90 ± 0.05	1.8	+0.1	m<0.30
78	26.3	2.90 ± 0.05	1.8	+0.1	m<0.30
78	27.5	3.20 ± 0.05	1.8	+0.1	m<0.45
152*	27.1	2.20 ± 0.05	1.4	<0.01	m<0.45
327	27.4	1.50 ± 0.07	0.9	<0.01	m<0.45
327*	27.0	1.40 ± 0.07	0.9	<0.01	m<0.45
1145*	27.5	0.24 ± 0.01	0.4	<0.01	m<0.35
78*	19.9	2.80 ± 0.05	1.8	+0.10	m<0.45
78*	10.1	2.90 ± 0.09	1.8	+0.4	m<0.45
131	10.0	3.30 ± 0.05	1.5	+0.1	m<0.45
328*	10.1	1.60 ± 0.05	0.9	<0.01	m<0.45
1145*	10.1	0.65 ± 0.03	0.4	<0.01	m<0.45
78*	1.9	3.00 ± 0.10	1.8	+0.45	m<0.40
152	2.0	2.65 ± 0.05	1.4	+0.05	50%<0.60
326	2.0	1.80 ± 0.10	0.9	<0.01	m<0.45
1145*	1.6	0.85 ± 0.03	0.4	<0.01	m<0.45

Note: The coefficient α corresponds to a fit of the data to $\eta(m)/\eta(0) = 1 + \alpha m^2$ for $m < 0.45$ (see text), the viscometer temperature being taken equal to that of the inner thermometer. Taking it instead equal to the theoretical estimate of table D.3 would result in the change $\Delta\alpha_{therm}$ of α. $\partial \ln \eta / \partial \ln T$, which measures the sensitivity of the viscosity to temperature changes, is estimated from our measurements in the unpolarized liquid. The range of polarization used to fit the data is indicated in the last column. The stars indicate the experiments in which the viscometer had an extra damping (cf. Section 6.3.3).

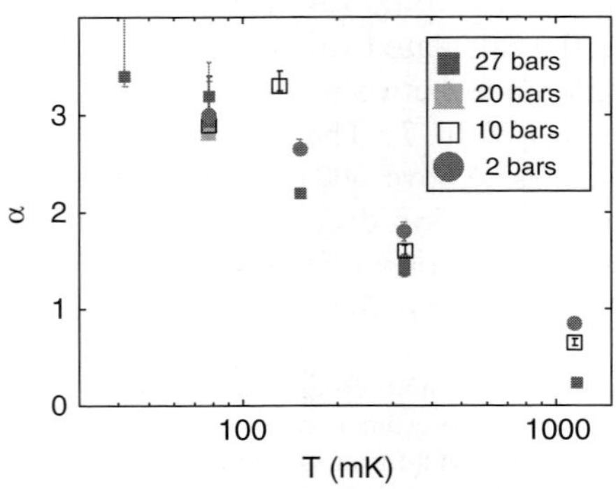

Fig. 24. Coefficient α as function of the temperature for different pressures.

6.3.2. At low temperature, the small variations of the coefficient α with the pressure observed at low temperature are within the error bars, an upper bound of these variations being 10%. On the other hand, the decrease of α with increasing temperature is clearly outside of the error bars. In the next section, we discuss the meaning of these results and compare them to the theoretical predictions given in Section 3. As the small dependence of α on pressure at low temperature is significant information, we stress that this result is not too sensitive to the problem of the determination of the viscometer temperature. This is illustrated in lower part of fig. 19d, where we chose the wall temperature as an estimate of the viscometer temperature, an extreme lower bound!

6.4. *Discussion*

We have shown that the viscosity is enhanced in spin-polarized ^{3}He over the whole range of pressure ($2 < P < 27$ bar) and temperature ($0.04 < T < 1$ K). Up to the highest polarization achieved ($m = 0.6$), the correction term is found to be nearly proportional to m^2. In the following discussion, we focus mainly on the value of the viscosity enhancement pre-factor α as function of the temperature and the pressure. Our results are compared with those of the previous experiments in fig. 25. Our measurements agree with those of Kranenburg et al. (1989) within the error bars up to $T = 200$ mK. However, their value at 340 mK and 27 bar ($\alpha = 4.9 \pm 1.5$) is substantially higher than ours ($\alpha = 1.5$–2). At low temperature, our data points lie between those of Vermeulen et al. (1988) and those of van Woerkens (1998). While the experimental result of Vermeulen et al. can be reconciled with ours (they

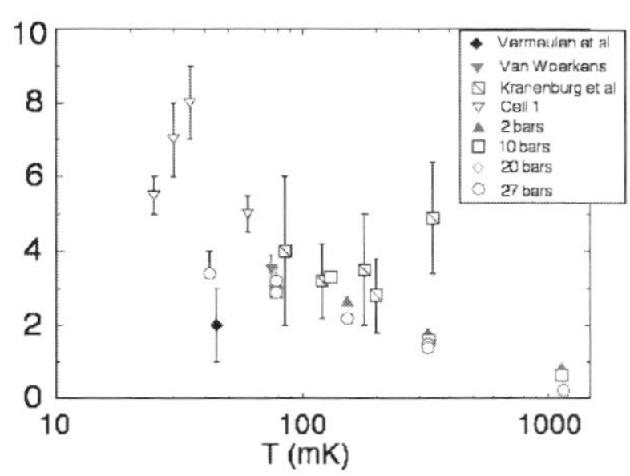

Fig. 25. Comparison of our results (closed symbols) to those of Kranenburg et al. (1989), Vermeulen et al. (1988), and van Woerkens (1998). The result of a preliminary experiment at 27 bar and 80 mK is also shown.

obtain $\alpha = 3 \pm 1$ with an alternative method), we did not find any explanation of the discrepancy between our results and those of van Woerkens. In particular, a fit of our viscosity curves $\eta(m)$ over a polarization range similar to van Woerken's gives values of α very close to those presented on fig. 24, since the deviations from $\Delta\eta/\eta \propto m^2$ are small.

We will first analyze our low-temperature data in light of the models describing degenerate ^{3}He. In a second section, we discuss the high-temperature results.

6.4.1. *Degenerate regime*

The main experimental facts we have observed at $T \leqslant 80$ mK are the increase of the viscosity with the polarization proportionally to m^2, and the weak pressure dependence of the pre-factor α. The models available to interpret these facts are the phenomenological Landau theory and the 'nearly ferromagnetic' model. Unfortunately, the 'nearly localized' model does not provide any prediction about the viscosity. We will therefore focus on the first two models.

6.4.1.1. *Landau theory.* Landau theory is the natural framework to interpret the experiments on liquid ^{3}He; unfortunately, its extension to spin-polarized systems does not give a simple answer about the polarization dependence of the viscosity. As we have seen in Section 3.1.2, the answer depends on the model introduced for the probability of collisions between the quasiparticles. In particular, the commonly accepted opinion that the viscosity should increase with the polarization because of the reduction of the number of collisions turns out to be wrong under certain assumptions. Here, we summarize the results of the different approximation schemes used to calculate the viscosity within the Landau theory. The computational details are given in Appendix D.

6.4.1.1.1. *s-wave limit.* The simplest assumption for the collision probability is to consider isotropic (s-wave) collisions, the Pauli principle requiring that they take place only between quasiparticles of unlike spin. As expected, this assumption leads to a monotonic increase of the viscosity with the polarization. It can be shown that the viscosity enhancement is a purely numeric factor, which does not depend on the collision probability. In particular, the viscosity enhancement does not depend on the pressure, in agreement with our observations. The order of magnitude of the effect at moderate polarization is consistent with our experimental results within our uncertainties (the leading theoretical correction for small polarization is actually $\propto |m|$ instead of the observed $\propto m^2$ behavior, but this linear term is dominated by the quadratic term above $m = 0.04$, see fig. D.1). The strong curvature predicted at large m, corresponding to a $\beta \approx 5$, is however clearly in disagreement with our experiments (we recall that for the 40 mK experiment of fig. 19a, $\beta \approx 2$. See also fig. 19d).

6.4.1.1.2. *s–p approximation.* The s–p approximation is expected to give more realistic predictions for the viscosity, since collisions of higher angular momentum are taken into account. However, the predicted behavior depends strongly on the model taken for the polarization dependence of the collision probabilities. As we have seen in Section 3.1.2, this issue is controversial: Hess and Quader, who used their own ansatz for the 'spin–flip' collision term in conjunction with the polarization-dependent Landau parameters given by the 'nearly metamagnetic' model predicted a strong initial decrease of the viscosity with the polarization, mainly reflecting an increase of the singlet scattering, due to the particular dependence assumed for the polarization-dependent Landau coefficients. This decrease is in contradiction with the experiments. Sanchez-Castro and Bedell (1989) argued that the anomalous decrease would not occur with a proper treatment of the phase space restrictions for the 'spin–flip' collisions, and that the polarization-dependent Laudau interactions, which form the basis of the 'nearly metamagnetic' model, would then lead to the correct behavior for the viscosity. However, as discussed in Appendix D, we believe that the phase space restriction on the number of spin–flip collisions have been correctly accounted for in Hess and Quader's calculation.

In an attempt to clarify the debate, we have undertaken our own calculations, based on the original paper of Anderson et al. (1987) (see Appendix D). If we neglect the polarization dependence of the probabilities of collisions and the effective mass, the s–p approximation cannot account for our experimental results: a first surprising feature is that the angular dependence of the probability of collisions (absent in the s-wave approximation) causes an initial decrease of the viscosity (although much smaller than in Hess and Quader's calculation), at variance with the experimental results (see fig. 33 in Section 7). Also, the effect of polarization on viscosity is much smaller than predicted by the s-wave approximation, and experimentally measured. If we take into account the polarization-induced decrease of the effective mass m^* (Section 8), this will result in a larger effect of polarization, due to the strong $(1/m^{*4})$ dependence of the viscosity. But this also implies strong deviations from a pure m^2 behavior, which, again, we do not observe (fig. 26b).

From our investigations, one may draw the following conclusions: the peculiar behavior calculated by Hess and Quader may arise partly from the use of the s–p approximation, and partly from the choice of the polarization dependence of the scattering amplitudes. So, one cannot draw conclusions on the validity of the 'nearly metamagnetic' model on these grounds. Furthermore, the use of the s–p approximation, where the Landau interactions $l \geqslant 2$ are neglected, seems unable to predict the right behavior for the

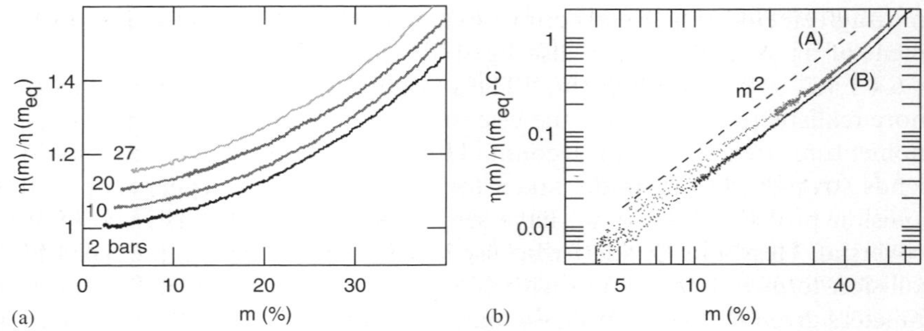

Fig. 26. The viscosity enhancement versus polarization m: (a) a linear plot for the four experiments of Fig. 14d does not show any initial linear decrease of the viscosity (for clarity, the different curves are vertically shifted by 0.05 from one pressure to the next). (b) A log–log plot exemplifies the m^2 behavior (dashed line) over the polarization range explored: (A) 40 mK, 27 bar; (B) 80 mK, 10 bar. The constant $C = \eta(0)/\eta(m_{eq})$ is adjusted to preserve the m^2 behavior at the lowest polarizations. It is 1.005 and 1.003, respectively, the difference from the expected values or $\alpha = 3$ (0.995 and 0.997) being possibly due to a small drift of the resonance width between the end of the relaxation and the calibration curve.

viscosity, whatever the choice for the polarization dependence of the Landau coefficients. The solution may be to include higher order Landau coefficients in the s–p approximation or to go beyond the s–p approximation itself. Even beyond the Landau transport theory, the observation of a regular m^2 behavior over such a wide range of polarizations ($0<m<0.6$), which covers both small and large values of the ratio $\mu_3 B/(kT)$ (for curve (A) of fig. 26, $\mu_3 B/(k_B T) = 1$ for $m \approx 25\%$) is to be emphasized. In our view, it represents a significant constraint on theories.

6.4.1.2. *Paramagnon theory.* The paramagnon theory correctly predicts the increase of the viscosity with the polarization and the initial m^2 behavior of the correction. The weak pressure dependence we observe, however, is in strong disagreement with the paramagnon theory: the predicted pre-factor increases by a factor of 2 between 2 and 27 bar, whereas we observe a variation smaller than 10%.

Strictly speaking, $T = 80$ mK is too high a temperature to apply the paramagnon theory, so we cannot compare directly our results at this temperature with the predictions of the 'nearly ferromagnetic' model (unfortunately, we have only one measurement at $T = 40$ mK and 27 bar; measurements at lower pressure at the same temperature would be difficult, due to the extra heat released by the decompression of the liquid). We may only conclude that the scattering of fermions by paramagnons does not account for the viscosity of ^{3}He at $T \geqslant 80$ mK. Although we cannot exclude that this mechanism becomes effective at lower temperature, it would be surprising from our point of view: the trend observed experimentally is that the lower the temperature, the smaller the pressure dependence.

6.4.2. *Non-degenerate regime*

The value of the pre-factor α decreases from 3.5 to <1 when the temperature is increased from 40 mK to 1 K. The decrease is monotonic, except for an 'anomalous' data point at 10 bar and 130 mK (this experiment, unfortunately, has not been repeated). The pre-factor measured by Kranenburg et al. (1989) at 350 mK is much higher than our results at the same temperature (see fig. 25), but we believe our values to be correct, due to our smaller error bars.

The pressure dependence of the pre-factor α is larger at high temperature than at low temperature. Equivalently, the characteristic temperature scale of the effect of polarization decreases with increasing pressure. This behavior is accounted for if we identify this characteristic scale with the 'magnetic Fermi temperature' T_{SF} (different for each pressure). Indeed, a plot of α as function of the reduced temperature T/T_{SF} (fig. 27), reveals a similar dependence for all pressures. However, more experiments would be needed to confirm the existence of a true scaling law as a function of T/T_{SF}. The decrease of the polarization-induced viscosity enhancement with the temperature shows that the effect of Fermi statistics decreases beyond the degenerate regime. This was not *a priori* obvious, as the scattering between

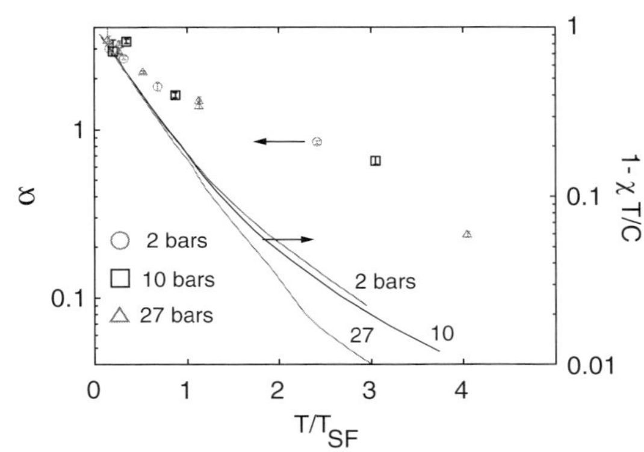

Fig. 27. Sensitivity to temperature of the effect of quantum statistics on thermodynamics and transport: for the various pressure studied, the effect of polarization of viscosity (measured by α), decays slower with temperature than the deviation $1 - \chi T/C$ of the susceptibility from the Curie law (continuous curves). The temperature scale is normalized by the spin fluctuation temperature T_{SF}.

bare ^{3}He atoms is always sensitive to the Pauli principle (see Section 3.1.2). According to Lhuillier (1991), this decrease can be understood in the following way: in the non-degenerate regime, the quasiparticles are not well-defined excitations of the system any more (their lifetime becomes extremely short), so spin-statistics effects could be expected to be washed out due to the large density of the liquid. This argument accounts for the fact that the magnetic susceptibility at $T > T_{SF}$ approaches the Curie-like behavior ($\chi = C/T$) rapidly, which indicates that the spin correlations induced by the Pauli principle become quickly negligible above the degeneracy temperature.

From this point of view, the surprising fact is not the decrease of the effect of polarization with temperature, but rather its persistence, well above the 'magnetic Fermi temperature'. This is illustrated in fig. 27, where we have plotted the deviation from the Curie law $1 - \chi T/C$ along with the pre-factor α; this comparison clearly shows that the spin-statistics effects on the viscosity are still present in a temperature range where the deviation of the susceptibility from the Curie law are small. This behavior is similar to that of a dilute Fermi gas where the transport properties show spin-statistics effects even in the non-degenerate regime. This similarity between liquid ^{3}He and dilute systems remains to be explained.

A final remark concerns the effect of polarization at 0.3 and 1.2 K. It is noteworthy that, for all pressures, the polarization dependence is nearly quadratic in m^2 over the full polarization range (up to $m \simeq 0.6$). One could have thought that, for such large polarizations, the very different Fermi energies of up and down spins could have resulted in a change of the behavior of the viscosity with the polarization, in a similar way to the phenomenon of 'semidegeneracy' for a Fermi gas (Meyerovich 1978). This does not seem to happen. The observed regular behavior also rules out the possibility of a ferromagnetic phase of ^{3}He at large temperature discussed by Bashkin (1984b).

7. Thermal conductivity

7.1. *Aim of the experiment*

As we just discussed in Section 6, the polarization dependence of the viscosity of liquid ^{3}He bears a strong similarity with that of a dilute Fermi gas, in that it persists up to high-temperature, into the non-degenerate regime. Another intriguing similarity is the low-temperature behavior: the viscosity increases linearly with m^2, with a slope independent of pressure and very close to the s-wave prediction. In contrast, the behavior predicted within the s–p approximation using the formalism developed by Anderson, Pethick and Quader (Anderson et al. 1987) falls far from the experiment.

This situation called for the measurement of the thermal conductivity. For s-wave scattering in degenerate systems, the increase in thermal conductivity is expected to be linear in polarization over a wide range (Bashkin and Meyerovich 1978, 1981, Mullin and Miyake 1983), with an increase of 60% at $m = 0.6$. As we discuss in Appendix D, this is a consequence of the phase space restriction induced by the conservation of momentum during a collision between unlike spins. The measurement of such an effect would be a test of s-wave behavior.

Measuring the thermal conductivity in transiently spin-polarized liquid ^{3}He is challenging, as the measurements have to be performed on the time-scale of a second, quite smaller than normally possible for ^{3}He (Greywall 1984). We were however able to use the setup of our viscosity experiment to meet this challenge (Sawkey et al. 2004). The idea is to use the viscometer not only as a fast thermometer (as in the specific heat experiment described in the next chapter), but also as a local heater. This chapter is organized as follows. We first discuss the principle of the method. Second, we show how measurements in the unpolarized liquid allow as to demonstrate and calibrate the sensitivity of the device to changes in the thermal conductivity of the surrounding ^{3}He. We then describe our experimental results for polarized ^{3}He, and finally discuss their theoretical implications.

7.2. *Principle of the experiment*

The idea is to heat liquid ^{3}He using the wire as a quasi 1D heat source, and to measure the increase of the local temperature around the wire (the HT temperature being fixed) through the induced change in the viscosity (fig. 28a). The effect is shown in fig. 28b for different temperatures at 27 bar. The advantage of the wire geometry is that a large part of the temperature increase can be expected to be controlled by the neighboring ^{3}He rather than by the silver sinter thermal conductivity, as would be the case for a plane heater, parallel to the slit. Two different methods can be used to heat the wire. One is to add to the AC excitation current a large DC current, which heats the resistive wire by the Joule effect. The second is to use a large AC excitation current, which mainly heats the liquid through the viscous dissipation[22] (the Joule effect also exists in this case, but is in practice much smaller). Typical numbers for the currents and powers used at 60 mK, and the corresponding temperature increase, are 50 μA, 45 nW and 5 mK for AC heating, and 250 μA, 3 μW and 6 mK for DC heating;

On the one hand, DC heating is less efficient as the heat is dissipated over the whole wire length (90 mm) rather than concentrated over its moving

[22]The heating power is computed from the in-phase-induced voltage at the wire terminals.

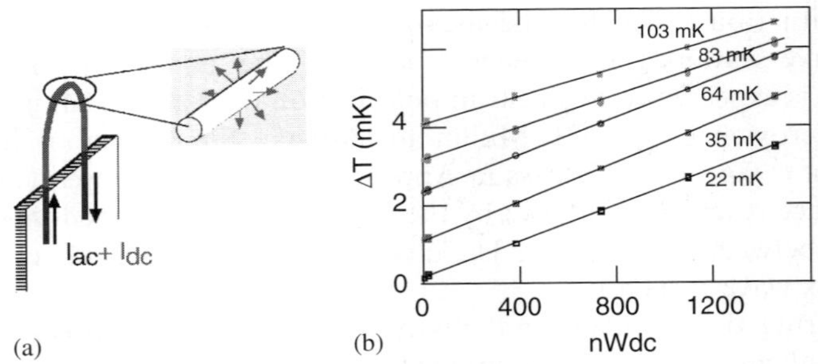

Fig. 28. Measurement of the thermal conductivity using the viscometer. (a) Using a large AC drive or adding a large DC current to the drive results in a radial heat flow round the wire and an increase of the temperature around the wire, which is probed through the induced change in viscosity. (b) Increase of the viscometer temperature as a function of the Joule power associated with the dc current, for different temperatures at 27 bar. For clarity, a different offset is used for each temperature. Note that the effect is linear over the explored range of powers.

portion (≈ 1 mm), where the temperature is measured. On the other hand, as discussed in (Buu et al. 2000a), AC viscous heating has the disadvantage that overdriving the wire causes an extra damping of the resonance, which partly offsets the temperature induced decrease of the viscosity, and makes imprecise the determination of the temperature increase above 100 mK. Furthermore, the large perturbation of the wire motion associated with the AC overdrive was found prone to induce changes in the viscometer state (a notion discussed in Section 6.3.3). For both reasons, DC heating was used in the experiments presented here.

In this case, one might worry about the possibility for the heat flux to be partly evacuated through the wire connections rather than through the ^{3}He. Under our conditions, this is not a concern. Taking from Lounasmaa (1974) the metal–^{3}He Kapitza resistance $R_{\mathrm{K,metal}}\, T^3 = 10^{-2}\,\mathrm{K}^4/\mathrm{m}^2/\mathrm{W}$, and a thermal conductivity of the wire $\kappa_{\mathrm{w}} = 0.1\,\mathrm{W/K/m}$, we find the characteristic length for thermal exchange, $2a(R_{\mathrm{K,metal}}\kappa_{\mathrm{w}})^{1/2}$, to be 1.5 mm at 50 mK, small compared to the total wire length. This implies that the Joule heating produced close to the center of the viscometer loop goes directly into the ^{3}He. This was confirmed by checking that the effects of AC and DC heating have a similar temperature dependence between 20 and 80 mK, which would not be the case if the metal – ^{3}He Kapitza resistance were to play a role.

7.3. *Sensitivity of the device from measurements in the unpolarized liquid*

As was the case for the effective resistance discussed in Section 6.3.3 and in Section 8, the thermal resistance R_κ associated with the present measurements

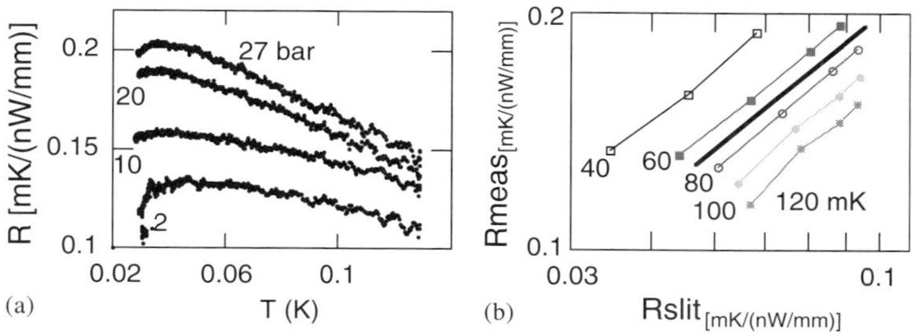

Fig. 29. (a) Temperature dependence of the total thermal resistance R_κ between the viscometer wire and the HT, for four different pressures (2, 10, 20, and 27 bar). These curves were measured by using a low-frequency square-wave modulation of the DC current fed into the viscometer (alternatively 0 and 250 μA every 30 s), during a slow cooling. Points in the 'cold' state give the temperature dependence of the viscosity η_{eq} at equilibrium polarization, while the comparison of the two states gives R_κ. (b) A plot of $R_\kappa(T, P)$ for selected temperatures (40–120 mK by steps of 20 mK from left to right) as a function of the expected ^{3}He thermal resistance for a slit with perfect thermal contact at the wall for the same temperatures and pressures (from bottom to top, 2 to 27 bar). The thick line corresponds to a slope 0.7, so that, at 60 mK, $R_\kappa \propto \kappa_3^{-0.7}$.

involves the thermal resistivities of the bulk ^{3}He in the slit and of the sinter, and the Kapitza boundary resistance. However, the 1D heating geometry makes the relative contribution of ^{3}He dominant. As a test, we varied the ^{3}He resistivity by changing the pressure at a given temperature, thus keeping the other parameters fixed.[23] Measurements of R_κ at equilibrium polarization from $T = 30$ to 110 mK at 2, 10, 20, and 27 bar, are shown in fig. 29a. A plot of the measured R_κ versus the expected ^{3}He resistance R_{slit} for a straight wire inside a slit with perfect thermal contact at the walls is given in fig. 29b on a logarithmic scale. Although R_κ is larger than R_{slit}, due to the non-zero boundary resistance and the sinter resistivity, it is quite sensitive to the ^{3}He resistivity. In the temperature range of the rapid melting experiments (60–80 mK), a 10% increase in the ^{3}He conductivity κ_3 decreases our measured R_κ by 7%, corresponding to $\kappa_3 \propto (R_\kappa)^{-1/0.7}$. This sensitivity is in reasonable agreement with a numerical calculation based on the parameters of Appendix C, which predicts a coefficient of 0.78 instead of 0.7. Despite the presence of the sinter, R_κ is thus a sensitive probe of the ^{3}He thermal conductivity (Sawkey et al. 2003).

[23]This makes the reasonable assumption – see section 8.3.2 – that the boundary resistance is independent of pressure, just as the analysis below of the experiments with the polarized liquid assumes that the boundary resistance is independent of polarization.

7.4. *Measurements in the polarized liquid*

Figure 30a shows the result of a typical rapid melting experiment, at 27 bar and a regulated HT temperature of 65 mK. The time evolution of the polarization and of the inner carbon resistor temperature are similar to those observed in the viscosity experiments described in Section 6. However, for the thermal conductivity measurements, we superimposed a 250 μA DC current on the AC drive of the viscometer. The DC heating current was square-wave-modulated with a period of 18 s, resulting in an oscillation of the resonance width between a 'cold' state and a 'hot' state. This allows us to measure the thermal resistance R_κ as a function of polarization provided the resonance width can be translated into a local temperature. This is the most delicate step of the analysis, due to the polarization dependence of the viscosity.

By a local fit of the viscometer signal versus the polarization squared over three consecutive states (see fig. 30b), we can determine the ratio of the viscosities in the 'hot' and 'cold' states, as a function of polarization. To proceed, we express this ratio as

$$\frac{\eta(m, T+\delta T)}{\eta(m, T)} = \frac{\eta_{\mathrm{eq}}(T+\delta T)}{\eta_{\mathrm{eq}}(T)} \frac{f(m, T+\delta T)}{f(m, T)}, \tag{7.1}$$

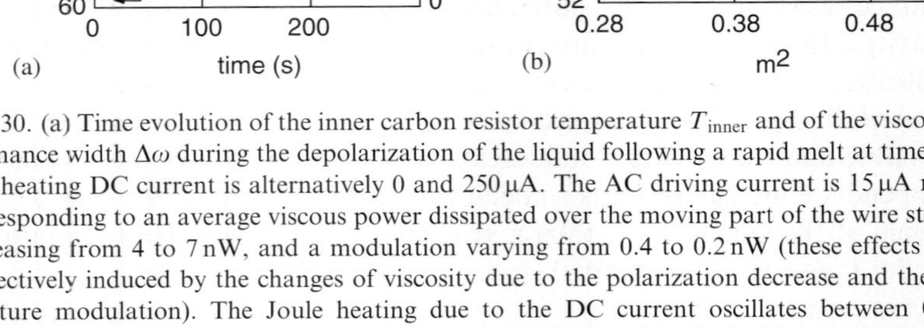

Fig. 30. (a) Time evolution of the inner carbon resistor temperature T_{inner} and of the viscometer resonance width $\Delta\omega$ during the depolarization of the liquid following a rapid melt at time zero. The heating DC current is alternatively 0 and 250 μA. The AC driving current is 15 μA r.m.s., corresponding to an average viscous power dissipated over the moving part of the wire steadily increasing from 4 to 7 nW, and a modulation varying from 0.4 to 0.2 nW (these effects being respectively induced by the changes of viscosity due to the polarization decrease and the temperature modulation). The Joule heating due to the DC current oscillates between 0 and $\approx$30 nW/mm, corresponding to a typical temperature change of 6 mK. (b) Fitting procedure for determining the ratio of the viscosity in the 'cold' and 'hot' states. For each sequence of three states (dark line), we eliminate the transients by keeping the last 4 s of each step (shown as a thicker line). The ratio $\eta(m, T+\delta T)/\eta(m, T)$ is determined such that, after correction by this ratio, the data (gray lines) lie on a single local parabola (thin line).

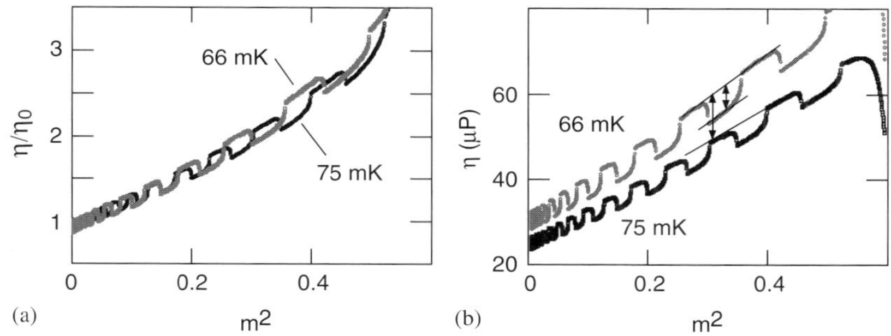

Fig. 31. (a) Polarization dependence of the viscosity enhancement factor $f(m, T_{\text{inner}}) = \eta(m)/\eta_{\text{eq}}(T_{\text{inner}})$ for two experiments with two different HT temperatures, 66 and 75.5 mK, respectively. (b) Polarization dependence of the viscosity for these two experiments: the arrows illustrate how we interpolate the viscometer 'hot' state temperature in the first experiment from the two viscometer 'cold' state temperatures (obtained from the value of the inner thermometer at the same polarization).

where $f(m, T)$ is the ratio, at the temperature T, of the viscosities in the polarized (η) and equilibrium (η_{eq}) states. Thus, extracting the temperature change from the experiment requires a knowledge of the temperature dependence of $f(m, T)$. We have determined this dependence by performing the viscosity measurement at two different temperatures, and found it to be weak, but not negligible. However, as we discuss below, such an approach requires a level of precision which might be unrealistic. Therefore, we will compare in the following the thermal conductivity obtained within this approach to that obtained assuming a temperature independent $f(m, T)$.

The first approach is illustrated in fig. 31a, which shows the polarization dependence of $f(m, T_{\text{inner}})$ for two experiments with two different HT temperatures, 66 and 75.5 mK, respectively. At $m^2 = 0.4$, f is about 5% smaller at the larger temperature, corresponding to a logarithmic derivative $d\ln(f)/d\ln(T) \approx -0.4$, not much smaller than $d\ln(\eta_{\text{eq}})/d\ln(T)(\approx -1.4)$. Hence, from eq. (7.1), this effect, albeit small, will significantly reduce R_κ with respect to the case of a temperature independent $f(m, T)$. However, the small temperature dependence of the enhancement factor $f(m, T)$ observed could well be due to experimental artifacts. As an example, we have seen in Section 6 that the measured polarization m is an average over the cell, rather than the polarization around the wire. Therefore, it might be that the observed change of α with temperature results from a 2% change of the polarization around the wire, at a given m, (due to the temperature dependence of both T_1 and spin diffusion) rather than from a true physical effect. A other uncertainty factor is the accuracy of the inner thermometer to measure the 'cold' state temperature of the viscometer (Section 6.3.3). In order to avoid biased conclusions, we have also computed the polarization dependence of R_κ assuming a temperature independent $f(m, T)$, as can be expected at sufficiently low temperature.

During depolarization, R_κ varies due to both the decay of polarization and the associated decrease of the slit temperature (see Section 6.3.1). For the experimental temperature decrease of 10 mK, the induced increase of R_κ is about 3%. We correct for this change by dividing $R_\kappa(m, T)$ measured during the depolarization by $R_\kappa(m_{\text{eq}}, T_{\text{inner}})$, where $R_\kappa(m_{\text{eq}}, T)$ is measured at the same time as $\eta_{\text{eq}}(T)$ (fig. 29a). Figure 32 thus presents $R_\kappa(m, T)/R_\kappa(0, T)$ for the results of fig. 30a, under the two assumptions above. The continuous line, which accounts for the possible temperature dependence of $f(m, T)$, is computed by interpolating from the viscosity and for any polarization, the temperature in the 'hot' state from the two 'cold' state temperatures (fig. 31b). The filled squares are computed from fig. 30a only, assuming $f(m, T)$ to be temperature-independent. As expected above, for the large polarizations, R_κ is smaller when the first hypothesis is used, the reduction being 25%. For $m < 0.4$, the difference between the two hypotheses is much smaller, as the temperature dependence of $f(m, T)$ is reduced at low polarization where $f \to 1$. We have checked that, in our experimental conditions, the modulated part of the AC drive heating effect (see legend of fig. 30a) had a negligible influence on the determination of R_κ (less than 1% at the maximal polarization, when the viscosity enhancement is the largest). The precision on $R_\kappa(m)$ then only depends on the DC power. For a relative change of the viscometer width of 10% induced by the DC

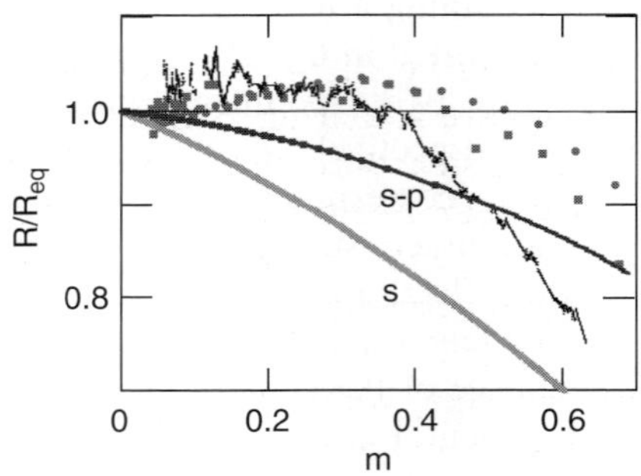

Fig. 32. Polarization dependence of the normalized thermal resistance $R_\kappa(m, T)/R_\kappa(m_{\text{eq}}, T)$ at 27 bar and a base temperature of 66 mK: the filled squares and circles correspond respectively to heating powers of about 30 and 100 nW/mm, resulting in a local temperature elevation of about 6 and 20 mK, and to the assumption of a temperature-independent enhancement of the viscosity by polarization. The continuous line corresponds to the result of the first experiment, when the interpolation scheme of fig. 31b is used instead of the latter assumption. The thick and thin lines are respectively the predictions for s-wave and the s–p approximation, assuming $R_\kappa \propto \kappa_3^{-0.7}$.

current (typical for a 250 μA DC current, at 65 mK), the precision is estimated to be of the order of $\pm 2\%$. In order to improve the signal to noise ratio, we also performed an experiment where a three times larger DC power was applied. The results, which assume a temperature independent $f(m, T)$, are shown as filled circles in fig. 32. Below $m = 0.5$, the results of both experiments analyzed using the same method agree within the experimental uncertainty, and show that, if anything, R_κ initially slightly increases with increasing polarization up to $m \approx 0.3$. At larger polarizations, R_κ decreases, but by less than 10% at $m \approx 0.7$, as compared to about 15% for the first experiment. This discrepancy might result from some bias in the determination of the ratio of viscosities due to the fast variation of the resonance width at short times, this bias having a smaller weight in the case of the larger heating power. Alternatively, the modulation of the viscometer temperature, which is about 30% of the absolute temperature, might be too large in the latter case.

The conclusions that we draw from these experiments and others is that, for polarizations smaller than about 40%, the thermal conductivity changes by less than 5%. For larger polarizations, the conductivity increases with polarization. The polarization dependence of the viscosity is a factor limiting the precision of the results.[24] For a constant $f(m, T)$, the thermal conductivity increase is less than $15\%/0.7 \approx 20\%$. This number is larger if this hypothesis is not made. We shall now show that, despite the resulting uncertainties, our experiments are still able to rule out the s-wave picture.

7.5. *Analysis*

Let us compare the results of fig. 32 to the theoretical prediction of Mullin and Miyake (1983) for s-wave scattering. The thick line corresponds to the hypothesis $R_\kappa \propto \kappa_3^{-0.7}$, consistent with our measurements in the unpolarized liquid (fig. 29). For both our two estimates, the effect of polarization is smaller than predicted from the theory. For $m < 0.4$, in which range the problem of the temperature dependence of $f(m, T)$ is not acute, the discrepancy is quite large. At $m = 0.4$, the measured R_κ is, within $\pm 2\%$, equal to its zero polarization value, while the predicted value is nearly 20% smaller. This difference is beyond our experimental uncertainties. Moreover, if we accept the validity of the more precise data at large DC power, R_κ increases slightly with polarization, up to $m \approx 0.4$, instead of decreasing.

Therefore, the thermal conductivity measurements, in contrast to those of the viscosity, show that pure s-wave scattering cannot account for the

[24] A solution based on the same geometry would be to deposit a resistive thermometer onto the wire, but the thermal coupling to the surrounding liquid may then be a problem.

polarization dependence of the transport properties of ^{3}He. Although the uncertainties mentioned above make it difficult to determine the precise analytical form of $\kappa_3(m)$ (i.e. linear or quadratic), it is clear that $\kappa_3(m)$ barely depends on m below $m = 0.4$. As we discuss in Appendix D, the linear increase of $\kappa_3(m)$ for s-wave scattering follows from the suppression of head-on collisions between unlike spins (see fig. D.1b). The absence of such an effect suggests that these collisions are unfavoured in liquid ^{3}He. Indeed, within the s–p approximation, the sign of the large constant and $\cos(\theta)$ terms in the scattering amplitude $T_{\uparrow\downarrow}(\theta, \phi)$ is the same (see eq. (D.14)), which makes these collisions less likely. However, the polarization dependence computed using Anderson et al. (1987) and the s–p approximation, although less marked than for s-wave, remains stronger than observed.

7.6. *Conclusions on the transport in polarized liquid ^{3}He*

We summarize the situation in fig. 33, where we compare the polarization dependence of both the viscosity and the thermal conductivity (computed using $\kappa_3 \propto (R_\kappa)^{-1/0.7}$) to the predicted behavior for s-wave and the s–p approximation (assuming the Landau parameters and the effective masses to be polarization independent, see Appendix D). For the viscosity, the predicted variation for s-wave correctly accounts for the results, while the variation predicted within the s–p approximation is too small. By contrast, for

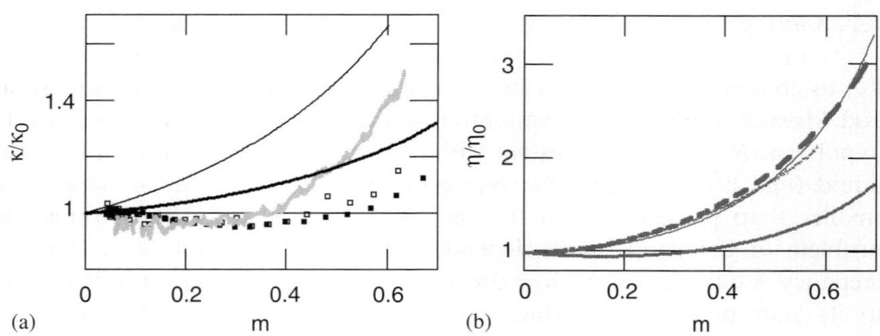

Fig. 33. Comparison of our results at 27 bar to theoretical predictions. Left panel: polarization enhancement of the thermal conductivity (assuming $\kappa_3 \propto (R_\kappa)^{-1/0.7}$) for the data of Fig. 32. The thin line is the prediction of (Mullin and Miyake (1983), the thick line is our prediction within the s–p approximation, using the formalism of Anderson et al. (1987) and polarization independent Laudau coefficients: Right panel: the same for the viscosity enhancement. Two sets of data are plotted: the thin line corresponds to the experiment of Fig. 19 at 80 mK and the dashed line to the slightly colder experiment of fig. 30 (the viscosity in the 'hot state' has been corrected back to the 'cold state' using the procedure of fig. 30).

the thermal conductivity, the s-wave overestimates the effect of polarization. Although the s–p approximation works better, the predicted variation at moderate polarization is clearly too large. Therefore, the overall behavior is more complex than for a simple gas. This is certainly not surprizing, but the difference between the viscosity and the thermal conductivity is striking. Whether this is just a coincidence or a consequence of some deeper physical property remains to be understood.

8. Polarization dependence of the ^{3}He specific heat

8.1. *Principle of the experiment*

As we have seen in Section 3, the polarization dependence of the specific heat of liquid ^{3}He is very different for the two 'extreme' models of liquid ^{3}He discussed in Section 2. Unfortunately, the measurement of the specific heat in polarized ^{3}He is difficult, due to its low thermal diffusivity, and also because the thermal effects associated with the rapid melting technique (see Section 4) render the use of the conventional adiabatic method impossible. The improvement in thermal diffusivity brought by confining the liquid inside a silver sinter opened the way to such measurements. A first attempt using the setup of our susceptibility experiment (Section 5) failed due to the dominant contribution of the HT specific heat (Bravin et al. 1994). Eventually, the key to the success was to use the weaker coupling between the HT and the sample, together with the ability to measure the ^{3}He temperature, which was provided by our viscosity cell (Buu et al. 2000c).

The method we use consists of measuring the thermal equilibration time of the sample following a temperature step during the magnetization relaxation. This thermal time constant can be shown (see Appendix C) to be proportional to the heat capacity of the sample. The proportionality constant is a thermal resistance which, in our experimental arrangement, is weakly polarization-dependent. We have observed that the thermal time constant measured in polarized liquid ^{3}He is shorter than at equilibrium. We interpret this result as a *decrease* of the heat capacity with the spin polarization. This result is discussed along with the theoretical models we have introduced in Sections 2 and 3. The technical details are given in Appendix C.

8.2. *Experimental procedure*

The specific heat measurements we present here have been carried out with the experimental setup we used for our viscosity measurements (Section 6). The production of polarized liquid ^{3}He by the method of rapid melting

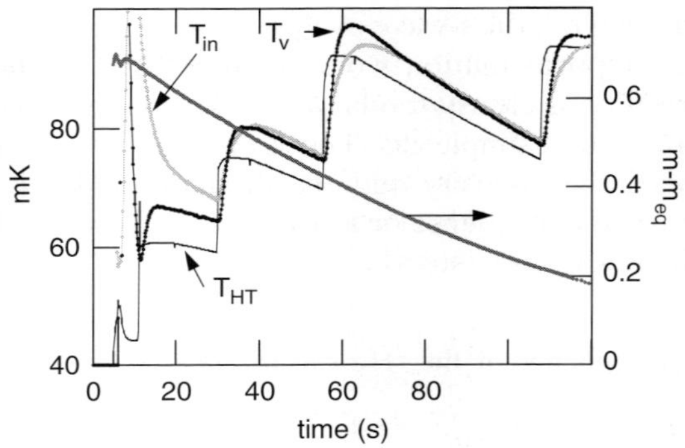

Fig. 34. Time evolution of the viscometer temperature T_{visco} (dark line) and of the inner thermometer temperature T_{inner} (gray line) following steps of the HT temperature T_{ht} (thin line). This experiment is performed during the relaxation of the ^{3}He polarization m, after melting the solid at zero time, by decompressing the filling line to 27 bar.

follows the procedure described in Section 5.4. For the viscosity measurement, the HT temperature was kept constant with thermal regulation during the polarization relaxation following melting. For the specific heat measurement, we use the thermal regulation to change the HT temperature by steps, and we monitor the subsequent evolution of the temperature at the center of the cell. We have performed the analysis using the temperature read by the viscometer rather than the inner thermometer, because of the anomalously long response of the latter (see below). The viscometer reading $\eta(m, T)$ is corrected for the polarization by using $\eta_0(T) = \eta(m, T)/(1 + \alpha m^2)$ (with $\alpha \approx 3$ at $T < 100$ mK according to Table D.4), before being converted into a temperature using a separate calibration of $\eta_0(T)$ carried out at equilibrium magnetization.

Figure 34 shows the temperature evolution during a typical experiment: after the decompression of the cell at time $t = 0$, the HT temperature T_{ht} raises to 50 mK while the inner temperatures T_{visco}[25] and T_{inner} increase up to 200 mK, due to the heat released by the melting process. At $t = 10$ s, T_{ht} stabilizes around 45 mK, while T_{visco} and T_{inner} relax toward this value (as noted in Section 6.3.3, the transient is slower for T_{inner}). At this point, we increase the HT temperature to 60 mK and measure the response of T_{visco} and T_{inner} during 7.5 s, at a polarization of 70%. The thermal regulation is then turned off and the system is let to cool down, owing to the contact with

[25]Here, we calculate T_{visco} using $\alpha = 3.4$ in order to ensure $T_{visco} \approx T_{inner}$ between the steps. However, the response time τ introduced below barely depends on this choice: choosing $\alpha = 2.4$ would modify its value by only 2%.

the mixing chamber. This pattern is repeated several times during the polarization decay. Between the steps, $T_{\rm visco}$ and $T_{\rm inner}$ lie above the temperature of the HT ($T_{\rm ht}$) because of the heat released by the magnetic relaxation. During the steps, the carbon thermometer lags behind the viscometer, as in the initial transient. This anomaly we have noticed during the viscosity measurements (Section 6.3.3) is discussed in Appendix C. A close examination of fig. 34 reveals that the response time of the cell is different for each step. This is due to the combined effects of polarization and temperature (see fig. C.2). In order to single out the effect of polarization, we carry out the same experiment in the unpolarized liquid. Starting with a cold sample, the initial temperature increase is obtained by applying a 3 s long heat pulse with the heater located at the bottom of the cell. The energy of the pulse (3 mJ) is chosen to obtain the same HT temperature ($\approx$ 45 mK) as in the melting experiment, and the same series of steps is then performed. One finds that, for a given temperature step, the response is faster in the polarized liquid. This is illustrated in fig. 35 (a and b) for the second step. At $m = 0.55$, both the viscometer and the inner thermometer response time are shorter by about 20%. This can be seen more clearly in fig. 35 (c and d) on the normalized signals (the normalization procedure is detailed in the next section). These observations are an indication that the specific heat decreases with the polarization. In the following, we focus on the response of the viscometer, and discard the inner thermometer, due to its anomalously slow response.

8.3. *Data analysis*

8.3.1. *Thermal response time*

In order to quantify the effect of polarization, we determine the response time τ of the viscometer temperature with respect to the HT, defined as the time integral of the response function $S(t)$ of the viscometer temperature (eq. (C.5)). In principle, the response time can be measured by the time integral of the normalized temperature difference between the viscometer and the HT following a temperature step. This analysis cannot be directly performed on the results of fig. 35 because of its sensitivity to, first, the temperature offset arising from the power released by the magnetic relaxation (which is constant on the timescale of the thermal response measurement), and second, small calibration uncertainties of the viscometer versus the HT thermometer.

In practice, we analyze the data in the following way: first, we fit the final part of the relaxation of the viscometer temperature (corresponding to the last 30% of the step height) to an exponential, which determines the final

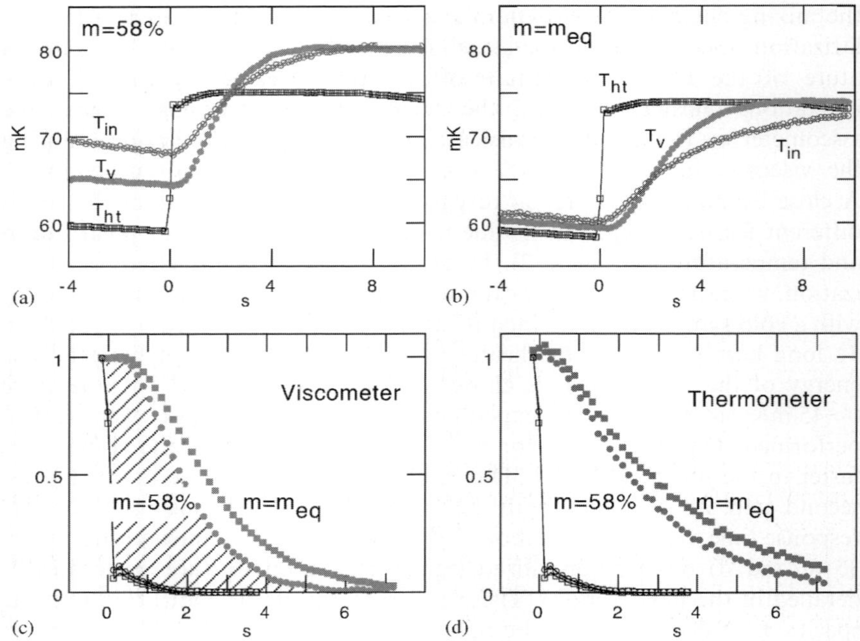

Fig. 35. A blow up (a) of the second step of Fig. 34 at $m = 0.58$, is compared to a similar step performed in unpolarized liquid (b). (c) (resp. (d)) shows the normalized variations of the viscometer (resp. inner thermometer) temperature, for both the polarized (circles) and unpolarized (squares) liquid. The carbon thermometer temperature relaxes more slowly than that of the viscometer, but, in both cases, the relaxation timescale is reduced by about 20% by the polarization. The hatched area in (c) gives the thermal response time τ.

temperature T_f by extrapolation. The initial temperature T_i is taken at the last data point before the step. The normalized response $(T_f - T_{visco}(t))/(T_f - T_i)$ is then integrated numerically over the time. The resulting quantity is taken as the viscometer characteristic time.[26] We apply the same procedure to the HT thermometer, which yields the effective time constant of the regulation system (of order 0.2–0.4 s, depending on the temperature and on the regulation parameters). The thermal response time of the cell τ is then given by the difference between these two times (represented by the hatched area in fig. 35c). This determination of the response

[26]For the first temperature step, a correction has to be applied, as part of the signal comes from the initial fast decrease of the viscometer temperature before the step. Owing to the linearity, this extra contribution can be suppressed by fitting the signal before the step to an exponential, and subtracting this exponential from the raw signal.

time suffers from an error due to the decrease in HT temperature before the step. This slow drift induces a small initial temperature offset between the viscometer and the HT. The relaxation of this offset after the step slightly biases the determination of τ. However, the resulting error is quite small: for a ideally sharp temperature step, the measured response time τ_{meas} is given by

$$\tau_{\text{meas}} = \tau \frac{1 - \tau_1/\tau_s}{1 - \tau/\tau_s} \tag{8.1}$$

with $\tau_1 = \int_0^\infty tS(t)\,dt / \int_0^\infty S(t)\,dt$ and $\tau_s = (T_f - T_i)/\dot{T}$, $\dot{T}$ being the cooling rate of the HT before the step. $S(t)$ is the response function of the viscometer temperature to a change in the HT temperature (see eq. (C.1)). For an exponential response function, $\tau_1 = \tau$, so that $\tau_{\text{meas}} = \tau$. In our case, we have $\tau_1/\tau \approx 0.7$, which gives an error of at most 2% for typical values of $\dot{T}$. Furthermore, because we compare experiments in the polarized and unpolarized liquids, which are performed with similar $\dot{T}$, the influence of this systematic error on our results is even smaller.

The results of our experiments at 27 bar are summarized in fig. 36, showing τ versus polarization, for steps of different initial and final temperatures. Also plotted are two points taken at 10 and 2 bar, which were obtained during the viscosity experiments of Section 6, from the temperature stabilization step (from $\approx$ 60 up to 80 mK, see fig. 17a). In each case, τ was

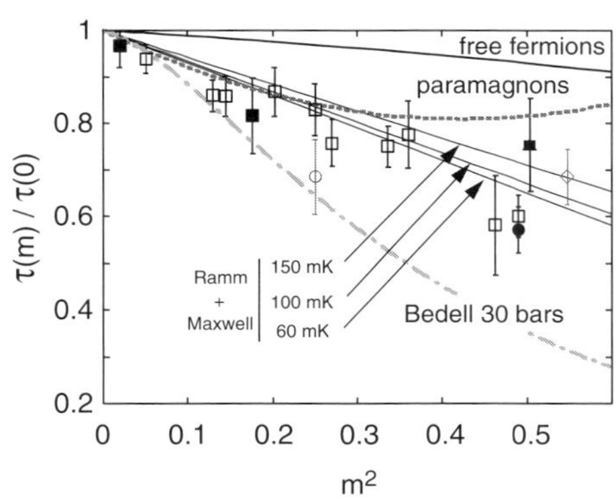

Fig. 36. Polarization dependence of the measured thermal relaxation timescale τ for different final temperatures at 27 bar • 60 mK, □ 75–85 mK, ■ 90–105 mK, ▲ 160 mK, and at a final temperature of 80 mK for 2 (○) and 10 bar (◇). The lines correspond to different predictions for the polarization dependence of the specific heat (free fermions, paramagnons, nearly metamagnetic model), and to the expected initial behavior, based on the experimental temperature dependence of the susceptibility (Ramm et al. 1970) and a Maxwell relation, for three temperatures.

normalized to its value measured in the unpolarized liquid with a similar temperature step. The error bars correspond to estimated errors of 0.05–0.15 s on τ in the polarized liquid (depending on the step height). All the experiments show a reduction of the response time τ with the polarization.

8.3.2. *Specific heat*

We interpret the systematic reduction of thermal response time we have measured in polarized ^{3}He as a decrease of the specific heat with the polarization. To support this interpretation, we have carried out a detailed analysis of the thermal response of the cell, described in Appendix C. Here, we just summarize the main results.

The thermal response time τ is the product of the specific heat of the ^{3}He sample[27] and an effective thermal resistance R_C distributed between the viscometer and the HT. In order to quantify the polarization dependence of the ^{3}He specific heat, we must assess the effect of the polarization on this resistance. As was the case for the resistance R_κ relevant for the conductivity experiment, R_C involves the ^{3}He thermal conductivity, the Kapitza resistance between ^{3}He and the silver sinter, the thermal conductivity of the silver sinter, and the metallic contact between the cell and the HT. However, due to the different spatial distribution of heating in both cases, these two resistances differ. Unlike R_{th}, which is controlled by κ_3, R_C is dominated by the contribution of the silver sinter, which does not depend on the ^{3}He polarization. Experimental evidence of the weak dependence of the response time τ on the ^{3}He thermal conductivity is the small pressure dependence of τ/C_3, which decreases only by 10% when going from 27 to 0 bar at 80 mK (in the non-polarized liquid), while the corresponding increase of κ_3 is 80%.

To investigate this point further, we have developed a model to calculate the thermal response time τ as a function of temperature. All the input parameters of this model are determined independently, except the Kapitza resistance which has to be adjusted. The model accounts for the thermal response time of the cell between 30 and 300 mK within 15%. Based on this model, we estimate that an increase of ^{3}He heat conductivity by a factor 2 would decrease the thermal response time by only 9%, a decrease of the Kapitza resistance by the same factor leading to a 6% decrease in τ. Such a large change in Kapitza resistance seems unlikely, if we assume it arises from the acoustic mismatch between ^{3}He and silver: measurements by Bonfait et al. have shown that the sound velocity increases by less than 10^{-3} at $m = 0.3$ (Bonfait et al. 1987). Concerning the thermal conductivity of ^{3}He, the

[27]The nuclear specific heat of the silver does not come into play on the timescale of measurements.

experiment described in Section 7 shows that it increases by at most 30% at $m = 0.6$ and 27 bar. Finally, another source of error could be a change in the number of moles contained in the cell, due to magnetostriction. On the basis of the pressure dependence of the low-field susceptibility, and Maxwell relations, this effect is negligible (Castaing and Nozières 1979, Matsumoto et al. 2000).

From this analysis, we may conclude that the observed 40% reduction in the thermal response time τ at a 70% polarization implies a reduction of the specific heat by at least 30%. In the following, we assume that *all* the polarization dependence of τ time stems from that of the specific heat.

8.4. *Comparison to models*

The observed decrease is much larger than for free fermions, demonstrating the role of interactions. Indeed, in the latter case, $C(m)/C(0) = (N_{\uparrow}(0) + N_{\downarrow}(0))/N(0)) = ((1+m)^{1/3} + (1-m)^{1/3})/2$, which decreases by only 8% between $m = 0$ and 0.7. The behavior of the specific heat then implies that the effective mass is reduced by 20–30% at a polarization $m = 0.7$. At low polarization, our measurements are consistent with the $\propto m^2$ decrease expected from the Maxwell relation $(\partial S/\partial M)_T = -(\partial B/\partial T)_M$ and the susceptibility data of Ramm et al. (1970). It is interesting to note that the extrapolation of this behavior accounts for the observed behavior up to the largest polarizations ($m \approx 0.6$) at 27 bar. Repeating the calculation for lower pressures leads to the conclusion that the effect should barely depend on pressure. This is not inconsistent with the data at 2 and 10 bar. In a next step, we can compare our results in fig. 36 to the predictions of the models we have described in Section 3.

The 'nearly localized' Vollhardt model predicts the occurrence of a metamagnetic transition. At $m > 0.15$, the paramagnetic phase should become unstable with respect to the fully polarized (and localized) state, the spinodal point being reached at $m \approx 0.4$. This transition gives rise to a divergence of the effective mass, which would show up in the specific heat. From fig. 35b, one would expect a 40% increase for $m^2 \approx 0.2$. This prediction is in strong disagreement with our results, which, consistently with the susceptibility measurements, show no sign of a metamagnetic transition. Hence, our results clearly rule out the appealing original Vollhardt model. This conclusion is all the more noteworthy as recent experiments on 2D fluid ^{3}He adsorbed on a crystalline substrate were interpreted as showing that the latter system undergoes a Mott transition as the filling approaches the density of the commensurate $\sqrt{7} \times \sqrt{7}$ phase (Mörhard et al. 1996, Saunders et al. 2000, Casey et al. 2003). There is however no contradiction, as in the

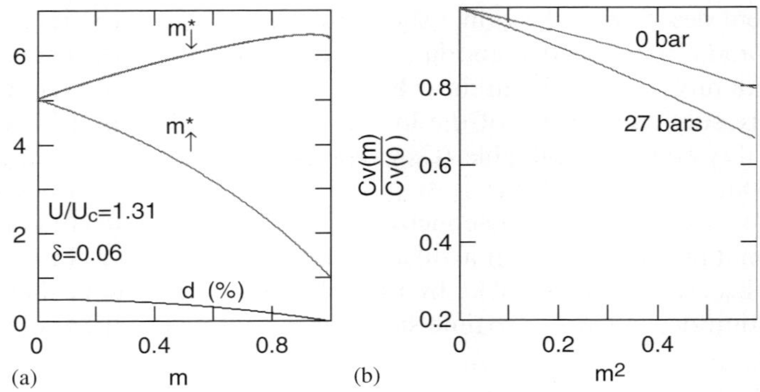

Fig. 37. (a) Polarization dependence of the effective masses (in units of the bare mass) in the Vollhardt, Wölfle and Anderson model, for a vacancy concentration of 6% (appropriate to 27 bar). The percentage of doubly occupied sites (d) is also shown. (b) Resulting polarization dependence of the specific heat at 0 and 27 bar.

2D system, there is a true underlying solid lattice, which is not the case for 3D ^{3}He. Beyond Vollhardt model, we argued in Buu et al. (2000c) that the measured decrease of the specific heat contradicts more generally the idea that, in ^{3}He, the effective mass enhancement results from a 'near localization' only. This assertion relied on the assumption that, in that case, spin polarization favors localization by decreasing the double occupancy of sites, and that this increases the effective mass. In fact, while both features hold within the original Vollhardt model, this is not generally so. As was pointed out by Laukamp and Vollhardt (1994), polarization can decrease both the double occupancy and the specific heat in the 'incompressible nearly localized' model of Vollhardt, Wölfle and Anderson (VWA), which we introduced in Section 2. This is illustrated in fig. 37, taken from Buu et al. (2002a), where we compute the polarization dependence of the effective masses and of the double occupancy in the framework of this model. Using a vacancy concentration $\delta = 6\%$, appropriate to 27 bar, and an interaction $U/U_c = 1.31$, as determined in Vollhardt et al. (1987) from the effective mass at zero polarization, we determine the double occupancy d by minimizing the total energy (eq. (2.14)), where the ε_σ are computed for an elliptical density of states. Due to the non-zero density of vacancies, $n = 1 - \delta$ in eq. (2.15) differs from 1, and the effective masses of up and down spins differ, unlike in the original Vollhardt model. As a result, although the number of doubly occupied sites decreases as was the case at half filling, the effective mass of majority spins decreases with polarization (fig. 37a). This makes the total specific heat decrease as shown in fig. 37b, where we used the relation $C(m) \propto \sum_\sigma N_\sigma/q_\sigma$, N_σ being the density of states per site. The computed behavior is quite close to that shown in fig. 36, and the pressure dependence is moderate, in line with what would be predicted from the Maxwell relation (compare to fig. 36).

At the same time, the finite concentration of vacancies suppresses the metamagnetic transition (Laukamp and Vollhardt 1994), in agreement with our experimental observations (Section 5). The same effect also occurs when the Gutzwiller approximation is replaced by a dynamical mean-field calculation, as in the 'Mott–Stoner' model introduced by Georges and Laloux (1997) and discussed in Section 5.6. It would be interesting to see whether in such a model where, as the authors point out, m^* is controlled by proximity to the Mott transition only, polarization also decreases the specific heat. If this is indeed the case, the debate between 'nearly localized' and 'nearly ferromagnetic' cannot be settled by our specific heat measurements only, and the confrontation between theories and experiment requires more quantitative comparisons than originally thought. From this point of view, the pronounced downward curvature of the magnetization curve (fig. 26) may be significant, as both the VWA and the Mott–Stoner models predict a highly linear behavior up to $m = 0.7$.

As for the phenomenological model of Bedell and Sanchez-Castro, which is based on an extension of Landau theory to polarized systems, a peak in magnetic susceptibility occurs at $m \approx 0.2$ (Sanchez-Castro et al. 1989). Sanchez-Castro et al. have interpreted this feature as an indication of a 'nearly metamagnetic' behavior in liquid ^{3}He. However, the 'pseudo transition' does not show up in the specific heat, the density of states being predicted to decrease monotonously with the polarization. As can be seen in fig. 36, the initial decrease is consistent with our data. This follows naturally from the fact that both our data and the model (by construction) are consistent with the temperature dependence of the susceptibility. However, at larger polarizations, the predicted deviation is significantly larger than that observed. Our experiment shows that the effective mass enhancement does depend on the polarization, since the decrease of the specific heat is larger than predicted for free fermions, but the choice of coefficients made by Sanchez-Castro overestimates this polarization dependence. As in the case of the magnetic susceptibility, a different choice for the coefficients entering the polarization dependence of the Landau parameters could possibly reconcile this model with our results.

In contrast, our observations are more in line with the 'nearly ferromagnetic' model. The agreement at small polarization is not surprising, since the paramagnon model correctly predicts the temperature dependence of the susceptibility (which, as discussed in Section 2, is related to the polarization dependence of the specific heat through a Maxwell relation). The deviation between the experimental data and the paramagnon prediction at $m > 0.4$ is possibly due to reaching the limit of validity of the expansion of the specific heat in powers of m (Eq. (3.7)). Even though we cannot conclude to the proximity of liquid ^{3}He to a ferromagnetic transition (see the susceptibility measurement Section 5.6), our specific heat measurements are consistent with the idea that spin fluctuations contribute significantly to the entropy (Greywall 1983, Clements et al. 1991, Krotscheck and Springer 2003). As such fluctuations are quenched in a magnetized system, the specific heat is expected to decrease with polarization, which is what is observed.

Finally, the observed decrease of the specific heat is also qualitatively consistent with microscopic model calculations, based on the induced interaction model (Bedell and Quader 1983), or on a Green function approach (Clements et al. 1991). A direct quantitative comparison is not possible as these calculations are restricted to fully polarized ^{3}He. However, they appear to overestimate the effect of polarization, as for both, the predicted decrease at full polarization is by a factor of 4, larger than a linear (in m^2) extrapolation of our 27 bar data to $m = 1$.

9. Conclusion

Some 25 years after the paper of Castaing and Nozières, it is satisfying that most of the basic properties – the specific heat, the magnetic susceptibility, the speed of sound, the viscosity, and the thermal conductivity – have been measured in polarized ^{3}He. However, none of these quantities show any sign of a phase transition at finite polarization: at 27 bar, that is reasonably close to solidification, the specific heat and the susceptibility are found to decrease monotonically with the polarization, while the speed of sound and the viscosity increase proportionally to m^2. Within the Hammel–Richardson model, the smooth dependence with polarization of the surface relaxation time is another indication that no irregular magnetic behavior of the liquid occurs, whatever the pressure. We stress that, while the specific heat and the magnetization curves are measured for ^{3}He confined inside a silver sinter, this is not so for the sound velocity, and the viscosity. This points to a negligible influence of the confinement on our results.

The picture which emerges from our measurements is at least partly consistent with the qualitative expectations of Bashkin and Meyerovich (Bashkin and Meyerovich 1981, Meyerovich 1990). These authors argued that, for fully spin-polarized ^{3}He, the suppression, due to the Pauli principle, of the odd harmonics of the scattering amplitude weaken the Fermi liquid interaction, leading to an increase of the transport coefficients, and a decrease of the effective mass. This is in agreement with our measurements of the viscosity, the thermal conductivity, and the specific heat. It is however not clear to us whether our measurements of the magnetization curve are also consistent with such an argument.

In the original expectation of Castaing and Nozières (1979), such measurements should have allowed one to discriminate between the 'nearly localized' and the 'nearly ferromagnetic' model. The present situation is however not so clear cut. The regular behavior of the thermodynamic quantities, and in particular the decrease of the specific heat, show indeed that liquid ^{3}He is *not* properly described by the 'nearly localized' model of Vollhardt. Nevertheless, we cannot rule out the possibility that ^{3}He is described by a more complicated version of such a model, with a finite density of vacancies, such as the 'incompressible nearly localized' model of Vollhardt, Wölfle and Anderson, or, possibly, the Georges and Laloux model.

On the other hand, our measurements also agree qualitatively with the paramagnon model, which describes systems with ferromagnetic tendencies: the observed downward curvature of the magnetization curve is comparable to that predicted with a simple Stoner model, and the polarization dependence of the specific heat is consistent with the idea that spin fluctuations play an important role on the system, a conclusion in agreement with Greywall's analysis. Quantitatively, however, the resolution of our magnetization curve and specific heat measurements is not good enough to probe the paramagnon model in the range of small polarizations where the calculations have been performed.

Hence, our results appeal for new, more detailed, theoretical calculations. On one hand, it would be interesting to extend the paramagnon model to large polarizations to see how well (or poorly) its predictions would fit our results. On the other hand, a full comparison of the predictions of the 'nearly localized' model (with vacancies) to the pressure and polarization dependences of the thermodynamic properties of liquid ^{3}He would be desirable. A number of other theoretical issues remain open as well. For example, could our thermodynamic results be consistently reproduced by a polarization-dependent Landau theory, similar to that introduced by Bedell and Sanchez-Castro, but with a modified set of parameters, or by a density functional theory? Although we do not know the answers to these questions, we may hope that our results will serve as a benchmark for future theoretical progress.

Disappointing as it may seem, the thermodynamic properties of liquid ^{3}He at finite polarization do not show any spectacular behavior. In our opinion, the surprise comes rather from the polarization dependence of the transport properties. At large temperature, our measurements of the viscosity show that degeneracy effects persist well above the 'magnetic Fermi temperature'. This feature, which is reminiscent of dilute Fermi systems, remains to be explained. At low temperature, the lack of pressure dependence of the effect of polarization on viscosity is also unexpected. In the framework of the paramagnon model, this would mean at the least that the viscosity is not due to scattering of quasiparticles by the spin fluctuations. Another intriguing feature is the weak deviation from a purely quadratic behavior in polarization, a point that we were unable to understand within the standard s–p approximation for the calculation of transport properties. In contrast, all these features resemble those expected for a dilute gas described by pure s-wave scattering, which is certainly surprizing for a dense, correlated, liquid. From this point of view, our measurements of the thermal conductivity are reassuring as its dependence on polarization is much smaller than expected for a gas. The open theoretical problem is now to understand whether the very simple behavior of viscosity is a mere coincidence or has a deeper significance.

Acknowledgments

This chapter owes a lot to a number of people. S.A.J. Wiegers, M. Bravin, A.C. Forbes, and D. Sawkey made essential contributions to the results reported here. We benefited from numerous stimulating discussions with a large number of colleagues, in France and abroad, among whom we are specially indebted to B. Castaing, A. Georges, R. Jochemsen, C. Lhuillier, J. Owers-Bradley, and G. Vermeulen. Last, but not least, P. Nozières had a large influence on us, both through fruitful exchanges of ideas and his Collège de France lectures he gave in Grenoble.

Appendix A. Kapitza resistance and surface magnetic relaxation of silver sinters

Our method for obtaining cold, polarized, liquid ^{3}He requires the cooling time to be much smaller than the magnetic relaxation time. The large surface area of a silver sinter reduces both times, so that the efficiency of confining ^{3}He in such a sinter depends on a subtle competition. When we started these experiments, we had to take a bet on their success, because of the lack of relevant literature data. The Kapitza resistance between ^{3}He and silver sinter had mainly been measured below 10 mK, in connection with the performances of dilution refrigerators, and the ability of nuclear demagnetization refrigerators to cool superfluid ^{3}He. The only data for pure ^{3}He around 100 mK were those of Andres and Sprenguer (1975) for 1 μm grain size silver powder, who found a specific Kapitza resistance R_K^A consistent with the acoustic mismatch theory (for a review, see Harrison, 1979), with $R_K^A T^3 \approx 5\,10^{-2}\,\mathrm{K^4 m^2/W}$. On the other hand, the specific Kapitza resistance between a saturated mixture, and the same 700 Å powder that we use, was about five times larger, as deduced from the analysis of the performances of a dilution refrigerator (Frossati et al. 1977)! As for the magnetic surface relaxation, it was a perilous exercise to guess its value from experiments at smaller magnetic fields and on different substrates. The final success of our experiments is due to a happy combination of a small Kapitza resistance (equal to, or smaller than that of Andres and Sprenguer (1975)), and a relatively long magnetic surface relaxation time, owing to the large applied field (Bravin et al. 1998). In this appendix, we describe our measurements of these quantities. As silver sinters are widely used in the low-temperature community, we hope that such a description is of general interest.

A.1. *Kapitza resistance of the heat tank silver sinter*

Measuring a Kapitza resistance around 100 mK is generally not easy, due to the low thermal conductivity of the liquid ^{3}He. In a situation where we would apply a heating power to a sinter immersed in the liquid, only a small fraction of the sinter area (the outermost part) will be involved in the heat transfer to the liquid, making it difficult to infer the true specific Kapitza resistance. The method we present here takes advantage of the short internal time constant of our HT to bypass this problem.

The idea is to measure the transient overheating of the HT silver walls, as a several seconds long heat pulse is applied to these walls (fig. A.1a). As the HT may be considered thermally insulated on the timescale of the measurements, the input heat flows directly into the ^{3}He inside the sinter, the ^{3}He temperature remaining initially the same, while, due to its low specific heat, the silver temperature T_{walls} increases instantaneously. After about 1 s, one reaches a stationary regime, where the temperature increases linearly with time, the rate of change of the temperature being uniform throughout the HT. Neglecting the ^{3}He conductivity, the temperature gradients in this regime can be computed using 2D conduction equations similar to eqs. (C.7) and (C.8) (with $\kappa_3 = 0$, and $\dot{Q}_{\mathrm{sinter}} = -C_3 \dot{T}_{\mathrm{walls}}$). The temperature before the pulse is switched off exceeds the final equilibrium temperature after the pulse by $R_K + (a^2/8\kappa_a')/\dot{Q}$, with $\dot{Q}$ the power applied per unit volume of the sinter, R_K the volume specific Kapitza resistance, a the radius of the sintered compartments, and κ_a' the effective thermal conductivity of the sinter (see Appendix C for details). At low enough temperature for the Kapitza resistance to dominate that of the sinter, the overheating of the silver walls

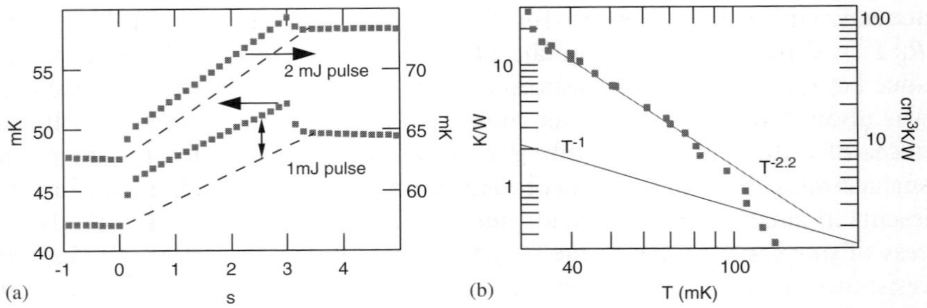

Fig. A.1. Measurement of the specific Kapitza resistance of the HT. (a) Principle: 3 s long heat pulses are applied to the HT. The transient overheating of the HT thermometer (shown by the vertical arrow), divided by the applied power, gives the total resistance between the silver walls and the ^{3}He at the considered temperature. (b) Below 100 mK, the resistance obtained is larger than the $1/T$ contribution of the sinter resistivity, i.e. is a measure of the Kapitza resistance. The right scale gives the specific Kapitza resistance in Kcm^3/W, obtained by multiplying the Kapitza resistance by the volume of ^{3}He ($\approx 4.3\,\mathrm{cm}^3$).

directly measures the specific Kapitza resistance. In our geometry ($a = 2.5\,\mathrm{mm}$) this 'low temperature' limit is $T < 100\,\mathrm{mK}$. At higher temperature, the sinter thermal resistance is no longer negligible, and our method determines only an upper bound of the Kapitza resistance.

Figure A.1b shows the specific resistance that is determined in K cm^3/W (i.e. the resistance times the volume of ^{3}He inside the sinter $-4.3\,\mathrm{cm}^3$), for the HT used in the viscosity experiment. The sinter effective resistance $a^2/8\kappa'_a$, calculated using a typical value $\kappa'_a/T = 0.7\,\mathrm{W/m/K^2}$ (Appendix C), is also shown. Below 100 mK, the resistance is dominated by the Kapitza resistance, which is found to be $\approx 0.009T^{-2.2}\,\mathrm{W/K}$, not too far from the expected T^{-3} behavior. This corresponds to a specific resistance $R_\mathrm{K} = 0.04 \times T^{-2.2}\,\mathrm{K\,cm^3/W}$, or (the area of our sinter being about $46\,\mathrm{m}^2$) $R^A_\mathrm{K} = 0.41 \times T^{-2.2}\,\mathrm{K\,m^2/W}$. Note that similar experiments performed on the HT of the susceptibility experiment yielded identical results. At 100 mK, $R^A_\mathrm{K}T^3$ is $0.065\,\mathrm{K\,m^2/W}$ (and is smaller below), consistent with the results of Andres and Sprenguer (1975) for a silver sinter of larger grain size. The resulting thermal relaxation time $R_\mathrm{K}C_3$, where C_3 is the specific heat of ^{3}He per unit volume, is of order of 0.5 s at 100 mK, much shorter than the magnetic relaxation time we discuss in the next section. This forms the basis of the method of polarization by rapid melting inside a sinter.

A.2. *The magnetic relaxation inside the sinter*

Our experiments allow us to investigate the magnetic relaxation of confined liquid and solid ^{3}He. A first series of experiments were performed using our susceptibility cell (Section 5). In the solid, the relaxation rate is found to increase spectacularly with polarization, reaching $40\,s^{-1}$ at 70% polarization, a yet unexplained phenomenon. We refer the reader to Bravin et al. (1998) for details. The point of interest with respect to our rapid melting method is that the corresponding T_1 is short enough to polarize the solid within a reasonable time, but still long enough to avoid a significant loss of polarization during the melting process. In the liquid, we find the relaxation time to increase with field as $B^{1.5}$, going from several seconds at 1 T up to a minute or so at 11 T. Such an increase, consistent with observations on other substrates at lower fields, is essential for our experiments.

Surface relaxation of pure liquid ^{3}He is usually described in the framework of the Hammel–Richardson (HR) model (Hammel and Richardson 1984). This model assumes a fast magnetization exchange between the bulk liquid and the solid layers adsorbed on the wall, and that all the relaxation takes place in the absorbed layer. In this context, a linear field dependence of T_1 is interpreted as resulting from a wide frequency spectrum of the exchange between the liquid and the adsorbed ^{3}He solid layer (Schuhl et al. 1987). From this standpoint, the field dependence of T_1 we measure supports the idea that, in our case also, the surface relaxation does involve the solid layer(s). Note that we do observe these layers by magnetization measurements, as shown in Section 5.5.1. Furthermore, the HR model is able to account for the observed dependence of the relaxation rate with polarization, which we now discuss.

The wide range of pressures and temperatures covered by the viscosity experiment allows us to study how the magnetic relaxation depends on these parameters. Our results are summarized in fig. A.2. At low temperature (< 100 mK), the relaxation is not exponential, the decay being slower at high polarization. Both the timescale and the shape of the relaxation curve at 27 bar and 80 mK are similar to those previously observed in the susceptibility cell in the same conditions (Bravin et al. 1998). This means that the magnetization diffusion from the slit to the sinter has a small influence on the behavior of the measured average polarization, which is then controlled by the relaxation inside the sinter. Our observations are consistent with the HR model. In this model, the relaxation time can be expressed as $T_1 = (M_{\text{liq}}/M_{\text{sol}})T_{1\text{sol}}$, where $T_{1\text{sol}}$ is the relaxation time in the solid layer, M_{liq} and M_{sol} are the total magnetization in the liquid and the solid layer. A key point is that, due to the fast exchange between the liquid and the solid, the effective field is the same in the two phases. Below 100 mK and at high polarization (i.e. strong effective magnetic field), the magnetization of the solid layer then tends to saturate, so that the relaxation is slower. At higher temperature, even at high polarization, the effective field is not strong enough to saturate the solid layer and the relaxation is exponential.

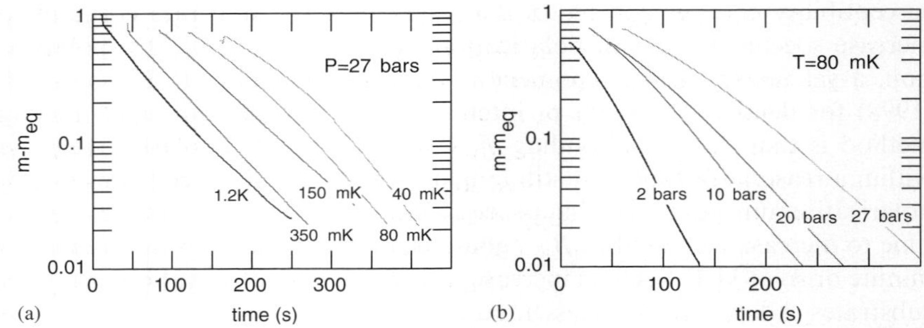

Fig. A.2. Polarization decay observed during rapid melting experiments. (a) As a function of temperature at 27 bar; and (b) as a function of pressure at 80 mK. The pressure is lowered at zero time. For clarity, the curves in figures (a) and (b) are successively shifted by 40 and 20 s, respectively. The final relaxation rate depends on pressure, and only weakly on temperature. At low temperature, the initial relaxation is slower. These features can be interpreted within the HR model.

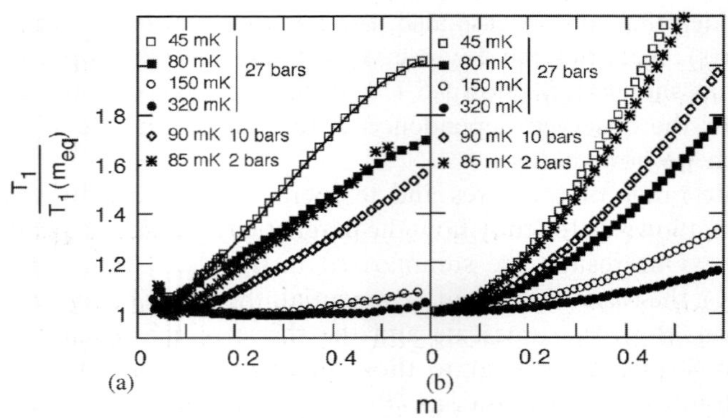

Fig. A.3. Polarization dependence of T_1: (a) observed and (b) predicted by the HR model.

Assuming the Curie law to hold in the solid, and the susceptibility of the liquid to be polarization independent, we can compute the expected polarization dependence of T_1 (Buu et al. 2000b). As shown by fig. A.3, the predicted behavior agrees reasonably well with the observations, supporting our interpretation. Conversely, the fact that we observe a regular increase with polarization is an argument against the occurrence of a metamagnetic transition at any pressure (at a given liquid polarization, this would reduce the solid magnetization and hence T_1).

Moreover, the HR model explains qualitatively the decrease in relaxation rate with increasing pressure: the observed changes can be accounted for by the increase in M_{liq} brought about by the change in number of atoms and in magnetic susceptibility of the bulk phase (the adsorbed layer density being mostly unaffected by the pressure, due to the large effective van der Waals pressure close to the surface). If we stick to the HR model to interpret our data, the weak temperature dependence we observe implies that the relaxation time T_{1sol} in the adsorbed layer should scale as $1/T$ to cancel the temperature dependence of M_{sol} (Buu et al. 2000b). As for the order of magnitude of the surface relaxation, T_{1sol} increases from 0.1 up to 0.5 s at 11 T, for temperatures from 1 K down to 40 mK, if we assume only one solid layer. On the basis of the magnetization measurements (Section 5.5.1 and Fukushima et al. 1992), a number of three layers seems more adequate, which would increase T_{1sol} by a factor of 3. These numbers are about one order of magnitude smaller than those reported by for a fluorocarbon substrate ($T_{1sol} \approx B$, where B is the magnetic field in T, for $B < 3$ T, which extrapolates to about 10 s in our 11 T field) (Schuhl et al. 1987).

Despite these smaller T_{1sol} values, the relaxation time T_1 remains long enough to carry out measurements, the lowest value being $\approx$30 s at 2 bar. Thus, the rapid melting method in a silver sinter is not limited to a particular range of pressure or temperature.

Appendix B. Effects of polarization gradients in the viscosity cell

Magnetization gradients arise inside the viscosity cell during the polarization decay due to the presence of the slit: because the magnetic relaxation inside the sinter ($T_1 = 30$–100 s) is much faster than the bulk relaxation ($T_1 > 1000$ s) (Van Steenbergen et al. 1998), the polarization inside the slit will decay by diffusion to the sinter and subsequent relaxation inside the sinter. This process creates time-dependent polarization gradients across the cell.

The purpose of this appendix is twofold: calculate the magnetization distribution during the polarization decay, and estimate the heat released by the magnetic relaxation. The knowledge of the magnetization profile is important, since the difference between the local magnetization in the slit (that alters the viscosity in the vicinity of the viscometer) and the average magnetization (measured by the magnetometer) contributes to the systematic errors in our measurements. Moreover, the heat released by the magnetic relaxation is an input of the thermal model we develop in Appendix C to calculate the temperature inside the slit.

B.1. *One-dimensional model*

We simplify the spin diffusion problem by using a one-dimensional model. We note x the distance from the center of the slit, $x = \pm\ell$ corresponding to the edge of the slit (ℓ = half-width of the slit), and $x = \pm(\ell + L)$ corresponding to the interface between the sinter and the silver walls ($\ell + L$ = cell radius). Inside the sinter, we use the same spin diffusion constant D_{spin} as in the bulk, i.e. we neglect the tortuosity. We neglect the bulk relaxation with respect to the relaxation in the sinter, which we characterize by a time constant T_1. Denoting m the deviation of polarization from its equilibrium value, the corresponding equations are

$$D_{\text{spin}} \frac{\partial^2 m}{\partial x^2} = \frac{\partial m}{\partial t}, \quad 0 < x < \ell, \tag{B.1}$$

$$D_{\text{spin}} \frac{\partial^2 m}{\partial x^2} = \frac{\partial m}{\partial t} + \frac{m}{T_1}, \quad \ell < x < \ell + L \tag{B.2}$$

with the boundary conditions $\partial m/\partial x|_{x=0} = 0$, $\partial m/\partial x|_{x=L+\ell} = 0$, $m(\ell^-) = m(\ell^+)$ and $\partial m/\partial x|_{x=\ell^-} = f\partial m/\partial x|_{x=\ell^+}$. The last equation expresses the continuity of flux at the interface between the slit and the sinter, $f = 55\%$ being the volume fraction of ^{3}He inside the sinter. We write the solution $m(x, t)$ as a superposition of eigenmodes in each domain:

$$m(x,t) = \begin{cases} \sum_{p=0}^{\infty} C_{1p} \cos(q_{1p} x) \mathrm{e}^{-D_{\text{spin}} q_{1p}^2 t}, & 0 < x < \ell, \\ \sum_{p=0}^{\infty} C_{2p} \cos(q_{2p}(L + \ell - x))\, \mathrm{e}^{-D_{\text{spin}} q_{2p}^2 t - (t/T_1)}, & \ell < x < \ell + L, \end{cases} \tag{B.3}$$

the wavevectors q_{1p} and q_{2p} are determined by the continuity equations at the interface, which imply

$$q_{1p}^2 = q_{2p}^2 + \Lambda_s^{-2}, \tag{B.4}$$

$$q_{1p} \tan(q_{1p}\ell) = -f q_2 \tan(q_{2p} L), \tag{B.5}$$

where we have introduced the spin diffusion length $\Lambda_s = \sqrt{D_{\text{spin}} T_1}$. In terms of q_{1p}, eqs. (B.4) and (B.5) have two series of poles, given by $(p + 1/2)\pi/\ell$ and $\sqrt{((p + 1/2)\pi/L)^2 + \Lambda_s^{-2}}$, p being a positive integer. Between two successive poles, eqs. (B.4) and (B.5) can be shown to have an unique solution, which is computed numerically using the method of dichotomy. Note that for $q_{1p} < 1/\Lambda_s$, q_{2p} is purely imaginary. The weighting factors C_{1p} and C_{2p}, determined by the initial condition $m(x,0) = 1$ (i.e. we assume that the polarization after melting is uniform) are obtained by projecting this initial condition onto the eigenmodes (using a scalar product modified to account for $f \neq 1$), and are given by

$$C_{1p} = \frac{2\sin(q_{1p}\ell)}{\left(\ell + Lf\,\frac{\cos^2(q_{1p}\ell)}{\cos^2(q_{2p}L)}\right) q_{1p}(1 - q_{1p}^2\Lambda_s^2) + \frac{\sin(2q_{1p}\ell)}{2}} \tag{B.6}$$

$$\frac{C_{2p}}{C_{1p}} = \frac{\cos(q_{1p}L)}{\cos(q_{2p}\ell)}. \tag{B.7}$$

B.2. *Evaluation of the polarization gradient*

The polarization profiles, calculated with the expressions above, are plotted in fig. B.1a. It is important to note that, in our geometry, the diffusion is fast in the slit and slow in the sinter ($\Lambda_s \approx 0.6$ mm, i.e. $\ell < \Lambda_s < L$). It implies that, during the magnetization relaxation, the polarization remains uniform in the slit. As the relaxation proceeds faster in the sinter, the excess magnetization which has built up in the slit propagates into the sinter towards the wall of the cell, but never reaches the wall on the timescale of our experiments ($\approx 3\ T_1$). Therefore, the relative gradients of magnetization are time dependent, i.e. one does not reach a stationary regime. As a consequence, the ratio of the polarization inside the slit to that far in the sinter grows during the whole experiment. This is illustrated in fig. B.1b.

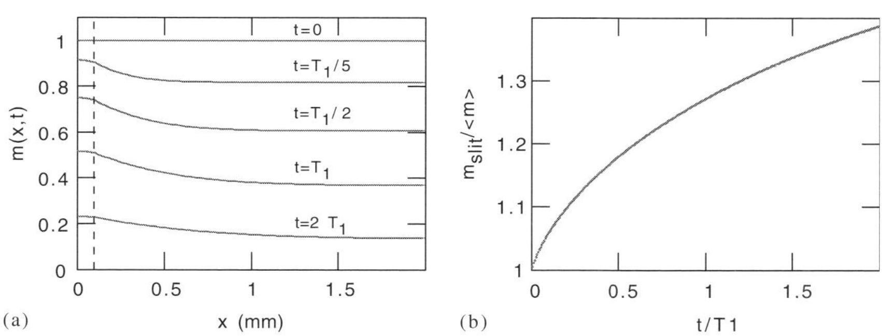

Fig. B.1. (a) Polarization profiles computed at different times during the magnetization decay. The parameters used here are: $\ell = 0.1$ mm, $L = 2$ mm, $f = 0.55$, $D_{\rm spin} = 4.10^{-3}$ mm^2/s (at 100 mK and 28 bar (Anderson et al. 1962), and $T_1 = 60$ s. The dashed line is the boundary between the slit and the sinter. (b) Polarization at the center of the slit ($x = 0$) normalized by the spatial average of m as function of the time.

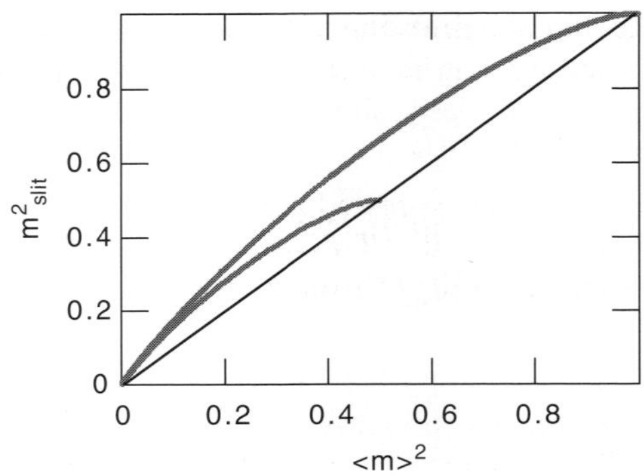

Fig. B.2. Theoretical effect of the finite spin diffusion: relationship between the squares of the average polarization $\langle m \rangle$ and the local polarization inside the slit m, for two experiments with initial uniform polarizations of 1 and 0.707, respectively.

Figure B.1b shows that the difference between the polarization in the slit and the average polarization may be significant during the magnetization decay. For example, at $t = T_1$, the difference between $\langle m \rangle$ and $m(x = 0)$ reaches 25%, which may cause large systematic errors. The bias between the polarization, which would be measured by the magnetometer and the polarization in the slit is plotted in fig. B.2. It shows that, if the viscosity enhancement is $\propto m^2_{\mathrm{slit}}$, we should observe a strong downward curvature in the plot η versus $\langle m \rangle^2$. The fact that we did not observe such a curvature (see Section 6.3.2) may imply that the actual relationship $\eta(m)$ includes a strong $\propto m^4$ compensating term, or that our diffusion model overestimates the polarization gradients. To test these two possibilities, we have carried out a series of experiments with different initial polarizations. If our model were correct, we should have observed large discrepancies between these experiments arising from the time-dependent polarization gradients, since data corresponding to the same average polarization would be recorded at different times during the magnetization evolution (see fig. B.2). The experimental curves, displayed in Section 6.3.3, show that it is not the case, and that the experiments at different initial polarizations are indeed consistent. We do not know the reason of the failure of our model (2D effects? Polarization-enhanced spin diffusion?, etc.)

B.3. *Relaxational heating*

The time-dependent magnetization profiles obtained with the one-dimensional diffusion model allow in principle to calculate the distribution of heat released during the magnetization relaxation. For an open system, the analog of the eq. (5.4) for the heat release is the local continuity equation

$$\dot{q} = -(B_T - B_a)\dot{M} - \operatorname{div} \vec{\jmath}_u, \tag{B.8}$$

all extensive quantities being expressed per unit of volume, the dots denoting the time derivative, and $B_T = B - T(\partial B/\partial T)_M$. $\vec{\jmath}_u$ is the energy flux, which is related to the spin flux $\vec{\jmath}_\uparrow$ and $\vec{\jmath}_\downarrow$ through

$$\vec{\jmath}_u = \vec{\jmath}_\uparrow \mu_\uparrow + \vec{\jmath}_\downarrow \mu_\downarrow, \tag{B.9}$$

where $\mu_\uparrow$ ($\mu_\downarrow$) is the chemical potential of the $\uparrow$ ($\downarrow$) spins. The difference in chemical potential is related to the effective field B by $\mu_\uparrow - \mu_\downarrow = 2\mu_3(B - B_a)$, μ_3 being the ^{3}He nuclear magnetic moment (Castaing and Nozières 1979). Taking the net mass flux to be zero ($\vec{\jmath}_\uparrow = -\vec{\jmath}_\downarrow$), the energy flux may be expressed as a function of the flux of magnetization $\vec{\jmath}_M$ by

$$\vec{\jmath}_u = \vec{\jmath}_M(B - B_a). \tag{B.10}$$

Using the continuity equation for the local magnetization, we can rewrite eq. (B.8) as

$$\dot{q} = (B_T - B_a)\frac{(M - M_{\text{eq}})}{T_1} - \vec{\jmath}_M \cdot \overrightarrow{\text{grad}}\, B + (B_T - B)\operatorname{div} \vec{\jmath}_M, \tag{B.11}$$

where T_1 is the magnetic relaxation time inside the sinter (which will differ from the observed T_1^{eff}, see appendix C).

To eliminate B, we neglect the field dependence of the magnetic susceptibility and the applied field. Equation (B.11) is then transformed into

$$\dot{q} = \frac{\mu_0 M_{\text{sat}}^2}{\chi_T}\frac{m^2}{T_1} + \mu_0 D_{\text{spin}}\left[\frac{|\overrightarrow{\text{grad}}\, m|^2}{\chi} - \left(\frac{1}{\chi_T} - \frac{1}{\chi}\right)\nabla^2 m\right], \tag{B.12}$$

where M_{sat} is the magnetization density at saturation, and $1/\chi_T = B_T/\mu_0 M = (1/\chi) - T\partial(1/\chi)/\partial T$. The released heat profiles obtained by substituting the polarization distribution computed with the one-dimensional model the expression (B.12) are shown in fig. B.3. With realistic parameters for ^{3}He at 100 mK and 28 bar, it turns out that the first term, i.e. the heat released by the magnetization relaxation, is much larger than the second one, i.e. the 'Joule' heating induced by the spin currents. For this reason, the heat released into the slit is small, and, in the sinter, expression (B.12) can be approximated by

$$\dot{q} = \frac{\mu_0 M_{\text{sat}}^2}{\chi_T}\frac{m^2}{T_1}. \tag{B.13}$$

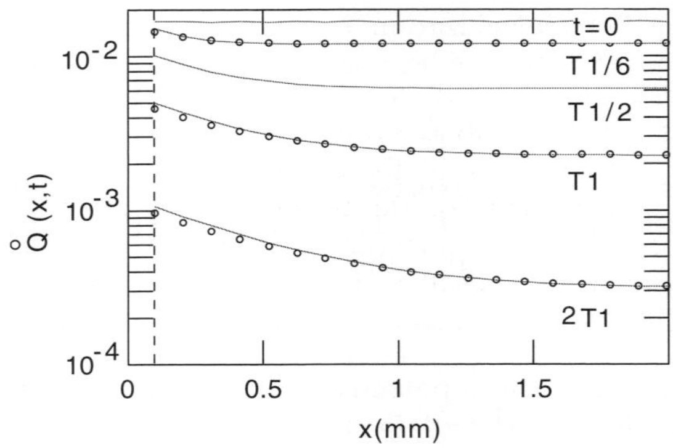

Fig. B.3. Successive profiles of the dissipation per unit volume, in units of $\mu_0 M_{\text{sat}}^2/\chi_T T_1$. The circles correspond to the contribution of the relaxation of magnetization only. The continuous lines also take into account the dissipation due to the spin diffusion. Inside the slit (i.e. on the left side of the dashed line), the dissipation is too small to be visible on this scale.

Appendix C. Thermal characterization of the viscosity cell

Determining the thermal resistance between the ^{3}He inside the slit and the silver walls is of prime importance for analyzing our experiments. Indeed, this resistance controls the temperature difference between the viscometer and the walls in the viscosity experiment, and contributes to the response time of the cell in the specific heat experiment. Different factors enter this resistance: the thermal conductivity of ^{3}He, that of the silver sinter, and the Kapitza resistance between ^{3}He and the sinter. The aim of this appendix is to determine experimentally the respective importance of these contributions by studying how the thermal response of the cell depends on temperature and pressure.

This appendix is organized as follows. First, we show how the thermal resistance can be measured by studying the response of the slit temperature to a change of the wall temperature. After having described such experiments, we develop a model to analyze the data, which allows us to determine the quantities sought after. The results allow us to justify that the polarization dependence of the response time is controlled by that of the specific heat only, a central assumption in our analysis of Section 8. Also, thanks to these results, we can estimate, in the viscosity experiment, the temperature of the viscometer during the magnetization relaxation and cross check the reading of the slit thermometer.

C.1. *Linear thermal response*

In the absence of any heat source in the cell, the temperature in the slit is determined by the temperature evolution of the HT. Here, we study the response of the temperature of the slit $T_{\rm slit}$ to a perturbation of the HT temperature $T_{\rm ht}$. We then relate this response to various physical situations.

Let us consider the response of the slit temperature to a HT temperature step of amplitude ΔT_0 applied at time $t = 0$, the cell being initially at a uniform temperature T_0. We define the response function of the cell $S(t)$ such that

$$T_{\rm slit}(t) = T_0 + \Delta T_0(1 - S(t)). \tag{C.1}$$

We further define the response time τ by

$$\tau = \int_{t=0}^{+\infty} S(t)\,\mathrm{d}t. \tag{C.2}$$

Physically, τ measures the timescale of the transient response of the slit temperature following a sudden change of the HT temperature. For an arbitrary change in the HT temperature, provided the variation of $T_{\rm ht}(t)$ is small enough for the response to be linear (that is for the function S not to vary), eq. (C.1) implies that the evolution of the slit temperature is given by (Carslaw and Jaeger 1959, p. 19)

$$T_{\rm slit}(t) - T_{\rm ht}(t) = -\int_{-\infty}^{t} S(t-t')\dot{T}_{\rm ht}\,\mathrm{d}t'. \tag{C.3}$$

We may use eq. (C.3) to relate the response time to the physical characteristics of the system: let us first consider the case where $T_{\rm ht}(t)$ increases linearly with time. Then the temperature difference between the HT and the slit is stationary, given by

$$T_{\rm slit} - T_{\rm ht} = -\dot{T}_{\rm ht}\tau. \tag{C.4}$$

In this situation, the temperature at any point in the cell varies at the same rate $\dot{T}_{\rm ht}$, corresponding to an effective uniform heating delivered to the ^{3}He given by $\dot{Q}_{\rm eff} = C_3\dot{T}_{\rm ht}$, where C_3 is the specific heat per unit volume of ^{3}He (the silver specific heat is negligible). On the other hand, in the situation where a uniform power $\dot{Q}_{\rm eff}$ would be released in the cell, the resulting temperature difference would be $T_{\rm slit} - T_{\rm ht} = R\dot{Q}_{\rm eff}V_{^3\rm He}$, where R is a distributed thermal resistance between the slit and the HT, and $V_{^3\rm He}$ the total volume of ^{3}He. We can therefore identify the thermal response time of the cell with the product of the distributed thermal resistance R by the total heat capacity of ^{3}He. This allows us to access R from a measurement of τ.

The method we use to measure τ consists in applying a temperature step to the HT and measuring the transient response of the slit. From eq. (C.3), the response time is given by

$$\tau = \frac{1}{\Delta T_0} \int_0^{\infty} (T_{\mathrm{ht}}(t') - T_{\mathrm{slit}}(t'))\,\mathrm{d}t', \tag{C.5}$$

which generalizes eq. (C.2) to the case where the HT temperature step is not ideally sharp. In the following, we first describe the measurements of this thermal response time, and then analyze them with a model.

C.2. *Experimental study of the thermal response of the cell*

We have measured τ as a function of the pressure and the temperature with a series of experiments in non-polarized liquid. Figure C.1 shows the responses of the different thermometers in non-polarized liquid ^{3}He at 27 bar following successive steps of the HT temperature. The temperature evolution for the viscometer and the carbon thermometer is clearly delayed with respect to the cell walls, demonstrating the thermal inertia of the cell. More surprisingly, the temperatures of the viscometer and the thermometer agree only for the shorter times ($t < 3$ s), the carbon thermometer lagging behind the viscometer at longer time. We focus on the response function measured by the viscometer, which, we will argue, is the true response function of the slit. The anomalous response of the inner thermometer is discussed in a later section. The normalized 'response function' of the viscometer is shown in

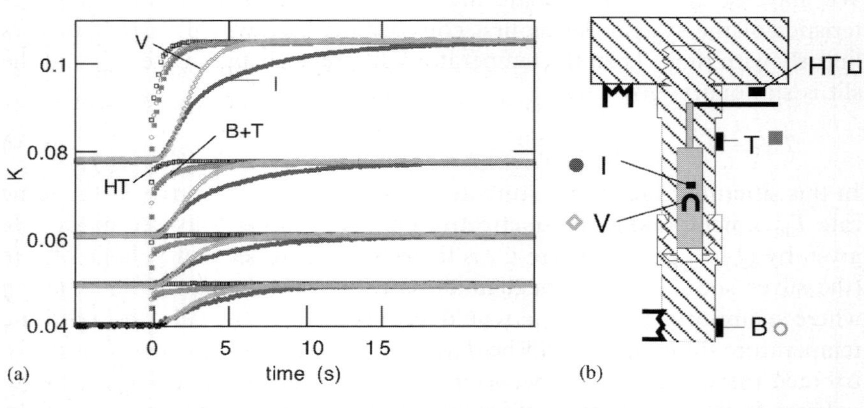

Fig. C.1. (a) Response of the different thermometers to successive steps of the heat tank temperature. The positions of the thermometers are shown in (b): HT; heat tank, B; bottom, T; top, I; inner, V; viscometer.

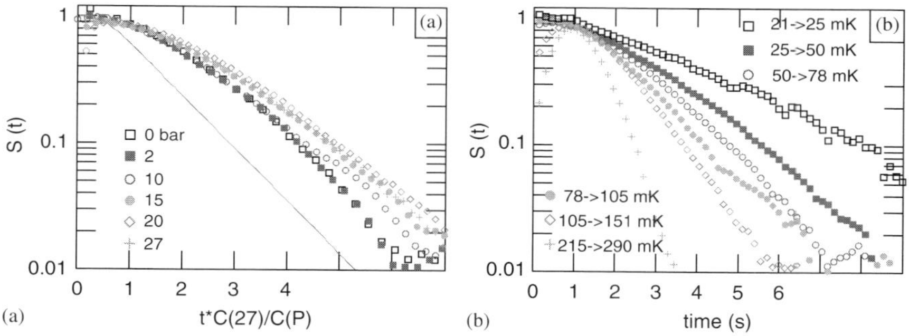

Fig. C.2. (a) The response function of the viscometer at 80 mK and different pressures. The pressure dependence of the specific heat is accounted by the normalization of the timescale. The remaining dependence is due to the ^{3}He thermal conductivity. The continuous curve is the expected behavior without slit, due to the sinter thermal resistivity only. (b) Temperature dependence of the response at a pressure of 27 bar. At the lowest temperatures, the response is mainly controlled by the Kapitza resistance.

fig. C.2a for steps of the HT temperature from 60 to 80 mK and pressures ranging from 0 to 27 bar. The response function does not start exactly from 1, because the step of the HT temperature is not ideally sharp, specially at large pressures. The pressure dependence of the response is nearly fully absorbed by rescaling the timescale by $C_3(P)/C_3(27)$ for each curve. The residual variations come from the pressure dependence of the heat conductivity of ^{3}He and of the Kapitza resistance. As the heat conductivity decreases by nearly a factor 2 from 0 to 27 bar, this proves that the distributed thermal resistance only weakly depends on the heat conductivity of ^{3}He. This characteristic is essential for the measurement of the polarization dependence of the specific heat from τ measurements (see Section 8). The response function is not a single exponential. This is only partly due to the distributed aspect of the thermal problem. As shown by the continuous line in fig. C.2a, the theoretical response function inside a cylinder has a similar long times decay rate (the diffusion constant D is computed from the ^{3}He specific heat and a sinter thermal conductivity $\kappa_a = 0.38T$ W/m/K, see below), but the deviation from an exponential is less marked. In fact, it appears from 1D exact calculations that the main reason for the observed slow decrease at short times is the thermal inertia of the liquid inside the slit.

Figure C.2b shows the response function of the viscometer for various temperature steps at 27 bar. Qualitatively, the relaxation becomes slower as the temperature decreases. This is quantified by the thermal response time τ, which we determine from the time integral of the normalized response. The results are summarized in fig. C.3. In the next section, we develop a model to analyze the experimental results.

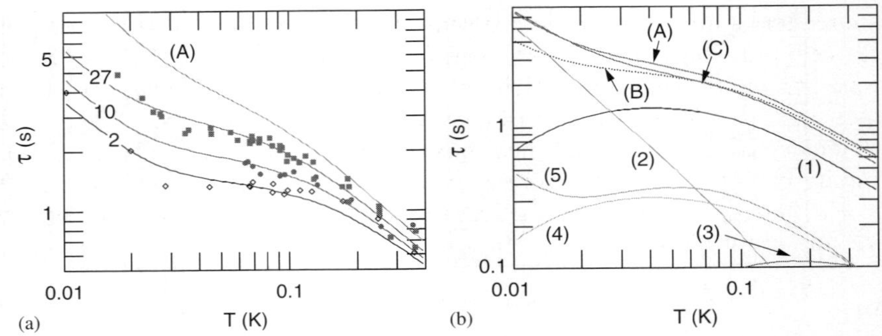

Fig. C.3. (a) Temperature dependence of the viscometer thermal response time in unpolarized ^{3}He: the thermal response time τ_{cell} of the viscometer with respect to the cell walls, measured at 2, 10, and 27 bar, is compared to the behaviour predicted from eqs. (C.12) and (C.13). Curve (A), calculated with the specific Kapitza resistance of the HT, increases too fast at low temperature. The other curves are obtained by dividing the Kapitza resistance by a factor 3 ($R_{\text{K}} = 0.013 \times T^{-2.2}$ K cm^3/W). (b) (1)–(5): breakout of the different contributions to τ_{cell}, using the same Kapitza resistance as for fig. C.3a (see text for the different labels). Above 60 mK, the sum (A) of these contributions is only weakly modified by dividing the Kapitza resistance by 2 (curve (B)), or doubling the ^{3}He thermal conductivity. (curve (C)).

C.3. *Analysis of the delay time*

C.3.1. *One-dimensional model*

As discussed in Appendix C.1, τ is the product of the ^{3}He specific heat by the distributed thermal resistance between the slit and the HT. In order to calculate the latter, we start with a one-dimensional model. We take the same convention as for the diffusion model of Appendix B.1, that is, x denotes the distance from the center of the slit, ℓ the half-width of the slit, and $(\ell + L)$ the cell radius. In the quasi-stationary regime, the temperature of the sinter T_a and the temperature T_3 of the ^{3}He are determined by:

$$\kappa_3 \frac{\partial^2 T_3}{\partial x^2} = -\dot{Q}_{\text{slit}}, \quad 0 \leqslant x \leqslant \ell, \tag{C.6}$$

$$\kappa_3 \frac{\partial^2 T_3}{\partial x^2} = \frac{T_3 - T_a}{R_{\text{K}}} - \dot{Q}_{\text{sinter}}, \quad \ell < x \leqslant \ell + L, \tag{C.7}$$

$$-\frac{\kappa_a}{f} \frac{\partial^2 T_a}{\partial x^2} = \frac{T_3 - T_a}{R_{\text{K}}}, \quad \ell < x \leqslant \ell + L, \tag{C.8}$$

where $f = 55\%$ is the volume fraction of ^{3}He inside the sinter, and the source terms $\dot{Q}_{\text{sinter}}$ and $\dot{Q}_{\text{slit}}$ are the powers injected per unit volume of ^{3}He. These source terms are equal in the problem of the thermal response function of the cell, but we will distinguish them for the purpose of eq. (C.5). κ_a is the macroscopic thermal conductivity of the sinter, κ_3 the heat conductivity of ^{3}He, and R_K the Kapitza resistance for one unit of volume of ^{3}He. The boundary conditions are: no heat flux in ^{3}He at $x = 0$ and $x = \ell + L$, continuity of flux and temperature at $x = \ell$, and no heat flux in the sinter at $x = \ell$. Solving eqs. (C.6)–(C.8) gives the temperature difference between the slit and the walls:

$$T_3(0) - T_a(\ell + L) = R_{\text{sinter}} V_{\text{sinter}} \dot{Q}_{\text{sinter}} + R_{\text{slit}} V_{\text{slit}} \dot{Q}_{\text{slit}}, \tag{C.9}$$

where V_{sinter} and V_{slit} are the respective volumes of ^{3}He in the sinter and in the slit, and

$$R_{\text{sinter}} V_{\text{sinter}} = \frac{1}{\kappa'_a + \kappa_3} \left[\frac{L^2}{2} + \kappa'_a R_K + \frac{\Lambda L}{\sinh L/\Lambda} \left(1 + \frac{\kappa_3}{\kappa'_a} \cosh \frac{L}{\Lambda} \right) \right], \tag{C.10}$$

$$R_{\text{slit}} V_{\text{slit}} = \frac{1}{2} \frac{\ell^2}{\kappa_3} + \frac{1}{f} \frac{1}{\kappa'_a + \kappa_3} \left[L\ell + \frac{\Lambda \ell}{\sinh L/\Lambda} \left(2 + \left(\frac{\kappa'_a}{\kappa_3} + \frac{\kappa_3}{\kappa'_a} \right) \cosh \frac{L}{\Lambda} \right) \right] \tag{C.11}$$

with $\kappa'_{\text{a}} = \kappa_{\text{a}}/f = 0.7\ \text{K/W}$ and $\Lambda^2 = R_K \kappa_3 \kappa'_a/(\kappa_3 + \kappa'_a)$. Λ is the thermal penetration depth, that is the distance needed for the heat injected from the slit into ^{3}He to flow into the silver sinter. If $\dot{Q}_{\text{sinter}} = 0$, the temperature difference $T_3 - T_a$ decays to zero away from the slit as $\exp(-x/\Lambda)$. The contribution R_{sinter} corresponds to the temperature difference caused by the heat injected directly into ^{3}He inside the sinter, while R_{slit} is the contribution arising from the heat injected into the slit which must cross the sinter to be evacuated into the HT. R_{sinter} is the sum of the thermal resistance of the sinter and ^{3}He in parallel and a contribution due to the Kapitza resistance between them. R_{slit} includes the thermal resistance of the slit itself, the thermal resistance of the sinter, seen from the slit and the 'interfacial' resistance between the slit and the sinter (proportional to Λ).

C.3.2. *Improvement of the model*

Despite the increase of the penetration depth with decreasing temperature (due both to the Kapitza resistance and the ^{3}He conductivity), Λ remains much smaller than the cell radius 2 mm for $T \geqslant 80$ mK ($\Lambda \leqslant 0.3$ mm[28]; $L = 2$ mm). This validates our use of a 1D model to calculate the contribution of the Kapitza resistance. On the other hand, the actual 2D geometry

[28]Computed with the Kapitza resistance estimated below.

modifies the contributions of the sinter effective resistivity to the resistance between the center of the cell and the walls: for a uniform $\dot{Q}_{\mathrm{sinter}}$, it decreases by a factor of 2 the corresponding term in R_{sinter}. As for the term in R_{slit}, it can be computed considering a heat source distributed along the slit (the heat injected from the slit into the sinter being concentrated in the neighborhood of the slit). The temperature at the center of the cell is then found by integrating the 2D Green's function $G(r,0) = G(0,r) \propto \log(L/r)$ of the problem along a diameter. This shows the contribution arising from the finite sinter conductivity in R_{slit} to be reduced by a factor $2/\pi$ with respect to the 1D case. Accordingly, eqs. (C.11) and (C.10) now can be written as

$$R_{\mathrm{sinter}} V_{\mathrm{sinter}} = \underbrace{\frac{L^2}{4(\kappa'_a + \kappa_3)}}_{a} + \underbrace{\frac{1}{\kappa'_a + \kappa_3}\left[\kappa'_a R_{\mathrm{K}} + \frac{\Lambda L}{\sinh L/\Lambda}\left(1 + \frac{\kappa_3}{\kappa'_a}\cosh\frac{L}{\Lambda}\right)\right]}_{b}, \tag{C.12}$$

$$R_{\mathrm{slit}} V_{\mathrm{slit}} = \underbrace{\frac{1}{2}\frac{\ell^2}{\kappa_3}}_{c} + \underbrace{\frac{1}{f}\frac{2L\ell/\pi}{\kappa'_a + \kappa_3}}_{d} + \underbrace{\frac{1}{f}\frac{1}{\kappa'_a + \kappa_3}\left[\frac{\Lambda\ell}{\sinh L/\Lambda}\left(2 + \left(\frac{\kappa'_a}{\kappa_3} + \frac{\kappa_3}{\kappa'_a}\right)\cosh\frac{L}{\Lambda}\right)\right]}_{e}. \tag{C.13}$$

From these equations, we compute the thermal response time τ of the slit with respect to the HT. In the case where the HT temperature is ramped at a constant rate, $\dot{Q}_{\mathrm{sinter}} = \dot{Q}_{\mathrm{slit}} = C_3 \dot{T}_{\mathrm{ht}}$, and the temperature difference between the slit and the HT is given by

$$\Delta T = C_3 \dot{T}_{\mathrm{ht}}(R_{\mathrm{sinter}} V_{\mathrm{sinter}} + R_{\mathrm{slit}} V_{\mathrm{slit}} + R_{\mathrm{contact}}(V_{\mathrm{slit}} + V_{\mathrm{sinter}})), \tag{C.14}$$

where R_{contact} is the resistance of the metallic contact between the cell walls and the HT. As, from eq. (C.4), $\Delta T = \tau \dot{T}_{\mathrm{ht}}$, we have $\tau = \tau_{\mathrm{sinter}} + \tau_{\mathrm{slit}} + \tau_{\mathrm{contact}}$, where the different contributions are given by $\tau_{\mathrm{sinter}} = R_{\mathrm{sinter}} V_{\mathrm{sinter}} C_3$, $\tau_{\mathrm{slit}} = R_{\mathrm{slit}} V_{\mathrm{slit}} C_3$, and $\tau_{\mathrm{contact}} = R_{\mathrm{contact}}(V_{\mathrm{slit}} + V_{\mathrm{sinter}}) C_3$. Equivalently, the effective resistance R_C used in Section 8.3.2 is given by $R_C(V_{\mathrm{sinter}} + V_{\mathrm{slit}}) = R_{\mathrm{sinter}} V_{\mathrm{sinter}} + R_{\mathrm{slit}} V_{\mathrm{slit}} + R_{\mathrm{contact}}(V_{\mathrm{sinter}} + V_{\mathrm{slit}})$.

C.3.3. *Thermal parameters*

We now compare the predictions of this thermal model to the measurements of the response time with non-polarized liquid ^{3}He. In the data plotted in fig. C.3, we have subtracted the contribution $\tau_{\mathrm{contact}} = 0.1$–$0.5$ s from the measured response time of the viscometer, to give the experimental time of thermal response of the slit with respect to the cell walls, τ_{cell}. The measurements are compared to $\tau_{\mathrm{sinter}} + \tau_{\mathrm{slit}}$, which we compute from eqs. (C.13) and (C.12).

τ_{contact} was calculated[29] from the contact resistance $R_{\text{contact}}T \approx 3.3\,\text{K}^2/\text{W}$, as measured by static experiments,[30] and the nominal specific heat of the cell ^{3}He (corresponding to $V_{\text{slit}} = 0.012\,\text{cm}^3$ and $V_{\text{sinter}} = 0.097\,\text{cm}^3$).

The input parameters in the thermal model are the heat conductivity and the specific heat of ^{3}He, the heat conductivity of the silver sinter, and the Kapitza resistance. The initial choices for these parameters were: the precision data of Greywall (1983, 1984) for κ_3 and C_3, $\kappa_a/T = 0.38\,\text{W/m/K}^2$ measured from the residual resistivity at 4.2 K with a sample of sinter prepared with the same silver powder in the same fashion as in the cell, and $R_{\text{K}} = 0.04 \times T^{-2.2}\,\text{K cm}^3/\text{W}$ as measured for the sinter of our HT at 0 bar between 30 and 100 mK (fig. A.1).

This choice of Kapitza resistance leads to a too fast increase of τ at low temperatures. This is shown by curve (A) of fig. C.3a for a pressure of 27 bar. A better agreement with experiment[31] requires us to reduce the Kapitza resistance by a factor of 3, keeping the same temperature dependence, and decrease κ_a/T by 10%. This allows us to reproduce the experimental data, including the 'plateau' around 80 mK (curves (B) of fig. C.3a). A possible explanation for this significant reduction is that the sinter might have different properties (e.g. specific area) in the vicinity of the slit and far from it, due to the method employed to create the gap in the cell (Section 6.2.1). As the Kapitza resistance entering τ is that within the small thermal penetration depth Λ from the interface, this may explain the observed change.

Putting these parameters into eqs. (C.13) and (C.12), we calculate the different contributions to the response time of the cell: in fig. C.3b, curves (1) and (2) are respectively the contribution of the diffusion across the sinter (first term 'a' of eq. (C.12)) and of the Kapitza resistance (last two terms 'b' of eq. (C.12)) to τ_{sinter}. As for the contributions to τ_{slit}, curve (3) comes from the conduction through ^{3}He inside the slit (first term 'c' of eq. (C.13)), curve (4) from the conduction across the sinter (second term of 'd' of eq. (C.13)), and curve (5) is the 'interfacial' (Kapitza) contribution (last term of 'e' of eq. (C.13)). It is clear from this plot that a large contribution to the thermal resistance seen by the slit arises from the sinter conductivity (curves (1) and (4)). These contributions control the decrease of τ_{cell} above 100 mK through the decrease of C_3/T in this temperature range. As the temperature decreases, C_3/T tends to a constant value, and (1) and (4)

[29]The thus obtained values were more precise than those directly obtained from the temperature difference between the walls and the HT during the response experiments.

[30]These experiments showed the thermal resistance of the silver walls to be negligible with respect to that of the contact.

[31]A similar agreement could as well be obtained choosing a T^{-3} temperature dependence of R_{K} with the appropriate pre-factor, to the expense of a slightly larger sinter thermal resistivity. This barely changes the results of Table 5.

saturate. This behavior persists until, below 50 mK, the ^{3}He parallel conductivity sets in and eventually bypasses the sinter conductivity, reducing (1) and (4). In the same temperature range, however, the Kapitza resistance becomes large enough for τ_{cell} to increase again. This is the mechanism giving rise to the bend observed around 80 mK.

Two conclusions are to be drawn from these results. First, although the slit contains only 12% of the total volume of ^{3}He, its effect on the thermal response time of the cell is significant ($\tau_{slit} = 0.3$–1 s), due to the different pre-factors in eqs. (C.12) and (C.13). On the other hand, the magnetic relaxation does not release any heat into the slit, as we shall see below. This will make it necessary to subtract the contribution of the slit from τ_{cell} in order to estimate the temperature gradient resulting from this relaxation. Second, above 60 mK, the thermal response time τ of the cell is mainly controlled by the thermal resistivity of the sinter. τ only weakly depends on R_K and κ_3, because these quantities only enter (5) through a square root, and because κ_3 is small compared to κ'_a. This is illustrated by curves (B) and (C) in the fig. C.3b, which have been calculated by dividing R_K and $1/\kappa_3$, respectively by a factor of 2. This weak dependence of τ on the Kapitza resistance and the heat conductivity forms the basis of our measurement of the polarization dependence of the ^{3}He specific heat.

C.4. *Anomalous thermometer response*

As we have seen, the response of the inner thermometer following a HT temperature step is slower than that of the viscometer (see fig. C.1). At long times (>3 s), the thermometer response function decays exponentially, but, except for the largest temperatures (>300 mK), significantly slower than the viscometer. The long-time decay rate barely depends on temperature, in contrast with the behavior observed at smaller times. The difference between the response of the inner thermometer and the viscometer may arise for two possible reasons: as the thermometer and the viscometer are not exactly at the same location, their respective temperature evolution may be different during the transient following a temperature step. Another possible explanation is an anomaly in the thermometer itself.

We believe that the second case holds: the cell has been designed to avoid any vertical temperature gradient, owing to the good thermal conductivity of the silver walls. We think that the problem does not stem from a defect in the sinter either, since the same anomaly was observed in two different cells, where the thermometer was respectively located close to the deep end or the surface end of the sinter (with respect to the compaction process). We are thus led to think that the thermometer response is intrinsically slow.

Although we do not know the exact cause of it, the anomaly is clearly related to the presence of liquid ^{3}He: first, the response function of the thermometer has a similar scaling in pressure as that of the viscometer (fig. C.2a),[32] and, second, the thermometer response is much faster in the solid phase (which has a smaller heat capacity and a higher thermal conductivity). We speculate that this anomaly might be due to a slow mechanic relaxation like in Bonfait (1987), or to a pocket of ^{3}He trapped in the core of the thermometer, because of the exfoliated structure of the carbon resistor.

In any case, the anomalous behavior of the thermometer could cast some doubts on the temperature measurements during the magnetization relaxation. In the following, we will show that, during the polarization decay, the thermometer, in spite of its slow response, reads a temperature which is close to the value predicted by our thermal model.

C.5. *Theoretical estimate of the temperature gradient generated by the magnetic relaxation*

We apply our analysis of the thermal response of the cell to calculate the temperature difference between the slit and the walls during the magnetization relaxation. This calculation allows us to cross-check the values read by the inner thermometer, whose anomalously slow response could have biased the results.

C.5.1. *Temperature difference between the slit and the wall*

The power released during the polarization decay varies over a timescale $\approx T_1/2$ ($T_1 = 30$–100 s), much slower than the thermal response time of the cell. Therefore, we can calculate the temperature difference ΔT between the slit and the cell walls in the stationary regime. We solve eqs. (C.8), (C.6), and (C.7) to calculate the contribution $\mathrm{d}T$ to ΔT, arising from a localized heat source $\dot{q}(x)$ at point x. The result, in the limit $\kappa'_a \gg \kappa_3$, is

$$\mathrm{d}T = \frac{\Lambda}{\kappa_3} \frac{\cosh\left(\dfrac{L-x}{\Lambda}\right)}{\sinh\left(\dfrac{L}{\Lambda}\right)} \dot{q}(x)\,\mathrm{d}x + \frac{(L-x)}{\kappa'_a}\left[\frac{1}{2}\dot{q}_\infty + \frac{2}{\pi}(\dot{q}(x) - \dot{q}_\infty)\right]\mathrm{d}x. \tag{C.15}$$

The factors $1/2$ and $2/\pi$ in the last term have the same origin as in eqs. (C.12) and (C.13), respectively, i.e. they account approximately for the 2D

[32] Also, polarizing ^{3}He decreases the thermometer response time, see Section 8.

character of the problem. Strictly speaking, the latter factor is valid only for times such that $\sqrt{D_{\rm spin} t} \ll L$, i.e. when the extra dissipation due to the polarization gradient, $(\dot{q}(x) - \dot{q}_\infty)$, is concentrated close to the slit (see fig. B.3).

In principle, we could use the full expression (B.12) with the magnetization profiles calculated with the 1D diffusion model developed in eq. (B.1). However, we have seen in eq. (B.2) that these profiles overestimate the polarization gradient. This would lead to an even stronger overestimate of the temperature difference between the slit and the walls, as the first term of eq. (C.15) is maximal in the vicinity of the slit ($x < \Lambda$), where the excess polarization is the largest (fig. B.3).

To allow a quantitative comparison with the experiment, we evaluate the power released into the sinter with the simplified expression (B.13), neglecting the magnetization gradients. The average magnetization $\langle m \rangle$ then decays exponentially with an effective time constant $T_1^{\rm eff} = T_1(1 + V_{\rm slit}/V_{\rm sinter})$, where $1/T_1$ is the decay rate in the sinter, $V_{\rm slit}$ and $V_{\rm sinter}$ are the ^{3}He volumes in the slit and in the sinter. From the above expression of $T_1^{\rm eff}$ and eq. (B.13), the power uniformly released into the sinter is, per unit of volume:

$$\dot{Q}_{\rm sinter} = \frac{\mu_0 M_{\rm sat}^2}{\chi_T} \frac{\langle m \rangle^2}{T_1^{\rm eff}} \left(1 + \frac{V_{\rm slit}}{V_{\rm sinter}}\right). \tag{C.16}$$

Using eq. (C.9) with $\dot{Q}_{\rm slit} = 0$, adding the contribution of the contact resistance between the HT and the walls of the cell, and expressing the thermal resistances in terms of the thermal relaxation times $\tau_{\rm sinter}$ and $\tau_{\rm contact}$ and the specific heat per unit volume C_3, the temperature difference between the slit and the HT is finally expressed as

$$T_{\rm slit} - T_{\rm ht} = \frac{\tau_{\rm relax}}{C_3} \frac{\mu_0 M_{\rm sat}^2}{\chi_T} \frac{\langle m \rangle^2}{T_1^{\rm eff}}, \tag{C.17}$$

where the time $\tau_{\rm relax}$ is given by

$$\tau_{\rm relax} = \tau_{\rm sinter}\left(1 + \frac{V_{\rm slit}}{V_{\rm sinter}}\right) + \tau_{\rm contact}. \tag{C.18}$$

Note that, because no heat is released into the slit, this temperature difference does not involve the contribution $R_{\rm slit}$ of the slit to the thermal resistance. This expression is a lower bound, since it neglects the magnetization gradients. For $t = T_1$, it underestimates the temperature difference by 30% with respect to a calculation using the full expression (C.15) and the magnetization profiles calculated by the magnetic diffusion model. Nevertheless, we think that the simple formula (C.17) gives a fairer estimate of the actual temperature difference during the magnetization decay, since the diffusion model overestimates the polarization gradients (Appendix B).

C.5.2. *Comparison with the experiment*

Expression (C.17) allows a direct comparison with the experiment, provided the contributions τ_{sinter} and τ_{contact} are known. The quantity directly measured by the thermal response experiments in the non-polarized liquid is $\tau = \tau_{\text{slit}} + \tau_{\text{sinter}} + \tau_{\text{contact}}$. We therefore determine the needed quantity τ_{relax} as $(1 + V_{\text{slit}}/V_{\text{sinter}})(\tau - \tau_{\text{slit}}) - \tau_{\text{contact}}(V_{\text{slit}}/V_{\text{sinter}})$, where τ_{slit} is calculated with our thermal model (curves (3)+(4)+(5) in fig. C.3b, and τ_{contact} is determined above (Section C.3.3).

Table 5 gives the measured τ, the calculated τ_{contact} and τ_{slit}, and τ_{relax}, for the different pressures and temperatures explored in the viscosity experiment. The error bars on τ correspond to the reproducibility of the experiments. The error bars on τ_{relax} also include the estimated uncertainty in the parameters R_K and κ'_a, amounting to ± 0.1 s for τ_{slit}.

As discussed in Section 6.3.3, the calculated temperature differences computed from τ_{relax} are equal to or larger than the values measured experimentally, however the difference remains consistent within the error bars. This implies that the carbon thermometer does not overestimate the slit temperature, despite the fact that its transient response is anomalously slow.[33] It also means that the actual temperature difference is not larger than the one predicted within our approximation of homogeneous polarization. Hence, the polarization gradient is smaller than expected from the diffusion model, in agreement with the discussion of Section 6.3.3.

Appendix D. The viscosity of ^{3}He within the Landau theory

The calculation of transport coefficients in Landau theory is based on a method similar to that used in the kinetic theory of gases: the current density of the transported quantity is expressed as a function of the deviation from the local equilibrium quasiparticle distribution function. This deviation is obtained as a function of the driving term (temperature gradient for the heat conductivity, velocity gradient for the viscosity, etc.) by solving the Landau kinetic equation (which is a Boltzmann equation for the quasiparticles). The corresponding transport coéfficient is the proportionality constant between the current and the driving term. For an unpolarized system, the problem

[33] In the pores-based explanation of the anomaly, this implies that these pores are much less relaxing than the pores of the silver sinter.

was first solved at a variational level by Abrikosov and Khalatnikov (1957), and exactly by Brooker and Sykes (1968), as described in detail in Baym and Pethick (1991). The transport coefficients are obtained as a function of various angular averages of the probability for collision between two quasiparticles. In the polarized case, one deals with a two-component system, with different Fermi momenta and effectives masses for the up and down spins. As in the unpolarized case, the problem has been solved at the variational level by Meyerovich (1983) and exactly by Anderson, Pethick and Quader (Anderson et al. 1987). In this appendix, we first summarize the approach for the unpolarized system, so as to introduce the basic concepts, then turn to a description of the calculation of Anderson et al. (1987), in terms of which we finally discuss our experimental results.

D.1. *Non-polarized system*

D.1.1. *Expression of the viscosity coefficient*

To calculate the viscosity, we assume that the average velocity $\vec{u}$ is in the x direction and varies only in the y direction. The momentum current density (that is the stress tensor) is written as

$$\sigma_{xy} = -\sum_{\sigma} \int \frac{d^3\vec{p}}{(2\pi\hbar)^3}\, p_x (v_p)_y \delta n_p^{\mathrm{le}} \tag{D.1}$$

(the discrete sum is over the spin index σ and the integral over the quasiparticle momenta $\vec{p}$). At the end of the calculation, σ_{xy} will be identified with $\eta(\partial u_x/\partial y)$ to find the viscosity coefficient. δn_p^{le} is the deviation from the local equilibrium distribution function which is a solution of the linearized kinetic equation. In this equation, the main problem is to deal with the collision integral, because the interactions between the quasiparticles are not known. However, the solution can be expressed in a compact way as a function of the collision probability between the quasiparticles. In a second step, one has to use an approximate expression for the collision probability to complete the calculation.

At low temperature, binary collisions dominate the interaction term. Moreover, due to energy conservation together with the Pauli principle, the quasiparticles undergoing scattering have to be taken close to the Fermi surface. Consequently, for two quasiparticles of incoming momenta $\vec{p}_1$ and $\vec{p}_2$, the conservation of momentum implies that the outgoing momenta $\vec{p}_3$ and $\vec{p}_4$ are fully determined by the angle θ between the two incoming quasiparticles, and the angle ϕ between the planes $(\vec{p}_1, \vec{p}_2)$ and $(\vec{p}_3, \vec{p}_4)$. The probability of collision between quasiparticles is then a function of

$W_{\sigma\sigma'}(\theta,\phi)$ (σ and σ' denoting the spins[34]). In the absence of magnetic field $W_{\uparrow\downarrow} = W_{\downarrow\uparrow}$ and $W_{\uparrow\uparrow} = W_{\downarrow\downarrow}$, and the relevant quantity is the 'total' probability of collision, $W(\theta,\phi)$ such that $W = W_{\uparrow\downarrow} + \frac{1}{2}W_{\uparrow\uparrow}$. To proceed, one separates the energy and the angular variables in $\delta n_p^{\rm le}$. For a small deviation, one writes

$$\delta n_p^{\rm le} = -(k_{\rm B}T)\frac{\partial n_p^0}{\partial \varepsilon_p}\,\gamma(\varepsilon)(\hat{p}\hat{x})(\hat{p}\hat{y})\frac{\partial u_x}{\partial y}, \tag{D.2}$$

where $\hat{p} = \vec{p}/|p|$, n_p^0 is the equilibrium distribution function, and $\gamma(\varepsilon)$ is the energy part of the deviation. Considering only the quasiparticles on the Fermi surface allows one to decouple the angular and energy parts of the collision integral. By integrating the kinetic equation over the angular variables, one finds the integral equation obeyed by $\gamma(\varepsilon)$:

$$-\frac{p_f v_{\rm F}\tau_0}{2\cosh(x_1/2)} = \left(1+\frac{x_1^2}{\pi^2}\right)\tilde{\gamma}(x_1) - \lambda\int_{-\infty}^{+\infty} {\rm d}x_2\,\frac{(x_1-x_2)}{\pi^2\sinh[(x_1-x_2)/2]}\tilde{\gamma}(x_2), \tag{D.3}$$

where $x = (\varepsilon-\mu)/k_{\rm B}T$ (μ being the chemical potential), $p_{\rm F}$ and $v_{\rm F} = p_{\rm F}/m^*$ are respectively the momentum and velocity at the Fermi level, with m^* the quasiparticle effective mass, $\gamma(x) = 2\cosh(x/2)\tilde{\gamma}(x)$, and τ_0 is a collision time averaged on the Fermi surface, given by

$$\frac{1}{\tau_0} = \frac{m^{*3}(k_{\rm B}T)^2}{16\pi^2\hbar^6}\int\frac{{\rm d}\Omega}{4\pi}\frac{W(\theta,\phi)}{2\cos(\theta/2)}. \tag{D.4}$$

In eq. (D.3), the averaging over the angular variables is contained in the coefficient λ, which is given by

$$\lambda = \frac{\int\frac{{\rm d}\Omega}{4\pi}\frac{W(\theta,\phi)}{2\cos(\theta/2)}[P_2(\hat{p}_1,\hat{p}_3)+P_2(\hat{p}_1,\hat{p}_4)-P_2(\hat{p}_1,\hat{p}_2)]}{\int\frac{{\rm d}\Omega}{4\pi}\frac{W(\theta,\phi)}{2\cos(\theta/2)}}, \tag{D.5}$$

where P_2 is the second-order Legendre polynomial. The coefficient λ contains angular averages of the probability of collision between quasiparticles, $W(\theta,\phi)$. Solving the kinetic eq. (D.3) for $\gamma(x)$, and calculating the momentum flux (eq. (D.1)) yields the viscosity η:

$$\eta = \frac{1}{15}N(0)p_{\rm F}^2 v_{\rm F}^2\tau_0 R^+(\lambda), \tag{D.6}$$

where $N(0)$ is the total density of states at the Fermi level. The dimensionless

[34] The total spin is conserved due to the weakness of the dipole–dipole interaction in ^{3}He.

factor $R^+(\lambda)$ is equal to

$$R^+(\lambda) = \sum_{\nu \text{ odd}} \frac{2\nu + 1}{\nu(\nu + 1)(\nu(\nu + 1) - 2\lambda)}. \tag{D.7}$$

D.1.2. *The forward scattering amplitudes*

A quantitative calculation of the viscosity coefficient from eq. (D.6) requires knowledge of the probability of collision $W_{\sigma\sigma'}(\theta, \phi)$. The probability of collision is related to the scattering amplitude $T_{\sigma\sigma'}(\theta, \phi)$ through

$$W_{\sigma\sigma'}(\theta, \phi) = \frac{2\pi}{\hbar} |T_{\sigma\sigma'}(\theta, \phi)|^2. \tag{D.8}$$

The scattering amplitude can be split into triplet and singlet terms:

$$T_{\uparrow\uparrow}(\theta, \phi) = T_{\mathrm{t}}(\theta, \phi),$$
$$T_{\uparrow\downarrow}(\theta, \phi) = \tfrac{1}{2}(T_{\mathrm{s}}(\theta, \phi) + T_{\mathrm{t}}(\theta, \phi)).$$

For the unpolarized liquid at low temperature, all momenta have the same absolute value p_{F}; simple geometric considerations show that, in the reference frame of the center of mass of two quasiparticles, the momentum of the 'reduced particle' is given by $p_{\mathrm{F}} \sin(\theta/2)$, while ϕ is the scattering angle. In the limit $\phi \to 0$, corresponding to small momentum transfer (forward scattering), the scattering amplitude can be expressed in terms of the Landau interactions. Furthermore, as $\phi = \pi$ corresponds to exchange the colliding quasiparticles starting from $\phi = 0$, the corresponding scattering amplitude is equal to plus or minus the forward scattering amplitude, the sign depending on the total spin. These considerations lead to the following sum rules:

$$T_{\mathrm{t}}(\theta, 0) = -T_{\mathrm{t}}(\theta, \pi) = N(0)^{-1} \sum_{l=0}^{\infty} (A_l^s + A_l^a) P_l(\cos\theta), \tag{D.9}$$

$$T_{\mathrm{s}}(\theta, 0) = T_{\mathrm{s}}(\theta, \pi) = N(0)^{-1} \sum_{l=0}^{\infty} (A_l^s + 3A_l^a) P_l(\cos\theta) \tag{D.10}$$

with $A_l^i = F_l^i(1 + F_l^i/(2l + 1))$ and $i = s, a$. In the general case, however, there is no direct relation between the scattering amplitude and the Landau coefficients. To proceed, one uses an approximate form of the probability of collision.

D.1.3. *Approximation schemes*

The scattering amplitude $T_{\sigma\sigma'}(\theta, \phi)$ can be expressed as a double Legendre polynomials expansion in terms of $P_l(\cos\theta)$ and $P_{l'}(\cos\phi)$. The Pauli principle implies that the triplet scattering amplitude contains only odd l' and the singlet contains even l' terms. The s–p approximation consists in truncating the $P_{l'}(\cos\phi)$ expansion to $l'<2$:

$$T_{\mathrm{t}}(\theta,\phi) \approx T_{\mathrm{t}}(\theta,0)\cos\phi,$$

$$T_{\mathrm{s}}(\theta,\phi) \approx T_{\mathrm{s}}(\theta,0).$$

This approximation amounts to assuming that, in the center of mass reference frame, the triplet scattering takes place in the p-channel, and the singlet in the s-channel. The results of this approximation, calculated with the Landau coefficients $l = 0, 1$ only, agree within a factor of two with the experimental values (cf. Table D.2). Numerically, it is interesting to note that the scattering amplitude between quasiparticles of unlike spins is larger than between quasiparticles of same spins:

at saturated vapor pressure:

$$N(0)T_{\uparrow\uparrow}(\theta,\phi) \approx (-1.43 + 1.16\cos\theta)\cos\phi, \tag{D.11}$$

$$N(0)T_{\uparrow\downarrow}(\theta,\phi) \approx (3.95 + 2.1\cos\theta) + (-0.71 + 0.58\cos\theta)\cos\phi. \tag{D.12}$$

At $P = 27\,\mathrm{bar}$:

$$N(0)T_{\uparrow\uparrow}(\theta,\phi) \approx (-2.2 + 0.9\cos\theta)\cos\phi, \tag{D.13}$$

$$N(0)T_{\uparrow\downarrow}(\theta,\phi) \approx (5.2 + 3.5\cos\theta) + (-1.1 + 0.45\cos\theta)\cos\phi. \tag{D.14}$$

More sophisticated approximation schemes to extrapolate the scattering amplitude from $\phi = 0$ to the entire Fermi sphere have been developed (for instance, see Ainsworth and Bedell 1987 and references therein). They generally give a better quantitative agreement for the transport coefficients, but are model-dependent.

D.2. *Polarized systems*

In a polarized system, the Fermi momenta p_σ, the densities of states at the Fermi surface $N_\sigma(0)$, the quasiparticle lifetimes τ_σ, and the distribution functions are different for each spin species $\sigma = \uparrow$ or $\downarrow$. To calculate the transport coefficients, there are now two kinetic equations which have to be solved. These equations are coupled through the collision terms, since the number of collisions undergone by (say) the $\uparrow$ quasiparticles depends on the distribution function of the $\downarrow$ quasiparticles. Anderson, Pethick and Quader (APQ) have provided exact analytical solutions for the set of kinetic equations and given the expressions of the transport coefficients in term of angular averages of the collision probability $W_{\sigma\sigma'}(\theta,\phi)$ (Anderson et al. 1987).

As in the case of the non-polarized system, these expressions have to be complemented by approximate forms of the collision probability to obtain quantitative results. In particular, we will see that the polarization dependence of the viscosity depends crucially on the choice of the expression for $W_{\sigma\sigma'}(\theta,\phi)$. In the following sections, we describe the main steps leading to a general expression for the viscosity coefficient; we then discuss the polarization dependence of the viscosity arising in various approximation schemes.

D.2.1. *General expression for the viscosity*

The kinetic equation for the energy part $\gamma_\sigma(\varepsilon)$ for each spin species has the same structure as eq. (D.3), with the collision term $\lambda\gamma(x)$ replaced by $\sum_{\sigma'} \lambda_{\sigma\sigma'}\gamma_{\sigma'}(x)$. This term couples the quasiparticles distributions of ↑ and ↓ spins. The matrix elements $\lambda_{\sigma\sigma'}$ are angular averages of the collision probability:

$$\lambda_{\sigma\sigma} = \frac{1}{\bar{W}_\sigma}\int \frac{\mathrm{d}\Omega}{4\pi}\,\frac{W_{\sigma\sigma}(\theta,\phi)}{2}\,\frac{[P_2(\hat{p}_1,\hat{p}_3)+P_2(\hat{p}_1,\hat{p}_4)-P_2(\hat{p}_1,\hat{p}_2)]}{2\cos(\theta/2)}$$
$$+\frac{(m^*_{-\sigma})^2}{(m^*_\sigma)^2\bar{W}_{-\sigma}}\int\frac{\mathrm{d}\Omega}{4\pi}\frac{W_{\sigma-\sigma}(\theta,\phi)}{2}\frac{p_{-\sigma}}{\ell}P_2(\hat{p}_1,\hat{p}_3). \tag{D.15}$$

$$\lambda_{\uparrow\downarrow} = \lambda_{\downarrow\uparrow} = \frac{1}{(\bar{W}_\uparrow\bar{W}_\downarrow)^{1/2}}\int\frac{\mathrm{d}\Omega}{4\pi}W_{\uparrow\downarrow}(\theta,\phi)\frac{(p_\uparrow p_\downarrow)^{1/2}}{\ell}$$
$$[P_2(\hat{p}_1,\hat{p}_4)-P_2(\hat{p}_1,\hat{p}_2)]. \tag{D.16}$$

where $\ell = |\vec{p}_1+\vec{p}_2| = ({p_\uparrow}^2+{p_\downarrow}^2+2p_\uparrow p_\downarrow\cos\,\theta)^{1/2}$, and m^*_σ is the quasiparticle effective mass of spin σ. $\bar{W}_\sigma$ is the average probability of collision for the spins σ given by

$$(m^*_\sigma)^2\bar{W}_\sigma = (m^*_\sigma)^2\int\frac{\mathrm{d}\Omega}{4\pi}\frac{W_{\sigma\sigma}(\theta,\phi)}{2}\frac{1}{2\cos(\theta/2)}+(m^*_{-\sigma})^2$$
$$\int\frac{\mathrm{d}\Omega}{4\pi}\frac{W_{\sigma-\sigma}(\theta,\phi)}{2}\frac{p_{-\sigma}}{\ell}. \tag{D.17}$$

These averaged probabilities correspond to collision times:

$$\frac{1}{\tau_\sigma} = \frac{m^{*3}_\sigma(k_\mathrm{B}T)^2\bar{W}_\sigma}{16\pi^2\hbar^6}. \tag{D.18}$$

In these equations, p_1 and p_3 refer to incoming and outgoing quasiparticles of same spin component, i.e. lie on the same Fermi sphere (consequently, the same holds for p_2 and p_4). Using momentum conservation, the arguments $(\hat{p}_1,\hat{p}_2)$, $(\hat{p}_1,\hat{p}_3)$, and $(\hat{p}_1,\hat{p}_4)$ of the Legendre polynomials in Eqs.

(D.15) and (D.16) can be expressed in function of the angles θ and ϕ, and $p_{\pm\sigma}/\ell$ (Anderson et al. 1987). For collisions between parallel spins, these expressions are identical to those for the non-polarized case and depend only on θ and ϕ. In contrast, for collisions between unlike spins, the expressions for the reduced particle momentum and the momentum transfer are no longer simply related to the angle θ and ϕ. The arguments of the Legendre polynomials now depend on $p_{-\sigma}^2/\ell^2$ for $(\hat{p}_1, \hat{p}_3)$ and $p_\sigma p_{-\sigma}/\ell^2$ for $(\hat{p}_1, \hat{p}_4)$. In a similar way, the $(2\cos(\theta/2))^{-1}$ term in the integrals, which arises from conservation of momentum, is replaced by $p_{-\sigma}/\ell$ in eq. (D.15) and $(p_\sigma p_{-\sigma})^{1/2}/\ell$ in eq. (D.16). Therefore, the integrals will depend on the ratio $p_\downarrow/p_\uparrow$. As we shall see, this gives rise to an implicit polarization dependence of $\lambda_{\sigma\sigma'}$ and the τ_σ.

In order to obtain the energy part of the deviations of the distribution functions $\gamma_\sigma(\varepsilon)$, the kinetic equations are decoupled by diagonalizing the matrix $\lambda_{\sigma\sigma'}$. The decoupled equations can then be solved separately using the procedure described in Section D.1 for non-polarized systems. The momentum flux, expressed with the original $\gamma_\sigma(\varepsilon)$ results in the following expression for the viscosity coefficient (Anderson et al. 1987):

$$\begin{aligned}\eta = \tfrac{1}{15}(&N_\uparrow(0)p_\uparrow^2 v_\uparrow^2 \tau_\uparrow R_{++} + N_\downarrow(0)p_\downarrow^2 v_\downarrow^2 \tau_\downarrow R_{--} \\ &+ 2(N_\uparrow(0)N_\downarrow(0)\tau_\uparrow\tau_\downarrow)^{1/2} p_\uparrow p_\downarrow v_\uparrow v_\downarrow R_{+-}).\end{aligned} \tag{D.19}$$

The coefficients R_{++}, R_{+-}, and R_{--}, which come from the diagonalization of the matrix $\lambda_{\sigma\sigma'}$, are given by

$$R_{++} = R^+(\lambda_1)\cos^2\xi + R^+(\lambda_2)\sin^2\xi,$$

$$R_{+-} = (R^+(\lambda_1) - R^+(\lambda_2))\cos\xi\sin\xi,$$

$$R_{--} = R^+(\lambda_2)\cos^2\xi + R^+(\lambda_1)\sin^2\xi,$$

where ξ is defined such that the eigenvectors associated with the eigenvalues λ_1 and λ_2 are $(\cos\xi, \sin\xi)$ and $(-\sin\xi, \cos\xi)$, respectively, and $R^+(\lambda)$ is the series expansion defined in eq. (D.7).

Expression (D.6) for non-polarized systems can of course be deduced from the general expression (D.16). We point out that unpolarized does not mean that the two components are uncoupled (as could be understood from Appendix A of APQ). Even though the kinetic equations for spins $\uparrow$ and $\downarrow$ are identical, the quasiparticles distribution functions $\gamma_\sigma(\varepsilon)$'s are still coupled through the collision term, and the matrix $\lambda_{\sigma\sigma'}$ is not diagonal in this case. As the absence of polarization implies $m_\uparrow^* = m_\downarrow^*$, $p_\uparrow = p_\downarrow$, $N_\uparrow(0) = N_\downarrow(0) = N(0)/2$ and $W_{\uparrow\uparrow} = W_{\downarrow\downarrow}$, we have $\lambda_{\uparrow\uparrow} = \lambda_{\downarrow\downarrow}$. This implies $\lambda_1 = \lambda_{\uparrow\uparrow} + \lambda_{\uparrow\downarrow}$, and $\xi = \pi/4$, so that eq. (D.19) finally reduces to the result of eq. (D.6), with

$\lambda = \lambda_1$, as would be trivially found by directly writing $\gamma_\uparrow(\varepsilon) = \gamma_\downarrow(\varepsilon)$ in the equations.

Equation (D.19) is the starting point to discuss the polarization dependence of the viscosity in a Fermi liquid. The polarization dependence arises from two different effects: the change in size of the Fermi spheres for the quasiparticles $\uparrow$ and $\downarrow$, and the polarization dependence of the effective masses m^*_σ and the probability of collisions $W_{\sigma\sigma'}(\theta, \phi)$.

As the Fermi momentum in a Fermi liquid is the same as that of the non-interacting system of same density (Landau 1956, Nozières 1964), the radii of the Fermi spheres are simply given by

$$p_\sigma / p_{\rm F} = (1 + \sigma m)^{1/3}. \tag{D.20}$$

Note that the change of Fermi momentum alone alters the phase space restrictions (see fig. D.1b) and thus induces a polarization dependence of the collision term (the matrix $\lambda_{\sigma\sigma'}$ implicitly depends on the ratio $p_\downarrow/p_\uparrow$) and the collision times τ_σ. It also trivially modifies the Fermi velocities and the densities of states.

The polarization dependence of the effective masses m^*_σ and of the probability of collisions $W_{\sigma\sigma'}(\theta, \phi)$ (which determines that of all the other factors, such as the densities of states, quasiparticle lifetimes, etc...), cannot be determined within the Landau theory. The choice of the approximation for the collision probability turns out to be of prime importance for the polarization dependence of the viscosity. Let us consider the various approximation schemes found in the literature:

D.2.2. *s-wave limit*

The usual approximation used for dilute systems is the s-wave limit, where there is no energy and angular dependence of the collisions between the quasiparticles (Meyerovich 1987), (Mullin and Miyake 1983). Due to the Pauli principle, isotropic collisions can occur only between quasiparticles of unlike spins ($W_{\uparrow\uparrow} = W_{\downarrow\downarrow} = 0$). The consequence of this restriction is the decoupling of the kinetic equations ($\lambda_{\uparrow\downarrow} = 0$). As $W_{\uparrow\downarrow}$ does not depend on the Fermi momentum, and hence on m, all the polarization dependence comes from the p^5_σ terms in eq. (D.19) and from the phase space restriction imposed by the conservation of momentum. One finds, for $m>0: \lambda_{\uparrow\uparrow} = 1 - 2(p_\downarrow/p_\uparrow)^2 + 6/5(p_\downarrow/p_\uparrow)^4$, $\lambda_{\downarrow\downarrow} = 1/5$, $\lambda_{\uparrow\downarrow} = 0$, $\tau_\uparrow = \tau_0 p_\uparrow/p_\downarrow$, and $\tau_\downarrow = \tau_0$. The relative viscosity enhancement for $m>0$ is (Mullin and Miyake 1983)

$$\frac{\eta(m)}{\eta(0)} = \frac{(1+m)^2 R^+[1 - 2(p_\downarrow/p_\uparrow)^2 + \frac{6}{5}(p_\downarrow/p_\uparrow)^4]/R^+(1/5) + (1-m)^2}{2(1-m)^{1/3}} \tag{D.21}$$

with $p_\downarrow/p_\uparrow = (1-m)^{1/3}/(1+m)^{1/3}$. Note that the relative viscosity enhancement does not depend on the strength of interaction, and therefore is pressure independent. The variational result of Bashkin and Meyerovich is obtained by retaining only the first term in the series expansion (eq. (D.7)) of $R^+(\lambda)$ (Meyerovich 1987).

As shown in fig. D.1a, both the variational result and the exact solution lead to a monotonous increase of the viscosity with the polarization. This may be thought as obvious, as the polarization reduces the number of (pure s-wave) collisions. However, the mechanism at play is more subtle than a mere counting argument, where the collision rate would be taken as the product of the densities of spins ↑ and ↓. The counting of the number of collisions would lead to an initial behavior $\propto m^2$ for the increase in viscosity. In contrast, eq. (D.21) implies a linear dependence at low polarization. Mathematically, this non-analyticity at $m=0$ arises because the various integrals involving polynomials in $\min(p_\uparrow, p_\downarrow)/\max(p_\uparrow, p_\downarrow)$. In particular, the collision time of the majority (say ↑) spin species is $\tau_\uparrow \propto p_\uparrow/p_\downarrow \approx (1+2/3|m|+\cdots)$, i.e. the scattering rate of these spins decreases linearly with the polarization. This reflects the fact that the $p_\downarrow/\ell$ term (for $m>0$) in eq. (D.17) decreases as the polarization increases. As stressed by Nozières (1998), the physical origin of the linear decrease of the scattering rate is quite clear in fig. D.1b: for $m>0$, the maximum momentum transfer during a collision is $2p_\downarrow$. The momentum conservation implies that the states with momentum p_3 in a region of the ↑ Fermi surface around $(\hat{p}_1, \hat{p}_3) = \pi$ are not accessible for a quasiparticle of spin ↑ and momentum p_1 after scattering by a quasiparticle of spin ↓. The unavailable portion of the ↑ Fermi surface is of angular extension of order $m^{1/2}$, hence is proportional to $|m|$, so that $\tau_\uparrow$ increases linearly with m. In contrast to the scattering rate, $\lambda_{\uparrow\uparrow}$ initially *increases* linearly with m, resulting from the fact that the phase space restriction is differently weighted in the numerator and denominator of eq. (D.15). In Bashkin and Meyerovich's calculation, the linear contributions of $\tau_\uparrow$ and $R^+(\lambda_{\uparrow\uparrow})$ happen to cancel, but this is not true in the exact solution, where a small linear contribution shows up (fig. D.1). In the same figure, we can see that, even at the variational level, the cancellation does not happen for the thermal conductivity, leading to a marked linear dependence up to $m \approx 0.2$. This emphasizes the central role of the phase-space reduction in the s-wave limit.

D.2.3. *s–p approximation*

The s–p approximation described in Section D.1.3 can be used in order to obtain a more realistic model for liquid ^{3}He. Such an approach was originally followed by Hess and Quader (1987). Starting from the general result of APQ, they used the s–p approximation to relate the collision probabilities

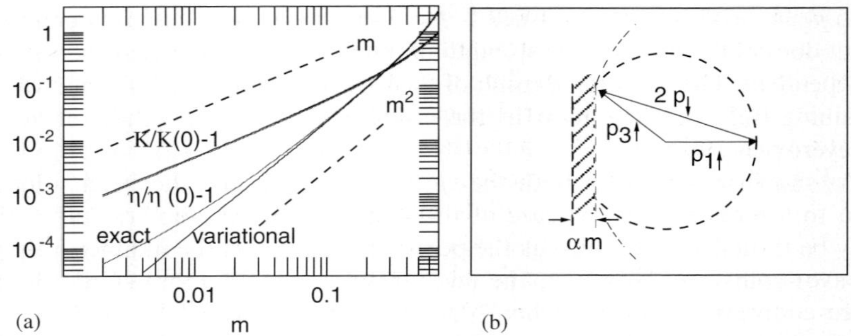

Fig. D.1. (a) Polarization dependence of the viscosity and the thermal conductivity in the s-wave limit; (b) for a polarized system, momentum conservation forbids the large angle scattering of the majority spins (hatched region). This gives rise to the *m* linear terms in Fig. D.1a.

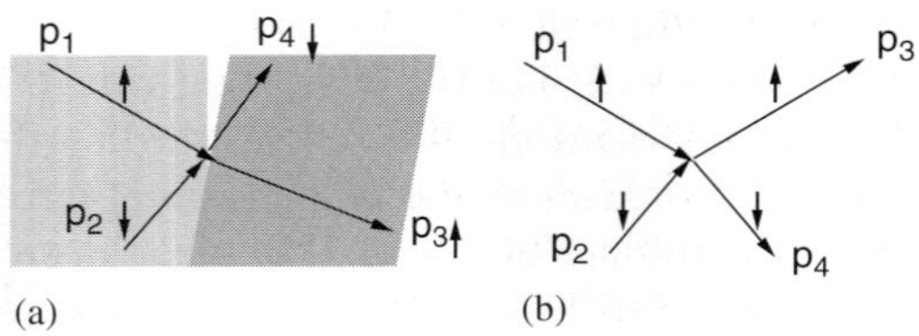

Fig. D.2. (a) A scattering event between unlike spins, where a given spin component is nearly forward scattered (corresponding to a small angle between the incidence and scattering planes); (b) a spin–flip scattering event in the close to forward direction can be viewed as a non-spin–flip scattering at larger angles.

to the $l = 0$ and 1 moments of the $\uparrow\uparrow$ ($\downarrow\downarrow$) and $\uparrow\downarrow$ forward scattering amplitudes in the polarized system (as for the unpolarized case, the higher moments were neglected). These amplitudes were expressed in terms of the polarization-dependent longitudinal Landau interactions given by the 'nearly metamagnetic' model of Bedell and Sanchez-Castro (Section 3.1) (Bedell and Sanchez-Castro 1986, Sanchez-Castro et al. 1989). However, an additional assumption was required for the $\uparrow\downarrow$ scattering amplitude, since it depends on the so-called 'spin–flip' interaction, a parameter which is not determined in the 'nearly metamagnetic' model. The spin–flip interaction involves the scattering process depicted by fig. D.2b, where a spin $\uparrow$ is scattered at low angle into a spin $\downarrow$. Alternatively, this process can be seen as a scattering at large angle with no spin–flip. Hess and Quader made the assumption that the spin–flip interaction $l = 0$, $\tilde{g}_0^{\downarrow\uparrow}$, behaves as a function of the Landau longitudinal interactions as $\tilde{g}_0^{\downarrow\uparrow} = (\tilde{f}_0^{\uparrow\uparrow} + \tilde{f}_0^{\downarrow\downarrow} - 2\tilde{f}_0^{\downarrow\uparrow})/2$, that is the very term responsible for the peak of the susceptibility (Section 3.1) in the 'nearly metamagnetic' model (Hess and Quader 1987). This expression gives the correct limit in the unpolarized case, and has the proper symmetry under time reversal. The spin–flip interaction $\tilde{g}_1^{\downarrow\uparrow}$ is not given in the original article, but Hess and Quader argued that it has a weak polarization dependence. Together with the polarization dependence of the $\tilde{f}_0$ terms of the Bedell and Sanchez-Castro model, this choice of $\tilde{g}_0^{\downarrow\uparrow}$ leads to a negative spin–flip interaction whose absolute magnitude increases with the polarization. The resulting singlet amplitude increases with m, which makes the viscosity decrease initially, before the phase space restrictions cause the viscosity to rise eventually. The maximal reduction, which occurs near $m = 0.35$, is of order 80%. Hess and Quader pointed out that an initial decrease was also observed if the 'spin–flip' scattering amplitude $\tilde{g}_0^{\downarrow\uparrow}$ was kept polarization independent, reflecting the fact that, even in this case, the singlet amplitude increases due to the contribution of the non spin-flip process shown in fig. D.2a. The maximal reduction still occurs at $m \approx 0.35$, but is only 20%.

This unexpected behavior, which is in disagreement with the currently accepted experimental results, was originally seen as evidence against the 'nearly metamagnetic' model (see the discussion in Béal-Monad 1988).

Sanchez-Castro and Bedell replied in a subsequent paper (Sanchez-Castro and Bedell 1989) that the initial decrease in viscosity was due to neglecting the phase space restriction in spin–flip scattering processes. In the first stage, this restriction reduces the absolute magnitude of the true spin–flip interaction $\tilde{g}_0^{\downarrow\uparrow}$. In the limit of full polarization, the total loss of phase space implies $\tilde{g}_0^{\downarrow\uparrow} = 0$. Hence, $\tilde{g}_0^{\downarrow\uparrow}$ does not become more negative with polarization, at variance with the first hypothesis made by Hess and Quader. However, as Sanchez-Castro and Bedell point out themselves, their calculation of $\tilde{g}_0^{\downarrow\uparrow}$ within the induced interaction model shows that $\tilde{g}_0^{\downarrow\uparrow}$ is approximately constant, up to $m \approx 0.9$, in agreement with the second hypothesis of Hess and Quader. According to Sanchez-Castro and Bedell, the failure of the Hess and Quader's calculation then comes from a second effect, the decrease of the number of spin–flip scattering processes caused by the phase space restriction. They argue that this effect alone would dominate the polarization dependence of the interactions, leading to a monotonic increase of the viscosity within the 'nearly-metamagnetic' model. We disagree on this point: as spin–flip scattering events at low angle can be viewed as large angle non spin–flip events, we believe that the phase space restriction is fully taken into account in APQ's calculation (as we discussed at length, this is the origin of the $p_\downarrow/\ell$ factors in Eqs. (D.15) and (D.16)) Consequently, it seems to us that the polarization dependence of the Landau interactions assumed by Hess and Quader does indeed lead, within the s–p approximation, to an initial decrease of the viscosity with increasing polarization, whether metamagnetism is present or not (i.e. whatever the precise behavior of $(\tilde{f}_0^{\uparrow\uparrow} + \tilde{f}_0^{\downarrow\downarrow} - 2\tilde{f}_0^{\downarrow\uparrow})/2$).

In retrospect, this debate has lost its significance, since our measurements of the magnetization curve do not show any metamagnetic behavior. Thus, there is no real point in trying to show that the observed increase of viscosity does not rule out the metamagnetism of liquid ^{3}He! In our opinion, the question now is to know how the decrease of viscosity predicted by Hess and Quader, which contradicts the experiments, is related to the peculiar polarization dependence they assume for the Landau interactions. To investigate this point, we have carried out calculations of the viscosity coefficient following a more phenomenological approach (Buu et al. 2002b): Following Hess and Quader, we start from the solution of the kinetic equation provided by APQ, and we use the s–p approximation expression for the collision probability. However, we choose simpler assumptions for the polarization dependence of the scattering amplitudes: In the first step, we assume the Landau interactions to be constant, equal to their value at $m = 0$ (the difference with the s-wave calculation is that we allow for the angular dependence of W). Although this choice is clearly unphysical (our experiments on the specific heat do show that the effective masses depend on the polarization), it tests the extent Hess and Quader's results depend on their particular choice of polarization dependent collision probabilities. As in the s-wave case, the angular integrals in the matrix elements $\lambda_{\sigma\sigma'}$ and the collision times τ_σ can be calculated analytically as polynomials in $\min(p_\uparrow, p_\downarrow)/\max(p_\uparrow, p_\downarrow)$. The results for s.v.p. and 27 bar are shown in figs. D.3a and b. Two features are worth pointing out: first, the initial behavior is linear in polarization. Second, in contrast to the *s*-wave case, $\lambda_{\downarrow\downarrow}$ and $\tau_\downarrow$ now depend on polarization. The resulting prediction for the viscosity at pressures of 0, 10 and 27 bar is compared in fig. D.4a to the *s*-wave result. The

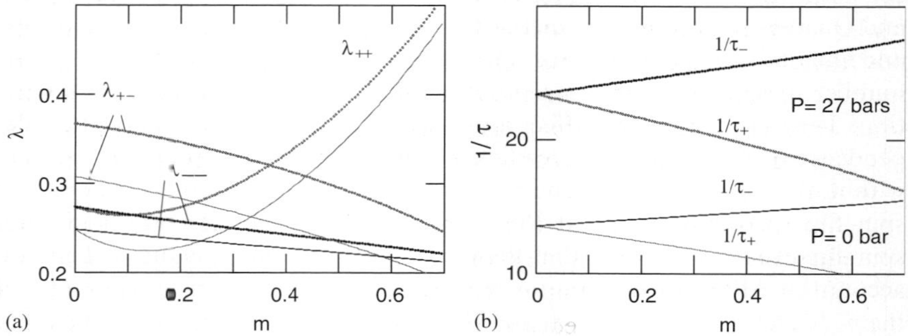

Fig. D.3. Calculated polarization dependence of (a) the coefficients of the matrix $\lambda_{\sigma\sigma'}$, and (b) the relaxation rates $1/\tau$ at s.v.p. (thin lines) and 27 bar (dotted lines). The s–p approximation was used, the probabilities of collision and effective masses being kept at their value at zero polarization.

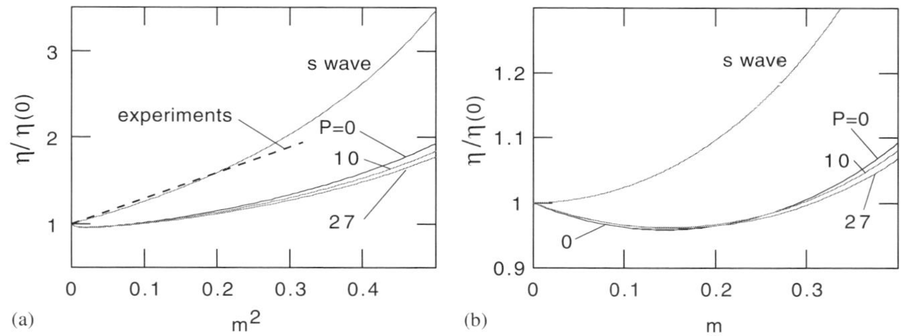

Fig. D.4. Theoretical polarization dependence of the viscosity plotted versus m^2 (a), and m (b). The different curves correspond to 0, 10, and 27 bar. The calculation is based on the s–p approximation, keeping the effective masses and collision probabilities W polarization independent. The difference with the s-wave behavior mainly comes from the θ and ϕ dependence of $W_{\uparrow\downarrow}$. The s-wave prediction and the experimental behavior at 80 mK are shown for comparison.

main result, is that, although we ignored the polarization dependence of the Landau interactions, the viscosity initially *decreases* (by 5%) before increasing at large polarization, this increase being slower than in the s-wave case. This behavior is quite insensitive to the choice of $W_{\uparrow\uparrow}$, reflecting that $W_{\uparrow\uparrow} \ll W_{\uparrow\downarrow}$. In fact, this behavior is not as surprising as it may seem. We already mentioned that, in the s-wave case, the linear contributions of $\lambda_{\uparrow\uparrow}$ and $\tau_{\uparrow}$ very nearly cancel each other. Therefore, it is conceivable that, by altering the angular dependence of $W_{\uparrow\downarrow}$, one can perturb this delicate cancellation in a way such that η initially decreases linearly with m. In fact, we can trace back the initial decrease to the negative sign of the $\cos\phi$ (p-wave) triplet contribution to the scattering amplitude. On the other hand, the positive $\cos\theta$ singlet contribution is such that it decreases the m^2 contribution with respect to the s-wave case. Both effects conspire to cause the observed minimum. As for the weak pressure dependence of the polarization effect, it originates in an approximate scaling of the dominant term $W_{\uparrow\downarrow}$ with the pressure.

A first possibility to explain this lack of agreement between the theory and the experiment would be to question the choice of the collision probability $W_{\sigma\sigma'}(\theta, \phi)$ made in the s–p approximation. We mentioned in Section 2.1.3 that this choice led to a predicted value for the zero polarization viscosity that is too small. In contrast, the total probability collision $W(\theta, \phi)$ computed by Ainsworth and Bedell (1987) within the induced interaction model accounted much better for the experimental data. According to these authors, the failure of the s–p approximation originates in its $W(\theta, \phi)$ being very different from that predicted within the interaction model (see fig. 9 of Ainsworth and Bedell 1987, which shows $W(\theta, \phi)\sin(\theta/2)$). In fact, we believe that this difference is overestimated by Ainsworth and Bedell. Using the same Landau parameters in the s–p approximation as these authors, we find a collision probability within the s–p approximation less singular than they do. However, this probability still differs from that given by the interaction model. In order to estimate whether this could explain the observed discrepancy concerning the polarization dependence, we changed the parameters in eq. (D.12), so as to obtain a behavior qualitatively similar to that predicted by Ainsworth and Bedell (1987). Although the initial decrease in the viscosity was somewhat reduced, we could not recover the experimental behavior.

This failure may be taken as an indication that the probability of collision *does* depend upon the polarization. To pursue the investigation, it is interesting to reintroduce some polarization dependence, for example through the effective mass (keeping $W_{\uparrow\downarrow}$ constant). We have calculated the polarization dependence of the viscosity using the empirical expression $m^*_\sigma(m) = m^*(0)(1 - m^2/2)$ for $\sigma = \uparrow, \downarrow$ from the experimental polarization dependence of the specific heat (Section 8). As shown in fig. D.5a, the viscosity is found to increase markedly faster than $\propto m^2$. This behavior stems mainly from the strong dependence of η upon the effective mass ($\propto 1/m^{*4}$).[35] The prediction now overestimates the effect of polarization on viscosity. Moreover, the minimum in viscosity is still present, although it is shifted to lower polarizations due to the increased m^2 dependence (fig. D.5b).

On the other hand, it is inconsistent to take into account the polarization dependence of the effective mass in eq. (D.19) while keeping the probability of collision constant. In particular, there is an implicit polarization dependence of $W_{\sigma\sigma'}$ through the density of states in Eqs. (D.10) and (D.9). If we assume a polarization dependence of the form: $W_{\uparrow\uparrow} \propto (m^* p_\uparrow)^{-2}$, $W_{\uparrow\downarrow} \propto (m^{*2} p_\uparrow p_\downarrow)^{-1}$ and $W_{\downarrow\downarrow} \propto (m^* p_\downarrow)^{-2}$, the effect of the polarization dependence of m^* on τ_σ will be partly offset. As shown by figs. D.5a and b, this leads to a behavior intermediate between the two preceding cases. However, the prediction remains far from the nearly pure $\propto m^2$ behavior observed experimentally.

From this study, one may conclude that it seems difficult to reproduce the experimental behavior within the s–p approximation with the scattering amplitude $l = 0$ and 1 only: if we neglect the polarization dependence of the interactions, suppressing the initial linear decrease would require changing the sign of $A_0^s + A_0^a$, an assumption which seems unreasonable. On the other

[35]The matrix $\lambda_{\sigma\sigma'}$ is not altered, since only the ratio $m^*_\uparrow/m^*_\downarrow$ enters in its expression. However, both τ_σ and v_σ in Eq. (D.19) are increased.

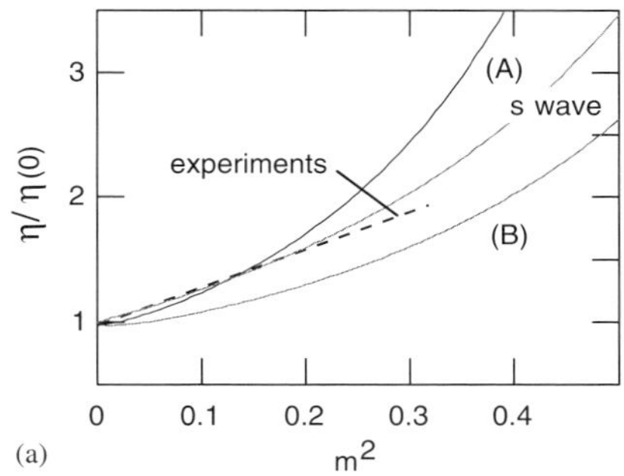

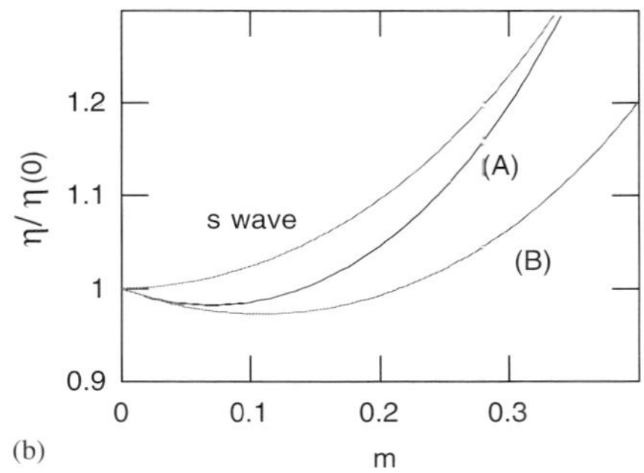

Fig. D.5. Theoretical polarization dependence of the viscosity, plotted versus m^2 (a) and m (b) at 27 bar. The curves labelled (A) correspond to a polarization dependence $m^* \propto 1 - m^2/2$, keeping the collision probabilities W's fixed as in Fig. D.4. The curves labelled (B) are obtained when the polarization dependence of the p_F and m^* is taken into account in the calculation of W. The s-wave prediction and the experimental behavior at 80 mK are shown for comparison.

hand, there is little chance that including a polarization dependence of the longitudinal Landau and spin–flip interactions would allow one to recover such a pure m^2 behavior, despite the large number of adjustable parameters. This may indicate that higher moments than $l = 0$ and 1 are needed to describe our data, or that the s–p approximation itself is not correct, either because it does not take into account high enough angular momentum collisions, or for a more fundamental reason.

References

Abrikosov, A.A. and I.M. Khalatnikov, 1957, Zh. Eksp. Teor. Fiz. **32**, 1083 [Sov. Phys. JETP **5**, 887].

Abrikosov, A.A. and I.M. Dzyaloshinskii, 1958, Zh. Eksp. Teor. Fiz. **35**, 771 [Sov. Phys. JETP **8**, 535].

Ager, J.H., A. Child, R. König, J.R. Owers-Bradley and R.M. Bowley, 1995, J. Low Temp. Phys. **99**, 683.

Ainsworth, T.L. and K.S. Bedell, 1987, Phys. Rev. B **35**, 8425.

Akimoto, H., J.S. Xia, E.D. Adams, D. Candela, W.J. Mullin and N.S. Sullivan, 2002, Int. J. Mod. Phys. **16**, 3117.

Akimoto, H., D. Candela, J.S. Xia and W.J. Mullin, 2003, Phys. Rev. Lett. **90**, 105301.

Anderson, A., W. Reese and J. Wheathley, 1962, Phys. Rev. **127**, 671.

Andersón, R.H., C.J. Pethick and K.F. Quader, 1987, Phys. Rev. B **35**, 1620.

Anderson, P.W. and W.F. Brinkman, 1975, in: The Helium Liquids, ed J.G.M. Armitage, and I.E. Farquhar (Academic Press, New York).

Andres, K. and W.O. Sprenguer, 1975, in: Proc. 14th Int. Conf. On Low Temperature Physics, eds M. Krusius and M. Vuorio, (North Holland, Amsterdam), Vol. 1, p. 123.

Andreev, A.F., 1978, J.E.T.P. **28**, 556.

Andreev, A.F. and Yu.A. Kosevich, 1979, [Sov. Phys. J.E.T.P. **50**], 1218.

Aziz, R.A., V.P.S. Nain, J.C. Carley, W.L. Taylor and G.T. McConville, 1979, J. Chem. Phys. **70**, 4330.

Barranco, M., E.S. Hernandez and J. Navarro, 1996, Phys. Rev. B **54**, 7394.

Bashkin, E.P., 1984a, Phys. Lett. A **101**, 164.

Bashkin, E.P., 1984b, Zh. Eksp. Teor. Fiz. **87**, 1948, [Sov. Phys. JETP **60**, 1122].

Bashkin, E.P. and A.E. Meyerovich, 1977, Pis'ma Zh. Eksp. Teor. Fiz. **26**, 696 [Sov. Phys. JETP Lett. **26**, 534].

Bashkin, E.P. and A.E. Meyerovich, 1978, Zh. Eksp. Teor. Fiz. **74**, 1904 [Sov. Phys. JETP **47**, 992].

Bashkin, E.P. and A.E. Meyerovich, 1981, Adv. Phys. **30**, 1.

Baym, G. and C. Pethick, 1991, Landau Fermi-Liquid Theory (Wiley, New York).

Béal-Monod, M.T., S.K. Ma and D.R. Fredkin, 1968, Phys. Rev. Lett. **20**, 929.

Béal-Monod, M.T. and E. Daniel, 1983, Phys. Rev. B **27**, 4467.

Béal-Monod, M.T., 1988, Phys. Rev. B **38**, 8801.

Béal-Monod, M.T. and E. Daniel, 1991, Comment in Phys. Rev. Lett. **68**, 3817; S.A.J. Wiegers, P.E. Wolf and L. Puech, 1991, Reply in Phys. Rev. Lett. **68**, 3818.

Bedell, K.S. and K.F. Quader, 1983, Phys. Lett. 96 A, 91.

Bedell, K.S. and C. Sanchez-Castro, 1986, Phys. Rev. Lett. **57**, 854.

Bertinat, M.P., D.S. Betts, D.F. Brewer and G.J. Butterworth, 1963, J. Phil. Mag., 977.

Black, M.A., H.E. Hall and K. Thompson, 1971, J. Phys. C: Solid State Phys., 129.

Bonfait, G., L. Puech, A.S. Greenberg, G. Eska, B. Castaing and D. Thoulouze, 1984, Phys. Rev. Lett. **53**, 1092.

Bonfait, G., L. Puech, W.P. Halperin and B. Castaing, 1987, Europhys. Lett. **3**, 489.

Bonfait, G., 1987. Ph.D. Thesis, University of Grenoble (France) (unpublished).

Bonfait, G., L. Puech and A. Schuhl, 1990. in: Helium Three, ch. 14, eds W.P. Halperin and L.P. Pitaevskii (Elsevier, Amsterdam).

Bravin, M., S.A.J. Wiegers, P.E. Wolf and L. Puech, 1992, J. Low Temp. Phys. **88**, 723.

Bravin, M., S.A.J. Wiegers, G. Remenyi, L. Puech and P.E. Wolf, 1993, Phys. Scripta T **49**, 322.

Bravin, M., S.A.J. Wiegers, L. Puech and P.E. Wolf, 1994, Physica B **197**, 410.

Bravin, M., O. Buu, A.C. Forbes, S.A.J. Wiegers, L. Puech and P.E. Wolf, 1998, J. Low Temp. Phys. **110**, 375.

Brinkman, W.F. and T.M. Rice, 1970, Phys. Rev. B **2**, 4302.

Brooker, J.A. and J. Sykes, 1968, Phys. Rev. Lett. **21**, 279.

Buu, O., A.C. Forbes, A.S. van Steenbergen, S.A.J. Wiegers, G. Remenyi, L. Puech and P.E. Wolf, 1998a, J. Low Temp. Phys. **110**, 311.

Buu, O., A.C. Forbes, L. Puech and P.E. Wolf, 1998b, J. Low Temp. Phys. **110**, 381.

Buu, O., A.C. Forbes, L. Puech and P.E. Wolf, 1999, Phys. Rev. Lett. **83**, 3466.

Buu, O., L. Puech and P.E. Wolf, 2000a, Physica B **284**, 184.

Buu, O., L. Puech and P.E. Wolf, 2000b, Physica B **284**, 182.

Buu, O., L. Puech and P.E. Wolf, 2000c, Phys. Rev. Lett. **85**, 1278.

Buu, O., L. Puech and P.E. Wolf, 2002a, J. Low Temp. Phys. **126**, 63.

Buu, O., L. Puech and P.E. Wolf, 2002b, J. Low Temp. Phys. **126**, 1553.

Buu, O., D. Clubb, R. Nyman and J.R. Owers-Bradley, 2002c, J. Low Temp. Phys. **128**, 123; Buu O., R. Nyman, D. Clubb and R.M. Bowley, 2002, J. Low Temp. Phys **128**, 143.

Candela, D., L.-J. We, D.R. McAllaster and W.J. Mullin, 1991, Phys. Rev. Lett. **67**, 330.

Candela, D., M.E. Hayden and P.J. Nacher, 1994, Phys. Rev. Lett. **73**, 2587.

Carless, D.C., H.E. Hall and J.R. Hook, 1983, J. Low Temp. Phys. **50**, 583.

Carslaw, H.S. and J.C. Jaeger, 1959, Conduction of Heat in Solids (Oxford University Press, Oxford).

Casey, A., H. Patel, J. Nyeki, B.P. Cowan and J. Saunders, 2003, Phys. Rev. Lett. **90**, 115301.

Castaing, B. and P. Nozières, 1979, J. Phys. **40**, 257.
Chapellier, M., G. Frossati and F.B. Rasmussen, 1979, Phys. Rev. Lett. **42**, 904.
Clements, B.E., C.W. Greef and H.R. Glyde, 1991, Phys. Rev. B **44**, 10239 and references therein.
Dutta, A. and C.N. Archie, 1985, Phys. Rev. Lett. **55**, 2949.
Dy, K.S. and C.J. Pethick, 1969, Phys. Rev. **185**, 373.
Fåk, B., K. Guckelsberger, R. Scherm and A. Stunault, 1994, J. Low Temp. Phys. **97**, 445.
Fomin, I.A., 1997, Zh. Eksp. Teor. Fiz. **65**, 717 [Sov. Phys. JETP **65**, 749].
Fomin, I.A. and G. Vermeulen, 1997, J. Low Temp. Phys. **106**, 133.
Frossati, G., H. Godfrin, B. Hébral and D. Thoulouze, 1977. in: Proc. ULT Hakone Symposium, Japan. Frossati, G., 1978, J. Phys., **39**, C6–1578.
Fukushima, A., S. Ogawa and Y. Okuda, 1992, J. Low Temp. Phys. **88**, 483.
Gatica, S.M., E.S. Hernandez and J. Navarro, 1998, Phys. Rev. B **58**, 12300.
Georges, A., G. Kotliar, W. Krauth and M.J. Rozenberg, 1996, Rev. Mod. Phys. **68**, 13.
Georges, A. and L. Laloux, 1997, Mod. Phys. Lett. B **11**, 913.
Glyde, H.R., B. Fåk, H. Godfrin, N.H. van Dijkn, K. Guckelsberger and R. Scherm, 2000, Phys. Rev. B **61**, 1421.
Godfrin, H., G. Frossati, A.S. Greenberg, B. Hébral and D. Thoulouze, 1980, Phys. Rev. Lett. **44**, 1695.
Greiner, M.O., Mandel, T. Esslinger, T.W. Hansch and I. Bloch, 2002, Nature **415**, 39.
Greywall, D.S. and M.A. Paalanen, 1981, Phys. Rev. Lett. **46**, 1292 These authors have measured in ^{3}He–^{4}He mixtures an increase of the second sound damping with polarization, which they interpreted in terms of a viscosity increase.
Greywall, D.S., 1983, Phys. Rev. B **27**, 2747.
Greywall, D.S., 1984, Phys. Rev. B **29**, 4933.
Hammel, P.C. and R.C. Richardson, 1984, Phys. Rev. Lett. **52**, 1441.
Hampson, T.M.M., R.M. Bowley, D. Brugel and G. McHale, 1988, J. Low. Temp. Phys. **73**, 333.
Harrison, J.P., 1979, J. Low Temp. Phys. **37**, 467.
Hensley, H.H., Y. Lee, P. Hamot, T. Mizusaki and W.P. Halperin, 1993, J. Low. Temp. Phys. **90**, 149.
Hess, D.W. and K.F. Quader, 1987, Phys. Rev. B **36**, 756.
Jeon, J.W. and W.J. Mullin, 1987, J. Low Temp. Phys. **67**, 421.
Kopietz, P., A. Dutta and C.J. Archie, 1986, Phys. Rev. Lett. **57**, 1231.
Kranenburg, C.C., L.P. Roobol, P.G. van de Haar, R. Jochemsen and G. Frossati, 1987, Japan J. Appl. Phys. **26**, 2133.
Kranenburg, C.C., S.A.J. Wiegers, L.P. Roobol, P.G. van de Haar, R. Jochemsen and G. Frossati, 1988, Phys. Rev. Lett. **61**, 1372.
Kranenburg, C.C., S.A.J. Wiegers, R. Jochemsen and G. Frossati, 1989, J. Low Temp. Phys. **77**, 371.
Krotkov, P.L., V.P. Mineev and G. Vermeulen, 2002, Phys. Rev. B **65**, 054304.
Krotscheck, E. and J. Springer, 2003, J. Low Temp. Phys. **132**, 281.
Laloux, L., A. Georges and W. Krauth, 1994, Phys. Rev. B **50**, 3092.
Landau, L.D., 1956, Zh. Eksp. Teor. Fiz. **30**, 1958 [Sov. Phys. JETP **3**, 920].
Landau, L.D., 1958, Zh. Eksp. Teor. Fiz. **35**, 97 [Sov. Phys. JETP **8**, 70].
Laukamp, M., 1994. Diploma Thesis, (Technishe Hochschule Aachen), M. Laukamp and D. Vollhardt (unpublished); Vollhardt D., 2001, private communication.
Leduc, M., P.J. Nacher, D.S. Betts, J.M. Daniels, G. Tastevin and F. Laloë, 1987, Europhys. Lett. **4**, 59.
Levin, K. and O.T. Valls, 1983, Phys. Rep. **98**, 1.

Lhuillier, C., 1983, J. Phys. **44**, 1.
Lhuillier, C., 1991. in: Les Houches Session LVI, Strongly Interacting Fermions and High T_c Superconductivity, eds B. Douçot and J. Zinn-Justin (Elsevier, Amsterdam).
Lhuillier, C. and F. Laloë, 1979, J. Phys **40**, 239.
Lhuillier, C. and F. Laloë, 1982, J. Phys **43**, 197.
Lieb, E.H. and F.Y. Wu, 1968, Phys. Rev. Lett. **20**, 1445. E.H. Lieb and F.Y. Wu, 1968, Phys. Rev. Lett. **21**, 192.
Lounasmaa, O.V., 1974, Experimental Principles and Methods Below 1 K (Academic Press, London).
Marchenkov, A., H. Akimoto, R. van Rooijen, R. Jochemsen and G. Frossati, 1999, Phys. Rev. Lett. **83**, 4598.
Matsumoto, K., S. Hasegawa and Y. Okuda, 2000, Physica B **284**, 194.
Metzner, W. and V. Vollhardt, 1989, Phys. Rev. Lett. **83**, 4598.
Meyerovich, A.E., 1978, Phys. Lett 69 A, 279.
Meyerovich, A.E., 1983, J. Low Temp. Phys. **53**, 487.
Meyerovich, A.E., 1985, Phys. Lett 107 A, 177.
Meyerovich, A.E., 1987. in: Progress in Low Temperature Physics, Vol. XI, Ch. 1, ed Brewer D.F., p. 1, Elsevier Science B.V.
Meyerovich, A.E., 1990. in: Helium Three, Ch. 13, eds W.H. Halperin and L.P. Pitaevskii, p. 757, Elsevier Science B.V.
Meyerovich, A.E. and K.A. Musaelian, 1993, Phys. Rev. B. **47**, 2897.
Mineev, V.P., 2004, Phys. Rev. B. **69**, 144429.
Modugno, G., F. Ferlaino, R. Heidemann, G.V. Roati and M. Inguscio, 2003, Phys. Rev. A. **68**, 11601.
Mörhard, K.D., C. Baüerle, J. Bossy, Yu. Bunkov, S.N. Fisher and H. Godfrin, 1996, Phys. Rev. B **53**, 2658.
Mullin, W.J. and K. Miyake, 1983, J. Low Temp. Phys. **53**, 313.
Mullin, W.J. and K. Miyake, 1986, J. Low Temp. Phys. **63**, 479.
Mullin, W.J. and J.W. Jeon, 1992, J. Low Temp. Phys. **88**, 433.
Nacher, P.J. and E. Stoltz, 1995, J. Low Temp. Phys. **101**, 311.
Nacher, P.J., F.O. Marion, B. Villard and G. Tastevin, 2002a, J. Low Temp. Phys. **126**, 85.
Nacher, P.J., N. Piegay, F.O. Marion and G. Tastevin, 2002b, J. Low Temp. Phys. **126**, 145.
Nelson, E.D. and W.J. Mullin, 1994, J. Low Temp. Phys. **97**, 251.
Nozières, P., 1964, Theory of Interacting Fermi Systems (Benjamin, New York) p. 102.
Nozières, P., 1986. Lecture Notes of the Collège de France (unpublished).
Nozières, P., 1998, private communication.
Owers-Bradley, J.R., P.C. Main, R.J. Church, T.M.M. Hampson, G. McHale and R.M. Bowley, 1988, Phys. Rev. Lett. **61**, 1619.
Owers-Bradley, J.R., 1997, Rep. Prog. Phys. **60**, 1173.
Owers-Bradley, J.R., R.M. Bowley, J.H. Naish, P.J. Richardson, A.P.J. Voncken and R. König, 1998, J. Low Temp. Phys. **110**, 321.
Owers-Bradley, J.R., O. Buu, C.J.M. Gloin, R.M. Bowley and R. König, 2000, Physica B, **284**, 190; Bowley R.M., O. Buu and R.A. Nyman, 2004, J. Low Temp. Phys. **137**, 1; Nyman R.A., R.M. Bowley, J.R. Owers-Bradley and O. Buu, 2004, J. Low Temp. Phys. **137**, 27.
Pickett, G.R., 1989. in: Proc. Third Conf. on Spin Polarized Quantum Systems, ed S. Stringari-ISI (World Scientific, Singapore).
Pines, D. and P. Nozières, 1966. Theory of Quantum Liquids, Vol. 1, 2nd ed. (Addison-Wesley, New-York).
Puech, L., 1987, Solid State Commun **64**, 8.

Puech, L., G. Bonfait, B. Castaing, C. Paulsen, J.J. Prejean and P.E. Wolf, 1989 in: Proc. Third Conf. on Spin-Polarized Quantum Systems, op. cited.
Quader, K.F. and K.S. Bedell, 1985, J. Low Temp. Phys. **58**, 89.
Ramm, H., P. Pedroni, J.R. Thompson and H. Meyer, 1970, J. Low Temp. Phys **2**, 521.
Regal, C.A., M. Greiner and D.S. Jin, 2004, Phys. Rev. Lett. **92**, 040403.
Rice, M.J., 1967, Phys. Rev. **162**, 189.
Rodrigues, A. and G. Vermeulen, 1995, J. Low Temp. Phys. **101**, 151.
Rodrigues, A. and G. Vermeulen, 1997, J. Low Temp. Phys. **108**, 103.
Roni, A. and G. Vermeulen, 2000, Physica B **284**, 87.
Saito, S., M. Okuyama and T. Satoh, 1985, Phys. Rev. Lett. **55**, 1757.
Sanchez-Castro, C., K.S. Bedell and S.A.J. Wiegers, 1989, Phys. Rev. B **40**, 437.
Sanchez-Castro, C. and K.S. Bedell, 1989b, J. Low Temp. Phys. **75**, 95.
Saunders, J., A. Casey, H. Patel, J. Nyeki and B. Cowan, 2000, Physica B **280**, 100.
Sawkey, D., V. Goudon, O. Buu, L. Puech and P.E. Wolf, 2003, Physica B 329–333, 118.
Sawkey, D., L. Puech and P.E. Wolf, 2004, preprint.
Schuhl, A., S. Maegawa, M.W. Meisel and M. Chapellier, 1987, Phys. Rev. B **36**, 6811.
Schumacher, G., D. Thoulouze, B. Castaing, Y. Chabre, P. Segransan and J. Joffrin, 1979, J. Phys. Lett. **40**, L143.
Seiler, K., C. Gros, T.M. Rice, K. Ueda and D. Vollhardt, 1986, J. Low Temp. Phys. **64**, 195.
van Steenbergen, A.S., S.A.J. Wiegers, P.E. Wolf, J.A.A.J. Perenboom and J.C. Maan, 1998, Phys. Rev. B **58**, 925.
Stipdonk, H.L., G. Frossati and J.H. Hetherington, 1985, J. Low Temp. Phys. **61**, 185.
Stokes, G.G., 1901, Mathematical and Physical Papers, Vol. 3. (Cambridge University Press, London), p. 38.
Stringari, S., M. Barranco, A. Polls, P.J. Nacher and F. Laloë, 1987, J. Phys. **48**, 1337.
Tastevin, G., P.J. Nacher, L. Wiesenfeld, M. Leduc and F. Laloë, 1988, J. Phys. **49**, 1.
Tastevin, G., 1992, J. Low Temp. Phys. **89**, 317.
Taylor, R.D. and J.G. Dash, 1957, Phys. Rev. **36**, 398.
Vermeulen, G., A. Schuhl, F.B. Rasmussen, J. Joffrin, G. Frossati and M. Chapellier, 1988, Phys. Rev. Lett. **60**, 2315.
Villard, B. and P.J. Nacher, 2000, Physica B **284**, 180.
Villard, B., P.J. Nacher and G. Tastevin, 2000, Physica B **284**, 178.
Vollhardt, D., 1984, Rev. Mod. Phys. **56**, 99.
Vollhardt, D., P. Wölfle and P. Anderson, 1987, Phys. Rev. B **35**, 6703.
Vollhardt, D., private communication.
Voncken, A.P.J., J.H. Naish and J.R. Owers-Bradley, 1998, J. Low Temp. Phys. **110**, 1105.
Wei, L.-J., N. Kalechofsky and D. Candela, 1993, Phys. Rev. Lett. **71**, 879.
Wheatley, J.C., 1975. in: The Helium Liquids, eds J.G.M. Armitage, and I.E. Farquhar (Academic Press, London), p 242–259.
van Woerkens, C.M.C.M., 1998, Ph. D. Thesis, Leiden University (The Netherlands) (unpublished).
White, R.M., 1970, Quantum Theory of Magnetism (McGraw Hill, New York).
Wiegers, S.A.J., P.E. Wolf and L. Puech, 1990, Physica B **165**, 139.
Wiegers, S.A.J., P.E. Wolf and L. Puech, 1991, Phys. Rev. Lett. **66**, 2895.
Wilks, J., 1967, Liquid and Solid Helium 1st ed., (Clarendon Press, Oxford), p 17–21 and 431.
Yawata, K., 2001. Ph. D. Thesis, University of Tsukuba (Japan) (unpublished).

CHAPTER 4

THE ^{3}HE MELTING CURVE AND MELTING PRESSURE THERMOMETRY

BY

E. DWIGHT ADAMS

Department of Physics, University of Florida, Gainesville, FL 32611-8440, USA

Progress in Low Temperature Physics, Volume XV

ISSN: 0079-6417 DOI: 10.1016/S0079-6417(05)15004-5

Contents

1. Introduction

Clearly, reasonably accurate determination of the temperature is essential in experimental research at low temperatures. As will be discussed in a particular example later, lack of accurate thermometry can lead to great difficulty in attempting to understand results and to considerable wasted effort trying to resolve differences arising from inaccuracies in temperature. Just such a situation played a prominent role in the development of ^{3}He melting pressure thermometry (MPT). In this chapter, the physics of MPT and its development from the initial suggestion by Scribner and Adams (1967) to adoption of the Provisional Low Temperature Scale based on it in 2000 (PLTS-2000) (Rusby et al. 2002) will be described in detail.

This chapter begins with the earliest prediction of the behavior of the melting pressure of solid ^{3}He by Pomeranchuk (1950) that was made even before sufficient quantities were available for experiments. Earliest efforts to detect the predicted minimum in the melting pressure by the blocked-capillary technique and the first success using an in situ resistive strain gauge are mentioned. This is followed by a description of the work of Scribner and Adams in which the highly sensitive capacitive pressure gauge was used for the first time to study the melting pressure, $P_m(T)$, with the realization that the melting pressure could be used as a highly sensitive thermometer.

Measurement of the melting pressure, $P_m(T, B)$, in a magnetic field, B, with the application of various thermodynamic relations have provided much useful information on both liquid and solid ^{3}He. In fact, both superfluidity in liquid ^{3}He (Osheroff et al. 1972a, b), and magnetic ordering in the solid (Halperin et al. 1974, 1978), were discovered through study of the melting pressure.

Development of the Pomeranchuk cooling technique for reaching ~1 mK and the discovery of superfluidity in ^{3}He gave impetus to melting pressure thermometry, with several scales being used by different laboratories. The thermodynamic scale of Halperin et al. (1974) used only the properties of the ^{3}He in determining $T(P)$ relative to the superfluid A-transition temperature T_A and gave a value of $T_A = 2.75$ mK.

One of the heated debates among different laboratories in the early 1980s was the value of m^*, the effective mass, for liquid ^{3}He obtained from the heat capacity. Various laboratories, using different values for T_A, found different values of m^*. The dilemma of what should be the correct value of m^* was resolved by Greywall (1986), who used a quite unconventional approach to the thermometry problem by requiring that his measured heat capacity follow the Landau Fermi-liquid linear dependence on T. This gave a value of $T_A = 2.49$ mK, significantly lower than 2.75 mK of Halperin et al. (1974) that had been generally accepted. Greywall's results were sufficiently convincing that the "Greywall scale" came into almost universal use.

Long-term efforts were begun at standards labs, National Institutes of Standards and Technology (USA), (NIST) (previously, National Bureau of Standards, NBS), and later at Physikalisch-Technische Bundesanstalt (Germany) (PTB), to measure the melting pressure versus a standard thermometer in order to establish the relation $P_m(T)$. A program to measure $P_m(T)$ with ^{60}Co nuclear orientation thermometry was carried out at the University of Florida in 1995. These three efforts culminated in a workshop in Leiden in September 1998 with the intent of coming to agreement on the relationship $P_m(T)$ and on the values for the various fixed points on the ^{3}He melting curve, T_A, and $T_{A\text{-}B}$, the superfluid transition temperatures, and T_N, the solid ordering temperature. Because the differences in values of the fixed points from PTB and Florida were sufficiently large, it was agreed that the scale adopted in 2000 would be "provisional" and designated PLTS-2000. (The work of NIST went to only 6 mK and did not include the three lowest fixed points.) The expression $P_m(T)$ and values of the fixed points that were adopted for PLTS-2000 (Rusby et al. 2002), are presented.

The features of ^{3}He melting pressure that make it the logical choice for defining the temperature scale between T_N and 1 K are described. Practical matters such as pressure sensors, capacitance bridges, pressure regulators, gas-handling systems, capacitance calibration, initial filling of the melting pressure cell, and procedures for observing the various fixed points are included. The intent is to provide first-time users of ^{3}He MPT with all the information needed to implement the technique.

2. Melting curve from Pomeranchuk to PLTS-2000

2.1. *The melting pressure minimum and the Pomeranchuk effect*

Experimental research on ^{3}He began only about 50 years ago, because of its scarcity of only about 1 part in 10^5 in natural helium. The lack of ^{3}He available for experiments changed dramatically in the late 1940s when both the US and USSR began producing tritium for development of the hydrogen bomb. The nuclear reaction to produce tritium is $^6\text{Li}+n = {}^3\text{H}+{}^4\text{He}$. Tritium, which decays into ^{3}He with a half-life of 12.3 years, may be separated from the ^{4}He before it decays, allowing ^{3}He with very little ^{4}He impurity to be produced. Further purification of the ^{3}He to better than 1 part per million (ppm) by techniques such as fractional distillation is possible (see, for example, Kirk and Adams 1974). Now ^{3}He is readily available in all countries that have developed fusion programs.

One of the first predictions of the unique behavior of ^{3}He that the melting pressure would have a minimum with a negative slope at low temperatures was by Pomeranchuk (1950). This fact is what makes melting pressure thermometry a viable technique. Pomeranchuk's prediction was based on the entropy of the solid in comparison with that of the liquid. The ^{3}He atom is a composite Fermion since the nucleus has a single neutron, resulting in a total spin $I = 1/2$. At high temperatures, the spins are completely disordered with spin entropy

$$S_{\mathrm{spin}} = R\ \ln(2I + 1) = R\ \ln 2 \tag{1}$$

where R is the gas constant (= 8.314 J/mol K). In addition, there is the phonon contribution, $S_{\mathrm{ph}}{\sim}T^4$. In considering interactions that would reduce S_{spin} of the solid, Pomeranchuk took into account only dipole–dipole interactions with energy $E_{\mathrm{D}}{\sim}\mu^2/a^3 \sim 10^{-7}$ K, where a is the interatomic spacing ~0.3 nm.Thus, his prediction was based on the solid entropy remaining at the "high-temperature" value of R ln 2 until $T{\sim}10^{-7}$ K. (Because of the exchange interaction in solid ^{3}He and the superfluid transitions in liquid ^{3}He, details of the entropies and melting pressure $P_{\mathrm{m}}(T)$ are significantly different than Pomeranchuk's prediction, especially in the few-millikelvin range and lower (see, for example, Osheroff et al. 1972a, b; Adams 2004b). On the other hand, the entropy of the liquid is that of a Fermi liquid and is approximately linear in T below ~0.5 K. Then, the entropies of the liquid and solid become equal at $T{\sim}0.32$ K, as shown in Fig. 1, with the unique situation that the solid has greater entropy than the liquid at lower temperatures. Application of the Clausius–Clapeyron equation then gives the slope of the melting pressure

$$\left(\frac{\mathrm{d}P}{\mathrm{d}T}\right)_{\mathrm{m}} = \frac{S_{\mathrm{L}} - S_{\mathrm{S}}}{V_{\mathrm{L}} - V_{\mathrm{S}}} \tag{2}$$

where $\Delta V = V_{\mathrm{L}} - V_{\mathrm{S}}$ is the molar volume difference between the liquid and solid, which becomes essentially constant at 1.314 cm^3/mol below $T{\sim}20$ mK (Halperin et al. 1974, 1978). The arrows illustrate compression of liquid ^{3}He along the melting curve to solidify it, thereby producing cooling—the Pomeranchuk effect.

Experiments on ^{3}He began in the late 1940s by Sydoriak et al. (1949) in Los Alamos, NM, where ^{3}He was being produced. Probably the first experimental work on solid ^{3}He was that of Osborne et al. (1951), who measured the melting pressure using the blocked capillary technique. With this technique, a capillary passed through the temperature where $P_{\mathrm{m}}(T)$ was to be determined and then to an external gauge. Upon increasing the pressure, once P_{m} was reached, there would be no further increase in pressure in the down-stream external gauge. However, it was realized that this technique

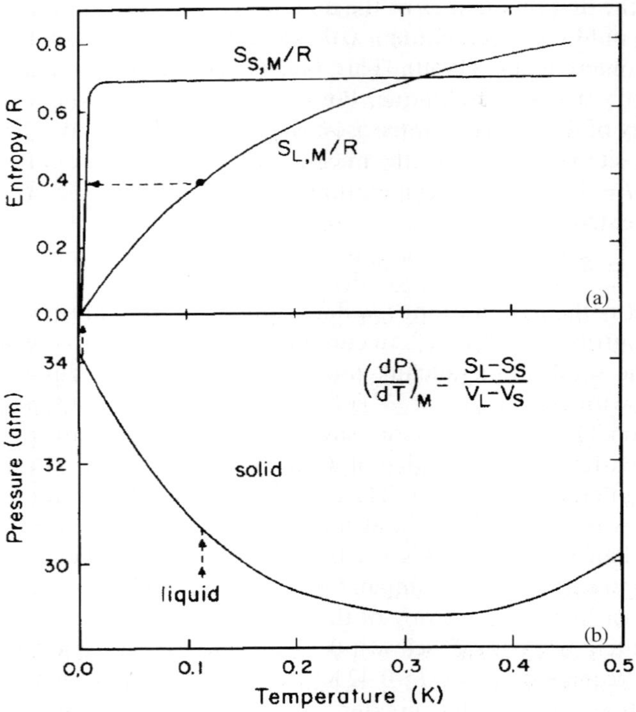

Fig. 1. Upper panel: schematic behavior of the entropies of liquid, S_L, and solid, S_S, of ^{3}He at melting pressure versus temperature. Lower panel: melting pressure, P, versus. temperature. The dashed arrows illustrate cooling by compression of liquid to form solid—the Pomeranchuk effect. From Trickey et al. (1972).

could not detect a minimum in the melting pressure since a block would occur first in the capillary at the position where $T = T_{\min}$. The first experiment to show the minimum in the melting pressure was that of Baum et al. (1959), who used a resistive strain gauge glued to their cell for detecting small motion in its wall with changing pressure.

In the early 1960s the highly sensitive capacitive pressure gauge was developed by Straty and Adams (1969) (see below). One of its first uses of this gauge by Scribner (1968) was to measure the melting pressure of ^{3}He. They saw that there was much greater resolution in the pressure measurement with the capacitive pressure gauge than in the temperature that they were measuring with a paramagnetic salt. Consequently, they suggested using the melting pressure $P_m(T)$ as the thermometer (Scribner and Adams 1967, 1970, 1972; Scribner, 1968). It was assumed that $P_m(T)$ would have to be determined first using another thermometer to give temperatures. Thus the melting pressure would provide a *secondary* thermometer, one that requires calibration by another thermometer. The suggestion that the melting pressure itself might provide its own *primary* thermometer, giving T directly without another thermometer for calibration, was made by Wheatley (1970), who pointed out that T in the Clausius–Clapeyron equation (2), for the slope dP_m/dT, is the absolute temperature. However, it does not appear that Wheatley saw how to make use of this fact to determine $P_m(T)$.

Although the demonstration of the melting pressure minimum showed the validity of Pomeranchuk's idea, the possibility of cooling by compression along the melting curve was not pursued for some time because of the concern that irreversible heating might offset the cooling available. Anufriyev (1965), using a double bellows arrangement with externally applied ^{4}He pressure to compress the ^{3}He, demonstrated that frictional heating could be kept small by reaching a temperature below 20 mK, having started at 50 mK (his actual final temperature was not known because of thermometry problems).

2.2. *Discovery of superfluid ^{3}He*

Following Anufriyev's success, others undertook Pomernachuk cooling and soon the method was routinely producing temperatures as low as 1 mK. Outstanding problems in quantum fluids and solids being investigated were the Fermi-liquid properties of liquid ^{3}He, its possible superfluidity, and magnetic ordering of the solid. The technique was being used by Osheroff et al. (1972a), who observed the two famous "glitches" in the trace of pressure versus time in the Pomeranchuk cell, as shown in Fig. 2. At the time, they attributed the glitches to the long-sought ordering in solid ^{3}He. However, additional experiments by Osheroff et al. (1972b) quickly revealed that the effect was not due to the solid in the cell, but was, in fact, due to two new superfluid transitions in the liquid at T_A and T_{A-B}, respectively. With this discovery, low-temperature labs around the world, with few exceptions, turned their attention to superfluid ^{3}He and two very productive decades of research in this area followed. Details of the discovery of superfluid ^{3}He may be found in the three Nobel symposium articles by Lee (1997), Osheroff (1987), and Richardson (1997).

It should be emphasized that the plot in Fig. 2 is $P_m(t)$ rather than $P_m(T)$, as will be discussed later (see Figs. 11 and 12). The glitches are a result of changes in the properties of the liquid as it undergoes the superfluid A and A–B transitions rather that features that can be seen in $P_m(T)$. At the A transition, which is second-order, there is a discontinuity in the heat

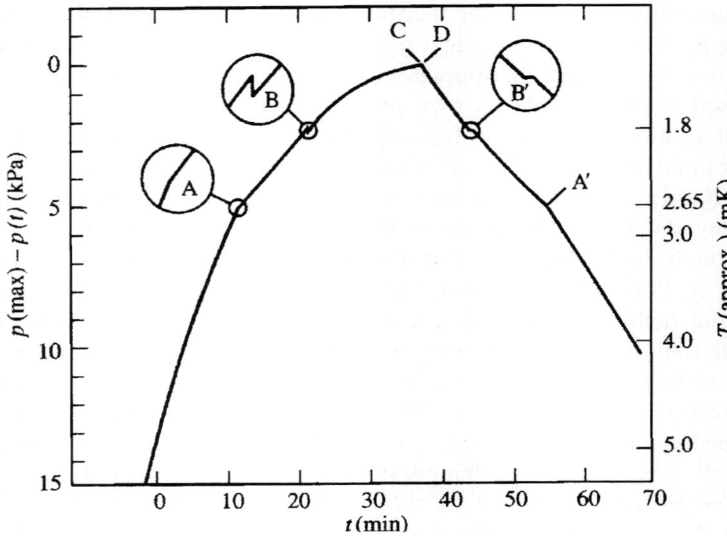

Fig. 2. The glitches in $P(t)$ in Pomeranchuk compressional cooling of ^{3}He that lead to discovery of the A- and A-B-superfluid transitions. From Osheroff et al. (1972a).

capacity of the liquid with the result that there is a decrease in the cooling rate of the Pomeranchuk compression cell and, thus, a discontinuity in $\mathrm{d}P_\mathrm{m}/\mathrm{d}t$. This fact, with the correct interpretation of the glitches as indication of superfluid transitions in liquid ^{3}He, was first pointed out by Vvedenskii (1972). However, as can be seen from eq. (2), $\mathrm{d}P_\mathrm{m}/\mathrm{d}T$ is continuous and the A transition produces no observable feature on the equilibrium $P(T)$ (see Fig. 11). The A–B transition is first-order with a very small latent heat (Halperin et al.1974, 1978). In Fig. 2, the small latent heat causes supercooling of the A–B transition and then a "backstep" when the transition occurs with the latent heat being absorbed by the system. In the decompression part of $P_\mathrm{m}(t)$ there is a small plateau at the A–B transition as the latent heat is removed from the liquid without superheating.

The appearance of $P_\mathrm{m}(T)$ at the A and A–B superfluid transitions (see Figs. 11 and 12), when the melting pressure cell is cooled by external means such as a nuclear demagnetization stage, is quite different from $P_\mathrm{m}(t)$ shown in Fig. 2, where cooling was by the Pomeranchuk effect. Neither the A nor A–B transition produces an observable effect on $P(T)$. Then, using T_A or $T_{\mathrm{A-B}}$ as fixed points in thermometry requires special attention, as will be discussed later.

In the few years following its discovery, several laboratories began research on superfluid ^{3}He. Most groups used the Pomeranchuk compression

TABLE 1
Various MPT Scales.

	T_A(mK)	T_N(mK)	Method	Primary T
Halperin et al.	2.752	1.100	Latent heat	$S_s \rightarrow R\ln 2$
Avenel et al.	2.75	1.08	Pt NMR ($K = 30.2$)	Osmotic pressure
Paulson et al.	2.625		LCMN	Noise $<v>$
Alvesalo et al	2.79		Pt NMR ($K = 29.9$)	^{54}Mn No
Greywall	2.491	0.931	LCMN	$C_v \sim T$
Fukuyama et al.	2.477	0.914	Pt NMR	
Fogle et al. (NIST)	$T > 6$ mK		LCMN	Noise $<v>$
Ni et al. (FL)	2.505	0.934	Pt NMR	^{60}Co NO
Schuster et al. (PTB)	2.41	0.88	Pt NMR, CMN	ITS – 90 (CMN at 1.5 K)

method of cooling, with ^{3}He melting pressure as the secondary thermometer and with various approaches to primary thermometry. These included osmotic pressure (Avenel et al. 1975), noise voltage (Paulson et al. 1979), and nuclear orientation of γ rays (Alvesalo et al. 1981). A tabulation of some features of the early scales, as well as more recent ones from different laboratories, is given in Table 1. Two of these scales in particular will be discussed briefly below.

2.3. *The thermodynamic scale and discovery of ordering in solid ^{3}He*

Two of the scales listed in Table 1 bear special attention since they rely only on the properties of ^{3}He itself for the primary thermometer. These are the scales of Halperin et al. (1974, 1978) and that of Greywall (1986). Halperin et al. (1974, 1978), in their experiment in which ordering in the solid was first observed, devised a procedure to determine the $P_m(T)$ relation from the latent heat of the solid. Their approach uses the Clausius–Clapeyron equation given above with $\Delta Q = \Delta n_S T(S_L - S_S)$ and $\Delta V = \Delta n_S(V_L - V_S)$, where Δn_S is the number of moles of solid produced in a heat pulse. While regulating the temperature of the Pomerachuk cell by controlling the pressure (see below), a pulse of heat ΔQ that could be measured was applied electrically. In the process, the cell was compressed automatically by the pressure regulator, with a measured quantity of liquid converted to solid, giving ΔV. Then by integration of dP_m/dT the expression for T is obtained

$$T(P) = T_o \exp\left[\int_{P_o}^{P} (\Delta V/\Delta Q)\, dP\right] \quad (3)$$

Halperin et al. provided a functional form $P_m(T)$ relative to T_A which they determined to be $T_A = 2.75$ mK by requiring that $S_{spins} = R \ln 2$ at "high T" (see eq. (1)). Thus, not only does ^{3}He provides its own cooling through the Pomeranchuk effect, it provides its own *primary* thermometry as well. The "Halperin scale" with $T_A = 2.75$ mK, taken as the reference point for comparison of results from different laboratories, was considered to be the most reliable for many years. Halperin and Adams (1998) have discussed possible reasons why $T_A = 2.75$ mK may be "too high" and have mentioned ways to improve this method.

Shortly after the results of Halperin et al. (1974), the same technique was used by Kummer et al. (1975) to study magnetic ordering in solid ^{3}He in an applied field B. In addition to the ordered phase that Halperin et al. had found, a second phase at fields above a critical field expected to be 21 T, which has yet to be observed (Osheroff et al. 1987; Yawata 2001). Providing a theoretical understanding of the two ordered phases has been a challenge that has led to the multiple-exchange model (see, for example, Roger et al. 1983; Roger and Hetherington 1990). Numerous studies of the melting pressure in applied fields have been used to provide quantities such as the entropy, magnetization, spin-wave velocity, and magnetic phase diagram of solid ^{3}He (see Ni et al. 1994b; Adams 2004b and references therein).

2.4. *The greywall scale*

The properties of liquid ^{3}He have been of great interest as an example of the Fermi-liquid theory. One of these, the heat capacity, is obtained from that of the non-interacting Fermi gas by replacing the ^{3}He mass, m_3, by an "effective mass" m^*. In the very-low-temperature limit the heat capacity has a linear dependence on T given by $C/nR = \gamma T$, where n is the number of moles of ^{3}He, and R is the gas constant. When interactions are included (Doniach and Engelsberg 1966), the functional form becomes

$$C/nR = \gamma T + \Gamma T^3 \ln(T/T_q), \tag{4}$$

where the second term in which Γ and T_q are constants results from the interactions. Quite different values of m^* were reported by various laboratories using their own temperature scales (see Table 1). Difference as large as 40% were reported in the measured values of the specific heat in the range of 10–40 mK.

Greywall (1983, 1986) measured the heat capacity of liquid ^{3}He over a wide range of temperatures down to the superfluid transition and took the usual step of adjusting the parameters in his thermometer calibration (see eq. (5), below) to yield a linear behavior for the specific heat in the range of a few millikelvin as shown Fig. 3. In this figure, $\gamma = C/nRT = 2.78$ is based on his higher T value (Greywall 1983), and T_c (0) is the value of T_A at $P = 0$. Although none of the curves in this figure is a straight line with zero slope, the curves in (b) with $T_A = 2.50$ comes close to these requirements. Further analysis of his data gave $T_A = 2.49$ mK, significantly lower than that for any of the other scales in use (see Table 1). Furthermore, Greywall found that by making simple adjustments in the other scales, he was able to bring all of the m^* values into agreement, with the conclusion that the "specific heat controversy was entirely associated with differences in thermometry." Greywall's results were sufficiently convincing that the "Greywall scale" with $T_A = 2.49$ mK came into widespread use.

Shortly after Greywall's results, Fukuyama et al. (1987) used Pt NMR as their thermometer and obtained $T_A = 2.48$ mK. However, this does not provide an independent measurement of T_A since their calibration of the Pt thermometer used the MPT scale of Greywall in the range 15–45 mK.

2.5. *Development of PLTS-2000*

^{3}He MPT has a number of features that make it the obvious choice for carrying the adopted international temperature scale in the range below ITS-90 that ends at 0.65 K. It can be considered to be analogous to vapor pressure thermometry in which temperature is determined from the vapor pressure of a pure liquid such as ^{3}He or ^{4}He. In contrast to vapor pressure thermometry, which is limited to T>0.3 K because of the vanishing vapor pressure, MPT has the distinct advantage of maintaining a high resolution in temperature down to the solid ordering at T_N. Other advantages of MPT are: (i) the superfluid A and A–B transitions and, especially, T_N, the ordering temperature of the solid, are easily identifiable fixed points; (ii) the pressure reference point at the pressure minimum, P_{min}; (iii) short times required for measurements; and (iv) relatively simple equipment and techniques. This latter point is not appreciated by some who have not worked extensively with liquid and solid He. For example, Spietz et al. (2003) have labeled MPT as of "limited practical use," "exotic," and "too complex and expensive for general use." Pekola (2004) refers to MPT as a "cumbersome instrument and method to measure temperature." However, the advantages of MPT for carrying the temperature scale to 1 mK are compelling. For anyone doing research on either liquid or solid ^{3}He, the technique is useful as an everyday working thermometer.

The "milestones" in the development of MPT from the introduction of the capacitive pressure gauge to the adoption of PLTS-2000 are listed in Table 2. Details of the "recipe" used for setting the values of fixed points

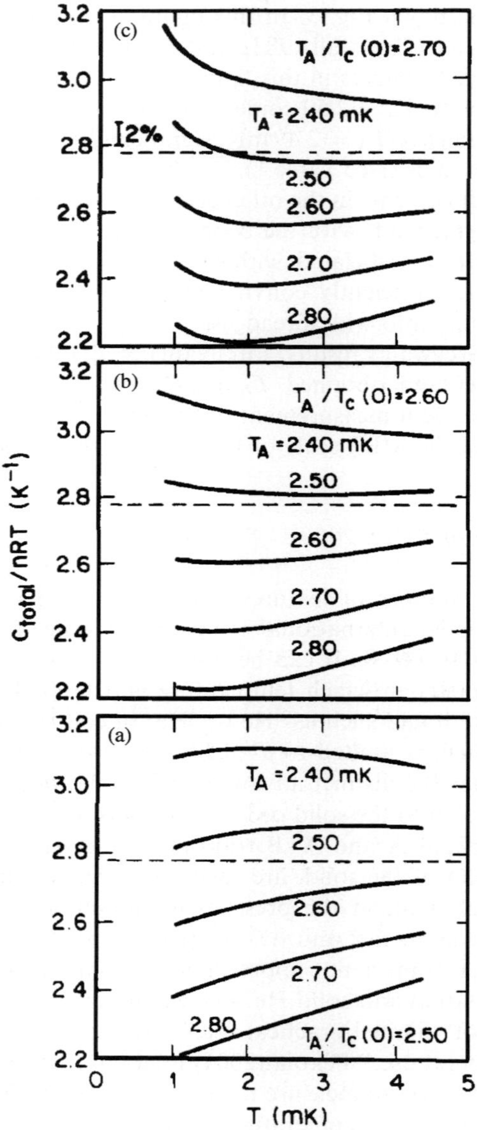

Fig. 3. Heat capacity C/nRT of liquid ^{3}He versus T for various values of T_A/T_C that lead to the "Greywall scale" of $P_m(T)$. From Greywall (1986).

TABLE 2
Milestones in the Development of MPT.

1965: Straty–Adams Pressure Guage
1967: First public suggestion of MPT, E. D. Adams, Crystal Mt. Gordon Conference
1968; 1970: First published reports on MPT, R A. Scribner, Ph.D. thesis, UF, Scribner and Adams, RSI
1970: First suggestion of thermodynamic $P_m(T)$, J. C. Wheatley, ULT Symposium NRL
1972: Nobel-prize "glitches," Osheroff, Richardson, Lee
1972–74: Many labs using PC Cooling & MPT, various scales
1974: First thermodynamic $P_m(T)$, $T_A = 2.75$ mK, W. P. Halperin, Ph.D. thesis, Cornell
1974–86: $C_V/{}^*m$ liquid ^{3}He controversy
1986: Use of $C_V \sim T$ to set $T_A = 2.49$ mK, resolution of $^*m$ controversy, Greywall
1992: NIST $P_m(T)$, noise and ^{60}Co thermometer, $T > 6.5$ mK
1994: Florida $P(T)$, ^{60}Co and Pt thermometer, $T > 0.5$ mK
1996: PTB, CMN and Pt thermometer, $T > 0.88$ mk
1998: Leiden Workshop to recommend $P_m(T)$ for extension of ITS-65
2000: PLTS 2000 Adopted

and $P_m(T)$ for PLTS-2000 are given by Rusby et al. (2002). An excellent review of MPT developments prior to 1997 is given by Soulen and Fogle (1997).

Although the Greywall scale with $T_A = 2.49$ mK had come into widespread use, the need for a scale based on a primary thermometer was recognized and undertaken at NIST by Fogle et al. (1990, 1992). Plans for this program were announced by R. J. Soulen at the 1984 Emil Warburg Symposium on Ultra-Low Temperature Physics, Bayreuth, Germany. The NIST work (see Fogle and Soulen 1998 and references therein) that extends from 0.7 K down to 6.3 mK used R-SQUID noise and ^{60}Co γ-radiation as primary thermometers below 22 mK. Unfortunately, this high-precision work did not go sufficiently low in temperature to include the fixed points T_A and T_{A-B} and T_N. Use of the melting pressure scale without these fixed points in temperature to serve also as *pressure fixed points* places a greater burden on measurement of the pressure, requiring highly accurate measurement of pressure differences relative to the pressure minimum at $T_{min} \approx 315$ mK.

In order to meet the need for determining $P_m(T)$ based on a primary thermometer that would include the fixed points at temperature T_A, T_{A-B}, and T_N, Ni et al. (1995) at the University of Florida (UF) made measurements from 25 mK down to well below T_N using a ^{60}Co γ-radiation primary thermometer along with a Pt NMR thermometer. The adsorption of β-particles in the ^{60}Co source produces self-heating of about 3 nW in the 5 μCi source of Ni et al. They were able to account for the self-heating by using the functional form for the transport of heat from the thermometer and that for the Pt susceptibility and then express the absolute temperature of the MPT in terms of that of the slightly hotter γ-ray thermometer. When the value of P_{min} of Ni et al. was adjusted to match that of Fogle et al. (1992), the two sets of measurements were in very good agreement in the region of overlap from 15 to 25 mK. Thus, Ni et al. combined their results with those of NIST to provide a scale, the UF scale, from T_N to 250 mK, with $T_A = 2.505$ mK and $T_N = 0.934$ mK.

Efforts to determine the ^{3}He melting pressure–temperature relation $P_m(T)$ as a thermometry standard were begun by Schuster et al. (1990) and by Hoffman et al. (1994) at PTB (the German Standards Laboratory) after the work at NIST had begun. The PTB work (Schuster et al. 1996, 1998) employed a single-crystal cerium magnesium nitrate (CMN) thermometer calibrated against the ITS-90 scale above 1.2 K and Pt NMR thermometer for the lowest temperatures. In addition, limited use was made of a Johnson junction noise thermometer. The CMN was in the shape of a sphere to allow use of the calculated Weiss constant $\Delta = 0.23$ mk (Hudson and Pfeiffer 1972), in the expression for the susceptibility:

$$\chi_{CMN} = A + B/(T + \Delta) \tag{5}$$

where A, a constant, is the off-set in the measurement at high temperature, and B is a constant determined in the calibration range.

Because of long thermal time constants of the spherical single crystal of CMN of a few hours below ~50 mK, its use by Schuster et al. was limited to temperatures above 40 mK, except for one measurement at 23 mK taken after a 16 h equilibrium period. The sensitivity of the Pt NMR thermometer limited its use to temperatures below 8 mK. Then, in order to bridge the gap between the two thermometers, the melting pressure relation $P_m(T)$, calculated in the range of the CMN thermometer, was extrapolated from 40 to 8 mK for calibration of the Pt NMR thermometer. The difficulty of lack of a thermometer between 40 and 8 mK was addressed by the limited use of the noise thermometer. However, heating effects in this thermometer gave rise to errors in temperature below 2.5 mK. The fixed-point temperatures of Schuster et al. (1996) are $T_A = 2.41$ mK and $T_N = 0.88$ mK, the lowest of any of the measurements (see Table 1).

An interesting comparison of several thermometers, including single-crystal CMN, with the PTB MPT scale has been made by Mohandas et al. (1998) at National Physics Laboratory, UK. This made use of one of the NIST superconducting reference material (SRM) fixed-point devices, recalibrated at PTB against their MPT scale and a ^{3}He MPT cell also constructed at PTB. Beginning below about 200 mK, the CMN temperature became systematically lower than the ^{3}He MPT, differing by as much as 1.0 mK at the lowest temperature of 10 mK reached by Mohandas et al. The data were then fitted to the Curie–Weiss law for the CMN susceptibility, eq. (5), with

B determined in the range from 1.2 to 1.5 K using a RhFe resistance thermometer. This fit gave a value of $\Delta = 1.27$ mK, which is 1.04 mK above the calculated value of Hudson and Pfeiffer (1972). Since Δ is additive to T, eq. (5), such a discrepancy between a calculated and fitted value would be a concern.

In the work of Mohandas et al., differences in temperature between a *powdered* CMN thermometer and the SRM device were $\lesssim 1$ mK for all the SRM fixed points. This suggests that the problem with the single-crystal CMN may be the long-thermal time constants, which are the result of using the single crystal for which thermal contact is problematic. The PTB scale relies on such an arrangement.

2.6. *Thermodynamic self-consistency*

In 1998, a workshop held at the University of Leiden was attended by 32 representatives of groups interested in MPT, with the purpose of adopting an agreed scale (Rusby et al. 2002). Although the differences in the various scales was not resolved, there was agreement that a provisional scale should be adopted and that the proposed scale should be analyzed for thermodynamic self-consistency which was being carried out by Reesink and Durieux (1999).

Details of the thermodynamic self-consistent calculation are given in the Ph.D. thesis of Reesink (2001). The starting point is the Clausius–Clapeyron equation (2). Since $\Delta V_{\rm m}$ is essentially independent of T below 25 mK (Halperin et al. 1974, 1978), by differentiation, this equation can be written as

$$\frac{d^2 P_{\rm m}}{dT^2} = \frac{1}{\Delta V_{\rm m}} \left(\frac{C_{\rm L}(T) - C_{\rm S}(T)}{T} \right) \tag{6}$$

where $C_{\rm L}$ and $C_{\rm S}$ are the heat capacities of the liquid and solid, respectively, at melting temperature (Greywall and Busch, 1987). Then, if the melting pressure and its slope are known at some starting temperature, T_0, along with the heat capacities, this equation can be integrated twice to give $P_{\rm m}(T)$ at lower temperatures:

$$P_{\rm m}(T) = P_{\rm m}(T_0) + (T - T_0)\left[\frac{dP_{\rm m}(T)}{dT}\right] + \frac{1}{\Delta V_{\rm m}} \int_{T_0}^{T} dT' \int_{T_0}^{T''} dT'' \left[\frac{C_{\rm L}(T'')}{T''} - \frac{C_{\rm S}(T'')}{T''}\right] \tag{7}$$

The results of this calculation are shown in Fig. 4 as PLTS calc, along with a comparison of various other scales that have been discussed in this chapter, plotted against T_{2000}, PLTS-2000, eq. (8) below, from Rusby et al.

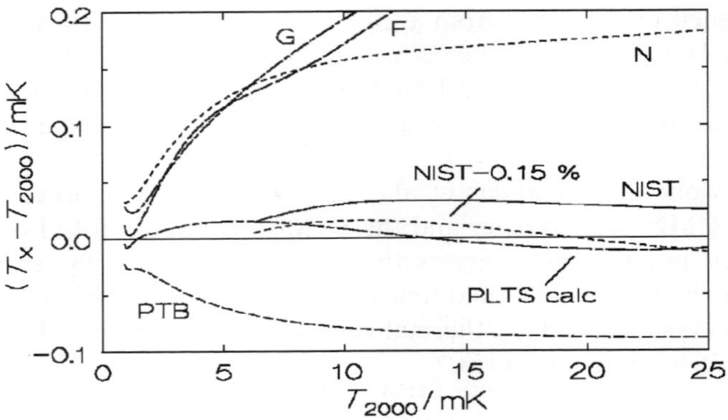

Fig. 4. Comparison of various temperature scales discussed in the text, including the self-consistent calculated PLTS, PLTS calc, with PLTS-2000 (T_{2000}), (Greywall 1986; Fukuyama et al. 1987; Ni et al. 1995) NIST, and NIST-0.15%, correction of NIST (see Rusby et al. 2002; PTB; Schuster et al. 1999. From Rusby et al. (2002).

(2002). As shown in Fig. 4, the departure of the self-consistent PLTS calc from T_{2000} (i.e. PLTS-2000) is <0.2 mK below 25 mK.

In January 2000, a meeting was held at NIST to derive a compromise scale that would be proposed to the Comité International des Poids et Mesures (CIPM) for adoption later in the year. The details of the compromise scale that was proposed by Reesink and Durieux are given by Rusby et al. (2002). Values of the fixed points T_A, T_{A-B}, and T_N, respectively, were taken as averages of those of PTB and UF after normalizing them to the NIST scale (see Rusby et al. 2002, for details) (Earlier determinations of the fixed points, as given in Table 1, were not considered.). Because of the lack of agreement on the values of these fixed points, the resulting scale is labeled "Provisional," with the anticipation that it will be revised when new determinations of these fixed points are available.

The scale was formally adopted by the CIPM in October 2000 (Rusby et al. 2002). Temperatures are given by a 12-term polynomial for $P_m(T)$ (P in MPa and T in K) in powers of T:

$$P = \sum_{i=-3}^{9} a_i T^i \tag{8}$$

where the coefficients a_i are: $a_{-3} = -1.3855442\text{E-}12$; $a_{-2} = 4.5557026\text{E-}9$; $a_{-1} = -6.4430869\text{E-}6$; $a_0 = 3.4467434$; $a_1 = -4.4176438$; $a_2 = 1.5417437\text{E}1$; $a_3 = 3.5789853\text{E}1$; $a_4 = 7.1499125\text{E}1$; $a_5 = -1.0414379\text{E}2$; $a_6 = 1.0518538\text{E}2$; $a_7 = -6.9443767\text{E}1$; $a_8 = 2.6833087\text{E}1$; and $a_9 = -4.5875709$.

Four fixed points in P (MPa) and T (mK) are

	P(MPa)	T(mK)
P_{min}	2.93113	315.24
A transition	3.43407	2.444
A–B transition	3.43609	1.896
Neél transition	3.43934	0.902

Rusby et al. also give a tabulation of the melting pressure and dP_m/dT in intervals of 1 mK up to 100 mK.

Although, PLTS-2000 is not defined below T_N, MPT may be extended to about 0.5 mK with reduced sensitivity by using the equation given by Ni et al. (1995),

$$P(T) - P_N = 0.1987 - 0.3002T^4 \tag{9}$$

where P is in kPa and T in mK. In using this equation, P_N must be identified in the experiment (see Fig. 12) and taken as a fixed point for obtaining pressure differences between $P(T_N)$ and $P(T)$.

3. Apparatus and procedures for the implementation of MPT

The advantages of MPT, mentioned above, make it the logical choice for extension of the temperature scale below ITS-90, which ends at 0.654 K. Although some not well-versed in the use of ^{3}He under pressure regard the method as complicated, in reality, it requires relatively sample equipment and techniques. The objective in this section will be to present sufficient details to make the technique readily accessible to those who may want to use it for the first time. A shorter version of some of the material in this section has been given elsewhere (Adams 2004a).

The techniques for using MPT are analogous to those for vapor pressure thermometry: a mixture of liquid and solid ^{3}He is cooled to the temperature to be determined, the melting pressure is measured, and then eq. (8) (or (9) for $T < 0.88$ mK) for $P_m(T)$ is solved for T. In the case of MPT, there is the slight complication that, because of the pressure minimum, the pressure must be measured in situ (for $T < T_{min}$). Although the apparatus and techniques for implementing MPT can be found in the original literature (see, for example, Adams 1993; Corruccini et al. 1978; Greywall and Busch 1982; Halperin et al. 1974, 1978; Ni et al. 1995; Scribner 1968), it is useful to give some details here for users not familiar with them. Transducers for in situ pressure measurement, the ^{3}He gas handling system for producing melting

^{3}He, procedures for calibrating the pressure transducer and for observing the various fixed points established in PLTS-2000 are described below. Simplicity in design without the need for hydrostatic-head corrections to the pressure calibration is stressed along with suggestions to avoid pitfalls for those not already familiar with the techniques.

3.1. *Sample cell and capacitance measurement*

The MPT cell must meet the following two requirements: (1) cool the mixture of liquid and solid ^{3}He to the temperature of interest, which may be below 1 mK, and (2) determine the melting pressure of the mixture with adequate resolution and accuracy. The first requirement is met by providing a large surface area of fine metal sinter within the cell to overcome the Kapitza resistance between the metal walls of the cell and ^{3}He. If the lowest temperature to which the MPT will be used is not less than about 10 mK, then sintered powder in not necessary to provide the require surface area. The second condition requires an in situ pressure transducer, which is usually a capacitive type patterned after the Straty–Adams design.

Since the design principles and several capacitive transducers have been described in the literature (Straty and Adams 1969; Greywall and Busch 1982; Adams 1993), only brief attention will be given to these here. The principle of operation is quite simple: the cell has a thin-wall diaphragm machined into it to which is attached a capacitor plate with a second plate held in close proximity. As the pressure changes, the diaphragm deflects slightly, causing changes in capacitance that are detected by suitable electronics (see below). The design of Straty and Adams, taken from Adams et al. (1991), is shown in Fig. 5. The most critical design parameters are the diameter and thickness of the diaphragm, and parallelism and spacing of the plates (for details, see Straty and Adams 1969). One of the simplest cells to construct and assemble is that of Greywall and Busch (1982), shown in Fig. 6, which is sealed with epoxy and has a unique way of achieving suitable spacing and parallelism of the capacitor plates. In these cells, thermal contact with the liquid is through the surface of packed powder "sinter" while the solid is cooled through the liquid. Sufficient open volume (space without sinter) must be left for the maximum quantity of solid that will be present at $P_{\min}$ (see below). Corruccini et al. (1978) have described a small demountable MPT cell that attaches to another cell that may contain a sample not at melting pressure. Miura et al. (1993) have constructed a gauge of titanium with a resolution in pressure $\delta P/P$ of 10^{-9} that is suitable for use in magnetic fields.

Cylindrical capacitors and others suitable for placing inside the cell have been described in the literature (Jarvis et al. 1968; Griffioen and Frossati

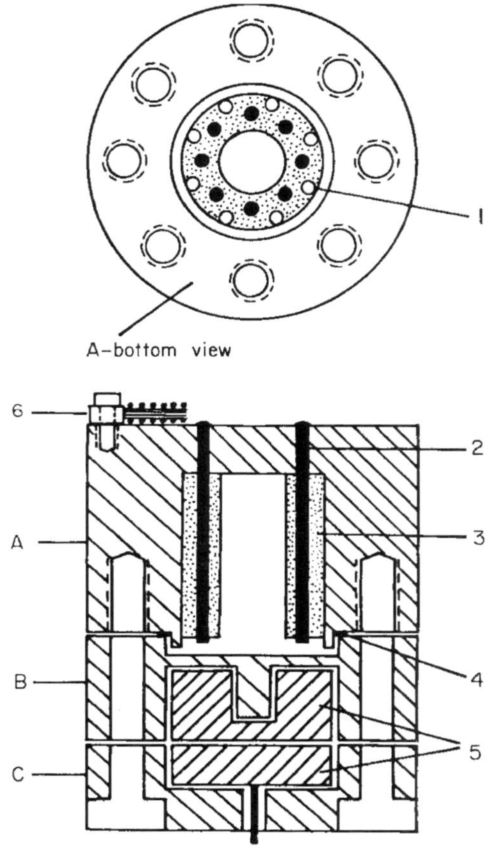

Fig. 5. Schematic view of a melting-pressure cell of the Straty–Adams design. A, heat exchanger with tapped holes at top for mounting; B, diaphragm and movable plate; C, fixed plate holder. 1, holes, 2, Ag wires welded to the body for thermal contact; 3, packed Ag powder; 4, indium seal; 5, capacitor plates; 6, Pt thermometer (for other use). For further details such as diaphragm design parameters and plate spacing, see Straty and Adams (1969). From Adams et al. (1991).

1985). See the review by Adams (1993) for descriptions of other gauges and references to the literature.

Two similar methods of measuring the capacitance of the pressure transducer with a capacitance bridge are in common use. Since the value of C and changes in it with pressure are relatively small compared with that of coaxial lines used to connect to instrumentation for measuring C, it is essential to

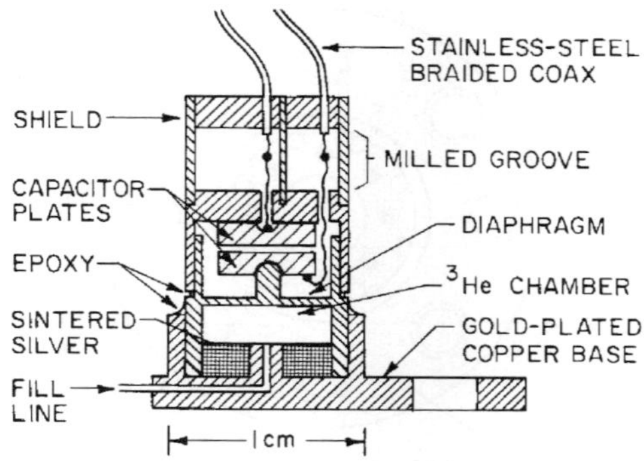

Fig. 6. The Greywall melting pressure cell. From Greywall and Busch (1982).

use a three-terminal method. With this method, each plate is connected to a separate coaxial line, with the third terminal being the ground. Then, variations in line capacitance do not affect the capacitance reading. A bridge with adequate resolution is commercially available (Andeen-Hagerling, Chagin Falls, OH), although heating at the lowest temperatures may be experienced with this bridge at the high-excitation voltage needed for maximum resolution.

The most commonly used bridge is the one assembled from commercial components, the key element of which is a precision ratio transformer (e.g. Tegam Incorporated, Geneva, OH), for comparing the unknown C to a reference standard. As shown schematically in Fig. 7, a variable resistor is included on one side of the bridge to balance the resistive, out-of-phase part of the signal. The balance point of the bridge is detected with a high-sensitivity lock-in amplifier, preferably with a separate low-noise preamplifier. In order to reduce temperature dependence of the reference standard, it is conveniently located in the cryostat, possibly incorporated into the pressure transducer as a third capacitor plate (Schuster and Wobler 1986). Corruccini et al. (1978) used a small silver-mica capacitor cooled to 77 K and situated in a shielded evacuated can for the standard.

With the three-terminal method, three coaxial lines are required with one being common to the reference capacitor and the unknown. Only the common coaxial line to which the detector is connected need be of high-quality with a low capacitance to ground. This is because the capacitance of this line, C_3, reduces the in-phase part of the out-of-balance signal,

$$v_0/v_i = [(R-1)C_1 + RC_2]/(C_1 + C_2 + C_3) \tag{10}$$

where C_1 and C_2 are the standard and the unknown capacitance, respectively, and R the ratio transformer setting (Swift 1980). Thus, C_3 should be

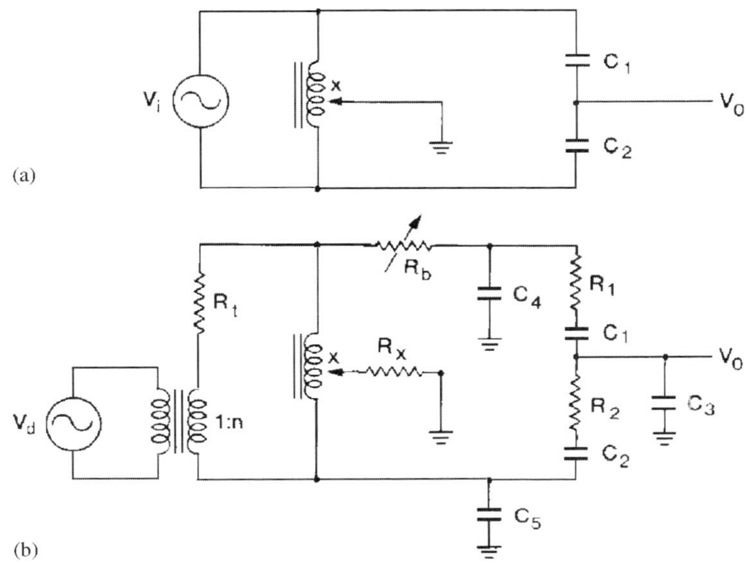

Fig. 7. Schematic diagram of a ratio-transformer-based capacitance bridge. (a) simplified diagram; (b) realistic diagram showing coaxial line capacitances and resistances. From Swift (1980).

kept as small as possible by locating the preamp at the top of the cryostat and with the line constructed to keep the capacitance to a minimum. The other two coaxial lines need not have low capacitance and may be constructed using a small CuNi capillary for the outer conductor with a thin copper wire for the inner one (copper is used to minimize resistance change with bath level). Vacuum grease pulled into the capillary as the wire is drawn in serves to heat-sink it along the capillary. (Twice the length of wire should be used so that one length without grease can be pushed through the first. Then the second half of the wire is pulled into the capillary by applying grease that can be heated to make it less viscous.) The capillaries should pass through the helium bath inside an evacuated tube (which may accommodate several miniature coaxes). The evacuated tube can be implemented easily by using hermetically sealed coaxial connectors at the top of the tube, which extends into the inner vacuum space and is open at the lower end with radiation baffles and heat-sinks for the coaxes. A single tube (~6 mm diameter) with "homemade" multiple coaxial connectors may accommodate several miniature coaxial lines.

Within the inner vacuum can, the coaxial lines connecting from 4 K to the lowest stage must be of low thermal conductivity and well heat-sunk at the various stages, while keeping C_3 small. Although miniature commercial coaxes are available, they can be constructed as described above using a CuNi capillary for the shield with a CuNi clad superconducting wire as the inner conductor. In order to reduce its capacitance, thread can be wound around the inner conductor to center it with vacuum grease to make thermal contact to the outer conductor. Prior to this assembly, strips of copper should be silver-soldered (use a "soft" flame and make sure that the copper is heated by the flame while not overheating the thin-walled CuNi) to the outer jacket at the appropriate points for bolting to the various stages of the refrigerator as heat sinks.

Another method of detecting changes in capacitance of the strain gauge is to incorporate the capacitor as a circuit element in an oscillator (Jarvis et al. 1968; Van Degrift 1975; Shvarts et al. 2003). Then, as the capacitance changes in response to pressure changes, the frequency of the oscillator changes and can be measured with high precision using a frequency counter. Oscillators of adequate stability require that the components be located within the cryostat, usually using a tunnel-diode located very near the strain gauge (see Adams 1993 and references therein). The oscillator method may be used in situations where one side of the capacitor must be grounded that would make the three-terminal capacitance measurement inoperable.

3.2. *Gas handling system and capacitance calibration*

The gas handling system used at the University of Florida, (Ni 1994a) for storing the ^{3}He gas and generating the pressure necessary for reaching the melting pressure is shown in Fig. 8. A similar gas handling system with fewer valves has been shown by Halperin et al. (1978), and others can be found in the literature.

A few liters of ^{3}He gas, with a ^{4}He impurity content of not more than 10 ppm, sufficient to fill the MPT cell and the gas handling system, is stored in a small storage cylinder (commercial or homemade). It should be capable of withstanding evacuation and an overpressure of a few MPa, and should be equipped with a valve in case it must be removed from the system. The valves labeled N are small low-pressure bellows-sealed vacuum valves (e.g. Nupro Co., Willoughby, OH) and the connection lines are 6 mm (1/4 in.) copper refrigeration tubing. The other valves are miniature high-pressure valves (e.g. High Pressure Equipment Co., Erie, PA) rated at 68 MPa and are connected with 1.5 mm steel tubing.

A low-pressure nitrogen-cooled charcoal-filled trap for removing impurities from the ^{3}He gas immediately follows the storage tank. The ^{4}He-temperature "dipstick" is a charcoal-filled container constructed from

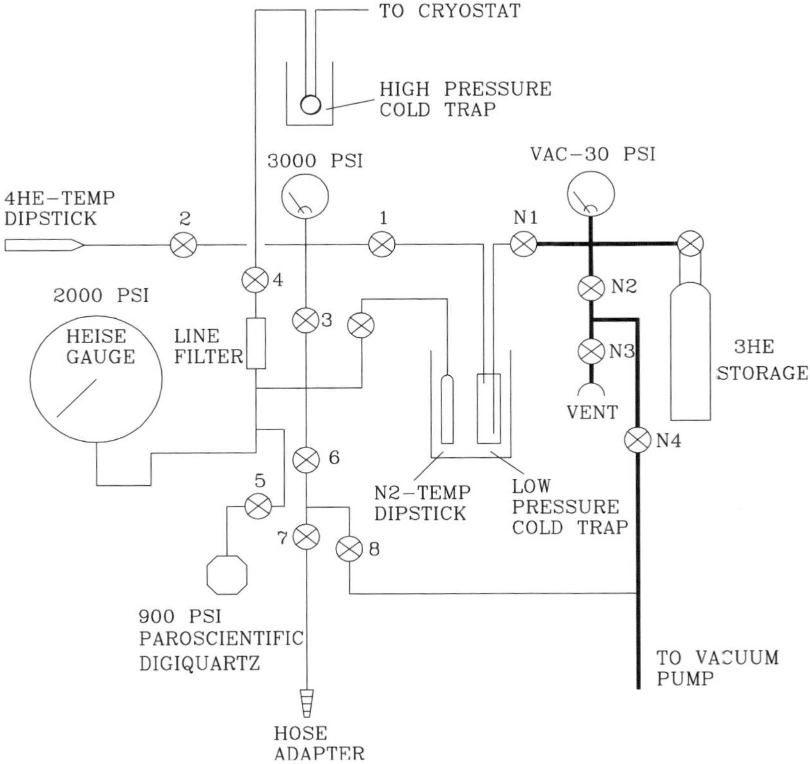

Fig. 8. Gas handling system for storing ^{3}He gas and producing the pressure to reach the melting curve. From Ni (1994).

12.5 mm heavy-walled stainless tubing about 20 cm in length connected to the system by about 2 m of 1.5 mm steel tubing, sufficient to reach to the bottom of a helium storage dewar. When this is slowly inserted into the helium storage dewar (being careful not to boil off helium rapidly), the volume of cold charcoal is sufficient to cryopump most of the ^{3}He gas from the storage tank. Then, the high-pressure valve to the tank (#1) is closed and the dipstick is raised slowly in the storage dewar to generate pressure as the ^{3}He is desorbed. A very small coarse pressure gauge (such as the one used on a pressure regulator) is used to monitor the pressure in the dipstick and to make sure that it does not become excessively high. ^{3}He gas from the dipstick is admitted into the portion of the system containing the pressure gauges for calibrating the strain gauge and monitoring the pressure during operation. A Bourdon dial gauge (e.g. Heise-Dresser Measurements, Shelton, CT) and a quartz gauge (e.g. Paroscientific, Inc., Redmond, WA) serve as pressure standards. Melting pressure thermometry may be performed without an expensive standard by using the pressure minimum as a pressure fixed point in combination with a reasonably good dial gauge, as is discussed further below (Corruccini et al. 1978).

The line leading to the cryostat may contain a particle filter and a high-pressure cold trap, which is just the length of the 1.5 mm tubing inserted into the same nitrogen dewar as the low-pressure trap. The 1.5 mm tubing ends at the top of the cryostat where it joins two parallel 0.6 mm i.d. CuNi capillaries that have separate valves where they join the larger tubing. The arrangement of filling capillaries extending to the MPT cell is not intended for calculation of the pressure head since this is unnecessary if the melting pressure minimum or one of the fixed points (P_A or P_N), is used to adjust the pressure calibration.

In order to prevent changes in bath level from causing large changes in pressure, it is essential that these capillaries pass through the helium bath inside an evacuated tube. This is most conveniently arranged by extending the tube through the 4 K flange into the inner vacuum can (IVC). The vacuum in the tube is conveniently provided through a small opening in its side just below the IVC top flange. The capillaries are joined together again at a homemade "T" just inside the IVC for connection to a single capillary extending to the MPT cell. Then before the cell is filled, the valve to one capillary is closed to allow removal of the ^{3}He if the other capillary becomes blocked.

The capillary extending to the MPT cell must have heat-sinks at the various cooling stages such as the 1 K pot, still, DR "coldplate", mixing chamber, and magnetic cooling stage (if there is one). These can be constructed by tightly winding about 30 cm of the capillary around copper posts to which it is then silver-soldered. About 30 cm of capillary should be included between each cooling stage for thermal isolation. This construction allows the entire filling line within the IVC to be constructed with only a few solder joints. (As a precaution against blockage, it is advisable to clean the inside of the capillary before using it for the first time by forcing a solvent through it under pressure, which is then blown out.)

Prior to cooling the cryostat, the cell should be evacuated, which may take a considerable length of time because of the long, thin capillary. After several hours of pumping, the cell can be flushed several times with dry N_2 gas. After the final evacuation of N_2, the cell may be flushed with ^{3}He. Since the volume of ^{3}He is only a few cubic centimeters, this can be discarded or returned to the storage tank through the cold trap. Continued pumping of the cell during the cooling of the cryostat is advisable.

In order to determine the pressure from the capacitance, the relationship $P(C)$ must be established by calibration of the gauge against a standard, which, in this case, is a quartz gauge. Since the capacitance of the gauge changes significantly upon cooling from room temperature, the calibration must be performed for each cooling from room temperature, where it then appears to be quite stable as long as it is kept cold. The calibration is carried out with the MPT cell at a temperature near 1 K where the filling line will not be blocked with solid in the pressure range of interest. Measuring the capacitance of the empty cell over the full range of temperatures below 1 K can check for any possible temperature-dependence below the temperature of the calibration. Carefully constructed capacitance gauges that do not have a temperature-dependent dielectric between the plates will show little temperature dependence below 1 K and the calibration can be assumed to be the same at lower temperatures.

Calibration of the gauge requires measurement of C versus P while the applied pressure is held at a series of values in the range from about 2.7 to 3.5 MPa.Usually there is a small hysteresis in C for increasing and decreasing pressures that can be reduced by "training" the gauge, i.e. successive raising and lowering the pressure. This process can be accomplished in a simple way by slowly raising the pressure to the maximum calibration range over a period of a few minutes, then slowly reducing it back to the low end of the calibration.

The next step is to take a series of readings of C (or of the ratio transformer, R) and P, while maintaining the pressure reasonably constant for sufficient time to take the reading. This can be accomplished readily by using the pressure regulation technique developed by Ihas and Pobell (1974), for which a block diagram is shown in Fig. 9. The out-of-balance signal from the MPT capacitance bridge is fed through an amplifier to a heater mounted on a small "bomb" in the ^{3}He filling line located in the IVC and weakly heat-sunk to the 1 K pot or to 4.2 K. Once a desired value of C (or P) is reached, the feedback loop is closed and maintained until the pressure stabilizes and is read on the external pressure gauge.

An alternative technique that works well is to observe the out-of-balance signal from the capacitance bridge and manually adjust the external pressure to hold C constant while the readings are taken. One or more of the open high-pressure valves in the filling line can serve as a small "displacement piston" for slight adjustments in pressure by screwing the valve handle in or out as needed. We fitted the calibration points P versus R to a polynomial of the form

$$P = b_0 + b_1 R + b_2 R^2 + b_3 R^3 \qquad (11)$$

which is used subsequently to calculate P from R.

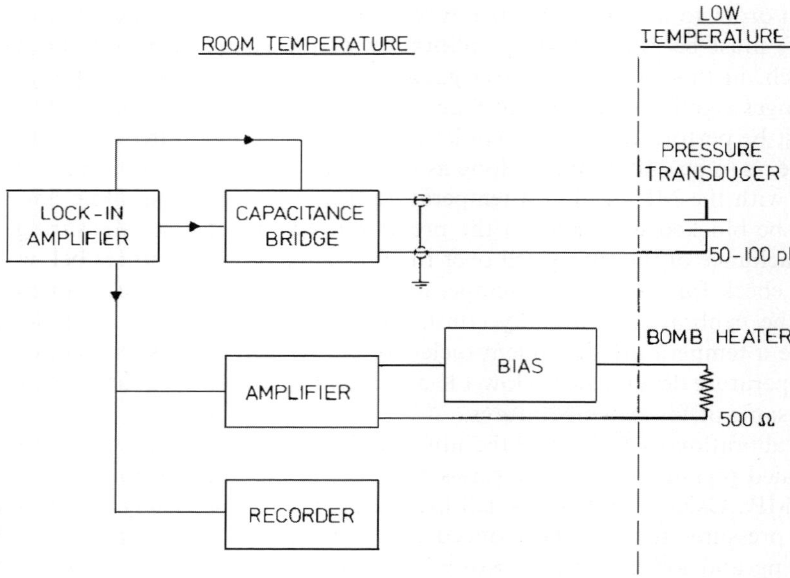

Fig. 9. Schematic of the electronics for using the output of the capacitance bridge to regulate pressure. From Ihas and Pobell (1974).

In using the bridge, it is frequently desirable to work with a small out-of-balance signal, with R set slightly off the balance point (see eq. (10)), and with v_0 recorded on a chart or electronically. Then it is not necessary to continually balance the bridge. Values of the balance point R can be read from the chart or electronic recording by reading off the out-of-balance signal and using small steps in R to convert v_0 into differences in R. This is particularly the case when $P(t)$ is needed to observe the fixed points (see below).

3.3. *Filling the cell and observation of the fixed points in pressure*

Once the capacitance gauge has been calibrated, the cell must be filled with liquid at the appropriate pressure. This can be seen by reference to $V(T)$ for liquid and solid ^{3}He and the melting pressure $P(T)$ along the melting curve (Grilly 1971; Halperin et al. 1978 ; Rusby et al. 2002), shown in Fig. 10. If the cell is filled with liquid at 3.38 MPa near 0.74 K, as shown by the point (P_i, T_i), in upper half of the figure, the molar volume will be $V = 25.5\,\text{cm}^3/\text{mol}$, shown by the dashed horizontal line in the lower half of the figure. Once

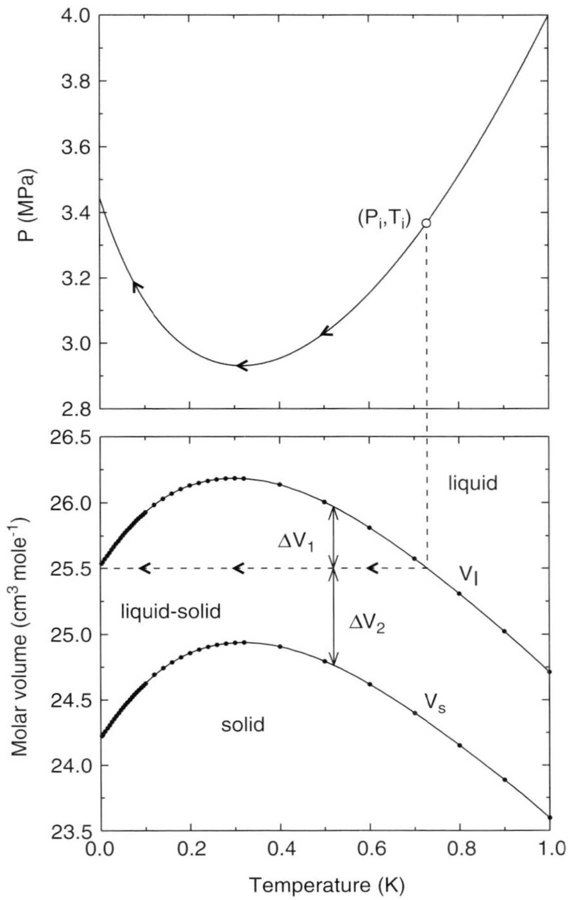

Fig. 10. Upper panel: melting pressure versus T, (eq. (8)). P_i, T_i are pressure and temperature for initial filling of the MPT cell. Lower panel: molar volumes of liquid V_L and of solid V_S at melting pressure versus T. The arrows show the paths followed as the cell cocls. Data from Halperin et al. (1978) and Grilly (1971).

cooling of the still blocks the capillary with solid, the cell will have a mixture of liquid and solid and therefore follows the melting curve to all lower temperatures, as shown by the arrows. For the shortest time constants, the cell must have a small solid fraction at the lowest temperature, which is provided by filling at these conditions. Since the thermal expansion of the liquid is negative at these temperatures, it is important to observe the pressure of the cell, though observation of C, as it cools until solid begins to form (the pressure drops), making sure that P does not increase above 3.38 MPa.This is especially important if the initial filling temperature is significantly above the freezing point $T(P_i)$.

At any temperature below the initial filling of the cell, the fraction x of ^{3}He in the cell that is solid is given by

$$x = [V_L(T) - V]/[V_L(T) - V_S(T)] \qquad (12)$$

This is largest near the melting pressure minimum (molar volume maximum) where it is about 55% for $V = 25.5\,\mathrm{cm}^3/\mathrm{mol}$. In order to observe the pressure minimum correctly (for use as a pressure fixed point), the liquid–solid interface must remain in the volume outside the sinter, which should be 55% or larger. A shorter time constant at the lowest temperature can be obtained by reducing the open volume in the cell. In that case, to observe the pressure minimum, the cell should be filled at a lower pressure so that the open volume will not fill with solid. Subsequently, the pressure should be set at about 3.38 MPa to maintain melting solid to millikelvin temperatures. This filling pressure will provide a few percent of solid at the lowest temperature.

As the cell cools, the pressure is lowest at $P_{\min}$. If the entire filling capillary is at a temperature above $T_{\min}$, slipping of the plug in the filling capillary may occur, resulting in too much ^{3}He in the cell. To avoid this problem, we cool the mixing chamber faster than the MPT cell by applying heat to the cell or by leaving the heat switch open (in a cryostat which employs one). Since a correct reading for $P_{\min}$ is essential, it is advisable to observe the minimum several times for both slowly warming and cooling the cell while maintaining the plug in the filling line by keeping the mixing chamber colder. This is accomplished most readily by leaving the bridge set with a slight imbalance and observing the point at which the out-of-balance signal v_0 goes through an extremeum. Then this value of v_0 can be converted into the value of R, and then to $P_{\min}$ for the balance point, as discussed above. Laborious small step-by-step determinations of $P_{\min}$ that have been reported (Schuster et al. 1998; Pitre et al. 2003), are unnecessary. During observation of $P_{\min}$, it is advisable to close the last valve in the filling line so that if the plug slips, only a small amount of extra ^{3}He can enter the cell. Afterwards, this valve should be kept open during the entire run so that the gas in the Bourdon gauge will act as a "constant pressure" source to prevent the pressure above the plug from dropping below $P_{\min}$. Adjustments in the external pressure may be necessary from time to time if changes are too large.

The accuracy required in the pressure calibration for acceptable temperature accuracy is a concern with MPT. Since the melting pressure slope $|\mathrm{d}P/\mathrm{d}T| \gtrsim 3.5$ kPa/mK in the few mK range, for $\Delta T/T < 1\%$ at $T = 2\,\mathrm{mK}$, an accuracy in P of $\Delta P \leq 70\,\mathrm{Pa}$, or 0.002% is required. The pressure standard may not have this accuracy and uncertainty in the hydrostatic head, ~700 Pa, may be a greater problem. However, the need for highly accurate pressures, which require determination of the hydrostatic head correction, can be avoided by using P_{min} (and the fixed points T_{N}, T_{A}, and $T_{\mathrm{A-B}}$ if they are accessible in the setup) as a pressure fixed point. For example, Corruccini et al. (1978) reported a resolution of 10 μK and an accuracy of 0.2 mK using only a Bourdon gauge with an accuracy of 1 in 10^3 and the pressure minimum as a fixed point (but not the lower-temperature fixed points) for adjusting the capacitance pressure calibration. With the extremely high resolution of the pressure gauge reported by Miura et al. (1993) of 1 in 10^9, a resolution in T of 1 nK is provided at about 1 mK.

If the low-T fixed points T_{A} or T_{N} are accessible in the experiment, the accuracy in temperature relative to PLT-2000 in the few-milikelvin range can be improved by using one of these to adjust the pressure calibration. If these fixed points are inaccessible in the setup, installation of one or more superconducting materials, such as tungsten, to serve as a fixed point in *pressure,* might be considered (Strom et al. 1998). Hechtfischer and Schuster (2003) have discussed the use of tungsten. Then, only accurate differences in pressure relative to the fixed point are required. Since the transition temperature of tungsten is ≈15.5 mK, accuracy in the lower temperature range is greatly increased.

Note that the two previous works with the highest reported accuracy in pressure determination (Halperin et al. 1978; Schuster et al. 1999) have somewhat different values for various pressure fixed points. For example, Halperin et al. found P_{min} and P_{N} to be 593 and 120 Pa, respectively, higher than that reported by Schuster et al. for these points. Near 2 mK, these differences in pressure correspond to differences in temperature of 8% and 1.5%, respectively. This illustrates the need to identify the low-temperature pressure fixed points for high accuracy in temperature in this range. If the pressure minimum is used as a pressure fixed point, the accuracy of T determined from the melting pressure obviously requires careful observation of this point. Differences between the various fixed points in pressure are given in Table 3. Comparing P_{N} and P_{min}, we see that Halperin et al. (1978) and Ni et al. (1995) are in close agreement, while Greywall's value is ~500 Pa lower than Halperin et al. and Ni et al., and that of Schuster et al. (1996) is ~400 Pa higher. Since the melting pressure slope in the few millikelvin range is $|\mathrm{d}P/\mathrm{d}T| \sim 3500$ Pa/mK, the difference in temperature δT corresponding to the pressure difference is only 0.1 mK.

Once P_{min} has been observed, the value of b_0 in eq. (11) is adjusted to a new provisional one until one of the low-T fixed points is observed. If the refrigerator is capable of reaching $T_{\mathrm{A-B}}$ and T_{N}, one of these, preferably T_{N}, should be used as a pressure fixed point for final adjustment of b_0 in the

TABLE 3
Pressure Difference between Fixed Points.

	$P_N - P_A$(Pa)	$P_A - P_{min}$(Pa)	$P_N - P_{min}$(Pa)
Halperin et al. (1978)	5,230 ± 20	502,600 ± 100	507,830
Greywall (1986)	5,252 ± 10	502,050 ± 100	507,302
Fukuyama et al. (1987)	5,272 ± 26		
Hoffmann et al. (1994)	5,210 ± 50	502,640 ± 80	
Ni et al. (1995)	5,234 ± 30	502,574 ± 100	507,808
Schuster et al. (1996)	5,270 ± 10	502,940 ± 50	508,210

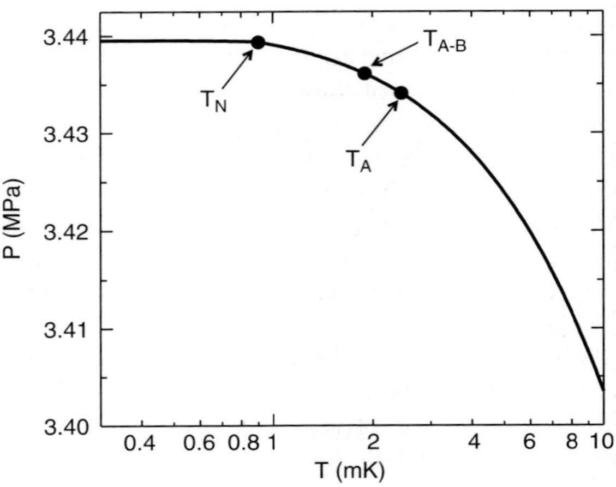

Fig. 11. The melting pressure, eqs. (8) and (9), versus T (log scale) for the region below 10 mK showing the reference points T_A, T_{A-B}, and T_N. As discussed in the text, only T_N is visible in the equilibrium $P(T)$.

calibration equation. To preserve the calibration at both P_{min}, and the lower fixed point, it may be necessary to introduce another adjustable parameter into eq. (2) by substituting $\alpha P'$ for P. Then α is determined by requiring eq. (11) to give the observed R at both P_{min}, and the lower fixed point.

Since the A superfluid transition is second-order, there is no discernible feature on the *equilibrium* melting curve $P(T)$. The A–B superfluid transition is first-order but with a very small latent heat that produces a relative discontinuity in slope of the melting pressure, $\delta|dP/dT|_m$, of only about 1.6 in 10^{-4}. This is also unobservable in $P(T)$ as shown in Fig. 11. Conversely, the transition at T_N is first-order with a large latent heat and large change in $|(dP/dT|_m$ and can be seen in the equilibrium $P_m(T)$. The two superfluid transitions can be seen only in $P(t)$ when they are traversed at a rate that

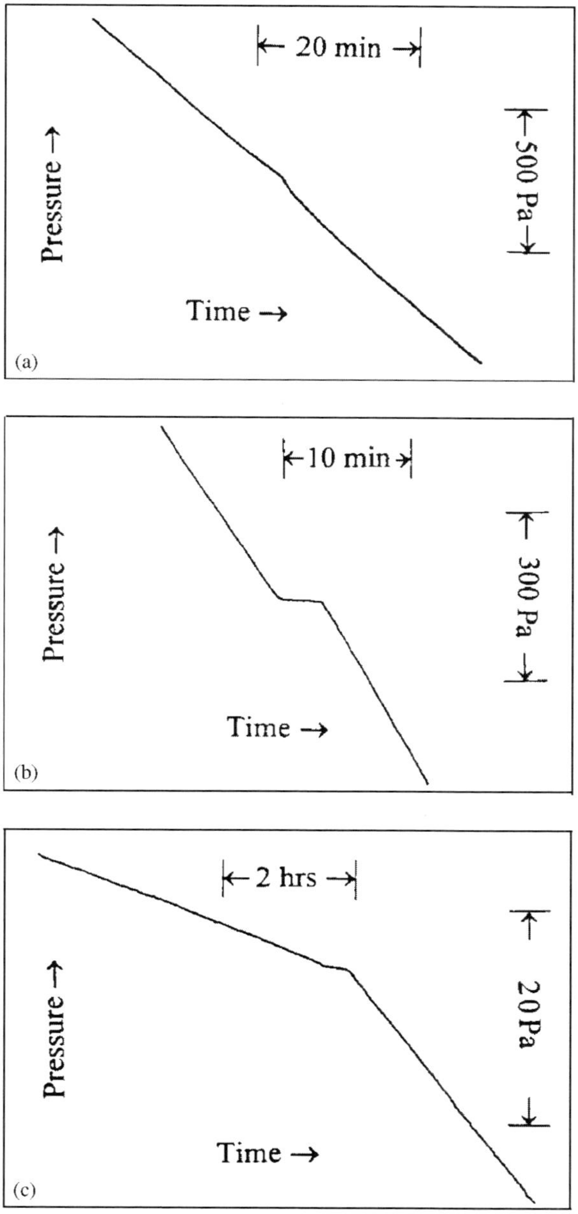

Fig. 12. Melting pressure versus time, showing the appearance of the reference points T_A, T_{A-B}, and T_N. Note the different chart speeds and sensitivities of the pressure gauge for observing these transitions. From Ni et al. (1995).

drives the cell and the ^{3}He out of equilibrium, as shown in Fig. 12. Observing the A or A–B transition for the first time can be quite challenging. This can be facilitated by calculating the expected position of the balance point based on the pressure calibration and the value of C at the pressure minimum. Then, the rate of cooling has to be adjusted carefully (see sensitivities shown in Fig. 12). Thermometry with the MPT may be facilitated by regulating the temperature at each desired point by using the out-of-balance signal from the capacitance bridge in a feedback loop to a heater. Then, each measured P is used in eq. (8) to determine T at that pressure.

3.4. *Possible future work*

By its very name, PLTS-2000 is a provisional scale, with the anticipation that revisions will be made once better determinations of the fixed points T_A, T_{A-B}, and T_N are available. Unfortunately, the program at NIST was discontinued several years ago with the shift of emphasis there as an agency to support US industry rather than one to develop new fundamental standards. At this point, PTB is apparently the only standard lab with facilities for doing work down to 1 mK, although there is fundamental research in this range being conducted by many low-temperature labs around the world. It would be useful if one or more of these labs would devote some effort to the determination of the fixed points more precisely and to the measurement of $P_m(T)$ based on these. There may not be a payoff in terms of spectacular results comparable to those achieved by Greywall (1986). However, we cannot know if discrepancies in the temperature scale will have a profound effect on results unless better measurements are made.

Acknowledgements

This work has been supported by the National Science Foundation through Grant No. DMR-0096791. The author thanks Dr. Mykola Omelayenko for assistance in preparing the figures for this chapter and Ms. Dori Faust for assistance in assembling the chapter.

References

Adams, E.D., 1993, Rev. Sci Instrum **64**, 601.

Adams, E.D., 2004a, D.C. Ripple, et al., ed., Temperature, its Measurement and Control in Science and Industry Vol. 7 (AIP, Melville, New York) Several other papers relating to PLTS-2000 appear in this volume.

Adams, E.D., 2004b, J. Low Temp. Phys **135**, 695.
Adams, E.D., Y.H. Tang and K. Uhlig, 1991, J. Low Temp. Phys **84**, 109.
Anufriyev, Yu.D., 1965, Zh. Eksperim. i. Teor Fiz.-Pis'ma Redakt. (6), 1 [translation: JETP Lett. **1**, 155 (1965)].
Avenel, O., M. Bernier, D. Bloyet, P. Piejus, E. Varaquaux and C. Vibet, 1975, M. Krusius, and M. Vuorio, eds, Proc. 14th Int. Conf. on Low Temperature Physics Vol. 4 (North-Holland, Amsterdam) 64.
Alvesalo, T.A., T. Haavasoja and M.T. Manninen, 1981, J. Low Temp. Phys **45**, 373.
Baum, J.L., D.F. Brewer, J.G. Daunt and D.O. Edwards, 1959, Phys. Rev. Lett **3**, 127.
Corruccini, L.R., K.R. Mountfield and W.O. Sprenger, 1978, Rev. Sci. Instrum **49**, 314.
Doniach, S. and S. Englesberg, 1966, Phys. Rev. Lett **17**, 750.
Fogle, W.E., J.H. Coldwell and R.J. Soulen, Jr., 1990, Physica B 165,166, 33.
Fogle, W.E. and R.J. Soulen, 1998, in: Proc. Leiden Workshop Towards an International Temperature Scale from 0.65 K to 1 mK, ed R.L. Rusby, and P. Mohandas (National Physical Laboratory, Teddington) 13.
Fogle, W.E., R.J. Soulen and J.H. Colwell, 1992, F. Schooley, eds, Temperature, its Measurement and Control in Science and Industry Vol. 6 (Am. Inst. Phys., New York) 91.
Fukuyama, H., H. Ishimoto, T. Razaki and S. Ogawa, 1987, Phys. Rev. B **36**, 8921.
Greywall, D.S., 1983, Phys. Rev. B **27**, 2747.
Greywall, D.S., 1986, Phys. Rev. B **33**, 7520.
Greywall, D.S. and P.A. Busch, 1982, J. Low Temp. Phys **46**, 451.
Greywall, D.S. and P.A. Busch, 1987, Phys. Rev. B **36**, 6853 and references therein.
Grilly, E.R., 1971, J. Low Temp. Phys **4**, 615.
Griffioen, W. and G. Frossati, 1985, Rev. Sci. Instrum **56**, 1236.
Halperin, W.P. and E.D. Adams, 1998, in: Proc. Leiden Workshop Towards an International Temperature Scale from 0.65 K to 0 mK, ed R.L. Rusby, and P. Mohandas (National Physical Laboratory, Teddington) 33.
Halperin, W.P., C.N. Archie, C.N. Rasmussen, R.A. Buhrman and R.C. Richardson, 1974, Phys. Rev. Lett **32**, 927.
Halperin, W.P., F.B. Rasmussen, C.N. Archie and R.C. Richardson, 1978, J. Low Temp. Phys **31**, 617.
Hechtfischer, D. and G. Schuster, 2003, D.C. Ripple, eds, Temperature, its Measurement and Control in Science and Industry Vol. 7 (AIP, Melville, New York) 149.
Hoffmann, A., J. Engert and W. Buch, 1994, Physica B 194–196, 19.
Hudson, R.P. and E.R. Pfeiffer, 1972, H.H. Plumb, ed., Temperature, its Measurement and Control in Science and Industry Vol. 4 (Instrum. Soc. Amer., Pitsburgh) 1279.
Ihas, G.G. and F. Pobell, 1974, Phys. Rev. A **9**, 1278.
Jarvis, J.F., D. Ramm and H. Meyer, 1968, Phys. Rev **170**, 320.
Kirk, W.P. and E.D. Adams, 1974, Cryogenics **14**, 147.
Kummer, R.B., E.D. Adams, W.P. Kirk, A.S. Greenberg, R.M. Mueller, C.V Britton and D.M. Lee, 1975, Phys. Rev. Lett **34**, 517.
Lee, D.M., 1997, Rev. Mod. Phys **69**, 645.
Miura, Y., N. Matsushima, T. Ando, S. Kuno, S. Inoue, K. Ito and T. Mamiya, 1993, Rev. Sci. Instrum **64**, 3215.
Mohandas, P., D.L. Head and R.L. Rusby, 1998, in: Proc. Leiden Workshop Towards an International Temperature Scale from 0.65 K to 1 mK, ed R.L. Rusby, and P. Mohandas (National Physical Laboratory, Teddington) 53.
Ni, W., 1994a. Ph.D. Thesis, University of Florida.
Ni, W., J.S. Xia and E.D. Adams, 1994b, Phys. Rev. B **50**, 336.
Ni, W., J.S. Xia and E.D. Adams, 1995, J. Low Temp. Phys **99**, 167.

Osborne, D.W., B.M. Abraham and B. Weinstock, 1951, Phys. Rev **82**, 263.
Osheroff, D.D., 1987, Rev. Mod. Phys **69**, 667.
Osheroff, D.D., H. Godfrin and R.R. Ruel, 1987, Phys. Rev. Lett **58**, 2458.
Osheroff, D.D., R.C. Richardson and D.M. Lee, 1972a, Phys. Rev. Lett **28**, 885.
Osheroff, D.D., W.J. Gully, R.C. Richardson and D.M. Lee, 1972b, Phys. Rev. Lett **29**, 920.
Paulson, D.N., M. Krusius and J.C. Wheatley, 1979, J. Low Temp. Phys **34**, 63.
Pekola, J., 2004, J. Low Temp Phys **135**, 723.
Pitre, L., Y. Hermier and G. Bonner, 2003, D.C. Ripple, eds, Temperature, its Measurement and Control in Science and Industry Vol. 7 (AIP, Melville, New York) 83.
Pomeranchuk, I.I., 1950, Zhur. Eksp. i. Teor. Fiz **20**, 83.
Reesink, L., 2001, Ph.D. Thesis, University of Leiden.
Reesink, L. and M. Durieux, 1999, in: Proc. 7th Int Symp. on Temp and Thermal Meas. in Ind. and Sci, ed J. Dubbeldam, and M. de Groot (NMi van Swinden Lab, Delft) 50.
Richardson, R.C., 1997, Rev. Mod. Phys. **69**, 683.
Roger, M., J.H. Hetherington and J.M. Delrieu, 1983, Rev. Mod. Phys **55**, 1.
Roger, M. and J.H. Hetherington, 1990, Phys. Rev. B **41**, 200.
Rusby, R.L., M. Durieux, A.L. Reesink, R.P. Hudson, G. Schuster, M. Kuhne, W.E. Fogle, R.J. Soulen and E.D. Adams, 2002, J. Low Temp. Phys **126**, 633.
Schuster, G., D. Hechtfischer, W. Buck and A. Hoffman, 1990, Physica B 165,166, 31.
Schuster, G., A. Hoffmann and D. Hechtfischer, 1996, Czech. J. Phys **46**, 481.
Schuster, G., A. Hoffman and D. Hechtfisher, 1998, in: Proc. Leiden Workshop Towards an International Temperature Scale from 0.65 K to 1 mK, ed R.L. Rusby, and P. Mohandas (National Physical Laboratory, Teddington) 27.
Schuster, G., A. Hoffmann, D. Hechtfisher and M. Kuehne, 1999, in: Proc. 7th Int. Symp. on Temp. and Thermal Meas. in Ind. and Sci, ed J. Dubbeldam, and M. de Groot (NMi van Swinden Lab, Delft) 129.
Schuster, G. and L. Wobler, 1986, J Phys **19**, 701.
Scribner, R.A., 1968, Ph.D. Thesis, University of Florida.
Scribner, R.A. and E.D. Adams, 1967, Comments at Crystal Mt. (Gordon Conf., WA).
Scribner, R.A. and E.D. Adams, 1970, Rev. Sci. Instrum. **41**, 287.
Scribner, R.A. and E.D. Adams, 1972, H.H. Plumb, eds, Temperature, its Measurement and Control in Sci. and Ind. Vol. 4 (Instrum. Soc. Amer., Pittsburgh) 37.
Shvarts, D., A. Adams, C.P. Slusher, R. Korber, B.P. Cowan, P. Noonan, J. Saunders and V.A. Mikheev, 2003, Physica B 329–33, 1566.
Soulen, R.J. and W.E. Fogle, 1997, Physics Today, p. 36.
Spietz, L., K.W. Lehnert, I. Siddiqui and R.J. Schoelkopf, 2003, Science **300**, 1929.
Straty, G.C. and E.D. Adams, 1969, Rev. Sci. Instrum. **40**, 1339.
Strom, A.J., C.M. van Woerkens, R. Jochemsen, G.J. Nieuwenhuys and W.A. Bosch, 1998, in: Proc. Leiden Workshop Towards an International Temperature Scale from 0.65 K to 1 mK, ed R.L. Rusby, and P. Mohandas (National Physical Laboratory, Teddington) 59.
Swift, G.W., 1980, Ph.D. Thesis, University of California, Berkeley.
Sydoriak, S.G., E.R. Grilly and E.F. Hammel, 1949, Phys. Rev **75**, 303.
Trickey, S.B., W.P. Kirk and E.D. Adams, 1972, Rev. Mod. Phys **44**, 668.
Van Degrift, C.T., 1975, Rev Sci Instrum **46**, 599.
Vvedenskii, V.L., 1972, JETP Lett **16**, 254.
Wheatley, J.C., 1970, in: Proc. 1970 Ultralow Temp. Symp, ed R.A. Hein, D.U. Gubser, and E.H. Takken (Naval Research Laboratory, Washington, DC) 99.
Yawata, K., 2001, Ph.D. Thesis, University of Tsukuba.

AUTHOR INDEX

SUBJECT INDEX